Management Science, Logistics, and Operations Research

John Wang
Montclair State University, USA

A volume in the Advances in Logistics, Operations, and Management Science (ALOMS) Book Series

BUSINESS SCIENCE Reference
An Imprint of IGI Global

Managing Director: Lindsay Johnston
Production Manager: Jennifer Yoder
Development Editor: Brett Snyder
Acquisitions Editor: Kayla Wolfe
Typesetter: Michael Brehm
Cover Design: Jason Mull

Published in the United States of America by
Business Science Reference (an imprint of IGI Global)
701 E. Chocolate Avenue
Hershey PA 17033
Tel: 717-533-8845
Fax: 717-533-8661
E-mail: cust@igi-global.com
Web site: http://www.igi-global.com

Library of Congress Cataloging-in-Publication Data

Library of Congress Cataloging-in-Publication Data

Management science, logistics, and operations research / John Wang, editor.
pages cm
Includes bibliographical references and index.
Summary: "This book examines related research in decision, management, and other behavioral sciences in order to exchange and collaborate on information among business, industry, and government, providing innovative theories and practices in operations research"-- Provided by publisher.
ISBN 978-1-4666-4506-6 (hardcover : alk. paper) -- ISBN 978-1-4666-4507-3 (ebook) -- ISBN 978-1-4666-4508-0 (print & perpetual access) 1. Management information systems. 2. Decision support systems. 3. Business logistics--Technological innovations. 4. Systems engineering. I. Wang, John, 1955- editor.

HD30.213.M3456 2014
658.4'034--dc23

2013028202

This book is published in the IGI Global book series Advances in Logistics, Operations, and Management Science (ALOMS) (ISSN: 2327-350X; eISSN: 2327-3518)

British Cataloguing in Publication Data
A Cataloguing in Publication record for this book is available from the British Library.

For electronic access to this publication, please contact: eresources@igi-global.com.

Advances in Logistics, Operations, and Management Science (ALOMS) Book Series

ISSN: 2327-350X
EISSN: 2327-3518

Mission

Operations research and management science continue to influence business processes, administration, and management information systems, particularly in covering the application methods for decision-making processes. New case studies and applications on management science, operations management, social sciences, and other behavioral sciences have been incorporated into business and organizations real-world objectives.

The **Advances in Logistics, Operations, and Management Science** (ALOMS) Book Series provides a collection of reference publications on the current trends, applications, theories, and practices in the management science field. Providing relevant and current research, this series and its individual publications would be useful for academics, researchers, scholars, and practitioners interested in improving decision making models and business functions.

Coverage

- Computing and Information Technologies
- Decision Analysis and Decision Support
- Finance
- Information Management
- Marketing Engineering
- Operations Management
- Organizational Behavior
- Political Science
- Production Management
- Services Management

IGI Global is currently accepting manuscripts for publication within this series. To submit a proposal for a volume in this series, please contact our Acquisition Editors at Acquisitions@igi-global.com or visit: http://www.igi-global.com/publish/.

The Advances in Logistics, Operations, and Management Science (ALOMS) Book Series (ISSN 2327-350X) is published by IGI Global, 701 E. Chocolate Avenue, Hershey, PA 17033-1240, USA, www.igi-global.com. This series is composed of titles available for purchase individually; each title is edited to be contextually exclusive from any other title within the series. For pricing and ordering information please visit http://www.igi-global.com/book-series/advances-logistics-operations-management-science/37170. Postmaster: Send all address changes to above address.

Titles in this Series

For a list of additional titles in this series, please visit: www.igi-global.com

Cases on Management and Organizational Behavior in an Arab Context
Grace C. Khoury (Birzeit University, Palestine) and Maria C. Khoury (Independent Researcher, Palestine)
Business Science Reference • copyright 2014 • 423pp • H/C (ISBN: 9781466650671) • US $175.00 (our price)

Management Science, Logistics, and Operations Research
John Wang (Montclair State University, USA)
Business Science Reference • copyright 2014 • 358pp • H/C (ISBN: 9781466645066) • US $185.00 (our price)

Outsourcing Management for Supply Chain Operations and Logistics Service
Dimitris Folinas (Department of Logistics, ATEI-Thessaloniki, Greece)
Business Science Reference • copyright 2013 • 596pp • H/C (ISBN: 9781466620087) • US $185.00 (our price)

Operations Management Research and Cellular Manufacturing Systems Innovative Methods and Approaches
Vladimir Modrák (Technical University of Kosice, Slovakia) and R. Sudhakara Pandian (Kalasalingam University, India)
Business Science Reference • copyright 2012 • 368pp • H/C (ISBN: 9781613500477) • US $185.00 (our price)

Fashion Supply Chain Management Industry and Business Analysis
Tsan-Ming Choi (The Hong Kong Polytechnic University, Hong Kong)
Information Science Reference • copyright 2012 • 392pp • H/C (ISBN: 9781609607562) • US $195.00 (our price)

Supply Chain Optimization, Design, and Management Advances and Intelligent Methods
Ioannis Minis (University of the Aegean, Greece) Vasileios Zeimpekis (University of the Aegean, Greece) Georgios Dounias (University of the Aegean, Greece) and Nicholas Ampazis (University of the Aegean, Greece)
Business Science Reference • copyright 2011 • 338pp • H/C (ISBN: 9781615206339) • US $180.00 (our price)

Quality Management for IT Services Perspectives on Business and Process Performance
Claus-Peter Praeg (Fraunhofer Institute for Industrial Engineering, Germany) and Dieter Spath (Fraunhofer Institute for Industrial Engineering, Germany)
Business Science Reference • copyright 2011 • 348pp • H/C (ISBN: 9781616928896) • US $180.00 (our price)

Managing Risk in Virtual Enterprise Networks Implementing Supply Chain Principles
Stavros Ponis (National Technical University of Athens, Greece)
Business Science Reference • copyright 2010 • 408pp • H/C (ISBN: 9781615206070) • US $180.00 (our price)

Table of Contents

Detailed Table of Contents

The nature of emerging business technologies, such automatic identification, and data capture innovations, such as smart cards, touch memory, and RFID, proves to be a difficult process to implement and achieve, despite recent efforts, even as implementation continues to lessen. The economy of scales associated with bar codes is a difficult barrier to overcome. Two Pittsburgh-based companies are showcased in this case study, namely Mobile Aspects, Inc., a recognized leader in the integration of RFID-related technologies in inventory management processes of large acute care, and Vocollect, Inc., which is especially noted for its voice-recognition software and small hardware platforms used in warehousing and partially automated inventory systems. The firms' goals for implementing inventory management, specific inventory recommendations and changes, comparison of inventory management processes, selected measures to ensure the quality and security of data transmitted via RFID-based technologies, and lessons learned are discussed.

An Integer Programming (IP) model is formulated and optimally solved for a real-life production planning problem. The model is used to determine the optimum production plan for the formation stage of battery manufacturing in a local battery-producing company. Battery manufacturing is a complicated process that involves several stages and several chemical and physical operations. The formation stage is considered the bottleneck of this process because it has the most critical limitations of time and production resources. In the formation stage, batteries are filled with acid and charged with electricity using different types of circuits. The objective of the model is to maximize the productivity of the critical formation stage by allowing the best utilization of the limited time and equipment resources. The model is able to consider a large number of battery sizes and types and also a large number of charging circuits with different capacities and charging speeds. The model-generated optimum production plans increase daily profits by 12% on average in comparison to manually-generated production plans.

Multimodal supply chains are characterized by multiple changes of transport modes, vehicles, and transport operators, and hence, the risks for theft, untimely delivery, and freight quality deterioration increase. There is a growing need to manage the security and efficiency of consignments. Tracking and tracing services, which are on the market, mainly concentrate on a single mode of transport, such as road vehicles, or transport units, such as containers. This chapter describes a concept for managing the security of multimodal supply chains using monitoring technology. A profound risk analysis was performed to identify threats and vulnerabilities during the different phases of the supply chain and key threat scenarios. A security service concept was developed to cope with the risks using monitoring technology. The technological solution depends on the needs of the supply chain actors and on the supply chain properties. The service concept was tested by monitoring different international shipments leaving from Finland. Intelligent monitoring devices, which were attached to the consignments, collected information on location and environmental parameters and transmitted in real-time to a background system. The chapter gives an overview of the framework, service concept, and the analysed results of multimodal shipments.

The use of RFID (Radio Frequency Identification) tracking in closed systems is replacing barcode as dominant tracking system in many industrial sectors because RFID enables reading multiple objects simultaneously without visual contact. To enlarge tracking systems to cover global supply chains, all aspects related to RFID, such as radio frequencies, data content, transmission protocols, and message sets, need to be standardised. By collecting, processing, and distributing information efficiently, organisations should be able to improve the efficiency of their transport logistics processes, lower their operational costs, and improve their portfolio of logistics services. This chapter describes the current perspectives, challenges, and benefits of RFID tracking applications in manufacturing industry. The perspectives derived from review of previous research are validated by using case study method.

While developing the inventory model with shortages under permissible delay in payments, it has been observed in the literature that the researchers have not considered the fact that the retailer can earn interest on the revenue generated after fulfilling the outstanding demand as soon as he receives the new consignment at the start of the cycle. Owing to this fact, the present study investigates the impact of interest earned from revenue generated after fulfilling the stock out at the start of the cycle on a single commodity inventory model with shortages for deteriorating item, in which the whole lot goes through an inspection process on arrival before entering into inventory system, under the conditions of permissible delay in payments. After inspection, the non-defective items are retained to fulfill the demand and the defective items are returned to the supplier. The results have been demonstrated with the help of a numerical example using the tools of Matlab7.0.1.

We can address innovation from different perspectives. In engineering practices we can look to changes resulting from attempts to discover ways of overcoming difficulties. How can we manage these innovative practices in engineering design projects? Furthering our perspective we use an actor-network way to look at change processes as chains of translations between heterogeneous actors that are enrolled in changes and where patterns of action are inscribed in durable manners. In an actor-network, the chains of inscriptions are stronger if the number of aligned actors is bigger, and this is the case of engineering projects where the cost of change builds up with time. Through the use of some stylized situations, the authors construct a Perturbation Index to obtain numeric values to assess the dynamics of innovations in engineering practices. The aim is that the application of this index to real situations could lead to meaningful descriptions of such innovation processes. Managing innovation in engineering design projects has to do with the management of project scope. The proposal extends scope management from its definition and planning phases through the control of changes along the execution.

This chapter provides a critical overview of Linear Programming (LP) from a manager's perspective. The main objective is to provide managers with the essentials of LP as well as cautionary notes and defenses on common modeling issues and software limitations. The authors illustrate the findings by solving a simple LP directly on the original decision variables and constraints space without adding new variables or translating the model to fit a specific solution algorithm. The aims are the unification of diverse set of topics in their natural states in a manner that are easy to understand and providing useful information to the managers. The advances in computing software have brought LP tools to the desktop for a variety of applications to support managerial decision-making. However, it is already recognized that current LP tools, in ample circumstances, do not answer the managerial questions satisfactorily. For instance, there is a costly difference between the mathematical and managerial interpretations of sensitivity analysis. LP software packages provide one-change-at-a-time sensitivity results; the authors develop the largest sensitivity region, which allows for simultaneous dependent and/or independent changes, based on the optimal solution. The procedures are illustrated by numerical examples including LP in standard-form and LP in non standard-form.

Governing extended networks is often associated with having the latest collaborative technology and business processes. However, current studies suggest that both are only part of the answer. This chapter argues that a holistic governing strategy needs to be adopted to achieve peak network performance. That strategy requires four conditions: 1) that each organization is ready to work with fellow business partners; 2) that network orchestrators recognize that all business relationships are not created equal and therefore need to be governed differently; 3) that a holistic governing strategy has to be adopted in order to integrate the people, technology, and processes in place; and 4) that people engagement hinges upon six factors (trust, beliefs, communications, culture, reward systems/metrics, and synergy). This chapter is a modification of an earlier work published by IGI Global.

Industrialization leaves no doubt that our planet is suffering from global warming, depletion of natural resources, pollution, waste, and other environmental concerns. Consequently, businesses, communities, and governments have been environmentally conscious and shown growing concern for sustainable development, particularly following the establishment of the United Nations Environmental Program (UNEP). This chapter provides further insights into sustainability and supply chain research through adopting a stakeholder perspective to understand drivers and consequences of supply chain sustainability in the United Arab Emirates. The results indicate that pressures enforced by champions and customers are positively related to supply chain sustainability. However, the relationship between government pressures and supply chain sustainability is insignificant. Moreover, the findings provide evidence that the impact of supply chain sustainability on the organizational sustainable performance is significantly positive. Supply managers can learn from these results in developing sustainable initiatives earlier along their supply chain through selecting and evaluating suppliers based on sustainability-related standards. In addition, environmental collaboration with customers and suppliers based on knowledge sharing and application may identify and reduce the total environmental impact.

Success in project management, and particularly in large, high-technology projects, continues to test the resources of organisations and their sponsors. This chapter revisits the conclusions of an earlier meta-study (Crosby, 2012a) that examined a large number of published case investigations and research efforts relating to the success and failure of projects. In that study, the success factors for general, and high-technology, projects were grouped and ranked as strategic success drivers for use prescriptively by project practitioners and approvers, and the principal drivers were examined closely to reveal any less obvious characteristics influencing project success. This chapter takes the original findings of ranked success drivers and investigates how these align with the experiences of three large contemporary high-technology projects. The conclusions show that, while the original set of drivers remains valid as predictors of project success, the ranking is likely to vary, even between projects that are technically and structurally similar. Two additional success factors are added as a result of the present study.

It is well over 30 years since the first (then called) Cockpit Resource Management (CRM) training, now called crew resource management was introduced. It is a shibboleth, a sacred cow as it were, despite many issues, concerns, and changes over the years. Some 21 years ago, 1992, an Air Transport Association (ATA)/Federal Aviation Association (FAA)-Sponsored Workshop was convened in an attempt to deal with some specific CRM issues. Yet the issues and needs as articulated in that workshop, and some newer ones, remain. Thus, this chapter is 21 years overdue, leading to the questions: Why now and is it still relevant? As said, some needs, issues, and concerns remain. The relevancy is that both a critique of civil aviation CRM on many levels and a comparison with current USAF, USCG, and USN CRM are presented. The proposed skeletal template for the long-overdue revision of civil aviation CRM, the R-MPM is shown. Next, a new model for an intelligent cockpit automated decision aid/advisory system, Event Response Integrated Decision Advisories (ERICA), is shown. ERICA came about from 2009-2012 work in automated decision-making tools for the cockpit and the realization that the Revised Mission Performance Model (R-MPM) and ERICA were interrelated.

Highly industrialized services add value to a company, but at the same time they can destroy it, should an almost identical proposal be made to two different clients who are seeking different experiences. The analysis of human subjectivity shows that the interaction with services goes beyond its mere use, as it is related to human beings' search for meaning, and it can potentially become a part of its users' biography and identity. The present chapter compiles some of the models that can contribute to getting over the paradox of service industrialization and which are here divided into two types: those that allow for a deep knowledge of the customer and those that are based upon designing experiences as a value proposition. The present chapter also suggests a research agenda that aims to get over the paradox of service industrialization.

This paper reports the findings of a questionnaire survey on the adoption and utilization of information and communication technology (ICT) – an important tool to enhance efficiency and responsiveness in modern-day supply chains – in the Chinese third-party logistics (3PL) industry. As a global manufacturing base, China has a burgeoning 3PL industry serving a large number of domestic firms as well as multinational corporations. With more and more organizations outsourcing their logistics function to 3PL firms, the latter have to make use of ICT for efficient communication with clients and coordination of activities in order to work hand-in-hand with customers to meet their day-to-day logistics needs. To investigate the status quo of ICT utilization in China, a questionnaire survey was conducted in 2009 to investigate the level of ICT adoption by the 3PL firms. For comparison, the findings of the Annual Third-Party Logistics Studies from 2007 to 2013 in this regard were also analyzed. The questionnaire survey results reveal that ICT is being widely adopted in the Chinese 3PL industry suggesting a high level of awareness of its significance and benefits to both the service providers and the clients. The ICT used ranges from low-tech telephone, facsimile, Internet access, to more advanced and sophisticated radio frequency identification technology and enterprise resource planning system. While small 3PL companies are using less expensive ICT at a tactical level mainly to cut costs and reduce errors in the day-to-day logistics operations, medium-sized and large firms have evolved to make use of more expensive ICT for planning and strategic purposes, such as business control, customer integration, and service differentiation. Transportation management system is regarded as the most important IT for business

Preface

INTRODUCTION

Globalization, competition, and increasingly sophisticated and informed customers are creating greater challenges for today's businesses. While achieving supply chain excellence in the face of these challenges is becoming more difficult, organizations that lead in supply chain improvement may be able to develop and maintain competitive advantages. The aim of Supply Chain Management is to produce, distribute, and deliver goods and services at the right location, right time, and right amount to satisfy customers, with minimum time and cost wastes. Technological innovations and applications bring new values to logistics and supply chain management. Newly developed technologies and techniques have substantially facilitated companies to achieve their goals to reduce costs, improve service, adding more values to their customers. New technologies, such as Geographical Information Systems (GIS) and Radio Frequency Identification (RFID), have changed traditional logistics and supply chain operations. In particular, RFID has emerged as part of a new type of inter-organizational intelligent system that aims to improve the efficiency of supply chains. It has become a new and exciting area of technological development, and is receiving increasing amounts of attention from both academicians and practitioners.

Green Supply Chain Management (GSCM) is an approach that targets at the overall optimization of material and information flows along the value chain. The main aspect is a stronger focus on ecological and sociological aspects when making organizational decisions. A related concept is corporate sustainability. Sustainability offers companies more opportunities to save costs, increase efficiency, and win new customers and suppliers. Even further, it incorporates the potential to gain a competitive advantage and to generate more profits by employing a variety of new practices and techniques such as reuse, recycle, reengineering. A common way to evaluate sustainability is through the lens of the Triple Bottom Line (3BL), which defines as: a) social responsibility, b) environmental stewardship, and c) economic viability. These pillars are sometimes termed "People, Planet, Profits." Social responsibility deals with such concerns as whether the firm provides a safe working environment. Environmental stewardship addresses how can a firm avoid depleting resources, prevent pollution, or otherwise reduce its ecological footprint. Economic viability concerns whether the firm is profitable, and can it be expected to grow and prosper, providing returns to investors? A firm that is deficient in any of these three facets is ultimately not sustainable. A company focusing on this triple bottom line through sustainability needs to investigate and apply changes to all upstream and downstream parties of its supply chain, as well as external stakeholders. This challenge includes questions at both strategic and operational levels that should be highlighted in its supply chain strategy.

In today's ever-growing competitive environment, effective supply chain management is progressively dependent on Third-Party Logistics (3PL) companies' functioning. Advancement in information technologies has brought all-encompassing values to Third-Party Logistics (3PL) companies. More often than not, three parties including the producer, the distributor, and the third-party logistics provider are involved in the supply chain. The occurrence of the 3PL provider in the entire supply chain has a noteworthy impact on its performance (Gotzamani, Longinidis, & Vouzas, 2010; Cai, et al., 2013). In order to stay competitive, all companies must identify the specific area(s) with high potential value, optimize their restricted resources, and enhance the most value to the whole supply chain, a value-added chain. E-procurement plays an ever-increasing role.

The purpose of this study is to explore some important issues in supply chain operations and discuss their applications, benefits, challenges, as well as future directions, providing some foundations for further discussions and research.

RFID INVENTORY CONTROL

Radio Frequency Identification (RFID) has become a hot topic and application in manufacturing and logistics industries for over a decade. RFID has emerged as part of a new pattern of inter-organizational intelligent system that targets to improve the efficiency of the processes in supply chains. It has become a new and exciting area of technological development, and is receiving increasing amounts of attention from both academicians and practitioners. Of the many organizations having implemented RFID technology, the one that received the most media attention is the retailing giant Wal-Mart. Wal-Mart has arguably one of the most efficient supply chains in the world, which include large selections of suppliers from many different industries, countries, and regions. Wal-Mart's RFID effort has the potential to cause a large wave of RFID implementations in many fields, ranging from manufacturing, packaging, to logistics and distribution. Wal-Mart's own estimated savings would reach $8.35 billion per year if it fully implemented RFID throughout its operations. Though this savings is quite substantial, Wal-Mart has taken its RFID implementation fairly slowly.

Wal-Mart and its suppliers quickly found several challenges in its RFID implementation – the UHF frequency they were using as a standard would not pass through many common products shipped to retail stores, such as water-based products and items shipped in metal packaging. Due to such challenges, by 2005, the top 100 suppliers had only tagged about 60 percent of their products. The development of RFID, such as the momentum provided by Wal-Mart along with the US Defense Department, and some European companies such as Tesco, has caused companies to take a hard look at what RFID can do for them and whether they should give further consideration to adopting the technology.

RFID plays a key role in bridging the information gaps among originations and is used to build up an "Internet of things" – a network that would allow companies to track goods through the global supply chain and run many applications simultaneously (Violino, 2005). Thonemann (2002) reports that after the deployment of RFID technologies, Procter & Gamble and Wal-Mart simultaneously reduced inventory levels by 70%, improved service levels from 96% to 99%. They also reduced administration costs by re-engineering their supply chains.

According to the market research firm IDTechEx (Das, 2005), the sales of RFID tags for the 60 years up to the beginning of 2006 reached 2.4 billion, with 600 million tags being sold in 2005 alone. It was expected that 1.3 billion tags and 500 million RFID smart labels would be needed in a range of areas, including retailing, logistics, animals and farming, library services, and military equipment. Similarly, Bagchi et al. (2007) forecast the prediction of RFID growth as from $1 billion in 2003 to $4 billion in 2008 to $20 billion in 2013. Though not without issues and challenges, RFID is a promising technology which analysts expect to become ubiquitous in the coming years, facilitating organizations to solve problems in supply chain management, security, personal identification, and asset tracking.

Along with the rapid development of RFID applications, RFID is an exciting area for research due to its relative novelty and exploding growth. RFID research has led to the emergence of a new academic research area that builds on existing research in an array of disciplines, including engineering, information systems, computer science, operations research, and business strategy, and there has been a significant increase in the number of papers on RFID in research journals, especially on the subject of production and inventory control.

Gunesekaran and Ngai (2005) highlight RFID technology as one of the important information technologies for Build-to-Order Supply Chain (BOSC) that increases enterprises efficiency and accuracy in production and inventory management. Similarly, Nemeth et al. (2006) present a review on RFID systems and the challenges and possibilities of the integration to supply chains. Chao et al. (2007) provide a review of the literature on trends and forecast of RFID technologies by a historical review method and bibliometric analysis. They focus on the RFID innovation, deployment by enterprises and market diffusion in supply chain management.

Inaccuracy issues in inventory management are important in supply chain management. Kang and Gershwin (2004) argue that although many organizations have automated their inventory control using information systems, inventory levels in information systems and the real physical inventory levels often do not match. Dehoratius and Raman (2008) report that 65% of the inventory records in retail stores were inaccurate. The result was obtained through a case study, by examining about 370,000 inventory records from 37 stores of a chain retailer. Delaunay et al. (2007) conduct a survey on the causes of inventory inaccuracy problems in supply chain management. Dolgui and Proth (2008) also present a literature review on RFID technology in supply chain. They focused on the advantages of this technology in inventory management. They analyze some problems and present ideas dealing with privacy this technology to supply chains is not only in increasing the efficiency of systems but also in supporting the reorganization of the systems that become more efficient.

The causes of inventory inaccuracy can be roughly categorized into four groups: transaction errors, shrinkage errors, inaccessible inventory, and supply errors. Iglehart and Morey (1972) first introduced transaction errors in inventory management. Transaction errors include shipment errors, delivery errors, scanning errors and also incorrect identification of items (Lee, Cheng, & Leung, 2005). Shrinkage errors (also named stock loss) include all types of errors that cause loss of products ready for sale. Hollinger and Davis (2001) report that shrinkage errors represent 1.69% of sales for retailers. Shrinkage errors include employee theft, shoplifting, administration and paperwork errors, vendor fraud and unavailable products for sale. Decrease shrinkage levels, increase profits.

Inaccessible inventory can be defined as products that are not in the correct place and are not available for customers. Inaccessible inventories have been studied widely. Raman et al. (2001) identify that misplaced items could reduce profits by 25%. Product quality, yield efficiency, and supply process can all affect inventory accuracy (Rekik, et al., 2007). RFID technologies provide better product traceability through its real time data capture abilities that enable improvements in the supply chains against these inventory inaccuracy issues. Although RFID cannot eliminate all errors, they can be detected quickly and they can be dealt with effectively. Zipkin (2006) shows that RFID is in particular very successful to eliminate transaction errors.

Replenishment policies are very important decisions for determining the size and timing of orders to lower ordering, holding and stock-out costs. There are several inventory policies available under continuous or periodic review inventory systems. Inventory replenishment decisions are made based on inventory levels in the information system. RFID technologies provide real-time inventory information and ensure the accuracy of inventory levels. The effects of RFID technologies on replenishment policies have been studied by Kok and Shang (2007), Lee et al. (2005), and Kang and Gershwin (2004) among others.

Inventory control could also benefit from RFID to improve inventory visibility through decrease shrinkage levels, increase profits. Customer service and the shopping experience can be enhanced by providing complementary applications enabled by RFID by manufacturer or supplier. Stock out cases can be decreased by increased inventory visibility. Decreased stock outs increase sales and ultimately, increase profits.

SUSTAINABLE SUPPLY CHAIN MANAGEMENT

Sustainable business models have received increasing attention from business practitioners and academic researchers in recent years. The growing importance of sustainable supply chain management is driven largely by the escalating deterioration of the environment, diminishing natural resources, expanded legislation, intensified global competition, and value-seeking initiatives. Sustainable supply chain management, green movement, along with reverse logistics, have their roots in both supply chain management and environmental literature. Adding the sustainability component to supply chain management involves addressing the influence and relationships between traditional supply chain management and the sustainable environment. Murphy et al. (1995) conducted a survey of more than 100 corporate executives and revealed that 60% of the group considered sustainability to be highly important and 82% expected that its importance would further increase in the years to come.

On the other hand, research and practice on sustainable supply chains have a relatively short history and issues at both the strategic and tactical levels are largely untouched. As sound reverse logistics networks serves as the foundation for efficient and effective sustainable supply chains where industrial ecology, profitability, and substantiality are primary concerns, a better understanding of the current situation of research on such topics is needed. Fundamentals of sustainable supply chain management as a competitive initiative are explained by Porter and van der Linde (1995), which argue that investments in sustainability lead to resource saving, waste eliminating, and productivity improving. As a consequence, sustainable supply chain management assists to lower the environmental impact of a firm, enhance its business efficiency, and possibly create major competitive advantages through operations and innovations.

CSCMP states logistics involves "the management of the flow of goods, services, and related information between the point of origin and the point of consumption, in order to meet customer requirements." Logistics deals with the movement and storage functions of the supply chain. The energy used in transport and in controlled climate storage is substantial. In particular, energy consumption across the US will increase by 1.5% annually in the commercial sector and 0.5% in the industrial sector. Between 1996 and 2006, the price of energy in the US more than doubled. This makes energy costs in the commercial industry have an average increase of close to 10% per year, which creates a substantial burden on most companies.

Sustainable supply chain management covers a variety of business activities, such as sustainable design, including product design for environmental compatibility and recovery, Environmental Conscious Design (ECD), and Life Cycle Assessment (LCA); sustainable operations, including green engineering, network design for reverse flow, green/sustainable manufacturing (e.g. production planning and scheduling, inventory management), and waste management (e.g. resource reduction and disposal). The commitment follows a "top-down" approach and requires both cross-functional (individual company) and cross-organizational (supply chain partners) collaboration.

While increased attention has been paid on sustainable supply chain management by industrial practitioners, it has also become a fast-growing topic in academic research. Kopicki et al. (1993) present three management approaches for managing sustainable supply chains: reactive, proactive, and value-seeking. In the reactive approach, to achieve sustainability, companies comply with environmental regulations by committing minimal resources. They start to procure goods with some recycled contents and use filters to lower the environmental impact of production. In the second approach, companies tend to expand the resources devoted to sustainable management. Recycling, reuse, and sustainable design are conducted by the firms to pre-empt new environmental regulations. Value-seeking is the most far reaching approach, where companies actively integrate sustainable activities into a business strategy and operate to reduce the impact on the environment as a strategic initiative.

Quantitative methods have been extensively applied in the research to improve sustainability in supply chains mathematical programming, such as Linear Programming (LP) and Mixed Integer Programming (MIP), is first widely used for modeling production lot sizing and routing problems in reverse logistics systems. A stream of research is particular focusing on the integrated production, inventory, and distribution problems with reverse flows. Del Castillo and Cochran (1996) study a production and distribution problem for products delivered in reusable containers. The model is defined for a two-stage network with a forward flow from a plant to a depot and a reverse flow of empty containers back to the plant. A hierarchical solution approach is proposed, where LP models are designed to determine the master production-distribution plan. Chen et al. (2007) explore a container vessel scheduling problem with bidirectional flows between a single origin and a single destination, where the forward flow sends customer orders packed in cargo containers from the origin port to the destination port, while the reverse flow brings empty containers back to the origin. The problem is solved through a linear program and a greedy heuristic method. Lei et al. (2009) study an integrated production and distribution problem motivated by the practice of a medical equipment leasing and service network that supplies, repairs, inventories, and distributes both new and refurbished medical devices to business customers. A partial LP relaxation-based heuristic approach is proposed to solve the problem.

Remanufacturing and inventory control are important areas. Inderfurth et al. (2001) develop a periodic review model for studying product recovery in stochastic remanufacturing systems with multiple reuse options. Mahadeven et al. (2003) focus on a product recovery and production-inventory control model. A remanufacturing facility that receives a stream of returned products according to a Poisson process is analyzed. Several heuristics based on traditional inventory models are proposed. Takahashi et al. (2007) consider a remanufacturing system with reproduction and disposal. Two control policies are provided and Markov analysis is used to evaluate the performance. Product take-back systems are analyzed in Klausner and Hendrickson (2000). An alternative take-back system would combine profitable remanufacturing and unprofitable materials recycling. A model that allows determining the optimal amount to spend on buy-back and the optimal unit cost of reverse logistics is established, which can be used to select a suitable reverse logistics system for EOL products.

Listes and Dekker (2005) present a stochastic programming based approach by which a deterministic location model for product recovery network design may be extended to explicitly account for the uncertainties. Stochastic models are used to understand a representative real case study on recycling sand. Recently, El-Sayed et al. (2010) develop a multi-period, multi-echelon, forward-reverse logistics network design under risk model. The network structure consists of three echelons in the forward direction (suppliers, facilities, and distribution centers) and two echelons in the reverse direction (disassembly and redistribution centers), with stochastic demands. The problem is formulated in a Stochastic Mixed Integer Linear Programming (SMILP) decision making form as a multi-stage stochastic program.

A variety of issues are addressed in the research on reverse logistics and sustainable supply chain management with more attention given to green operations. Both quantitative and qualitative approaches are used, where research concerned with strategic issues are conducted by using qualitative methods such as case study and conceptual descriptions, and tactical and operational decisions are analyzed by using quantitative techniques such as linear programming and stochastic programming. It can be observed that there has been a trend to use mathematical models to analyze green supply chain problems in recent years.

THIRD-PARTY LOGISTICS (OUTBOUND LOGISTICS)

Using *Agency Theory*, Whipple and Roh (2010) assess the likelihood of quality fade in buyer-supplier relationships and prescribe contractual mechanisms for reducing quality fade. A 2×2 matrix is proposed that contrasts outcome measurability with outcome uncertainty to illustrate buyer and supplier vulnerability (quality fade) and to suggest contractual mechanisms that can be used to mitigate vulnerability for both parties. A case of zero-inventory production-distribution systems under pool-point delivery has been studied by Geismar, Dawande, and Sriskandarajah (2011). Including multiple trucks, a modifiable production rate, and alternative objectives, the authors develop intuition into this case. Combining theoretical analysis and computational experiments, they gain insights into optimizing the total cost of a production-delivery plan by understanding the trade-off between production and transportation.

In order to increase customer loyaty, Jiang, Wang, and Ding (2013) bring disruption management idea into Vehicle Routing Problem (VRP) and present a delivery service priority method. Both a single- and a multi-vehicle delivery delay disruption management recovery models with service priority are constructed to ensure all the key customers' service level with distribution delay. They used an improved Genetic algorithm to solve the problem. Applying traditional Multiple Attribute Decision-Making (MADM) for selecting 3PL providers has many shortcomings, especially for NP-complete problems. Based on Fuzzy Analytic Network Process (FANP) and Preemptive Fuzzy Integer Goal Programming (PFIGP), Wong (2012) propose a Decision Support System (DSS) for a critical issue of selecting logistics outsourcing providers under the trends of globalization. The model takes into consideration flexible resource and interactions among providers. After obtaining experts' scores of the providers, the author integrates the scores into PFIGP solves it via genetic algorithms.

From a provider's perspective, Large, Kramer, and Hartmann (2011) investigate the potential impact of customer-specific adaptations by Third-Party Logistics (3PL) providers on the success of 3PL-relationships. Based on research in 3PL, relationship marketing and transaction cost theory, the authors present a document analysis, develop and test hypotheses with structural equation modelling and causal analysis with partial least square. Their study provides evidence that customer-specific adaptation by providers is an important prerequisite to 3PL-performance. Building upon the basic notions of service theory, competence research and the resource based view, Prockl, Pflaum, and Kotzab (2012) develop a general frame of reference addressing the value proposition and the value creation architecture that leads to generic business model configurations for contract logistics services. Combining the dimensions of integration power and intangible knowledge creation, the authors specify generic types of contract logistics services. From the service provider's point of view, the authors deliver a set of distinct business models for 3PL services reflecting the customer's and addresse specific aspects of the generation/production of required services.

Supposing that logistics capabilities play a mediating role, Evangelista, Mogre, Perego, Raspagliesi, and Sweeney (2012) explore the relationship between IT and 3PLs' performance. Based on a questionnaire survey, the authors test a conceptual Resource-Based View (RBV) framework linking IT adoption, logistics capabilities and firm performance with factor analysis and Ordinary Least Square (OLS) regression analysis. Multidisciplinary approach combining management of information systems, strategy, logistics and supply chain management have been used. Rajesh, Pugazhendhi, and Ganesh (2011) devise a four-stage model for the development of a Generic Taxonomy Components Framework (GTCF) to implement Knowledge Management (KM) solution for 3PL service providers. The main components of the four-stage model are modified Q-sort method and Delphi analysis. The authors identified KM components and derived the hierarchical structure of the taxonomy. Using a novel neighborhood rough set approach, Bai and Sarkis (2013) introduce a performance evaluation of third party reverse logistics providers model and evaluate its robustness with various factors.

Reverse Logistics (RL) has attained more importance among practitioners and academicians due to growing environmental legislations recently. Govindan and Murugesan (2011) use fuzzy extent analysis to solve a Third-Party Reverse Logistics Provider (3PRLP) selection problem. The authors found that for a company to manage outbound logistics (Accenture supply chain management practice) takes 1/12 as many steps to process returns as it maintains the separate RL system. It stimulates many companies to outsource the RL activities or functions through 3PRLPs. As a highly multi-faceted methodology, fuzzy extent analysis requires more numerical calculations. Using structural equation modelling from a survey, Juga, Juntunen, and Grant (2010) investigate how perceived service quality influences both a shipper's satisfaction and subsequent loyalty in 3PL outsourcing relationships. The authors identified critical service dimensions and their impact on satisfaction and loyalty. A theoretical model is in turn examined empirically. Logistics service providers recognize the importance of service quality in outsourcing relationships.

Zhong and Zhou (2011) investigate the application of GPS-based information technologies to optimize operations of companies providing 3PL service. The authors discuss how these technologies help to enhance the effective and efficient management of their businesses and analyze the interaction of GPS implementation and several key characteristics of the logistic distribution context. Hingley, Lindgreen, Grant, and Kane (2011) investigate benefits of and barriers to the use of Fourth-Party Logistics (4PL) management as a catalyst for horizontal collaboration. The authors show that barriers to such integration are created by power plays among lead stakeholders in grocery retailing that inhibit horizontal collaboration. However, Sheffi (2012) noticed the problem of unemployment among less educated and less trained workers.

Trappey et al. (2010) apply a two-stage clustering approach combined with Ward's minimum-variance method and the K-means algorithm and help the logistics companies to prioritize their services to better satisfy groups of customers with specific preferences. The authors model logistic services as customizable services and develop a data system methodology to define the profiles of automobile manufacturers and their preferredlogistic services. Drawing on types of third party collections of returned products, Qian, Han, Da, and Stokes (2012) propose a 0-1 mixed integrate linear programming (0-1MILP) model for reverse logistics networks in e-business and determine the market demands and returns. Facing returned products from e-business customers, the authors identify possibilities for the application of the reverse logistics network models. Lin and Pekkarinen (2011) develop a framework of QFD (Quality Function Deployment)-based logistics service design to integrate the HOQ (House of Quality) technique and modular logic to help in designing logistics services with high quality and a large service variety.

E-PROCUREMENT SYSTEM/OUTSOURCING

In line with Barahona and Elizondo (2012), due to its characteristics, size, and impact, e-procurement has a strategic importance not only for public administration but for e-government, since its implementation necessarily crosses many institutional barriers and paradigms of many public managers. However, the realization of these savings requires user compliance. Using survey data, Brandon-Jones and Carey (2011) examine the extent to which user-perceived E-Procurement Quality (EPQ) influences both system and contract compliance and found a strong positive relationship between user-perceived EPQ and both system and contract compliance. Practitioners should focus on managing different dimensions of perceived quality as they may have totally different effects on both contract and system compliance. Also, Kersten, Vahidov, and Gimon (2013) notice that concession-making behavior is an essential process in negotiations and auctions and has critical impact on the outcomes of an exchange since it may achieve Pareto-optimal agreements.

According to McCue and Roman (2012), worldwide governments seem to identify and peddle technology as a way to transform how they govern. Public procurement is at the forefront of most reform efforts since it plays a significant role in promoting accountability and transparency. The authors found that digitalized public procurement has not yet led to significant transformative changes due to the many obstacles, such as unsuitability of software platforms, organizational resistance, lack of strategic systems' integration and failure to involve public procurement professionals in the design of e-procurement systems, etc. In order to capitalize on the potentially transformative nature of ICT in procurement, decision makers must take an active role in both the design of the software and its adoption process dealing with political, institutional and behavioral domains. Starting to discuss the evolution of e-procurement systems within the chemical industry as a whole, Taube and Bryant (2010) examine the use of Electronic Procurement within the Supply Chain Management systems of industrial firms. The authors found that both administrative acquisition costs and total purchase cost for all components, basic ingredients, and raw materials are cut down.

Rather than working on the macro-level impact of e-government on citizens' trust in government, Smith (2010) provides a deeper understanding of the complexity of the interaction between e-services and citizens' trust in government. A comparative case study of two e-services was conducted. The author presents an integrated conception of trust, institutional trust, and a comparative analysis of citizens' perceptions and interpretations of "successful" e-services. Griffiths and Payab (2010) examine many issues and solutions to the implementation of e-procurement systemswithin supply chains. The authors place these challenges into five categories: lack of standardisation, integration of e-commerce-driven transactions with other systems, immaturity ofe-procurement-based services, security fears and end-user resistance and maverick purchasing. Also, the authors offer some solutions to these challenges, such as providing appropriate infrastructure for e-procurement, effective supplier and contract management, the control of end-user behaviour in e-procurement activities and product differentiation and strategic alliances.

Quesada, González, Mueller, and Mueller (2010) examine the impact of electronic procurement technologies on Procurement Practices (PPR) and Procurement Performance (PP). A model of the relationships between E-Procurement Technology (EPT) usage, PPR, and PP was proposed, tested and validated. By providing an empirical test of the impact of EPTs on perceptions of PPR and performance, the authors' contribution is obvious. Hong, Dobrzykowski, and Vonderembse (2010) examine the use of specific supply chain Information Technologies (IT) for e-commerce, e-procurement, and Enterprise Resource Planning (ERP), when implementing lean practices to achieve Mass Customization (MC) performance with different focuses. After collecting data via survey, the authors utilize exploratory factor analysis, hierarchical multiple regression, and T-tests to analyze relationships between lean practices, IT use, and MC performance in aggregated and bifurcated samples of product and service focused manufacturers. The authors confirm that e-commerce use is a better predictor of performance than e-procurement or ERP for service focused manufacturers. Best practices related to lean practices, e-commerce, e-procurement, and ERP emerge among high MC performers. This paper is the first study to examine these three IT approaches in the context of lean practices and supply chain MC performance.

Azadegan and Teich (2010) develop propositions on the adoption of electronic data interchange and online auctions, two distinct types of E-Procurement Technologies (EPT). The authors highlight three prominent adoption theories explaining innovation adoption and show how benchmarking of innovation adoptions and EPT can be made more effective by applying a multi-dimensional perspective. Accounting for the economic planning and development of a nation, E-procurements are the major activities in G2B. Cloud computing provides the most possible solutions to the emerged transformations. Wang (2012) try to catch the current e-procurement flow and its bottlenecks if any. The author explore the necessary services and display an architecture of publice-procurement under the cloud. García et al. (2013) propose a new suppliers' evaluation-and-selection model based on a FDSS (Fuzzy Decision Support System). The proposed method is robust enough to improve the main shortcomings of more simplistic methods.

CONCLUSION

Globalization, competition, and increasingly sophisticated and informed customers are creating greater supply chain challenges for today's businesses. While achieving supply chain excellence in the face of these challenges is difficult, companies that lead in supply chain improvement may be able to build competitive advantages. Our study provides information on the current use of RFID technology in supply chain operations and its impacts on supply chain management systems. RFIDs have demonstrated tremendous opportunities for increasing value, reduce inventory costs, eliminate stock errors, and allow companies to update their logistics and inventory databases. As RFID technology can provide important business benefits, the results of this research deliver a better understanding of current problems and issues in RFID technology introduction and show which factors influence the level of success.

We also address the important issue of sustainable supply chain management. With a substantial large number of variables and constraints, the challenges with which we are dealing in sustainable supply chain systems are much more complicated than those in traditional settings. The inherent complexity of environmental issues poses serious challenges to business managers and researchers. There are certainly rewards for being an environmentally responsive company. Not only should the positive effect on a company's image be taken into consideration, there are also long-term benefits to a company's financial bottom-line. The primary needs from today's businesses are not tactical or operational in nature. In fact, they tend to be strategic and unstructured in general with more requirements in the areas of intra- and inter-firm collaboration and diffusion of best practices. Many changes in policies, practices, concepts, and technologies can be expected in the years to come, which can bring fresh opportunities for meaningful academic research in sustainable supply chain management.

For gaining competitive advantage, companies need to enhance value chain velocity, intensify customer service, and decrease total logistics costs, including consolidation, transportation and inventory carrying costs. Inbound and outbound logistics consists of many multi-party processes. Managing orders, appointments and shipments follows the data exchange on multiple networks. Certainly, the removal of the barriers between inbound and outbound operations is the first step for moving to a demand-driven value network.

John Wang
Montclair State University, USA

REFERENCES

Azadegan, A., & Teich, J. (2010). Effective benchmarking of innovation adoptions. *Benchmarking*, *17*(4), 472–490. doi:10.1108/14635771011060558

Bagchi, U., & Guirida, A. ONeill, L., Zeng, A., & Hayya, J. (2007). The effect of RFID on inventory management and control. Berlin: Springer.

Bai, C., & Sarkis, J. (2013). Flexibility in reverse logistics: A framework and evaluation approach. *Journal of Cleaner Production*, *47*, 306–318. doi:10.1016/j.jclepro.2013.01.005

Barahona, J. C., & Elizondo, A. M. (2012). The disruptive innovation theory applied to national implementations of e-procurement. *Electronic Journal of E-Government*, *10*(2), 107–119.

Brandon-Jones, A., & Carey, S. (2011). The impact of user-perceived e-procurement quality on system and contract compliance. *International Journal of Operations & Production Management*, *31*(3), 274–296. doi:10.1108/01443571111111928

Cai, X., Chen, J., Xiao, Y., Xu, X., & Gang Yu, G. (2013). Fresh-product supply chain management with logistics outsourcing. *Omega*, *41*(4), 752–765. doi:10.1016/j.omega.2012.09.004

Chao, C.C., Yang, J.M., & Jen, Y, W. (2007). Determining technology trends and forecasts of RFID by a historical review and bibliometric analysis from 1991 to 2005. *Technovation*, *27*(3), 268–279. doi:10.1016/j.technovation.2006.09.003

Chen, Z., Lei, L., & Zhong, H. (2007). Container vessel scheduling with bidirectional flows. *Operations Research Letters*, *35*(2), 186–194. doi:10.1016/j.orl.2006.03.010

Das, R. (2005). *RFID tag sales in 2005 - How many and where*. IDTechEx Ltd.

Dehoratius, N., & Raman, A. (2008). Inventory record inaccuracy: An empirical analysis. *Management Science*, *54*(4), 627–641. doi:10.1287/mnsc.1070.0789

Del Castillo, E., & Cochran, J. (1996). Optimal short horizon distribution operations in reusable container systems. *The Journal of the Operational Research Society*, *47*(1), 48–60. doi:10.1057/palgrave.jors.0470105

Delaunay, C., Sahin, E., & Dallery, Y. (2007). A literature review on investigations dealing with inventory management with data inaccuracies. In *Proceedings of RFID Eurasia*. RFID.

Dolgui, A., & Proth, J. M. (2008). RFID technology in supply chain management: State of the art perspectives. In *Proceedings of the 17th World Congress the International Federation of Automatic Control*, (pp. 4465-4475). Academic Press.

El-Sayed, M., Afia, N., & El-Kharbotly, A. (2010). A stochastic model for forward-reverse logistics network design under risk. *Computers & Industrial Engineering*, *58*(3), 423–431. doi:10.1016/j.cie.2008.09.040

Evangelista, P., Mogre, R., Perego, A., Raspagliesi, A., & Sweeney, E. (2012). A survey based analysis of IT adoption and 3PLs' performance. *Supply Chain Management*, *17*(2), 172–186. doi:10.1108/13598541211212906

García, N., Puente, J., Fernández, I., & Priore, P. (2013). Supplier selection model for commodities procurement. Optimised assessment using a fuzzy decision support system. *Applied Soft Computing*, *13*(4), 1939–1951. doi:10.1016/j.asoc.2012.12.008

Geismar, H. N., Dawande, M., & Sriskandarajah, C. (2011). Pool-point distribution of zero-inventory products. *Production and Operations Management*, *20*(5), 737–IX. doi:10.1111/j.1937-5956.2010.01190.x

Gotzamani, K., Longinidis, P., & Vouzas, F. (2010). The logistics services outsourcing dilemma: Quality management and financial performance perspectives. *Supply Chain Management*, *15*(6), 438–453. doi:10.1108/13598541011080428

Govindan, K., & Murugesan, P. (2011). Selection of third-party reverse logistics provider using fuzzy extent analysis. *Benchmarking*, *18*(1), 149–167. doi:10.1108/14635771111109869

Griffiths, G., & Payab, H. (2010). Supply on the heels of demand: Issues of e-procurement. *Journal of Global Business Issues*, *4*(2), 29–35.

Gunesekaran, A., & Ngai, E. W. T. (2005). Build-to-order supply chain management: A literature review and framework for development. *Journal of Operations Management*, *23*(2), 423–451. doi:10.1016/j.jom.2004.10.005

Hingley, M., Lindgreen, A., Grant, D. B., & Kane, C. (2011). Using fourth-party logistics management to improve horizontal collaboration among grocery retailers. *Supply Chain Management*, *16*(5), 316–327. doi:10.1108/13598541111155839

Hollinger, R. C., & Davis, L. (2001). *National retail security survey, report (Technical Report)*. Gainesville, FL: Department of Sociology and the Center for Studies in Criminology and Law, University of Florida.

Hong, P. C., Dobrzykowski, D. D., & Vonderembse, M. A. (2010). Integration of supply chain IT and lean practices for mass customization. *Benchmarking*, *17*(4), 561–592. doi:10.1108/14635771011060594

Iglehart, D. L., & Morey, R. C. (1972). Inventory systems with imperfect asset information. *Management Science*, *18*(2), 388–394.

Inderfurth, K., Kok, D. A., & Flapper, S. (2001). Product recovery in stochastic remanufacturing system with multiple reuse options. *European Journal of Operational Research*, *133*(1), 130–152. doi:10.1016/S0377-2217(00)00188-0

Jiang, L., Wang, H., & Ding, B. (2013). Disruption management recovery model of distribution delay with service priority. *Asian Social Science*, *9*(2), 170–179. doi:10.5539/ass.v9n2p170

Juga, J., Juntunen, J., & Grant, D. B. (2010). Service quality and its relation to satisfaction and loyalty in logistics outsourcing relationships. *Managing Service Quality*, *20*(6), 496–510. doi:10.1108/09604521011092857

Kang, Y., & Gershwin, S. B. (2004). Information inaccuracy in inventory systems – Stock loss and stockout. *IIE Transactions*, *37*(4), 843–859.

Kersten, G. E., Vahidov, R., & Gimon, D. (2013). Concession-making in multi-attribute auctions and multi-bilateral negotiations: Theory and experiments. *Electronic Commerce Research and Applications*, *12*(3), 166–180. doi:10.1016/j.elerap.2013.02.002

Klausner, M., & Hendrickson, C. (2000). Reverse logistics strategy for product take-back. *Interfaces*, *30*(3), 156–165. doi:10.1287/inte.30.3.156.11657

Kok, A. G., & Shang, H. K. (2007). Inspection and replenishment policies for systems with inventory record inaccuracy. *Manufacturing and Service Operations Management*, *9*(2), 185–205. doi:10.1287/msom.1060.0136

Kopicki, R., Legg, L., Berg, L., Dasappa, V., & Maggioni, C. (1993). *Reuse and recycling: Reverse logistics opportunities*. Oak Brook, IL: Council of Logistics Management.

Large, R. O., Kramer, N., & Hartmann, R. K. (2011). Customer-specific adaptation by providers and their perception of 3PL-relationship success. *International Journal of Physical Distribution & Logistics Management*, *41*(9), 822–838. doi:10.1108/09600031111175807

Lee, Y. M., Cheng, F., & Leung, Y. T. (2005). *A quantitative view on how RFID will improve a supply chain* (Technical report). IBM Research Report.

Lei, L., Zhong, H., & Chaovalitwongse, W. (2009). On the integrated production and distribution problem with bidirectional flows. *INFORMS Journal on Computing*, *21*(4), 585–598. doi:10.1287/ijoc.1080.0311

Lin, Y., & Pekkarinen, S. (2011). QFD-based modular logistics service design. *Journal of Business and Industrial Marketing*, *26*(5), 344–356. doi:10.1108/08858621111144406

Listes, O., & Dekker, R. (2005). A stochastic approach to a case study for product recovery network design. *European Journal of Operational Research*, *160*(1), 268–287. doi:10.1016/j.ejor.2001.12.001

Mahadeven, B., Pyke, D., & Fleischmann, M. (2003). Periodic review, push inventory policies for remanufacturing. *European Journal of Operational Research, 151*(3), 536–551. doi:10.1016/S0377-2217(02)00627-6

McCue, C., & Roman, A. V. (2012). E-procurement: Myth or reality? *Journal of Public Procurement, 12*(2), 221–248.

Murphy, P., Poist, R., & Braunschweig, C. (1995). Role and relevance of logistics to corporate environmentalism. *International Journal of Physical Distribution and Logistics Management, 25*(2), 5–19. doi:10.1108/09600039510083916

Nmeth, P., Toth, L., & Hartvanyi, T. (2006). Adapting RFID in supply chain. In *Proceedings of ICM 2006 IEEE 3rd International Conference on Mechatronics*, (pp. 263-266). IEEE.

Porter, M., & van der Linde, C. (1995). Green and competitive. *Harvard Business Review, 73*(5), 120–134.

Prockl, G., Pflaum, A., & Kotzab, H. (2012). 3PL factories or lernstatts? Value-creation models for 3PL service providers. *International Journal of Physical Distribution & Logistics Management, 42*(6), 544–561.

Qian, X., Han, Y., Da, Q., & Stokes, P. (2012). Reverse logistics network design model based on e-commerce. *The International Journal of Organizational Analysis, 20*(2), 251–261. doi:10.1108/19348831211227864

Quesada, G., González, M. E., Mueller, J., & Mueller, R. (2010). Impact of e-procurement on procurement practices and performance. *Benchmarking, 17*(4), 516–538. doi:10.1108/14635771011060576

Rajesh, R., Pugazhendhi, S., & Ganesh, K. (2011). Towards taxonomy architecture of knowledge management for third-party logistics service provider. *Benchmarking, 18*(1), 42–68. doi:10.1108/14635771111109814

Raman, A., Dehoratius, N., & Ton, Z. (2001). Execution: The missing link in retail operations. *California Management Review, 43*(3), 136–152. doi:10.2307/41166093

Rekik, Y., Sahin, E., & Dallery, Y. (2007). A comprehensive analysis of the newsvendor model with unreliable supply. *OR-Spektrum, 29*(2), 207–233. doi:10.1007/s00291-005-0025-0

Sheffi, Y. (2012). Driving growth and employment through logistics. *MIT Sloan Management Review, 54*(1), 20–22.

Smith, M. L. (2010). Building institutional trust through e-government trustworthiness cues. *Information Technology & People, 23*(3), 222–246. doi:10.1108/09593841011069149

Takahashi, K., Morikawa, K., Takeda, D., & Mizuno, A. (2007). Inventory control for a Markovian remanufacturing system with stochastic decomposition process. *International Journal of Production Economics, 108*(1), 416–425. doi:10.1016/j.ijpe.2006.12.023

Taube, L. R., & Bryant, M. (2010). The evolution of electronic procurement within supply chains. *The Business Review, Cambridge, 16*(1), 74–79.

Thonemann, U. W. (2002). Improving supply-chain performance by sharing advance demand information. *European Journal of Operational Research, 142*(1), 81–107. doi:10.1016/S0377-2217(01)00281-8

Trappey, C. V., Trappey, A. J. C., Ai-Che, C., & Huang, A. Y. L. (2010). Clustering analysis prioritization of automobile logistics services. *Industrial Management + Data Systems, 110*(5), 731-743.

Violino, B. (2005, November/December). Leveraging the Internet of things. *RFID Journal*, 1-2.

Wang, H. (2012). The planning of the architeture of a public e-procurement environment under the cloud - The case of Taiwan. *Journal of Global Business Management*, *8*(2), 134–143.

Whipple, J. M., & Roh, J. (2010). Agency theory and quality fade in buyer-supplier relationships. *International Journal of Logistics Management*, *21*(3), 338–352. doi:10.1108/09574091011089781

Wong, J. (2012). DSS for 3PL provider selection in global supply chain: Combining the multi-objective optimization model with experts' opinions. *Journal of Intelligent Manufacturing*, *23*(3), 599–614. doi:10.1007/s10845-010-0398-z

Zhong, H., & Zhou, B. (2011). Using information technology to optimize operations of third-party logistics provider. *International Journal of Management and Information Systems*, *15*(1), 115–121.

Zipkin, P. (2006). OM forum the best things in life were free: On the technology of transactions. *Manufacturing and Service Operations Management*, *8*(4), 321–329. doi:10.1287/msom.1060.0113

Chapter 1
Case Studies of RFID Practices for Competitive Inventory Management Systems

Alan D. Smith
Robert Morris University, USA

ABSTRACT

The nature of emerging business technologies, such automatic identification, and data capture innovations, such as smart cards, touch memory, and RFID, proves to be a difficult process to implement and achieve, despite recent efforts, even as implementation continues to lessen. The economy of scales associated with bar codes is a difficult barrier to overcome. Two Pittsburgh-based companies are showcased in this case study, namely Mobile Aspects, Inc., a recognized leader in the integration of RFID-related technologies in inventory management processes of large acute care, and Vocollect, Inc., which is especially noted for its voice-recognition software and small hardware platforms used in warehousing and partially automated inventory systems. The firms' goals for implementing inventory management, specific inventory recommendations and changes, comparison of inventory management processes, selected measures to ensure the quality and security of data transmitted via RFID-based technologies, and lessons learned are discussed.

DOI: 10.4018/978-1-4666-4506-6.ch001

INTRODUCTION

Operational and Strategic Issues Associated with RFID-Related Technologies

Currently, one of the latest packaging innovations and, perhaps most controversial, in retail stores is radio frequency identification, or RFID. RFID-related technologies, if properly implemented, have the potential to provide many advantages to both the service and manufacturing industries, despite the relatively large cash outlays initially required. Supplying more information than the standard barcode, eliminating the potential for inventory stock outs and reducing theft occurrences are some main concerns of users of such technology, but these concerns can be addressed and resolved. The global RFID market has been estimated to consists of US$5 billion per year and is entering a phase of rapid growth due to huge orders by China for its national ID card scheme, the adoption of e-passport in over 70 countries, and MasterCard for its Paypass™ program (Harrop, 2010; "RFID industry solutions," 2010; Sirico, 2010), with general purpose of identification and tracking via the utilization of radio waves. RFID-related technologies quickly gained attention because of its ability to track moving objects and has been used by thousands of companies for a decade or more (What is RFID?" 2010).

Purpose of Present Study

It is important for both service and manufacturing firms to develop and implement successful strategies that integrate RFID-embedded technological initiatives and information/knowledge practices and procedures, technological innovations, and customer relationship building of trust if they are to survive and, hopefully, prosper. To accomplish this task, this task, two major Pittsburgh, PA-based companies will be explored, through the selected company's description, background, a brief introduction to an industry overview in terms of information technology needs, partnering solutions, and product and/or service offerings as they related to RFID-based applications. The discussion that follows in the case study section tries to incorporate a customer relationship management (CRM) approach (Anton and Petouhoff, 2002; Smith, 2005a-b) in describing the types of customer-based problems and solution that RFID technologies are designed to address, both in terms of B2C (business-to-customer) and B2B (business-to-business) transactions.

As for the basis of the present study, the purpose of an RFID system is to enable data to be transmitted by a portable device, called a tag, which is read by an RFID reader and processed according to the needs of a particular application. The data transmitted by the tag may provide identification or location information, or specifics about the product tagged, such as price, color, date of purchase, to name a few applications. In particular, asset tracking, manufacturing, supply chain management, retailing, payment systems, and security and access control are illustrations of common uses of RFID (What is RFID?" 2010).

IMPLEMENTATION CONCERNS AND CUSTOMER SATISFACTION

Strategic Support for RFID-Related Technologies

RFID has been referred to as "not a science, but an art," by Neco Can, who spearheaded item-level RFID trials at U.S. clothing retailers Gap and Abercrombie & Fitch (Collins, 2006). In 2001, Can was the director of Gap's project management office, leading an RFID project that provided an inventory accuracy of up to 99.6%. Despite the success of the pilot in both improving in-store inventory and increasing sales, numerous factors caused the company to choose not to invest immediately in RFID tagging. This experience led Can to believe that any RFID project must be supported at the highest levels of management, but management needs to get the strongest support from its major stakeholders, in particular the support of the CEO. RFID is believed to have a slow start, which is inherent in the newness of the technology and ideas surrounding it. However, many proponents of RFID believe that in a few years this will become an industry standard. Through Can's experience in RFID retail implementation, he maintained that tags don't have to be much cheaper than they are now to deliver a return on investment. For Gap, it was 17-cents and at Abercrombie & Fitch, it was 24-cents per tag. Despite the proven benefits of this new technology, many retailers are still cautious about deploying the project; this is partially due to the many variables that must be managed to ensure adequate tag-read rates.

Besides the cost of RFID tag, the readers require substantial capital outlays as well, especially when compared to the unit profit of some higher volume items. When you are selling purses or shoes that are over US$800, for example, the retail industry would generally not have a problem placing at US$2 RFID tag on the merchandise; however, if that retailer would want to read this tag at a certain distance in order to have even less of a chance of theft, they would also need to purchase a stronger reader or active tags versus passive tags. As the distances the reader is supposed to cover increase, the more costly this technology can become.

With the use of RFID technology, there are some business and technical problems and issues that can arise. Some of these problems include: data sharing, data quality, costs and benefits, security and privacy, and RFID standards. Divergent factors can affect the perceptions of the internal and external stakeholders within an organization in the process of adopting RFID. Before attempting to share and use RFID data, an organization must consider potential costs in mastering collaborative planning and implementation with its partners. RFID researchers and practitioners were asked to provide their viewpoints on the application of RFID technologies in the study by Koong and Lin (2007). The most important factor involved in the adaptation of RFID was the cost consideration, particularly cost of RFID tags and hardware. Industry evaluators were extremely concerned about integrating their existing hardware and new RFID systems. Most organizations would prefer to retain their current system to avoid compatibility problems. Other criteria that fell under the cost consideration category were the software, system integration, operations, and human resources. The second most important concern was the system applicability factor. Was the product simplistic in nature? Did any interference occur with existing product materials? Security and privacy issues were also placed under the category of system applicability. Finally, data quality ranked third, in RFID-related systems implementation, data sharing is the use of a standardized data format to communicate between RFID supply chain suppliers.

When considering the application of RFID technologies, companies must be assured that the data are collected in a format that both the organizations and suppliers can interpret. Koong and Lin (2007) suggested that the obstacles of RFID can be minimized by reducing the total costs of RFID; resolving the interference problems; improving the identification accuracy; protecting intellectual property rights; establishing international standards; and developing better software supports.

RFID-Related Applications beyond Retail

Besides the apparent usage of RFID tags in the retail arena, there is much discussion of its advantages and disadvantages outside of stores. Some manufactures have been accused of using RFID tags to monitor the habits of its customers beyond store walls (Roberti, 2007; Schneier, 2006). Since finding this information, some buyers have shied away from the brand, uneasy about how they are gaining information about their buyers (what information are they collecting, how far the tags track, to name a few). Another item that has gained attention for the company is its Nike+Ipod Sport Kit™ combination, which combines Nike shoes with the Ipod nano. These products are supposed to work together to help the consumer gain information about their workout-time, distance, pace, and overall calories burned. Although this has become a popular kit for running enthusiasts, it has gained the attention of computer science researchers across the nation.

Although Nike has denied that it uses RFID-related applications for its Nike+Ipod Sport Kit™, researchers have studied the product and have concluded that the system basically uses radio waves and a unique identification to ensure one runner does not receive information from another consumer's sneakers (Roberti, 2007). This personally identifiable information may lead some to wonder how others could use the information from this product. Well-known security technologist and author, Schneier (2006), has voiced his concern of this item after researchers from the University of Washington developed a surveillance system that tracked people through the Nike/Apple product. With their program, which only cost them about US$250 to create, they were able to track the individual with the kit using Goggle Maps™. Schneier used the Nike/Apple product to further suggest that this demonstration can be used to persuade even skeptical individuals that RFID chips can be used to track people. "It's a good example because the chips have no personal identifying information, yet can still be used to track people. As long as the chips have unique IDs, those IDs can be used for surveillance" (Schneier, 2006, p. 2). It is certainly early enough for researchers and developers to standardize rules and regulations for what the capabilities of these tags should be. Since RFID tags have not technically created an issue for personal surveillance problems, lawmakers are somewhat reluctant to create regulations, although some popular press items and books leverage the concepts for invasion of public privacy.

Wal-Mart's previous mandate requiring their top 100 suppliers to utilize RFID by January 2005 has made them one of the leading drivers of this new technology ("Wal-Mart CIO to speak on RFID," 2003). The packaging industry has already begun to see drastic improvements in security, inventory control and the amount of information that can be monitored by placing tags on individual items, cases and pallet loads. The suppliers are learning new information capabilities of the tags and how they can play an integral role in future shopping experiences (Smith and Offodile, 2007, 2008, 2009). Human error is being reduced, inventory stockouts are lessening, and theft is decreasing; however, due to the many implementation problems the retailer has encountered, Wal-Mart has greatly relaxed some of its requirements for their 100 suppliers to utilize the new technology (Collins, 2005, 2006; Roberti, 2005a-c). RFID-related technologies are instead being used in multiple clothing stores, since the technical challenges are substantially reduced when the tag is applied to clothing ("Retailer tests RFID on garments," 2005).

RFID applications are expanding quickly and are still considered a relatively emerging technology in the retail world as a medium for information storage and communication (Smith, 2005, 2008). Whether or not RFID-related technologies are utilized as a location device or storage bank for a list of product information, the retail world is slowly but surely recognizing the benefits of this technology. RFID is being utilized in various venues such as electronics, clothing, produce, and much more. The tag incorporates information unique to the product. Retailers benefit from its functionality the minute the products using the technology enter the loading docks, until it leaves the front door of the store. With such a vast number of products and large amounts of information being exchanged between retailers at multiple locations in varying quantities and times, it becomes apparent that an efficient system of product information organization and communication be utilized to give retailers an edge over the competition ("Wal-Mart tackles out-of-stocks," 2005).

RFID information exchanges are particularly attractive alternatives to the more traditional barcodes, simply because of its efficiency, information accessibility, and individual accountability ("Fifty uses of RFID in retail" 2003). Information transfer will no longer require individual attention from a hand held scanner and barcode, but now only require a pass through an energy field sent from a handheld or mounted antenna. From here the required from the RFID tag will be read and stored in an appropriate computer software system. Once read, information such as manufacturing dates, transit duration, distribution center location, previous handler, price, and method of payment, date of expiration, warranty, and much more can be accessed and employed by retailers before even placing the product on the shelf.

Using the information collected through such tags and applying it to management strategies may also give retailers and consumers specific "on-shelf" product information. Examples of such information would include visibility of replenishment requirement, expiration dates, alerts, tracking, security, price changes, medicinal information, consumption trends and more (Kay, 2007; Lindsay, Reade, and Roth, 2003). Food and clothing retailers might receive specific information about ingredients or materials within the package that may cause an allergic reaction to those who use the item. Based on the profile of the purchaser, for example, an alarm or warning may be triggered to notify the consumer of potential harm at the point-of-sale. Similarly, drug retailers utilize a system in which an alert is sent when specifically tagged items are approaching or have passed expiration. In this instance, notification is made before the purchase. Smaller products that are easily concealed may provide tracking information after the product is removed from the shelf until the point of purchase. With this information, retailers eliminate time spent on manual data collection and entry, improve store efficiency and reduce costs throughout the store (Green and Khermouch, 2005).

With the advantage of RFID technological applications and its information exchange capabilities, some customers may feel that the exchange of too much information causes controversy between retailer and customer relationships (Anton and Petouhoff, 2002; Smith, 2005a). Certain RFID devices are able to hold significant amounts of information that is irrelevant to the product but valuable to the retailer. A primary security concern surrounds RFID technology, as with this technology, organizations can obtain unsolicited tracking of consumer location and analyze the shopping habits or behaviors of its customers. Therefore, assuring customers that retailers are not obtaining and exploiting personal information is necessary during customer and tag interaction. To combat this information intrusion, some companies have developed smart labels that include directive commands that restrict after-sale information collection ("RFID in retail-the Consumer in Control," 2005). By obtaining information through real-time customer purchases, retailers will be able to use RFID technologies to prompt in-store promotions, savings and discounts, allowing customers a technological advantage.

A basic RFID system consists of three components, namely an antenna or coil, a transceiver (with decoder), and a transponder or tag that usually is electronically programmed with unique information (What is RFID?" 2010). The antenna emits radio signals to activate the tag and to read and write data to it, while a typical RFID tag consists of a microchip attached to a radio antenna mounted on a substrate, which typically can store as much as 2 or more kilobytes of data. The reader emits radio waves in ranges from one inch to 100 feet or more, depending upon its power output and the radio frequency used. When an RFID tag passes through the electromagnetic zone, it detects the reader's activation signal. The reader decodes that data encoded in the tag's integrated circuit (silicon chip) and that data is passed to the host computer for processing.

Most RFID tags contain at least two parts, namely an integrated circuit for storing and processing information, modulating and demodulating a RFID signal, and other specialized functions, and an antenna for receiving and transmitting the signal. There are generally three types of RFID tags with an active RFID tag which contain a battery or power source and can transmit signals autonomously being the most expensive option. The other types, such as a passive RFID tag, which has no battery and requires an external source to provoke signal transmission, and a battery-assisted passive tag, which requires an external source to activate the forward link, are generally less expensive and may serve particular applications in a more cost sensitive manner (Smith and Offodile, 2009).

General Growth and Concerns of RFID-Related Technologies

The birth of RFID resulted from early-1900 predecessor technologies, such as IFF transponders in aircraft or covert listening devices. Mario Cardullo's U.S. Patent in 1973 was the first true ancestor of modern RFID; a passive radio transponder with 16-bit memory, followed by the first patent to be associated with the abbreviation RFID was granted to Charles Walton in 1983The basic patent covers the use of radio frequency, sound, and light as transmission media. The original use was for transportation, security, and financial services. Since then, RFID has found numerous uses as automobile vehicle identification, automatic toll system, personnel and medical identification.

Privacy and customer advocacy organizations have expressed concerns in the context of ongoing efforts to embed electronic product code RFID tags in consumer products. The use of RFID technology has engendered considerable controversy and even product boycotts by consumer privacy advocates. The major privacy concerns regarding RFID are that since the owner of an item will not necessarily be aware of the presence of an RFID tag and the tag can be read at a distance without the knowledge of the individual, it becomes possible to gather sensitive data about an individual without consent. The other concern is that if a tagged item is paid for by credit card or in conjunction with use of a loyalty card (Cowles, Kiecker, and Little, 2002; Devaraj, Fan, and Kohli, 2002), then it would be possible to indirectly deduce the identity of the purchaser by reading the globally unique ID of that item ("An Internet of objects-the future of RFID technology" 2009; Newland, 2005; O'Connor, 2004a-b). These privacy concerns need to be considered when choosing RFID as a tracking device in business operations since a brand name's reputation is sensitive to customers' perception and loyalty. Although some of the selected frequencies used for RFID in the U.S. are currently incompatible with those of Europe or Japan, suggesting that there is no emerging standard, RFID will probably not completely replace barcodes in the immediate future, but will complement them.

The general advantages of RFID-related technology only make sense where it provides advantages over barcodes. Some of the important advantages that RFID has over barcodes include the following (Harrop, 2010; Smith, 2005a, 2008):

1. Speed, since a RFID reader can read tags faster than a typically barcode scanner can scan barcodes. For example, RFID readers designed for supply chain operations can perform up to 1,500 read operations per second.
2. Read distance, as it is common to read RFID tags at distance of at least three meters between the tag and the antenna and the read field is three-dimensional.
3. Simultaneous as opposed to sequential scanning, since RFID readers can identify multiple tags within the read field.

4. Non-line of sight, as radio waves can penetrate most materials depending on the frequency utilized.
5. Durability, since RFID tags can operate in extremely rugged working conditions and can be encased in plastic or even embedded directly into the products manufactured.

However, bar codes can be easily printed or applied directly on the products it is designed to identify as extreme accurate read rates and very low costs.

Inventory and Operational Decisions and Cost Analysis

As illustrated in the case studies in the present study, the use of RFID tags in an organization can provide significant benefits to operations managers involved in inventory management and related supply management applications. Every business must balance the desire to cut costs through reduced inventory with the need to maintain safety stock as insurance against supply shortages. Inventory-related challenges include ensuring stock arrives at the right place in production at the right time, that the correct quantity is delivered, and shrinkage is reduced. By enabling operations managers to automatically and accurately track the flow of crucial materials through the facility, RFID tags can streamline processes and reduce costs, leading to a leaner organization (Hu, Wang, Fetch, and Bidanda, 2008; Jain, Benyoucef, and Deshmukh, 2008; Kennedy and Widener, 2008; Scherrer-Rathje, Boyle, and Deflorin, 2009; Smith and Offodile, 2009).

As previously mentions, RFID tags are a more expensive inventory control device than more common barcodes and visual labeling systems and should be reserved for high-priced items or those critical to the production process. A common way to determine which inventory items should receive RFID tags is via the ABC inventory analysis. This method is derived from the Pareto principle which states there are a critical few and trivial many (Heizer and Render, 2008). ABC analysis initially divides inventory into three categories usually based upon dollar volume, though other criteria such as anticipated engineering changes, quality problems, or high unit costs can be factored in as well.

Once the criteria are selected for the ABC analysis, on-hand inventory items are divided into their three classes, A, B, and C, in descending order of importance. Though items in category A, the most critical category in terms of potential impact to the business, comprise a small percentage of total inventories, these units can be responsible for 70 to 80% of the total dollar value of the manufactured product and should, therefore, could be the recipients of RFID technology (Heizer and Render, 2008; Smith and Offodile, 2009).

Since RFID tags and their associated equipment are more expensive on a per-unit basis than most inventory control options commonly in use today, managers must realize the cost benefits of RFID are not associated with the tags and systems itself, but rather through process improvement. One of the fastest paybacks from RFID-related technologies is the reduction in direct labor costs (Gaukler, Özer, Hausman, 2008; Smith, 2005b, 2008). This is achieved when RFID technologies are used to eliminate redundant manual tasks such as barcode scanning, data entry, and manual paperwork. These advanced inventory control devices enable a facility to process more materials and products with the same or reduced manpower, thereby increasing process throughput. Cost savings may be realized during periods of peak demand, as the firm will be able to continue operations without hiring additional personnel ("Building a case for active RFID …" 2006).

RFID technology provides further cost benefits by exposing process inefficiencies. For example, suppose a manufacturer receives a pallet of raw materials tagged with RFID on a timely basis from a reliable supplier. The pallet is delivered to a central receiving depot where the tag is read by the scanning devices, thereby automatically adding the materials to on-hand inventory. A forklift driver then moves the pallet to manufacturing building B, where the RFID station records its new location. One week later, work-in-process product from manufacturing building A arrives at building B and the material on the pallet is introduced to the production process. At this time, the RFID system automatically notifies the MRP system to move the raw material from on-hand inventory to work-in-process inventory. This cycle repeats itself each month.

In the above example, RFID identified two areas of waste, transportation and storage. Operations managers made the following decisions based on data from the RFID system. First, the order for the raw material in question will be delayed by one week to reduce on-hand inventory, an operational liability, and storage costs. Second, the supplier will deliver the product directly to building B where it is needed; thereby freeing resources in receiving as well as eliminating the need for internal transportation. This simple example is a small demonstration of the benefits of RFID information systems. Rather than obtaining data through observation, the system automatically gathers operational data and presents it to management for further analysis. For complex operations spanning various locations, an RFID system may be the only feasible method of performing this type of process scrutiny.

RFID-Related Theft Prevention Measures

Each year, as suggested by O'Connor (2004a-b), billions of dollars are lost due to fraudulent merchandise returns and exchanges to retailers. In general, inventory shrinkage resulting from employee theft, shoplifting, administrative and/or related errors, and vendor fraud is a major retailer concern that RFID initiatives can directly address (Green and Khermouch, 2005; Hackman, 2004). Since one of the major goals of implementing RFID is to reduce counterfeiting and theft, many retailers are continuing to invest in this technology in order to enhance its ability to decrease product theft.

Within the supply chain, a main concern is protection of merchandise and store property. For the past thirty years, barcodes and Universal Product Codes (UPC) have been supplying retailers with the necessary information for product recognition, yet theft is still a major issue. RFID can aid in product security because the tags contain more detailed, unique information, which is connected to the item that is being represented. The technology presents manufacturers and retailers with the power to know the "amount of time the item was in transit, the location of the distribution center holding the item, name of the last person to handle to item, and the amount for which the item was sold," increasing total product visibility ("Fifty uses of RFID in retail" 2003, p. 2). The opportunity for a loss of sale is decreased as both manufactures and retailers are aware of the products at all times throughout the supply chain and keep better management of their inventories.

These tags are used on an individual level, case or pallet load, and at times all three may be utilized. In sectors where inventory accuracy is suspect, tagging on an item level aids in reducing inventory shrinkage. An instance where retailers would find individual coding most beneficial would be with high-valued products, such as clothing, purses, and high-end shoes. The ability to monitor and track movement of these goods decreases the potential for item theft and misplacement. It is also important to consider those items that are most at risk within a particular store. Product and package size, item availability and store visibility are common factors considered when deciding which items to steal. As noted by Hachman (2004), the Gillette Co. in its European operations is using RFID tags inside packages of several products, such as razorblades, to minimize theft. Along with razorblades, the European Metro Group has utilized these tags to track Pantene Shampoo™ and Philadelphia Cream Cheese™.

Not only will RFID bring benefit to retailers, consumers experience some of the rewards as well. When consumers' return previously purchased goods, via the reverse logistics process, the return process is typically extensive, time consuming, and expensive (Smith, 2005a). Many stores require consumers to provide verification that the items have been purchased, mainly through paper receipt. RFID helps to eliminate this stage because the tag links and supplies date of purchase along with other pertinent information to the retailer computer program. Hence, the advantages of implementing RFID in retail stores at the item level is that tagging every item allows the stock employees to know when shelf replenishment is necessary. Knowing when to refill the shelves, product is always available for the customer to purchase thus eliminating stock-outs. Another advantage of RFID implementation is that inven tory storage is more organized, lessening clutter; due to stores requiring fewer inventories on site. RFID tags are read at the receiving docks and also during transit to the store shelves, during which a computer software program records the decrease in inventory and alerts the supplier to send additional product.

In theory, as noted by Kearney (2005), many retailers see obvious benefits in the primary areas of reduced inventory (through a one-time cash savings estimated at 5% of total inventory), annual benefit from a reduction in store and warehouse labor expenses (typically about 7.5%), a significant reduction in out-of-stock items (estimated in an annual benefit of US$700,000 per US$1 billion in annual sales for retailers who implement such technologies to upgrade their current shelf-fulfillment processes. The ability to quantify the actual benefits that RFID provides will help reinforce to retailers that the innovative technology will save them time and money. Informing companies of the potential return-on-investment encourages suppliers to begin RFID implementation at the store, warehouses, and distribution center levels.

When inventory is delivered to the store through the shipping dock, the antenna can scan the RFID tags to know what exactly is on a pallet. Electronic data entry eliminates the potential for human error from inventory discrepancies. Since RFID tags collect real-time data, time is saved allowing stockers to unload the pallet and replenish store shelves, as noted by Roberti, (2005a-c), reducing the dependency on to record on a handheld computer what has been picked in the back room and subject to frequent entry mistakes. Relying on a handheld computer to scan incoming deliveries may cause a threat to the inventory, should the worker miss a read or over-scan. Placing antennae throughout the line-of-transit eliminates the human error element, increasing accuracy and efficiency. In general, RFID data allow management to actually measure the amount of time it takes a product to move through the supply chain ("Wal-Mart tackles out of stocks," 2005). Allowing retailers to track how long a case or pallet is at one particular location reduces the chance for item stock outs. The retailer can view when pallets or cases were shipped from the manufacturer and delivered to the store, when product is placed on the shelves and up to point of purchase. Tracking product at each stage during transit decreases the possibility for empty store shelves, resulting in a loss of sale, but also reducing money wasted from too much unaccounted for store inventory.

To reduce shopping time, many consumers create lists stating what items they intend to purchase; though during shopping, impulse buying is likely to occur. During instances when a particular item is not on store shelves, a transaction does not occur, which results in a loss of sale. In addition, the customer may decide to leave the store; consequently, other purchases the consumer may have bought would be lost. Customer dissatisfaction may cause a broken bond between the consumer and retailer, ending future visits. On-the-contrary, when the consumer can quickly locate the desired items, it benefits both the retail store and the consumer. As noted by Collins (2005a-b), the significant reducing out-of-stocks and the improving product availability are probably the greatest potential consumer benefit from the use of RFID-related technologies, since having a constant supply of product available for shoppers enhances their shopping experience and increases the chance for reoccurring purchases.

RFID-Related Implementation Concerns

It is a common school of thought in the RFID community that item-level RFID will move only as far as retailers choose to take it (Koong and Lin, 2007; Smith and Offodile, 2008, 2009). Many retailers still need to "see to believe" the benefits of RFID before they are willing to experiment with it. In order to fully prepare for an implementation as large and detailed as this, a cross functional team should be assembled well in advance to actively investigate a particular retail area. The first important focus should be on item-level deployments, since this is where the payoff will be quickest and largest. The sooner a retailer can see the direct results of the implementation, the more trust they will instill in the process going forward (Chen and Dubinsky, 2003; Chiou, 2004; Collier and Bienstock, 2006).

CASE STUDIES OF RFID-BASED PRODUCT AND/OF SERVICE DESIGN AND ADAPTABILITY

Methodology

As the previous studies reviewed on the tactical and strategic importance of inventory management systems in its various forms and its associated implementation problems, the author of the present study decided to review such measures from important representatives from the general service and manufacturing sectors. The two firms that were selected were Pittsburgh, PA-based, with global operations and reach capabilities, which were within relatively easy access and could be reviewed for basic principles of the strategic, financial, informational, and operational viewpoints. Such organizations are mainstream economic drivers of the Midwestern U.S. and provided an opportunity to review a spectrum of service and manufacturing-related firms that closely linked its strategy with its service marketing of quality and its internal product/service design managerial approaches.

While these firms are quite dissimilar in nature and scope in terms of clientele, in both situations, there were considerable managerial commitments to use technology in its inventory practices to improve its service to its clientele. Each firm faces unique challenges in utilizing its reputation for quality assurance and expertise in a highly competitive and cost-sensitive environment, with significant consequences for getting it wrong in a recessionary economy (Hu, Wang, Fetch, and Bidanda, 2008; Kamhawi, 2008). Commonly established case study procedures associated with quality initiatives and improvements were followed in the present study (Nonthaleerak and Hendry, 2008; Smith and Offodile, 2007).

Sample Selection

The two relatively large organizations analyzed from a case study-perspective in terms of its inventory management practices, especially in terms of its RFID-related systems implementation and security practices. The Pittsburgh-based companies showcased include Mobile Aspects, Inc., a recognized leader in the integration of RFID-related technologies in inventory management processes of large acute care, especially those caring for children; and Vocollect, Inc., which is especially noted for its voice-recognition software and small hardware platforms that are especially convenient for users in handling partially automated inventory systems. The firms were chosen because of their leadership roles in the utilization of RFID technologies in the inventory field and, part, due to their willingness to open up some of their processes for research and review.

Through a combination of personal interviews of upper-to-middle management, as well as comments from convenient samples of employees was used to gather perceptions of the accuracy of inventory management systems and their implementations. A variety of sources, including personal experiences, were obtained either directly from interviews of management and/or from the firms' websites or a combination of both sources and used for the purposes of these case studies.

The following sections begin with a brief introduction to the companies and its general operating environments. These sections are followed by sections describing the firms' goals for implementing inventory management, specific inventory recommendations and changes, comparison of inventory management processes, and selected measures for ensure the quality and security of data transmitted via RFID-based technologies. Discussion of the practical applications of lessons learned from the case studies follow these sections.

COMPANY CASE STUDIES

Company Case Study 1: Mobile Aspects, Inc. (Mobile)

Company Description and Background

Mobile Aspects, Inc., headquartered in Pittsburgh, PA, was founded in 2000. Mobile Aspects is a leader in the integration of technologies, in particular RFID, in the inventory management processes of large acute care hospitals and hospitals that specialize in the care of children. More specifically, the company aims to assist hospitals with eliminating manual inventory processes through the automation of supply, equipment, drug, and patient management. Presently, the automation solutions provided by Mobile reside primarily in the following areas of a hospital, namely cardiac catheterization laboratories, interventional radiology suites, anesthesia, and surgical specialty areas.

Healthcare Industry Overview in Terms of Information/ Knowledge Technology Needs

Similar to most other knowledge and technology-based fields, the healthcare industry is one that is rapidly evolving. Acceptable practices, processes, and technologies have changed quite a bit over the past 10 to 20 years (Scanlon, Swaminathan, Lee, and Chernew, 2008; Swaminathan, Chernew, and Scanlon, 2008). To this point, Mobile and other healthcare suppliers and service providers must deal with some of the many relevant changes and future chaotic conditions of the medical environment. For example, in terms of facility layouts, it was not long ago the hospitals were arranged into wards, where several patients (up to six) could be located in the same room, with only curtains providing separation and only one bathroom to share. Today, hospitals are moving in the direction of providing a hotel-like atmosphere for their patients, with a trend towards offering single-patient rooms. Clinical care providers, such as nurses, technicians, and physicians were traditionally aware of the cost associated with providing use day-to-day care. Today, many hospitals and health systems strive for transparency in their data for the leadership and decision-makers of the organization, for the government and regulatory agencies, and also for the public. This transparency also provides valuable information to and about the caregivers who directly impact the cost of care on a regular basis.

Inventory management used to occur through a manual effort as the medical staff may have been asked to remove a sticker from a medical supply or equipment and add it to the patient's chart or check of a box when something was used in the care process for the patient. This sticker or checkbox provided a means to track the inventory used for billing and ordering purposes; but, understandably this manual process was prone to errors. Many times, supplies tracked this way were misplaced, forgotten, or even applied to the wrong patient chart. Although flawed, there were little to no ramifications when working in this environment where the hospital was being paid a percent of the charges that were billed. Inventory management is critical as supply costs have grown considerably, and good documentation is extremely important to receiving proper payment for the care provided.

In the past 10 to 20 years, nurse-to-patient ratios have decreased significantly, some through state-mandate (for example, Pennsylvania has mandated certain standards for nurse-to-patient ratios for trauma and critical care services) and others through professional society recommendations. This reduction in ratios increases the cost to provide care, thus emphasizing the need to achieve greater efficiencies in order to reduce costs. Currently, there are shortages of clinical care providers, particularly with nurses who are capable of caring for a growing patient population as baby boomers are now crossing the 50 to 60 years-old threshold. In the past, a nurse could have graduated from a technical school or received a nursing school diploma to practice. Today, practicing nurses need at least an Associate's Degree, along with state licensure, and there is an effort underway to move the minimum requirement to a Bachelor's degree.

A simple back surgery used to require a minimum of a three-day inpatient stay. Now, it is considered an outpatient procedure where the patient returns home the same day. Open heart surgery patients used to arrive at the hospital several days prior to the procedure and the average length of stay was 9 to 12 days. Patients routinely can come in the day of the procedure, and their average length of stay has been reduced to about four to six days. Healthcare costs in the U.S. have raised steadily over the last 30 years, with US$253 billion spent in 1980, US$714 billion in 1990, and US$2.3 trillion in 2008 with little indication that these cost increases will slow down (Ranji, Lundy, Salganicoff, 2010). It can be hoped that these costs will slow a bit with the passage of the comprehensive healthcare legislation, but industry and technology are always placed in the forefront to solve cost-related problems (Gaukler, Özer, Hausman, 2008). As supply costs rise as well, it is increasingly important to effectively manage a hospital's inventory. Increasing use of technology is now being promoted by all types of leadership in the United States. President Obama's administration has already allocated US$19 billion in federal funding to uniformly upgrade information technology in healthcare (Ranji, Lundy, Salganicoff, 2010). RFID spending in healthcare was estimated to total US$90 million in 2006, and is expected to grow to US$2.1 billion by 2016 (Harrop, Das, and Holland, 2009). In the examples listed above, it should be apparent that the cost of patient care continues to rise, at least in the short term. Although it may not be as prevalent in healthcare, inventory theft continues to occur in situations where poor controls are in place to effectively manage high cost inventory, as previously mentioned.

Rising costs further emphasize the need to run an efficient operation. As can be found in other industries, healthcare continues to move from a paper-based environment with manual processes to one that is more automated, with reliance on computers and other paperless methods that should decrease the time and cost associated with repetitive activities. With this move comes additional implementation cost, though the opportunity for long-term cost reductions exists as well. Based on these and other industrial practices, the healthcare consumer also wants the care to happen faster, with better outcomes, and at a reasonable price. As technologies and research continue to advance, so do the costs of care. If a healthcare provider can work to increase efficiencies and reduce waste and scrap (thus, costs) within its existing systems, then it may successfully work to meet the consumers' demands. Technology advances provided by Mobile Aspects are positioned to assist a provider with doing just that.

Company's Healthcare Information Solutions

Since its inception, Mobile has worked to develop its "One System of CARE" portfolio of solutions. Some of the major solutions that include RFID-related technologies are centered on using technology to ensure better quality, faster throughput, and lower costs, as demanded by both B2B and B2C customers. For example, iRISupply™ is an RFID-enabled inventory management solution supporting the storage and management of supplies used in cardiac catheterization laboratories, interventional radiology suites, and surgical operating rooms. Other related technological and knowledge-based systems include iRISecure™ is an RFID system to store and manage implantable tissues and other biologic materials used in surgical operating rooms; iRIScope™, which is an RFID tracking system that is designed to improve the utilization, cleaning, reprocessing, and storage of endoscope equipment; iRISMed™, a real-time RFID system for drug supply tracking during anesthesia processes of surgical procedures; iRISite™, which is an RFID embedded patient management system; iRISynergy™, a web-based system management software that provides reporting, analytics, and system management tools for the "One System of CARE" clinical resource management solutions; while VistaTrak™ is an RFID-automation solution for contrast media management. Each of these patented solutions assists their client base in efforts to increase patient safety, positively impact clinical outcomes, and enhance clinical staff efficiency in the care processes. In addition to their innovation in healthcare technologies, the company has made a commitment to quality through its certification under the ISO 9001:2008 Quality Standard.

Although Mobile offers several solutions to its client base, the author of the present study was allowed to their primary solution, iRISupply™, through a demo. At a closer glance, it was easy to see the benefits that a hospital could realize through the implementation of the solution, since it was very easy to use as the ease-of-use factor is critical to the adoption of any new technology (Davis, 1989). Basically, iRISupply™ is closed cabinetry equipped with Mobile's proprietary software and RFID-embedded technology. The basic procedure to properly activate the RFID-embedded cabinetry in a clinical care process, a member of the medical staff working in the procedural area where the equipment is based would simply need to login to the system at the start of a procedure, and identify the patient in the room. Once logged in, the staff member (assuming their credentials allow for it) can access the patient care supplies necessary for the procedure. It should be noted that the system can be set up so certain team members only have access to certain supplies/equipment. When the supplies are accessed by opening the cabinet door and removing them from the shelf or rack, the supplies are automatically charged to the patient's account for billing. If supplies are returned to the cabinet, then they are automatically removed from the account and returned to inventory. Data collected from the system were used to complete the patient's bill and medical record, as well as to assist materials management in purchasing decisions and supply management. This system effectively minimizes or possibly eliminates waste and theft that can otherwise occur in hospital procedural rooms. In addition, data from the system can also alert staff to supplies that are approaching the expiration date of their useful life, and assist management in understanding the cost per case by procedure, by physician, or by other variables.

As one would observe through a demonstration, the technology works to simplify the job of clinicians involved in the care process. Rather than having to remember to document what was used in each patient's case, the potential for manual error is taken out of the process through this automation. In addition, accountability is now incorporated into the process, especially since a clinician, typically a physician, nurse, and/or technician, is now effectively signing out the use of each supply taken from the cabinetry for the patient case. In instances where supplies go unused and aren't returned to the cabinet, they can now be traced back to a user.

Collaboration to Develop Solutions

Although Mobile Aspects is a privately-held, independent firm, they have a track record of collaborating with other organizations in the development and enhancement of their customer solutions. Three such examples of collaboration are showcased in it iRISecure™ partnering, as the company has worked with the Children's Hospital of Boston to develop the industry's first RFID-enabled tissue tracking system. Through this strategic partnering, an integration of knowledge-based systems, the company significantly improved inventory management automation and application of various RFID technologies by leveraging Children's Hospital of Boston's existing expertise in the related clinical care processes. This leveraging was also apparent with VistaTrak™, as Mobile worked with Bayer HealthCare Pharmaceuticals to develop this contrast media management system. Through this partnering activity, Bayer worked to modify Mobile's existing RFID-based inventory management platform to address its requirements for storage, utilization, and tracking of contrast media. VistaTrak's™ development leveraged each organization's RFID knowledge, via through Mobile's management solutions and Bayer HealthCare's RFID tagged products, to develop this specialized product offering.

Other recent examples of collaboration to improve clinical throughput via information and knowledge-based systems involved Mobile's partnering with Cerner Corporation to enhance workflow as it relates to the clinical documentation processes for implantable devices, tissues, and related items. This solution provides real-time communication of device and supply utilization to the clinical information system. This development was focused on Cerner's Millenium© clinical information system, but likely can be applied to any other such system. Such co-produced solutions, Mobile has found value in developing collaborative relationships with other organizations in healthcare or in technology fields. To that end, the company's has documented success in its corporate relationships, including Cerner Corporation, Novation, Bayer HealthCare Pharmaceuticals, Texas Instruments RFID Systems, Association for Automatic Identification and Data Capture (AIDC) Technologies, CIMIT, and EPCglobal. Each example of collaboration focused on inventory management, utilization, and/or the reduction or elimination of missed charges. Through the elimination of manual processes in the inventory management system, the company hopes to improve a hospital's charge capture, minimize the risk of expired inventory, and eliminate waste in the hospital's inventory management systems.

Challenges Facing the Organization

Although Mobile is in a seemingly good position on the forefront of a wider adoption of the RFID innovation, their future outlook is not without its challenges. To start, the implementation of this technology comes with a relatively high cost of adoption. Many providers are already working with an inventory management system today that may be using bar codes or another means to track inventory. The return-on-investment to move from the current system to an RFID-enabled system can be a tough sell to many companies, especially healthcare providers ("Providing a more advanced automation ..," 2007). In reviewing Mobile's sample clients, and through a conversation with members of their management team, it's clear that the company is focused on larger or specialty providers (like Children's hospitals) for the implementation of their solutions. At this point, it does not seem to be a cost-effective solution for smaller, community-based hospitals.

As part of a cost-benefit analysis of an RFID-embedded solution, a healthcare provider would also need to evaluate the number of suppliers that currently provide RFID tagging. It is our understanding that this tagging has not been widely adopted by all healthcare vendors. In addition, the tagging that is done may not follow a uniform standard, thus necessitating addition tagging to be performed by hospital staff, often a time-consuming and tedious task. The increased man-hours required for this additional, typically manual, step are an example of just one of the incremental costs associated with this inventory automation option.

Although there has not been any incidents reported to the U.S. Food and Drug Administration (FDA) at the time of the present study, research by Seidman, Brockman, and Lewis (2010) indicated that the potential for electromagnetic interference of RFID with some common implantable cardiac devices do exist. Although the researchers concluded that their findings did not pose an urgent health risk, the study findings are of special concern to companies such as Mobile, due to a significant portion of its target market involving cardiac catheterization laboratories and surgical operating rooms where such devices may be implanted. It will be important for management to stay abreast of this situation in order to ensure its business model is well-positioned to adapt their technology if this interference is deemed a threat to the health of cardiac patients.

Company Case Study 2: Vocollect, Inc. (Vocollect)

Company Description and Background

Throughout the past two decades, Vocollect, Inc., headquartered in Pittsburgh, PA, has developed a worldwide reputation for developing products that increase productivity, accuracy, cost reductions. Vocollect has strived to develop voice-recognition software, and small hardware platforms that are convenient for individuals to utilize in all industries. The founders of Vocollect knew the potential uses for their software were almost limitless. They had to find investors that were willing to invest in an unproven fledgling technology. Larger engineering companies, such as Westinghouse, may have slowed down the development of their voice-recognition software development, and the founders of Vocollect wanted to take advantage of the market's potential. Throughout the 1990s the voice-recognition marketplace did not grow in large increments as early adopters, such as General Mills and Ford, had the capital to invest in the technology, but smaller organizations did not implement the technology due to costs and technological resistance.

The original private investors and employees at Vocollect were fully committed to the idea of taking voice-recognition products to mainstream marketplaces. A large breakthrough came when a nationwide hardware company approached Vocollect to develop a system for their distribution network. Vocollect has developed a wearable computer with voice-recognition software. This product was well received by their client, and their employees. A strategic decision was made to direct company resources to focus on distribution products rather than manufacturing products. The company began to double in size year after year as a result of its wearable computers and currently the company records over US$100 million annually in sales.

Currently, Vocollect's products are routinely used on six continents and their voice-recognition technology is used in 22 languages. In the future, Vocollect will probably need to increase its collaboration and strategic partnering activities, as management at Mobile has experienced, with other companies to create advanced technology for the ever-changing marketplace. Linking their voice-recognition software to RFID-related technologies is certainly a potential that Vocollect may continue to have a competitive advantage over its competition.

Creation of Partnerships

Both voice-recognition products and RFID-based products have benefits on their own, but in the future linking these technologies will undoubtedly open new opportunities for all related industries. RFID-related technologies have the ability to develop larger information systems that are easily navigated by companies. In many applications, as previously outlined, RFID-related technologies can significantly increase usable information volume, decrease errors, and create cost savings. In the current market, expenses that are incurred by using RFID technologies can be justified in some circumstances, but many companies cannot implement RFID sue to high initial costs. Low-sale price items that are mass-produced may not justify the cost of RFID implementation. As noted, only around 25% of businesses that have been surveyed believe RFID technology will add value to their business at some point in time ("Building a case for active RFID …," 2006). Larger, more premium-priced consumer goods, such as cars and electronic items, can immediately benefit from RFID technology implementation because the higher prices of these items justify the cost of individual RFID tags. Linking voice-recognition software that the company develops as well as their wearable computer platforms should generate larger value to companies that implement RFID-embedded technologies.

Applications of voice-recognition software can aid corporate employees in overall inventory management, such handling and locating products, where to place or move products, as well as any other information a company wants its frontline employees to have knowledge concerning inventory procedures. RFID-related technologies, on-the-other-hand, still supplies a product's basic information as the voice-recognition software will supply the advanced information to increase the capabilities of a distribution network. These technology links have the potential to create an intelligent and conversational RFID tag that should increase information flow exponentially. The uses for linking these technologies may immediately increase the ability several departments and older forms of information flow. Inventory systems can instantly be impacted through the larger amounts of information and the ability of the systems to process that information. Companies that have departments who focus on discounting items, recalling items, and expired items will benefit from this technology in many ways. The information that Vocollect's systems can generate has the potential to vastly increase the efficiency and effectiveness of these departments as well as generate significant cost savings.

Partnerships with hardware providers and RFID-technology developers should increase the number of potential customers that Vocollect can reach. One such partnership was with Psion Teklogix, a developer of mobile-computing systems, RFID-related technologies, and wireless data collection has benefited both companies greatly. The partnership helped the companies develop a mobile computer called the Walkabout Pro Speech. This handheld computer combines voice-recognition software that the company developed along with other software and the hardware platform that Psion Teklogix developed. Solid partnerships that help develop new products will increase the ability of all companies involved to expand their business ("Pison Teklogix teams up with Vocollect …," 2006).

Developing global partnerships to expand Vocollect's brand is also a main strategic goal of the company. Vocollect recently signed a partnership agreement with Schmidt & Co., Limited in Hong Kong ("Vocollect celebrates more than two decades in Pittsburgh ..." 2010). Schmidt & Co, Limited is Asia's leading systems integrator of voice-recognition software business solutions.

This partnership should help Vocollect gain access to two more countries in Asia in the near future. Both companies will benefit from their ability to reach more markets and companies that need their services as Schmidt & Co. obtained the status of Total Solution Provider, which the highest level of participation in the global Vocollect partner program. In particular, significant increases for the demand of technology are occurring in emerging markets around the world, such as China, India, and Latin America. These markets are supplying more developed nations with goods of all kinds and the management at Vocollect, as well as many other information and knowledge-based organizations, are trying to strategically position its assets throughout the world to take advantage of these growing markets.

Future Directions at Vocollect, Inc.

Vocollect has enjoyed more than 20 years of success in the Pittsburgh, PA area, and they have surpassed US$100 million dollars in global sales. They were the first company in the voice-recognition industry to reach this sales mark. Management at Vocollect strongly believes that developing global markets will increase competition levels in all industries, and this will in turn drive the need for more voice-recognition software applications in knowledge-based organizations, in particular, and inventory applications, in general. In the near future, RFID-related technologies should continue to gain traction in the warehousing aspects of all businesses. Vocollect has already linked their software products with RFID-related software products, and hardware tag readers. They are on the forefront of technology development and are leading the way in developing new products to make operations more efficient in distribution systems. Healthcare systems are also a main focus for Vocollect, as with Mobile, and they are working with healthcare providers to develop technology that will reduce costs and eliminate errors. Voice-recognition technology and RFID technology will increase the ability of healthcare systems to provide their employees and patients with more efficient and effective services. Vocollect's CEO, in an interview, suggested that the company has the ability to reach yearly sales of US$500 million worldwide over the next decade.

Implementing the newest technology should aid knowledge-based organizations in their quest for global competitiveness. It is a major strategic goal at Vocollect to continue to develop and provide their global customers with the most advanced and proven voice-recognition systems that link with other technologies, such as RFID, to create advanced information systems. These advanced information systems will help develop successful industries on a global scale, and companies such as Vocollect and Mobile should become a global contender as long as management continually enhance their technology-based product and service offerings by providing customers with high value for their investments. Vocollect especially has strategically position their advanced voice-recognition products for their customers to become a global leader in the field of such technological innovations.

FUTURE USES OF RFID-RELATED TECHNOLOGIES

As the previous sections of this paper have pointed out, RFID-related technologies have been extremely important to supply chain and inventory management advancements, especially in relation to the healthcare industry. However, the applications of RFID are not limited to healthcare and the areas that Vocollect and Mobile service. In fact, as this section will illustrate that RFID is and will be useful various business segments, especially in knowledge-based organizations, in the near future as well as new applications of RFID are being routinely developed and tested in such organizations. However, there are perceived flaws with regards to RFID-related technologies that must be minimized.

Perhaps, RFID-related technologies will truly be ubiquitous and transparent in many transaction involving technology ("The future of RFID," 2009). Therefore, in order to take advantage of RFID-related efficiencies, early adoption is imperative (Davis, 1989). RFID-based solutions offer many companies methods to automate tasks that are currently completed manually in a more operational effective and cost-sensitive manner. For example, a Norwegian food company has developed a program, referred to as the "Smart Flow of Goods," in which they are heavily investing capital to develop intelligent packaging with RFID chips and sensors ("An Internet of objects ...," 2009). RFID technologies, in this case, allow the company to track where the food has been and whether it is in good condition. Moreover, the sensors in the packaging will indicate the pressure, temperature and movement of the food from supplier to retailer. Such use of RFID should be greatly beneficial in the early detection of contaminated food, thus alerting food recalls earlier and identifying the necessary recalled food using the RFID information. There are also plans for many retailers to integrate RFID into their everyday inventory process. For instance, instead of having employees constantly monitor incoming and outgoing inventory, RFID should allow companies to eliminate these non-value added positions by using the technology to track these movements. Ideally, when inventory reaches a store, and is properly tagged, the inventory can enter the facility and be uploaded into the inventory management systems within seconds. Companies will be able to focus more time on value-added operations, in turn, increasing performance.

An example of a RFID advancement that can by used in the above mentioned situation is a chip developed by the Japanese multinational, high technology company, Hitachi, whom is a major player in the future of RFID-related technologies. They have developed a microscopic RFID chip they coined the Mu-chip (or μ-chip) ("Overview of RFID solutions, 2010). The Mu-chip is the world's smallest RFID integrated circuit, barely viewable by the naked eye. It allows a company to discreetly track, traces, and monitor an item's costs effectively. Furthermore, the chip enhances security, protects brand image and can offer high visibility throughout supply or value chain or lifecycle of the tagged item.

There are also green-RFID applications that are taken seriously by several companies. For example, Bechtel Jacobs, an environmental management contractor for the U.S. Department Energy's Oak Ridge Office, has been working on the cleanup of over 300,000 cubic yards of waste materials ("Radioactive waste cleanup project ..." 2010). The company is charged with packaging, transporting and disposing of the waste materials. One interesting caveat is this waste is not your typical landfill waste, for it is radioactive. Therefore, the proper tracking and disposal of the waste is important. In fact, transporting and disposing of the contaminated and hazardous waste products required compliance with numerous transportation, health and safety, radiological, physical security, classification and nuclear material control and accountability requirements. In order to accomplish this task, Bechtel Jacobs has employed a RFID-based system, comprised of EPC Gen 2 RFID tags and interrogators, in conjunction with GlobeRanger's iMotion© platform. Their RFID process is based on a RFID tag that is attached to a truck, a transportation specialist in the field then uses a handheld-RFID system to select shipping information, a visual inspection occurs, project-shipping data is encrypted and written to the vehicle's RFID tag. As the transport vehicle leaves the disposal zone, their tags are read using interrogators and the truck's weight is recorded, and this information is sent to company's waste transportation management system. The company is able to use this information to validate their procedures to confirm that the waste is being transported correctly.

Since implementing this system, the company has seen a reduction in costs in many operating areas. Since its inception, the company has seen reduced consumption of fuel by 60,000 gallons, nitrogen dioxide emissions reduced by 6,500 pounds, carbon dioxide emissions reduced by 350,000 pounds; since the system is paperless process, about 100 pounds of air pollution and two tons of paper were avoided ("Radioactive waste cleanup project …" 2010).

Although RFID-related technologies generally provide much efficiency, there are skeptics predicting some potential weaknesses of RFID. Most notably, RFID is relatively expensive in its initial capital outlay. Although, as previously discussed, chips themselves cost approximately between 10 to 15-cents, depending on volume, a RFID reader usually costs between US$500 to 2000; hence large scale operations, these costs can add up quite quickly. However, the governing body of RFID, EPCglobal, have suggested that due to projected increased volume of production, these chips would ultimately cost only US$0.05 a piece. Nonetheless, as previously discussed, the direct cost of RFID should not be the only consideration ("Frequently asked questions," 2010).

The increase in data storage could also pose a potential problem with the implementation of RFID-related technologies. Wal-Mart alone conservatively has the potential to produce seven terabytes of operational RFID information per day (Collins, 2005, 2006). Existing computer systems would be hard pressed to house such data volumes without adding significant memory storage and computation abilities. Therefore, an increase in operational databases would need to take place. Along with data storage incapacities, information security and privacy must also be considered, as RFID tags may be manipulated, thus putting sensitive company/personal information in the hands of thieves. Since large scale RFID-related systems must have massive capabilities, so much information can be gained from tags. Consequently, companies like RSA Laboratories are designing products to combat these potential risks. They have a suite of authentication and access-management products that help businesses validate RFID readers and secure back-end access to their RFID infrastructure. What is more, RSA Laboratories is persistently developing new security products to mitigate the latest risks as well ("RFID privacy and security," 2010).

GENERAL CONCLUSIONS AND IMPLICATIONS

RFID-Related Managerial Implications

As much of the paper has emphasized, RFID is becoming a more widely used tool for supply chain and inventory management, as well as an essential tool for knowledge-based organizations. Companies are able to leverage the RFID-embedded technologies to make their operations more and more efficient and effective. Although many corporate stakeholders know the fundamentals of RFID, what the strategic decisions that must be made to implement RFID and the associate cost-benefit analysis and trade offs used when weighing the option of RFID are still very much debatable subjects. As described in the case study of the two companies headquartered in Pittsburgh, PA are currently offering RFID-based applications to their customers. In particular, both companies, namely the healthcare industrial provider Mobile Aspect, Inc. and voice-activate warehouse and inventory provider Vocollect, Inc., see great potential in these technological innovations. Management at both companies are continuing to improve their technology to keep up with the ever changing improvements to enhance the capabilities of RFID, but reducing its initial cash outlays.

It is apparent that product communication is important to both retailers and consumers. At times it is difficult to introduce new technologies because consumers, manufacturers and retailers are unaware of its benefits and/or resistant to innovation. By fusing information with technology, RFID technologies can be successfully integrated to existing software and hardware applications. These technological innovations are potentially rewarding to various industries that have begun quantifying its abilities. Information transference, reduction in theft, and eliminating stock outs are some of the main attributes RFID provides to industry. However, in few instances, such as Wal-Mart's initial push into RFID, management may have been so excited about the cost and operational benefits of the new technology that they did not properly anticipate the financial risks and the disadvantages to using such technology. However, in part to Wal-Mart's initial push for RFID implementation, other industries have become aware and the retail industry, on particular, was able to utilize the technology to their completive advantage.

RFID-Related Technical Implications

An RFID technological deployment involves the essential areas of investment in hardware, middleware, and services. The hardware costs include not only the RFID tags themselves, but also the equipment and software required to read the information and transmit data into the firm's MRP system. The middleware investment includes upgrades to the MRP software, such as SAP or Oracle, to accept and act upon data received from the RFID system. Finally, the services cost includes all of the labor associated with the installation and maintenance of the system, vendor support, and employee training ("Building a case for active RFID ..." 2006). Prior to making the decision to deploy RFID-related technologies, several questions must be asked and sufficiently answered. Perhaps the most important question is whether the entire organization will get on-board with the initiative or not. Many modern businesses with multiple locations operate in a decentralized capacity, as is particularly the case with global enterprises. If RFID cannot be implemented throughout a process from start to finish, the benefits may not outweigh the costs as there will be insufficient data to expose waste in processes.

Once it is established that the organization will internally implement RFID-related systems across its own processes, it must be determined which materials and products to track and if the associated suppliers are willing to participate in the RFID initiative. Even if the lowest-cost passive RFID tag is selected, an approximate 15-cents cost each, this initiative can add up to a significant up-front expense across an entire operation. For this reason, only critical or expensive items, such as those identified via the ABC analysis, should be tagged; inventory items of lesser importance can continue to be tracked through conventional means. Once participating items are selected, the suppliers can be contacted to verify participation in the RFID program.

Once these key questions have been answered, management can decide whether to implement an RFID tracking system or not. The system should be implemented if there is enough participation by internal and preferably also external parties to expose potential inefficiencies in a process and increase throughput time via expanded automation. Though it is possible for this to occur completely internal to an organization, higher involvement from external suppliers, including transportation providers, will significantly increase the system's rate of return.

In general, a RFID system is not just about cost-effective inventory tagging, but rather the facilitation of an inventory and work-in-process tracking system that reduces the overall cost of production. In order to be successful, an RFID implementation must be used on a high enough number of units and processes to identify sources of waste. This includes all entities involved in production such as suppliers, transportation providers, and internal departments. Though the up-front investment in this type of tracking system will initially appear more expensive than traditional barcode or labeling mechanisms, the long-term increase in automation, reduction in manpower, and improved throughput time will more than offset setup and implementation costs associated with RFID.

RFID-Related Customer Satisfaction Implications

In general, when used properly, RFID utilization can increase customer satisfaction, allowing important data that can be immediately used to ensure that a constant shelf supply is readily accessible. Once tags are scanned during transit from the back room to the store floor, the data are obtained, which notifies retailers when product reorders and restock are necessary. RFID can also decrease some customer satisfaction if the customer feels that his or her privacy is being invaded. Other customers feel that it is a nuisance. When they are shopping close to the 'gates' by the front of the store, sometimes the readers will go off, because they think someone is shoplifting. This can cause the shopper to become uncomfortable.

It is important to remember that "RFID is a journey that will take time for retailers and suppliers to learn how to deploy it successfully and for RFID vendors to improve their products" (Roberti, 2005a, p. 1). Whether or not RFID-based technologies are the most appropriate or reliable solution for organizations is still open for exploration.

ACKNOWLEDGMENT

The author wishes to thank most heartedly for the valuable contributions by the reviewers for their input into the final paper. Peer reviewing and editing are commonly tedious and thankless tasks.

REFERENCES

About Mobile Aspects. (2010). *Mobile aspects.* Retrieved February 4, 2010, from www.mobile-aspects.com

An Internet of Objects-the Future of RFID Technology. (2009). Retrieved November 23, 2009, from http://www.cellular-news.com/story/37900.php

Building a Case for Active RFID ROI: Cost-Benefit Analysis. (2006). Retrieved November 18, 2012, from http://active-rfid.com/download/Building_a_Case_for_Active_RFID.pdf

Chen, Z., & Dubinsky, A. J. (2003). A conceptual model of perceived customer value in e-commerce: A preliminary investigation. *Psychology and Marketing, 20*(4), 323–347. doi:10.1002/mar.10076

Chiou, J. S. (2004). The antecedents of customer loyalty toward Internet service providers. *Information & Management, 41*(6), 685–695. doi:10.1016/j.im.2003.08.006

Collier, J. E., & Bienstock, C. C. (2006). Measuring service quality in e-retailing. *Journal of Service Research, 8*(3), 260–275. doi:10.1177/1094670505278867

Collins, J. (2005). School studies RFID's effect on Wal-Mart. *RFID Journal.* Retrieved November 22, 2009, from http://www.rfidjournal.com/article/articleview/1514/1/9/

Collins, J. (2006). RFID implementation is an art. *RFID Journal.* Retrieved November 18, 2009, from http://www.rfidjournal.com/article/articleview/2427/1/1/

Cowles, D. L., Kiecker, P., & Little, M. W. (2002). Using key informant insights as a foundation for e-retailing theory development. *Journal of Business Research, 55*(8), 629–636. doi:10.1016/S0148-2963(00)00203-4

Davis, F. D. (1989). Perceived usefulness, perceived ease of use and user acceptance of information technology. *Management Information Systems Quarterly*, *13*(3), 319–340. doi:10.2307/249008

Devaraj, S., Fan, M., & Kohli, R. (2002). Antecedents of B2C channel satisfaction and preference: Validating e-commerce metrics. *Information Systems Research*, *13*(3), 316–333. doi:10.1287/isre.13.3.316.77

Gaukler, G. M., Özer, O., & Hausman, W. H. (2008). Order progress information: improved dynamic emergency ordering policies. *Production and Operations Management*, *17*(6), 599–614. doi:10.3401/poms.1080.0066

Green, H., & Khermouch, G. (2005). Bar codes better watch their backs. *BusinessWeek*. Retrieved December 8, 2012, from http://www.businessweek.com/magazine/content/03_28/b3841063.htm

Hackman, M. (2004). *RFID could hack retail fraud*. Retrieved February 4, 2012, from http://www.eweek.com/article2/0,1759,1628696,00.asp

Harrop, P. (2010). *HF RFID-the great leap forward*. Retrieved February 4, 2012, from http://www.rfidwizards.com.html

Harrop, P., Das, R., & Holland, G. (2009). RFID in healthcare: 2009-2019. *IDTechEx.com*. Retrieved February 4, 2010, from http://www.idtechex.com/research/reports/rfid_for_healthcare_and_pharmaceuticals_2009_2019_000146.asp

Heizer, J., & Render, B. (2008). *Principles of operations management* (7th ed.). Upper Saddle River, NJ: Prentice Hall.

Hu, G., Wang, L., Fetch, S., & Bidanda, B. (2008). A multi-objective model for project portfolio selection to implement lean and Six Sigma concepts. *International Journal of Production Research*, *46*(23), 6611–6648. doi:10.1080/00207540802230363

Jain, V., Benyoucef, L., & Deshmukh, S. G. (2008). What's the buzz about moving from 'lean' to 'agile' integrated supply chains? A fuzzy intelligent agent-based approach. *International Journal of Production Research*, *46*(23), 6649–6678. doi:10.1080/00207540802230462

Kamhawi, E. M. (2008). System characteristics, perceived benefits, individual differences and use intentions: A survey of decision support tools of ERP systems. *Information Resources Management Journal*, *21*(4), 66–83. doi:10.4018/irmj.2008100104

Kay, M. (2007). Closing the RFID knowledge gap in apparel and footwear. *RFID Journal*. Retrieved November 18, 2012, from http://www.rfidjournal.com/article/articleview/3267/1/82/

Kearney, A. T. (2005). *RFID will bring great benefits for retailers*. Retrieved November 24, 2009, from http://retailindustry.about.com/cs/it_rfid/a/bl_atk111003.htm

Kennedy, F. A., & Widener, S. K. (2008). A control framework: Insights from evidence on lean accounting. *Management Accounting Research*, *19*(4), 301–319. doi:10.1016/j.mar.2008.01.001

Koong, L., & Lin, C. (2007). Evaluating the decision to adopt RFID systems using analytic hierarchy process. *Journal of American Academy of Business*, *11*(1), 72–77.

Lindsay, J., Reade, W., & Roth, L. (2003). Retail RFID systems without smart shelves. *RFID Journal*. Retrieved November 18, 2009, from http://www.jefflindsay.com/rfid1.shtml

Mobile Aspects. (2007). *Is RFID ready for healthcare? Yes, if RFID is focused on addressing healthcare's needs* (White Paper). Retrieved November 24, 2009, from www.mobileaspects.com

Mobile Aspects. (2007). *Providing a more advanced automation strategy for medical device and supply utilization in preoperative care through RFID* (White Paper). Retrieved November 18, 2009, from www.mobileaspects.com

NCR Corporation (2003). *Fifty uses of RFID in retail*. Retrieved November 18, 2009, from http://www.usingrfid.com/features/read.asp?id=3

Newland, G. (2005). *The potential impacts of RFID on the retail enterprise*. Retrieved November 18, 2012, from http://www.nrf-arts.org/presents/2004/retail/graham.pdf

Nonthaleerak, P., & Hendry, L. (2008). Exploring the six sigma phenomenon using multiple case study evidence. *International Journal of Operations & Production Management*, *28*(2), 279–303. doi:10.1108/01443570810856198

O'Connor, M. C. (2004a). All CVS store to offer express pay. *RFID Journal*. Retrieved November 23, 2012, from http://www.rfidjournal.com/article/articleview/1297/1/11/

O'Connor, M. C. (2004b). Ending retail scams with RFID. *RFID Journal*. Retrieved November 23, 2012, from http://www.rfidjournal.com/article/articleview/1260/1/20/

Overview of RFID Solutions. (2010). Retrieved February 4, 2010, from http://www.hitachieu.com/mu/

Pison Teklogix Teams Up with Vocollect to Provide Voice Recognition to Work About Pro VIP Product Line. (2006). Retrieved February 5, 2010, from http://www.logisticsit.com/absolutenm/templates/articlevoice.aspx?articleid=2317&zoneid=39

Ranji, U., Lundy, J., & Salganicoff, A. (2010). *US healthcare costs: A background brief*. Retrieved February 4, 2010, from http://www.kaiseredu.org/topics_im.asp?imID=1&parentID=61&id=358

RFID and Voice Enhanced Productivity on Pick-to-Tote Operations. (2008). Retrieved November 23, 2012, from http://www.rfidproductnews.com/reader/83/articles/index.php

RFID in the Retail Industry. (2004). Retrieved November 18, 2012, from http://www.microsoft.com/industry/retail/RFIDoverview.mspx

RFID Industry Solutions. (2010). Retrieved November 18, 2012, from http://www.symbol.com/category.php?category=720

RFID Journal. (2003a). *RFID for your shopping cart*. Retrieved November 23, 2012, from http://www.rfidjournal.com/article/articleview/489/1/20/

RFID Journal. (2003b). *Wal-Mart CIO to speak on RFID*. Retrieved November 23, 2012, from http://www.rfidjournal.com/article/articleview/419/1/26/

RFID Journal. (2003c). *Wal-Mart details RFID requirement*. Retrieved November 23, 2012, from http://www.rfidjournal.com/article/articleview/642/1/1/

RFID Journal. (2005). *Wal-Mart tackles out-of-stocks*. Retrieved November 18, 2012, from http://www.rfidjournal.com/article/articleview/1551/4/160/.

RFID Journal. (2005a). *RFID in retail - The consumer in control*. Retrieved November 23, 2012, from http://www.semiconductors.philips.com/news/publications/content/file_1283.html

RFID Journal. (2005b). *Retailer tests RFID on garments*. Retrieved November 22, 2012, from http://www.rfidjournal.com/article/articleview/491/1/20/

RFID Journal. (2009). *The future of RFID*. Retrieved December 8, 2009, from http://www.rfidjournal.com/article/view/7268

RFID Journal. (2010a). *Frequently asked questions*. Retrieved February 4, 2010, from http://www.rfidjournal.com/faq#20

RFID Journal. (2010b). *Radioactive waste cleanup project becomes more efficient, greener*. Retrieved February 4, 2010, from http://www.rfidjournal.com/article/view/7435/3

RFID Privacy and Security. (2010). Retrieved November 22, 2012, from http://www.rsa.com/rsalabs/node.asp?id=2115

Roberti, M. (2005a). Retailers say RFID will take time. *RFID Journal*. Retrieved November 22, 2009, from http://www.rfidjournal.com/article/articleview/1344

Roberti, M. (2005b). Wal-Mart begins RFID process changes. *RFID Journal*. Retrieved November 18, 2012, from http://www.rfidjournal.com/article/articleview/1385/1/20/

Roberti, M. (2005c). Wal-Mart to expand RFID tagging requirement. *RFID Journal*. Retrieved November 18, 2012, from http://rfidjournal.com/article/articleview/1930/1/9/

Scanlon, D. P., Swaminathan, S., Lee, W., & Chernew, M. (2008). Does competition improve health care quality? *Health Services Research*, *43*(6), 1931–1944. doi:10.1111/j.1475-6773.2008.00899.x PMID:18793214

Scherrer-Rathje, M., Boyle, T. A., & Deflorin, P. (2009). Lean, take two! Reflections from the second attempt at lean implementation. *Business Horizons*, *52*(1), 79–85. doi:10.1016/j.bushor.2008.08.004

Seidman, S. J., Brockman, R., & Lewis, B. M. (2010). In vitro tests reveal sample radio frequency identification readers inducing clinically significant electromagnetic interference to implantable pacemakers and implantable cardioverter-defibrillators. *Heart Rhythm: The Official Journal of the Heart Rhythm Society*, *7*(1), 99–107. doi:10.1016/j.hrthm.2009.09.071 PMID:20129290

Sirico, L. (2010). *An RFID primer for manufacturers*. Retrieved February 4, 2010, from http://www.rfidwizards.com.html

Smith, A. D. (2005a). Reverse logistics and their affects on CRM and online behavior. *VINE: The Journal of Information and Knowledge Management*, *35*(3), 166–181.

Smith, A. D. (2005b). Exploring the inherent benefits of RFID and automated self-serve checkouts in a B2C environment. *International Journal of Business Information Systems*, *1*(1-2), 149–183. doi:10.1504/IJBIS.2005.007405

Smith, A. D. (2008). Evolution and acceptability of medical applications of RFID implants among early users of technology. *Health Marketing Quarterly*, *24*(1-2), 121–155. doi:10.1080/07359680802125980 PMID:19042524

Smith, A. D., & Offodile, O. F. (2007). Exploring forecasting and project management characteristics of supply chain management. *International Journal of Logistics and Supply Management*, *3*(2), 174–214.

Smith, A. D., & Offodile, O. F. (2008). Gauging web portals impact on supply chain management concepts. *International Journal of Logistics and Management*, *4*(2), 184–206.

Smith, A. D., & Offodile, O. F. (2009). The perceived importance of major RFID-related technology initiatives among retail store managers. *International Journal of Services and Operations Management*, *5*(4), 520–547. doi:10.1504/IJSOM.2009.024583

Swaminathan, S., Chernew, M., & Scanlon, D. P. (2008). Persistence of HMO performance measures. *Health Services Research*, *43*(6), 2033–2051. doi:10.1111/j.1475-6773.2008.00890.x PMID:18783460

Vocollect Celebrates More Than Two Decades in Pittsburgh, Surpasses $100 Million in Revenue, Sees Future With New Markets. (2010). Retrieved February 4, 2010, from http://www.pghtech.org/news-and-publications/teq/article.aspx?Article=1855

What is RFID? (2010). Retrieved February 4, 2010, from http://www.aimglobal.org/technologies/rfid/what_is_rfid.asp.html

Chapter 2
An Integer Programming Model to Maximize Battery Manufacturing Productivity

Hesham K. Alfares
King Fahd University of Petroleum and Minerals, Saudi Arabia

ABSTRACT

An Integer Programming (IP) model is formulated and optimally solved for a real-life production planning problem. The model is used to determine the optimum production plan for the formation stage of battery manufacturing in a local battery-producing company. Battery manufacturing is a complicated process that involves several stages and several chemical and physical operations. The formation stage is considered the bottleneck of this process because it has the most critical limitations of time and production resources. In the formation stage, batteries are filled with acid and charged with electricity using different types of circuits. The objective of the model is to maximize the productivity of the critical formation stage by allowing the best utilization of the limited time and equipment resources. The model is able to consider a large number of battery sizes and types and also a large number of charging circuits with different capacities and charging speeds. The model-generated optimum production plans increase daily profits by 12% on average in comparison to manually-generated production plans.

DOI: 10.4018/978-1-4666-4506-6.ch002

INTRODUCTION AND BACKGROUND

Battery production is complex process that involves several products, many components, and multiple manufacturing stages. The manufacturing of automotive batteries for cars, trucks, and boats is done through a unique manufacturing process that includes several physical and chemical operations. These operations include lead strip manufacturing, steaming, chemical curing, acid filling, formation, electrical charging, and final assembly. The distinctive nature of the battery manufacturing process lead to special considerations and objectives in planning and scheduling of battery production activities. This chapter presents an integer programming model for achieving the maximum profitability in the formation stage of the battery manufacturing process at a mid-size battery making company. The model is formulated and applied to actual production problem using real data from the Middle East Battery Company (MEBCO) in Saudi Arabia. Optimum production plans generated by the model have shown significant improvement in profitability over the previously used manually generated production plans.

Middle East Battery Company (MEBCO) is a joint venture between a group of Saudi Industrialists and business leaders and Johnson Control International (JCI). MEBCO produces AC Delco, JCI and Toyota branded maintenance free batteries. The plant is located in 2nd Industrial City in Dammam, Kingdom of Saudi Arabia. JCI is a 49% equity partner in MEBCO and has management responsibility for the plant operation. The plant is the first Delphi battery plant built specifically to serve the after-market and is Delphi's first manufacturing operation in the Middle East producing batteries for comprehensive vehicle market, including European, Asian, North American, as well as marine, commercial and industrial applications.

The plant started its test production in November 1997 and had the first shipment to customers on January 14, 1998. MEBCO's primary market is the Gulf Cooperation Council states: Saudi Arabia, United Arab Emirates, Kuwait, Bahrain, Qatar and Oman. Other markets include Egypt, Lebanon, Jordan, Syria, Asia Pacific, North America and South Africa. The production has been consistently growing and passed the 2.5 million mark in October 2006. The plant produces batteries for all vehicle types, as well as marine, commercial and industrial applications. MEBCO's primary market is the Middle East, but it also has growing markets in Asia, Europe, and North America. Other potential markets and part numbers are continuously explored to suit the market demands and application requirements.

MEBCO's battery manufacturing plant is divided into 4 main areas; X-Met, Green Group, Formation, and Final Assembly. The Formation area's main function is the electrical charging of batteries for periods that range from 14.5 hours to 24 hours, depending on the type of battery. Because of the limited time per day, and also because of space limitations that restrict the number of charging circuits, the Formation area is considered to be the bottleneck of the battery production capacity.

According to the theory of constraints, the throughput rate of any multi-component system is controlled by the slowest component (called the constraint or the bottleneck). In order to maximize the productivity (profitability) of the battery production line, the company needs to optimize the production planning output in the formation stage. As shown in the following section, a variety of optimization and heuristic techniques have been used to solve battery production planning problems. In this chapter, an integer programming (IP) model is used to maximize the profitability of the formation area. This IP model has been practically applied to optimally schedule the daily production of 74 battery models. The model considers different battery sizes, charging times, and daily demands, as well as limitations on times, capacities, and numbers of different types of charging circuits. Compared to manually generated production plans, the IP-optimized production plans increase daily profits by 12% on average.

The remaining sections of this chapter are arranged as follows. First, previous literature on battery production planning is reviewed. Next, the manufacturing process, especially in the formation step, is described in more detail. Subsequently, the integer programming model for battery production planning is presented, and then the model's performance is compared to the current manual approach. Finally, results are analyzed, and conclusions are drawn.

SURVEY OF RELEVANT LITERATURE

Previous approaches to battery production planning include simulation, network flow models, EOQ-type techniques, mathematical programming, and various heuristics. Turnquist (1991) described a software system for battery production planning developed by Cornell University for the Battery Strategic Business Unit of Delco Remy. The PC-based system includes three main components: (1) a forecasting module, (2) a multi-plant product allocations module, and (3) a scheduling module. The production scheduling module employs a network flow model to dynamically balance inventory and overtime costs given limited capacities and fluctuating demands for each plant.

Frost (1999) discussed technological advances in the design and production of lead-acid batteries, especially those made by the Advanced Lead–Acid Battery Consortium (ALABC). He described some of the most likely developments and their expected impacts, and suggested three strategies to minimizing any adverse effects. Li et al. (2011) developed a model for the automatic configuration and equipment selection of assembly systems for manufacturing automotive batteries. A recursive algorithm is used to determine task sequences and groupings, and assign them to equipment, in order to minimize the total cost.

Elimam and Sartawi (1986) integrated statistical quality control techniques with process and production control methods to improve production schedules and quality in a battery plant. Their methodology consists of scheduling operations for a level production rate, while applying quality control tools to track and control quality aspects during production. Relying on simple practical solutions, quality-related problems were reduced, and significant improvements were obtained in reliability and performance. Elimam and Udayabhanu (2011) developed a mixed integer programming (MIP) model to generate least-cost battery production plans. The total cost is obtained by summing the costs of material ordering, inventory holding, and production setup, as well as the cost of defects. As the optimum solution of the model is difficult, a heuristic based on material requirements planning (MRP) was used to produce efficient solutions.

Yenradeea (1994) combined simulation, optimized production technology (OPT), and simple scheduling rules to schedule a four-stage battery production line. The OPT production plans successfully minimize inventory while maximizing the throughput rate. Using simulation experiments, these plans were shown to outperform both the push and the pull policies. Khadem and Ali (2008) developed a simulation model to optimize the cost effectiveness of a car battery manufacturer. The model was used to represent and analyze the dynamics of the battery assembly line, and also to make several recommendations for improving the line's cycle time, productivity, and quality.

Sharma (2003) integrated the principles of lean production with the practices of Six Sigma to improve daily work life in a battery company on a continuing basis. The advantages realized by the company included significant cost reduction, more efficient manufacturing process, and higher customer satisfaction. Al-Turki (2000) presented a heuristic multi-product batch scheduling algorithm to schedule battery batch production with the minimum set-up time. The effects of set-up time/frequency and batch size on the production performance were investigated in a real-life battery manufacturing company. César et al. (2007) described a case study implementation of intermediate stock management in the production process for automotive batteries, highlighting the main positive aspects and the deficiencies.

Huston (1996) used a three-stage process for production planning at the International Fuel Cell Corporation (IFC) in the US. First, a spread-sheet model determines daily production and workforce size. Next, an EOQ model determines optimal lot sizes. Finally, a process improvement model increases bottleneck production by reducing setup time and/or cycle time. Elimam and Udayabhanu (1999) presented two procedures for planning the production of lead-acid batteries: a simple MRP-based heuristic, and a mixed integer programming optimization model. In a later paper, Elimam and Udayabhanu (2001) modeled battery production as a multi-product, multi-period, and multi-component supply chain process. The trade-offs in this process – including the costs of holding, ordering, and defects – were tested in a small battery manufacturing plant.

Dunstall and Mills (2002) presented a mixed-integer programming (MIP) model to optimally schedule cyclic production of batteries for Exide Technologies in Australia. A production schedule is developed for 42 periods in a cyclic workweek. The objective is to maximize the throughput of a multi-product battery charging and finishing facility while maintaining a given product mix. Dunstall and Mills (2007) slightly modified their earlier MIP model and also used simulation experiments to evaluate its performance compared to two simple scheduling rules: shortest processing time (SPT), and random sequencing (RND). The simulation results confirmed that the optimal MIP schedules consistently yield higher throughputs than simple on-line dispatching rules.

Several authors discussed reverse supply chains of batteries, where the objective is to take back old used batteries, and properly dispose or recycle them. Other authors analyzed closed-loop supply chain models that include both a forward path for new batteries and a reverse path for used ones. Tsoulfas et al. (2002) applied life cycle analysis (LCA) to assess the environmental impact of different phases of the reverse supply chain process in the recovery of used batteries. Schultmann, et al. (2003) combined a closed-loop supply chain optimization model with a recycling simulation model for used batteries. Results indicated the potential for complete recycling of used batteries for use in the German steelmaking industry. Daniel et al. (2003) studied lead recovery from used batteries using life cycle inventory analysis (LCIA). Each stage of the battery reverse supply chain, including collection, processing, and recycling, is associated with a different environmental problem. A model of the reverse supply chain of is developed to evaluate the environmental trade-offs in different stages of the battery life cycle.

Kannan et al. (2010) formulated a mixed integer linear programming (MILP) model for a closed loop battery supply chain network to optimize material procurement, production, distribution, recycling and disposal. A genetic algorithm (GA) based heuristic was developed and used to obtain high-quality solutions in short computational time. Ramkumar et al. (2011) proposed revisions to eliminate certain inadequacies in the Kannan et al. (2010) model. Sasikumar and Haq (2011) developed a closed-loop distribution supply chain (CLDSC) network model for lead-acid batteries, considering used batteries as a source of recycled lead. Applying a fuzzy multi-criteria decision-making (FMCDM) approach, a mixed integer linear programming model was used to balance total cost and on-time delivery. The model optimizes decisions involving battery production, distribution, recycling, and material procurement.

The formation-stage production planning problem at MEBCO, which is described in detail later, does not involve any stochastic or nonlinear aspects. Although the problem has integer variables, it is fairly small and thus easy

Figure 1. Flow chart of the manufacturing process

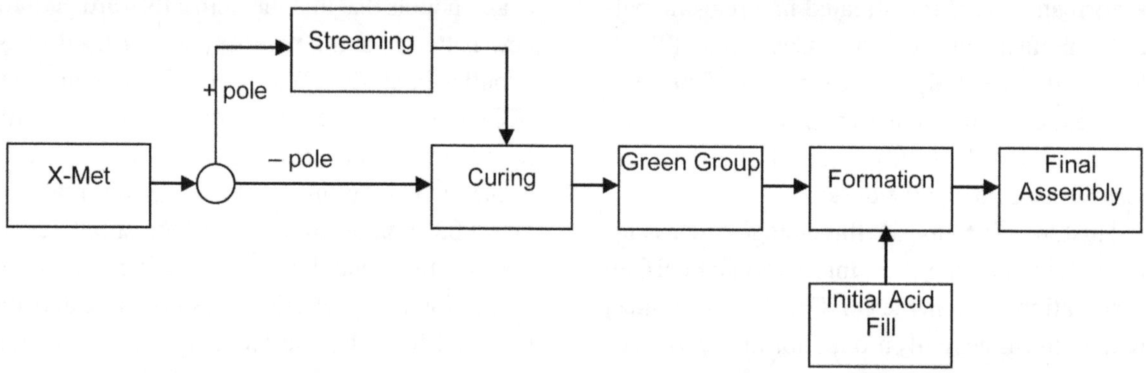

to solve. Therefore, optimization by integer linear programming is the most suitable approach for this problem. Compared to Dunstall and Mills (2002) MIP model, this chapter presents a pure-integer programming model with substantially different decision variables, objective function, and constraints. Before introducing the model, the battery manufacturing process is described in the following section.

STAGES OF BATTERY MANUFACTURING

The manufacturing space in the plant is divided into 4 main areas; X-Met, Green Group, Formation, and Final Assembly. The main processes in the X-Met area are lead strip manufacturing, oxide manufacturing, paste mixing, X-Met line, steaming, and curing. The Green Group main sections are encapsulation/collation, case preparation, battery assembly, cast-on-strap, side terminal weld, extrusion fusion weld, heat seal, and top terminal weld. The main processes in the Formation stage, which is described in more detail later in this section, are initial acid fill and formation. The Final Assembly's main processes are formation acid dump, final acid fill, final heat seal, labeling, and packing. Figure 1 illustrates the flow of the manufacturing process.

First, green group assemblies are transported to the initial fill line where they are placed on the line. The first operation in the formation area is filling the batteries with a lower-specific-gravity acid called the forming acid. This weak acid contains more water than acid to facilitate and promote the chemical conversion of the plates. This forming acid also contains a specific ingredient called formax. Controlling the amount of formax, acid gravity, and acid volume in the acid mixing process is very important in achieving the right battery formation. In addition to acid gravity and formax control, acid volume placed in the batteries is another critical factor in attaining a proper formation. Currently, automotive batteries are targeted at having a fill level just below the case to cover joining point.

Once the Green Group units have been filled with forming acid, they are transferred by conveyors to the charging row area. Each row has its own charging panel, which is the source of the electrical energy put into the batteries. The table tops are a plastic material with designed in space openings to allow for the pulling of air down through the platform thus providing cooling to the batteries while on charge. The pull of air is accomplished through the use of large air washers (motor driven devices designed to create a vacuum and pull air). Once the Green Group batteries have been filled with forming acid, they are transferred by conveyors to the charging area.

As the batteries are placed on the charging rows, they are aligned in groups of up to 18 each to match the circuit output points of the chargers. Such groupings are known as circuits. As was noted previously, each charger has 36 output points, with 18 batteries per output point (circuit). Each row could potentially hold 648 units; however, due to the variation in sizes of the batteries (8, 9, 10 and 13 inch), circuits are rarely fully utilized because of the spacing limitations. Therefore, depending on the battery sizes, the number of batteries grouped together actually ranges from 12 to 18 in each circuit. In addition to being organized in groups, individual batteries in the circuits are hooked together (positive of one battery to negative of next battery) by the use of charging straps.

Once the batteries are loaded on to a row in the circuits, they are subjected to a one-hour rest or pickle. During this one-hour time frame, strap and circuit connections are completed, and the batteries are spaced apart to attain proper air flow cooling.

The charging area, whose layout is shown in Figure 2, has 10 tables, 4 of which have 20-Ampere chargers while the remaining 6 have 30-Ampere chargers. On the charging tables, batteries are arranged in groups (known as circuits) to match the circuit output points of the chargers. Each charger has 36 output points (circuits), with up to 18 small-size batteries per circuit. However, due to space limitations, only 15 medium-size or 12 large-size batteries can be connected to each circuit.

Now that the batteries have been filled with forming acid, placed on a row, grouped in circuits, spaced properly, strapped up and allowed to pickle for one hour, they are ready to have electrical charge applied. At this stage, batteries are given two steps of electrical charges for a specific duration and ampere rates. Between the two steps, a rest time of 3 hours is provided. The electrical charges are applied in order to turn the green sulfated plate

Figure 2. Layout of the charging tables in the formation area

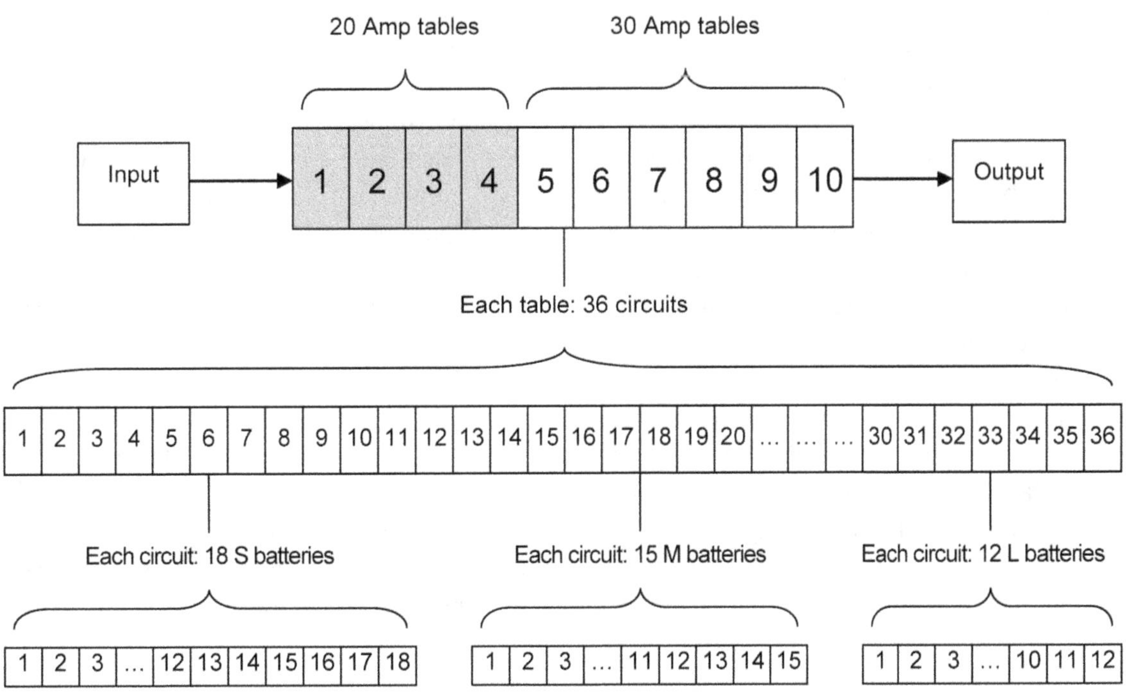

Figure 3. Detailed flow chart of the formation stage

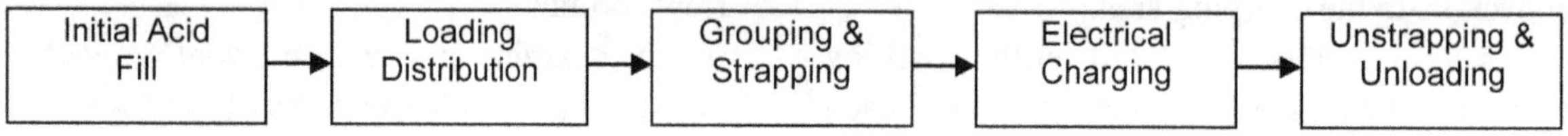

material into formed plates capable of storing and delivering electrical charge.

The electrical charging cycle of the batteries varies with the size of the battery. Smaller automotive batteries require less Ampere-hour (Amps-charger output multiplied by time on charge) charging than larger automotive batteries. Extra-large heavy duty batteries tend to have the longest charging cycles. Once the formation is completed, the batteries are unstrapped from each other and from the charging circuits. Finally, the batteries are off loaded from the charging rows onto conveyors that transfer the units to the finishing line. The steps of the formation process are depicted in Figure 3.

Though the on charge amperage rates and duration of charge (time) vary depending upon the size of the battery, the formation schedules of all batteries do have certain elements in common. Most notably, they are divided into three main phases: acid filling, loading and strapping, and electrical charging.

Sometimes during the peak season, the formation stage becomes a bottleneck that limits the battery production capacity because it involves many constraints and obstacles. These include long formation times, limited number of charging tables, other capacity constraints, lack of scheduling system, and inexact demand forecasting. In the following section, an integer programming optimization model for production planning is formulated for maximizing the productivity (profitability) of the formation stage.

PRODUCTION PLANNING INTEGER PROGRAMMING MODEL

In this section, we present the assumptions, decision variables, objective function, and constraints of an integer programming production planning model that is formulated to maximize the profitability of batteries produced from the formation area.

Model Assumptions

1. There are 10 tables for charging batteries. The number of tables is currently limited due to scarce space availability in the formation area.
2. There are six 30-Amps and four 20-Amps charging tables. This is probably a sub-optimum allocation of the 10 tables among the two types, because more battery models can be charged on 20-Amps circuits than on 30-Amps circuits.
3. Each table has 36 charging circuits. This number is fixed by the table size and the space requirements of the different battery models and charging circuits.
4. Each circuit consists of 18, 15 or 12 batteries, depending on battery size.
5. There are three battery sizes:
    - Size 1 (small) 18 batteries per circuit.
    - Size 2 (medium) 15 batteries per circuit.
    - Size 3 (large) 12 batteries per circuit.
6. In any charging circuit, we cannot have any mix among the three battery sizes.

7. Some batteries can be charged only on 20-Amps circuits, some others can be charged on 30-Amps, while the rest can be charged on both.
8. If a battery can be charged on either 20-Amps or 30-Amps circuits, then the charging time is shorter on the 30-Amps.
9. The larger the battery size, the longer the charging time required.
10. The total number of battery models produced by the company is 74. Out of these, 35 models can be charged on 20-Amps circuits, 31 models can be charged on 30-Amps circuits, while 8 models can be charged on both types of circuits.
11. The maximum demand for each battery model is given per day, based on the overall weekly demand.

In reality, there are several other considerations that are not explicitly included in the model because they do not affect the model's output. For example, storage time constraints require batteries are not kept for more than 96 hours after formation. In addition, batteries taking longer times are not loaded when there is an urgent need for finishing certain quantities of other models. Finally, where demand is normal, batteries with longer charging duration will be put on charge first.

Model Parameters

d_i = demand of model i battery, $i = 1, ..., 74$
p_i = profit per unit of model i battery, $i = 1, ..., 74$
t_{ia} = formation time of model i battery if charged on a 20-Amps table
t_{ib} = formation time of model i battery if charged on a 30-Amps table
S_k = set of indices (model numbers) of batteries of size k, $k = 1, ..., 3$

Model Variables

X_i = number of model i batteries charged on 20-Amps tables per day, $i = 1, ..., 74$
Y_i = number of model i batteries charged on 30-Amps tables per day, $i = 1, ..., 74$
A_k = number of 20-Amps circuits assigned to batteries of size k, $k = 1, ..., 3$
B_k = number of 30-Amps circuits assigned to batteries of size k, $k = 1, ..., 3$

Model Objective

The objective of the integer programming model is to maximize the total daily profit of batteries processed in the formation stage of the battery manufacturing process:

$$\text{Maximize } Z = \sum_{i=1}^{74} p_i(X_i + Y_i) \quad (1)$$

Model Constraints

The maximum daily demand for each battery model cannot be exceeded:

$$X_i + Y_i \leq d_i, \; i = 1, ..., 74 \quad (2)$$

The number of batteries charged in the 20-Amps (30-Amps) tables is limited by the available respective circuit capacity:

$$A_1 + A_2 + A_3 \leq 4(36) \quad (3)$$

$$B_1 + B_2 + B_3 \leq 6(36) \quad (4)$$

For each of the three sizes, the total number of hours used for charging 20-Amps batteries cannot exceed the total available circuit-hours per day:

$$\sum_{i \in S_1} t_{ia} X_i \leq 18(24) A_1 \quad (5)$$

$$\sum_{i \in S_2} t_{ia} X_i \leq 15(24) A_2 \quad (6)$$

$$\sum_{i \in S_3} t_{ia} X_i \leq 12(24) A_3 \quad (7)$$

For each of the three sizes, the total number of hours used for charging 30-Amps batteries cannot exceed the total available circuit-hours per day:

$$\sum_{i \in S_1} t_{ib} Y_i \leq 18(24) B_1 \quad (8)$$

$$\sum_{i \in S_2} t_{ib} Y_i \leq 15(24) B_2 \quad (9)$$

$$\sum_{i \in S_3} t_{ib} Y_i \leq 12(24) B_3 \quad (10)$$

X_i, Y_i, A_k, $B_k \geq 0$ and integer, $i = 1, \ldots, 74$, $k = 1, \ldots, 3$ (11)

MODEL APPLICATION CASE STUDY

All of the data required for the above model were collected from the MEBCO Product Standards Manual for all 74 battery models produced by MEBCO. Some of these models can be charged either on 20-Amps or on 30-Amps charging circuits. Table 1 shows standard data (formation time, charge table type, and battery size) for a sample of 15 battery models. In addition, current unit profits and typical daily demands are shown for these 15 battery models.

For each work day, the above integer programming model was solved for different daily demands d_i, $i = 1, \ldots, 74$, using the Solver optimization tool in Microsoft Excel. Compared to the current manual schedules, the optimum solutions found by the above model produced significantly higher daily profits. The increase in daily profit ranged from \$3,500 to \$6,200, and averaged around \$4,600. In relative terms, the increase in daily profit ranged from 8% to 16%, and averaged around 12%. Table 2 compares the daily profits of the model's optimum solution and the company's manual solutions for 5 typical work days.

CONCLUSION AND FUTURE EXTENSIONS

This chapter presented the formulation and solution of an applied optimization model for production planning of automotive batteries in a mid-size company. Battery manufacturing is a complex, multi-stage, and multi-operation process with several specific requirements and constraints. The formation stage, in which batteries are filled with acid and charged with electricity, is the most critical and constrained, and thus is considered as the bottleneck of the production process. An integer linear programming model is formulated whose objective is to maximize the total profitability of the battery formation stage, The constraints of the model include demand values for each battery type, limited availability of charging circuits, and charging time requirements. The ILP model determines the optimum quantities of different types of batteries, and assignment to different types of charging circuits.

The integer programming model for battery production planning presented above has several advantages and also a number of limitations. The most important advantage of the optimization model is the significant improvement profitability, amounting to an average increase of 12% in daily profits. The second advantage is its flexibility, because it allows changes in the production planning problem to be incorporated easily, whether in input parameters or in decision variables. For example, the model can easily adapt to different demand values and consider new battery types. The third advantage of the model is the use of Microsoft Excel Solver, which is virtually available on all personal computers around the world.

Table 1. Typical data for a sample of 15 battery models

Model	Formation time (Hr)	Table type	Battery size	Profit ($)	Daily demand
1	17.5	20 Amps	1	3.52	748
4	21	20 Amps	1	4.59	218
	17	30 Amps			
5	24	30 Amps	2	6.72	4
20	22.5	30 Amps	3	6.72	180
21	23.5	20 Amps	2	5.28	75
22	23.5	20 Amps	2	5.28	225
28	24	20 Amps	1	5.23	0
29	24	30 Amps	2	11.12	0
30	21	30 Amps	3	7.15	72
31	16	20 Amps	1	3.52	192
	15.5	30 Amps			
32	23	30 Amps	3	7.87	0
33	15.5	20 Amps	1	3.52	256
	16	30 Amps			
34	24	30 Amps	3	10.40	9
35	18	20 Amps	1	3.04	0
36	16.5	20 Amps	1	3.04	0
37	20.5	30 Amps	2	9.20	144

Table 2. Optimal solution profit versus company schedules' profits for five work days

Day	Company profit ($)	Optimum Profit ($)	% Profit Increase
1	37,387	43,200	**15.55**
2	41,605	46,160	**10.95**
3	32,803	37,099	**13.10**
4	41,369	45,077	**8.97**
5	35,973	39,767	**10.54**

This makes the model a convenient and low-cost tool for optimum battery production planning in many locations and for most budgets. On the other hand, Microsoft Excel itself limits the model size (number of variables and constraints) that can be handled due to the limited capability of the standard built-in Solver tool.

Other limitations of the model are imposed by the given assumptions and constraints, reflecting the company's current practice. These include space and time constraints and the limited number of charging tables.

Several opportunities for extending the model could be pursued by MEBCO in order to improve performance further. First, daily demands for different battery types could be estimated more carefully in order to take full advantage of the model. Second, the optimum number of 20-Amps and 30-Amps tables could be determined under space and budget constraints, subject to varying daily demands for individual battery types. Third, mixing different battery sizes on the same charging circuit could be allowed to increase the flexibility to meet demands. Fourth, the IP model could be combined with facility layout techniques in order to maximize the space utilization and increase the number of charging tables.

Many promising and interesting directions for future research in the area of battery production planning based on the integer programming model presented in this chapter, including the following:

1. Generalizing the integer programming model of the battery production planning problem to consider multi-component, multi-period, and multi-plant production planning situations.
2. Combining the battery production planning problem with inventory control, simultaneously considering the costs of material ordering, inventory holding, and production setup.
3. Integrating the battery production planning model with statistical quality control and inspection models, and modifying the objective function to include the costs of inspections and defects.
4. Applying simulation models to analyze a stochastic and dynamic version of the battery production planning problem.
5. Integrating the battery production planning problem with workforce assignment and scheduling, in addition to equipment allocation

REFERENCES

Al-Turki, Y. A. Y. (2000). Multi-product batch scheduling for lead-acid battery assembly line. In *Proceedings of the 11th Annual Production and Operations Management Society (POMS) Conference*. San Antonio, TX: POMS.

César, P., Rodrigues, C., & Oliveira, O. J. (2007). Intermediate stock management in the production process for automotive batteries: A case study. In *Proceedings of the 18th Annual POMS Conference*. Dallas, TX: POMS.

Daniel, S. E., Pappis, C. P., & Voutsinas, T. G. (2003). Applying life cycle inventory to reverse supply chains: A case study of lead recovery from batteries. *Resources, Conservation and Recycling*, *37*(4), 251–281. doi:10.1016/S0921-3449(02)00070-8

Dunstall, S., & Mills, G. (2002). Scheduling the charging of batteries. In *Proceedings of the Mathematics in Industry Study Group*. Canberra, Australia: University of South Australia.

Dunstall, S., & Mills, G. (2007). Robustness of cyclic schedules for the charging of batteries. *ANZIAM Journal*, *48*(4), 475–492. doi:10.1017/S1446181100003163

Elimam, A. A., & Sartawi, M. M. (1986). Quality-control practices in lead-acid battery manufacturing to improve quality, cost, reliability. *IEEE Transactions on Reliability*, *35*(4), 369–374. doi:10.1109/TR.1986.4335471

Elimam, A. A., & Udayabhanu, V. (1999). Production planning for lead-acid batteries. In *Proceedings of IIE Industrial Engineering Research'99 Conference*. Phoenix, AZ: IIE.

Elimam, A. A., & Udayabhanu, V. (2001). Multi-level supply chain in lead-acid battery manufacturing. In *Proceedings of the First International Conference on Logistics and Supply Chain Management*. Coimbatore, India: IEEE.

Elimam, A. A., & Udayabhanu, V. (2011). Optimal and heuristic production planning in battery manufacturing. *California Journal of Operations Management*, *9*(1), 1–18.

Frost, P. C. (1999). Developments in lead–acid batteries: A lead producer's perspective. *Journal of Power Sources*, *78*(1–2), 256–266. doi:10.1016/S0378-7753(99)00042-7

Huston, T. W. (1996). *Production system design and cycle time reduction in a fuel cell manufacturing operation*. (Unpublished Master Thesis). MIT, Cambridge, MA.

Kannan, G., Sasikumar, P., & Devika, K. (2010). A genetic algorithm approach for solving a closed loop supply chain model: A case of battery recycling. *Applied Mathematical Modelling*, *34*(3), 655–670. doi:10.1016/j.apm.2009.06.021

Khadem, M., & Ali, A. (2008). Modeling and simulation for car battery manufacturing for cost effectiveness. In *Proceeding of the Intelligent Systems and Control Symposia: Computational Biology and Bioinform*. Orlando, FL: IEEE.

Li, S., Wang, H., Hu, S. J., Lin, Y.-T., & Abell, J. A. (2011). Automatic generation of assembly system configuration with equipment selection for automotive battery manufacturing. *Journal of Manufacturing Systems*, *30*(4), 188–195. doi:10.1016/j.jmsy.2011.07.009

Middle East Battery Company. (2006). *MEBCO product standards manual*. Dammam, Saudi Arabia: MEBCO.

Ramkumar, N., Subramanian, P., Narendran, T. T., & Ganesh, K. (2011). Erratum to a genetic algorithm approach for solving a closed loop supply chain model: A case of battery recycling. *Applied Mathematical Modelling*, *35*(12), 5921–5932. doi:10.1016/j.apm.2011.05.026

Sasikumar, P., & Haq, A. N. (2011). Integration of closed loop distribution supply chain network and 3PRLP selection for the case of battery recycling. *International Journal of Production Research*, *49*(11), 3363–3385. doi:10.1080/00207541003794876

Schultmann, F., Engels, B., & Rentz, O. (2003). Closed-loop supply chains for spent batteries. *Interfaces*, *33*(6), 57–71. doi:10.1287/inte.33.6.57.25183

Sharma, U. (2003). Implementing lean principles with the Six Sigma advantage: How a battery company realized significant improvement. *Journal of Organizational Excellence*, *22*(3), 43–52. doi:10.1002/npr.10078

Tsoulfas, G. T., Pappis, C. P., & Minner, S. (2002). An environmental analysis of the reverse supply chain of SLI batteries. *Resources, Conservation and Recycling*, *36*(2), 135–154. doi:10.1016/S0921-3449(02)00016-2

Turnquist, M. A. (1991). University/industry cooperation on an integrated production planning software system. In *Proceedings of Technology Management: The New International Language*. Portland, OR: IEEE. doi:10.1109/PICMET.1991.183691

Yenradee, P. (1994). Application of optimized production technology in a capacity constrained flow shop: A case study in a battery factory. *Computers & Industrial Engineering*, *27*(1-4), 217–220. doi:10.1016/0360-8352(94)90274-7

Chapter 3
Improving Security and Efficiency of Multimodal Supply Chains Using Monitoring Technology

Johan Scholliers
VTT Technical Research Centre of Finland, Finland

Sirra Toivonen
VTT Technical Research Centre of Finland, Finland

Antti Permala
VTT Technical Research Centre of Finland, Finland

Timo Lahtinen
T&RH Magazine, Finland

ABSTRACT

Multimodal supply chains are characterized by multiple changes of transport modes, vehicles, and transport operators, and hence, the risks for theft, untimely delivery, and freight quality deterioration increase. There is a growing need to manage the security and efficiency of consignments. Tracking and tracing services, which are on the market, mainly concentrate on a single mode of transport, such as road vehicles, or transport units, such as containers. This chapter describes a concept for managing the security of multimodal supply chains using monitoring technology. A profound risk analysis was performed to identify threats and vulnerabilities during the different phases of the supply chain and key threat scenarios. A security service concept was developed to cope with the risks using monitoring technology. The technological solution depends on the needs of the supply chain actors and on the supply chain properties. The service concept was tested by monitoring different international shipments leaving from Finland. Intelligent monitoring devices, which were attached to the consignments, collected information on location and environmental parameters and transmitted in real-time to a background system. The chapter gives an overview of the framework, service concept, and the analysed results of multimodal shipments.

DOI: 10.4018/978-1-4666-4506-6.ch003

INTRODUCTION

Logistics is the part of the supply chain process that plans, implements, and controls the efficient, effective flow and storage of goods, services, and related information from the point of origin to the point of consumption in order to meet customers' requirements (Ballou, 2004). The aim of logistics is to execute the delivery process productively, economically securely and safely, taking into account the client's requirements. A disturbance free supply chain is efficient, reliable, visible, resilient and economical. This means that the supply chain is able to deliver the shipments as planned and not deteriorating the quality of goods. The vulnerability of one point will decrease the total performance of the supply chain and it should be viewed in the overall context of the business needs (CIPS 2006).

The configuration of transport networks leads to the need for security measures in different forms, depending on the exact function and appearance of each node and link. Multimodal supply chains are characterised by changes in transport modes, vehicles and transport operators. This poses risks on both security and the timely delivery of the goods, and challenges on tracking of door-to-door shipments. The different actors in the supply chain make their decisions, on e.g. the actual timing and route, independently, without communicating them to the other actors in the supply chain. Hence, often no actor has a complete view of the actual supply chain, and hence on the control of the security. The availability of fast and accurate information exchange plays hence a key role in improving efficiency and security (Zhenxin Yu & al., 2001; Nilsson & Waidringer, 2010).

The product damages are often caused by interaction of various factors and damage mechanisms may be complex. "Undesirable" events include late (or too early) delivery, non-delivery (the chain breaks), delivery of faulty, perished, damaged, wrong or wrong amount of the products or delivery at the wrong location. Intelliflex estimates that the cost of losses of perished food in the cold supply chain in the U.S amounts to 17 Billion $ each year (Intelliflex 2013). The frequency of security incidents as well as their consequences is increasing. The total loss of value caused by theft of cargo and freight vehicles is about 8.2 Billion Euro each year in Europe, or about 6.7 Euro per loaded trip (van den Engel & Prummel, 2007). The management of supply chain security is a topic which is hence high on the agenda of logistics operators and governmental agencies (Salmela & al., 2010; Urciuoli, 2010; Voss & al., 2009).

The use of sensor and communication technologies allows monitoring consignments in real time, and hence provides the possibility to take immediate actions to mitigate the supply chian risks.

However, the best solution for improving security and monitoring, both in terms of information exchange, performance and economics, depends on the specific needs of the actors involved. Different models can be identified regarding the amount of information in supply chains exchanged between the different stakeholders (Pilli-Sihvola 2012):

- **Black Box Model:** No information is shared between the supply chain partners during transport or other logistics processes. Problems are noticed and actions are taken with a significant delay.
- **Notification Model:** In which information is exchanged at nodal points of the supply chain, where goods are handled.
- **ID Model:** This model uses identification technologies such as OCR, barcode and RFID to enable collection and transmission of tracking and monitoring information.
- **Continuous Monitoring Model:** This model uses real time wireless communication together with sensor technologies enables significant development of the supply chain monitoring and tracking solutions.

In order to manage the security in supply chains, knowledge for building monitoring and emergency systems capabilities should be implemented. The constant changes of processes, business practices and economic power in the global business environment have generated the need for the security management approach to enterprise resiliency (Ritter & al., 2007).

Tracking and tracing services assist both in improving the efficiency and security of supply chains. Increased supply chain visibility offers possibilities for reducing transportation and expedition costs and for increasing the on-time delivery level, and also assists in transportation planning (Goel, 2010). Frost&Sullivan estimates that the government regulations regarding asset tracking and monitoring in pharmaceutical, nuclear, and hazardous chemicals supply chains will drive the market. (Frost&Sullivan, 2012)

As solutions to track and trace vehicles and trailers with stationary mounted devices are already steadily used, the use of tracking devices for consignments is still scare. The main reasons are that the business model is not seen as profitable from the operator's point of view or that the technology is seen as to be immature or expensive for global supply chain purposes. Another issue related to tracking and tracing services is that often there are no mechanisms defined in order to identify and react on exceptional situations.

In the SCIE (Supply Chain Integrity and Efficiency) project a holistic service framework for the management of the security and efficiency of multimodal consignments. The service includes a alarm centre, which monitors shipments on a 24/7 basis, and reacts on identified exceptions.

The value of the increased security measures and monitoring of the supply chain can be found e.g. in improved product safety and supply chain processes, in increased supply chain visibility and delivery speed as well as in resilience improvements, which include better anticipation and reaction to the undesired occurrences in the supply chain (Peleg-Gillai & al., 2006).The paper will give an overview of the supply chain critical stages vulnerability and risk assessments, consignments tracing devices requirements, service concept framework and analyse the results of a multimodal shipment in Europe.

CRITICAL STAGES IN MULTIMODAL SUPPLY CHAINS

There are obvious phases and stages along the supply chain that are more critical or carry more risks than others. Typically the most vulnerable phases in multimodal supply chains are slowing, stopping and parking en route on the road transport, moving through dangerous geographical areas, loading and unloading, change of transport vehicles or transport modes and static points along the routes (warehouses, terminals, ports, borders, etc.) (Burges, 2012; FreightWatch, 2012; Europol, 2009). Craddock and Stansfield (2005) have assessed incidents that can take place during a container journey. These threats have been generalized for consignments (parcels, transport units) and are presented in Table 1.

In many existing supply chain security applications the transport vehicle is monitored and exceptions and deviations are reported on real time basis. However, without effective response measures that allow to normalize the situation or at least minimize the damage without unnecessary delays, the overall impact remains limited.

RISK ASSESSMENT

Supply chain risk assessment's objectives are to identify and assess those risks and vulnerabilities in the supply chain process and route that can affect the quality of the delivery process. Several security risk assessment and management methods for supply chains and container freight have been developed in recent years (Boukachour & al., 2011; Papas, 2011; Salmela & al., 2010). The

Table 1. Threats and incidents that may occur at different stages of the supply chain journey

	Incident								
	Consignment integrity			**Consignment contents**				**Consignment movement**	
Stage in supply chain	**Unauthorized opening of transport unit**	**Transport unit intrusion**	**Transport unit damage (handling...)**	**Theft of contents**	**Unauthorized addition**	**Unexpected movement due to inappropriate fixtures**	**Content damage (shock, temperature...)**	**Unexpected movement of whole consignment**	**Route deviation, delay**
Warehousing				X	X		X	X	
Consignment stuffing/ stripping into transport unit				X	X	X	X	X	
Land transport move			X			X	X	X	X
Stationary during transport	X	X		X	X		X	X	
Transport unit handling and mode transfer	X		X	X	X	X	X	X	X
Sea transport move						X	X	X	X
Customs check						X	X	X	

ways of assessing risks are selected according to the required objectives. The risk assessment's objectives in the project were to assess the risks of the supply chains in a more general level, to establish policies for rapid risk assessment for supply chain monitoring purposes and to assess the capability of the monitoring device to identify risks during the supply chain. Supply chain vulnerability and risk assessments were based on supply chain modelling data. Databases on incidents can provide information on risk probabilities. The different handling and transport processes to which the shipment is subjected are modelled and analysed.

The risk factors were classified into five entities: human factors, malfunction, crime, environment and processes. Scenarios regarding the risks are developed to support the analysis, and prioritised together with the main stakeholders. The different potential sensor systems able to resolve, detect or mitigate the risks are then listed. The effectiveness of monitoring devices to deal with the risks and scenarios is then evaluated. The analysis is discussed with the customer and cost-benefit assessment performed prior to final system selection. It is also important to identify the vulnerabilities and risks (or incidents) that are impossible or difficult to identify with the sensor technologies. It is also important to be prepared to interpret the data caught during the transport and take appropriate measures when needed. More accurate information is gathered when the assessments are done for special deliveries, taking into account actual operators and risk factors.

TECHNOLOGY FOR INTEGRITY PROTECTION

There are lots of potential sources of information, which can provide value and added value to the integrity monitoring and which can transmit revealing data about the situation of the consignment. These include (Craddock & Stansfield, 2005)

- data from devices, which are installed in or attached to the vehicle, the transport unit or the consignment;
- documents regarding the consignment, such as the cargo manifest, packing list, dangerous goods information, expected route and timing. Information may be included on timing, routing and geofencing areas;
- environmental data, such as meteorological data, road traffic and road weather conditions, news (special events, strikes,…).

The only way to obtain automatically real-time information from consignments is through attached sensors. There are many technologies on the market for tracking and tracing. RFID (Radio Frequency Identification) is an affordable and standardised technology, which allows automatic reading in free flow from 4 to 6 meters at discrete spots (Permala & al., 2012). A major drawback in tracking is that there is no information if the tag does not pass near the reading units.

Tracking devices include location, communication and possibly MEMS sensors such as temperature, pressure, humidity, shock. The main technology used for positioning is satellite positioning (GNSS), especially GPS (Bruneteau & al., 2010). A drawback of this technology is that it does not work well when there is no line of sight to the satellites. Alternative technologies are cellular and wireless network based technologies, which use the position of wireless or cellular network access points to determine location.

The location and sensor data are transferred in real-time through cellular or satellite communication to a background system, allowing monitoring the complete supply chain from origin to destination.

The selection of the asset, to which the devices are attached and the movement of which is monitored, depends on the needs and the possibilities of the actors involved. In order to track single goods of high value, the tracking unit should preferably be directly and unobtrusively attached to the asset. As an example, FreightWatch has brought a tracking product on the U.S. market which fits in a pharmaceutical pill bottle (FreightWatch, n.d.). For consignments containing many different parcels, the transport unit may be a more appropriate asset to follow. For freight containers, security technology has been developed, such as electronic seals (McCormack & al., 2010) and CSDs (Container Security Devices) (Meyer-Larsen, 2009; Davydenko, 2012). These devices are attached to the container doors; CSDs transmit regularly information on their position through cellular networks or to satellite.

The actor in charge of security may however not have the possibility to attach the equipment to the asset of interest, for instance if he does not have access to the asset, or due to size or performance limitations of suitable devices (e.g. battery limitations, transport in metal enclosure).

SECURITY SERVICE CONCEPT

The goal of the specified security service is to support customers in a comprehensive manner in protecting and developing their supply chains. To a large extent, any well-planned and executed operation also has a lot of in-built security and integrity. The approach for improving supply chain security starts from the requirements of the different stakeholders, regarding security, efficiency and transparency of the supply chain is shown in Figure 1. Starting from the stakeholder requirements, a

Figure 1. Phases in monitoring system development

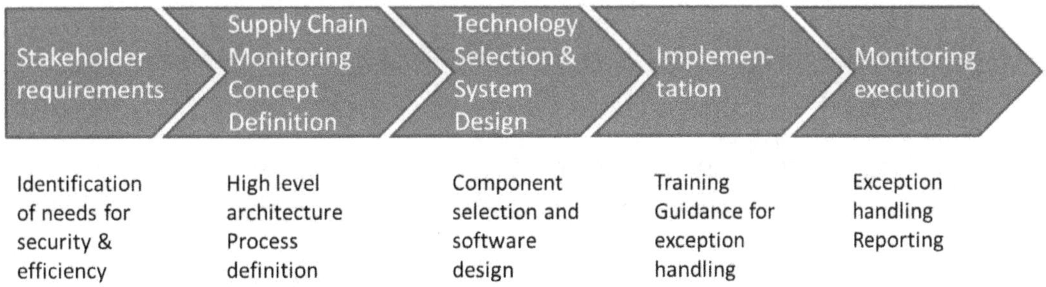

high level architecture regarding the information to be collected and to be exchanged between the different stakeholders is developed. The suitable technology is selected and the complete system is designed, including design or tailoring of the software and, if needed, required organisational changes and co-operation contracts. In the next phase the system is implemented, training is provided to the personnel involved, as well as concrete guidelines for service personnel on how to react on the information provided on exceptions by the monitoring system.

Monitoring is mostly reactive by nature, meaning that most measures taken are initiated by exceptions and deviations or technical alarms detected and forwarded to the service centre.

Figure 2 shows the conceptual architecture of the monitoring service. Alerts can be generated by the monitoring equipment on the shipment (e.g. opening of a container door) or by the back-end system (e.g. long delay with respect to plan), dependent on system design. The service centre collects information from monitoring devices, from partners and operators along the supply chains, and from other data sources, such as weather and traffic information. The service centre reacts on identified exceptions. Value-added services in the back-end system provide additional information to stakeholders, e.g. more accurate data on the expected time of arrival (ETA) allows optimizing resources.

In addition to the operation of the service centre, services are provided for planning of the service (risk assessment, device selection, alerts and reactions specification, guideline development and operator training) and for management of the devices (device maintenance, programming, installation and activation) and for return logistics of the devices.

Critical issues in the monitoring of consignments moving through the supply chain are the identification of exceptions and the efficient reaction on exceptions (e.g. search of misplaced products, attempt to recover stolen products). As logistics systems consist of a complex network of organizations and processes, exception monitoring is critical for the success of logistics services. Efficient detection of exceptions requires not only monitoring technologies, but also a proper understanding of the exceptions and the actions needed to manage them. Exception management consists of monitoring, diagnosis, and resolution that are involved in the process of keeping track of logistics activities, checking the cause of the exception, and applying the solution for the exception (Xua, 2011).

Tracking equipment may generate a vast amount of data, but only exceptions are of importance for security personnel. Since the cost of reactive measures (e.g. police involvement) is high, a clear diagnosis of the event should be made before any reactive action. The number of false alerts should be as low as possible, to allow

Figure 2. Concept for real-time monitoring of consignments

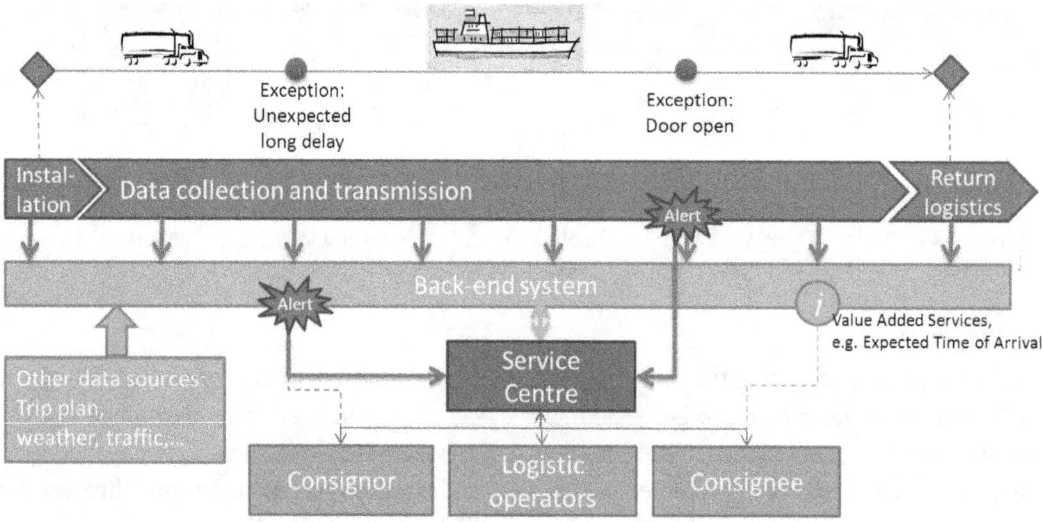

all partners to have confidence in the technology. A high amount of false or non-relevant alerts may cause the security personnel to ignore also relevant alerts.

For the correct identification and diagnosis of exceptions, the information from monitoring equipment has to be fused with other available data. By combining all relevant data, knowledge is gained on the actual status of the consignment (location, transport mode and process, such as loading and discharging…) If for instance a geofencing alert regarding deviation from prescribed route is received, this can be due since the driver is forced to take another route due to road works, or due to theft of the goods/vehicle. Also the lack of information can raise an alert, e.g. if the back end software notices that no location information is received, this may be due to an obstruction of the tracking device (e.g. moved to a location where there is no cellular network), to failure of the tracking device or to jamming of the communication network by criminals. In general the biggest restraints for the real time location systems market are high costs, a lack of standardization, and too low end-user awareness. (Frost&Sullivan, 2012)

PILOTS

Multimodal Transport between Finland and Italy

During the SCIE project multimodal shipments between Finland and the rest of Europe (Sweden, Estonia, Germany, Italy) with metal coils and metal parts were monitored. In total 9 multimodal shipments were monitored. This paper describes a shipment of metal coils from Tornio (Finland) to Castellano near Milan (Italy) between 16.4 and 3.5.2010. The objectives of the shipment were threefold: to provide more information on the actual delivery of the shipment and the environmental conditions during the shipment, to test the performance of the monitoring devices and to test the new concept for the security service.

In the project the dSeal intelligent mobile device from 4TS, a partner in the SCIE project, and the 4TS monitoring solution were used. The dSeal includes multiple sensors (temperature, humidity, air pressure, shock, fall, movement and light for integrity checking), multi positioning and real-time SMS and GPRS communication (4TS, 2013). The location is determined from GPS or

GlobalMPS, a cellular network based positioning technology developed by 4TS, which allows to achieve indoor accuracy levels of less than 30 m (4TS, 2013), with less power consumption than GPS. The device is commercially used for monitoring transport of valuable assets like electronics, perishable goods like pharmaceuticals and food or anything sensitive for damages. The device sends alerts when a sensor value exceeds a pre-set value.

The devices and their functions are controlled by profiles, in which the sensors used, the sensor alert ranges, and the frequency of measuring and geofencing areas are defined. Geofencing areas can be programmed in the device, and the device sends alerts when entering and/or leaving the geofenced areas. The profiles are set by the user and transmitted wirelessly to the device. The monitoring profiles can be changed during the transport if necessary, for instance due to condition deviations, route changes or changes in predefined schedules. The event data (measurement values, alerts, positions, and times) can be monitored in real-time in the 4TS monitoring internet service. Reports of the events can be made for example for insurance claims and for customers (e.g. quality certification).

The devices come standard with an internal 1130 mAh rechargeable battery. The battery lifetime depends mainly on the frequency of communication and positioning quality. In standard usage the battery lasts for 4-7 days (without communication about 17 days). For longer transports external batteries must be linked to the monitoring device.

At the manufacturing plant four monitoring devices were attached to product packaging, including a standard 4TS dSeal monitoring device (Figure 3), a 4TS dSeal with external battery, and temperature and acceleration logger. Geofencing areas were defined around the major terminals and nodes for the rail transport, in order to allow monitoring how the consignment is proceeding. The temperature alert range was set between -5 and 25 degrees for testing purposes, in order to assure that some alerts were generated.

Figure 3. dSeal Monitoring device attached to consignment (© 2011 Turvatiimi Oyj. Used with permission)

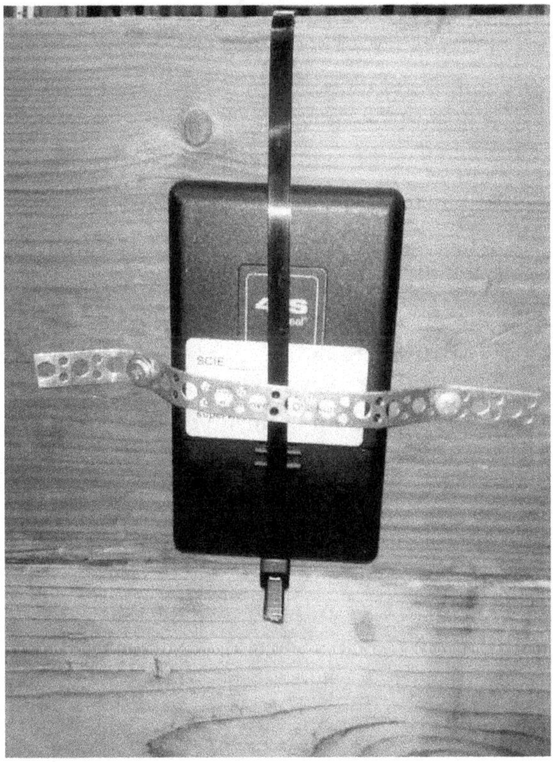

The metal coils were transported by road from the manufacturing plant in Tornio (Netherlands) to the nearby port, where they were loaded in a vessel. The vessel took the load to Terneuzen (Netherlands), where they were stored in a warehouse next to the port and from there transported by rail, through Belgium, Germany and Switzerland, to Desio nearby Milan (Italy). The last leg from the rail yard in Desio to the final destination in Castellano was by road.

The security service could follow the moving shipment through the collected location information, which was transmitted daily, and in real-time through the geofencing alerts, which were generated round the major points along the planned route (harbours, rail yards, final destination).

Geofencing alerts were sent by SMS to the security service persons monitoring the transport. Figure 4 shows the monitored path. For the sea leg, no location data were received, as the consignment was under a metal deck, and hence neither GPS location nor cellular communication was available. Location and sensor data were received according to the device profile at regular time intervals after unloading in Terneuzen warehouse and during the rail part. Geofencing alerts were received for the complete track. Alerts were received every time the shipment entered and left predefined geofencing areas.

The actual transport lasted longer than the planned transport. For instance, the train leg lasted one day longer than planned. The shipment remained also longer than expected in the Terneuzen warehouse, at the Antwerp rail yard (1 day) and at Desio rail yard, where shipment remained for 3 days over the weekend before being delivered to the customer the next Monday 3.5.

Figure 5 shows the measured temperature and humidity during the sea transport, the storage in the Terneuzen warehouse and the train transport. During the sea transport (from 16.4 to 21.4) the temperature remains constant and humidity decreases from the initial value. From 21.4 to 26.4 the load is in a local warehouse, and from 26.4 onwards on a train. The temperature in the train wagon increases due to warm weather and sunshine, and becomes higher than 35 degrees. The temperature and humidity conditions of the load were compared with meteorological data from weather stations in the neighbourhood of the track.

Temperature alerts were generated when the measured temperature exceeded the alert value (25 °C). Due to sunny warm weather, alerts were received once a day when temperature in the train wagon surpassed/reached the alarm level. No alarms were however received for the part in Italy, when the weather was bad according to the meteorological station nearby Milan.

The information of the delays and temperature alarms was analysed in the service centre but no action was considered necessary. During the test, information was sent to the stakeholders during

Figure 4. Monitored path for Tornio-Castellano shipment; Courtesy of GPSVisualizer.com

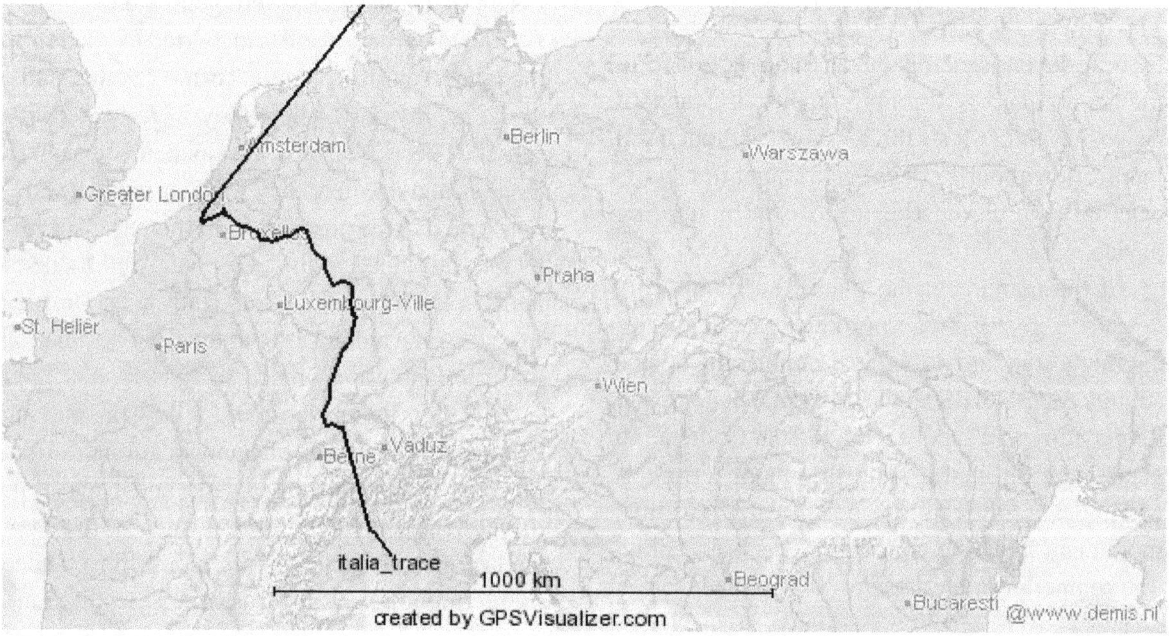

Figure 5. Measurements during the transport from Tornio over Terneuzen to Italy

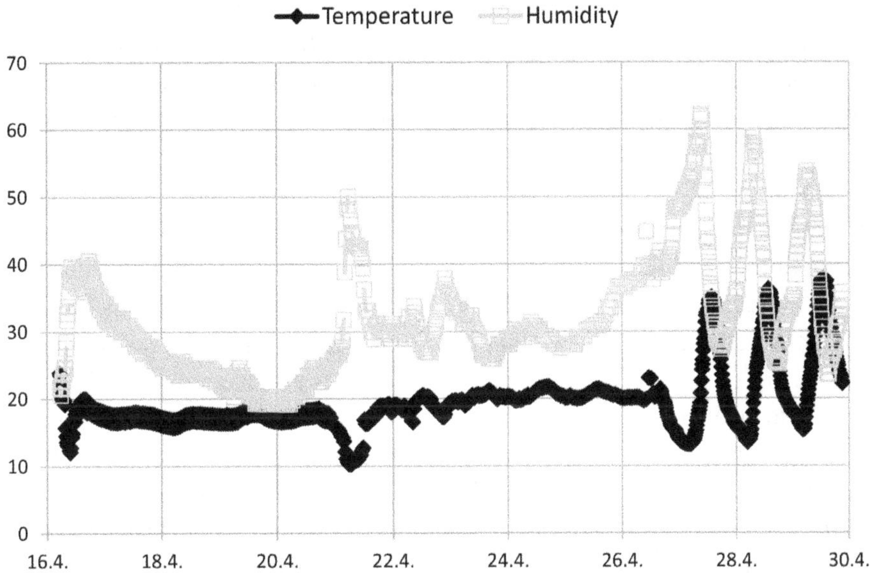

working hours on 3.5 and after the consignment arrived at the destination.

Shipments from Southern Finland to Estonia

Six shipments of wooden cases filled with metal products between Hyvinkää in Southern Finland and Estonia were monitored. The transport involved the following phases: collection of the consignment by transport operator, and transport to a warehouse in Helsinki. The cases are loaded in a truck, which boards in the port of Helsinki on a RoRo-vessel to Tallinn, and is delivered at a consignee near Tallinn. The average duration of a shipment is about 24 hours.

The GPS interval was set at 45 minutes. In all shipments, the battery capacity was sufficient. If the device is deactivated after receipt, the capacity is also sufficient for the return trip. During the transport on the vessel, no location data were transferred. The devices awake after arrival in the port, but in 3 of the 6 cases the first location after arrival was the final destination.

During 2 shipments, alerts for shock were set a low level (1.2g). Shock alerts were received during the handling phases, as well as during the collection phase. As in many other devices, the shock alert is sent together with the last measured location. This makes interpretation of the alert more difficult: e.g. when an alert is received after the sea transport, it is not clear where the shock alert was actually generated.

Container Shipments between Europe and Asia

Container Security Devices (CSD) consists of intrusion and location sensors as well as communication possibilities and it is installed on the container door before shipment. In the SMART-CM project (Smart Container Chain Management) (Davydenko, 2012), which received funding from the European Commission European Union's Seventh Framework Programme (FP7/2007-2013), the possibility of using CSDs for enhancing security and the efficiency of global supply chains, was researched. In the project devices from different manufacturers were tested. The devices send information to the device manufacturer's server, which transfers the data in a uniform format to a "neutral layer", where it is combined with information from other sources such as AIS. An additional layer provides value added services, such as ETA (Estimated Time of Arrival) estimation and geofencing alerts.

During the project, more than 100 container transports between Europe and Asia (Abu Dhabi, India, China) were successfully monitored. The cost per trip was estimated at 300 € (Davydenko, 2012).

Based on the evidence gathered during the evaluation activities, it emerges that, by addressing user needs and requirements, the SMART-CM system may deliver significant benefits to the stakeholders, the potential users, and the general public, such as cost reduction, throughput time reduction, productivity and reliability improvement, flexibility and added value and quality monitoring.

The SMART-CM solutions were proven to fulfil to a great extent the users' needs by providing real time information on the location and the integrity status of containers, and this is considered by the industry as a unique and desirable operational element. It also provides a single source of information resulting in better communication across companies, and gives the necessary data for fast customs processing offering ultimately a reduced risk profile for the containers that are covered by the SMART-CM system.

CHECKLIST FOR DEVICE SELECTION

A major prerequisite for being adequately able to identify incidents is the availability of accurate transport plan and execution information, so as to be able to distinct exceptions from normal situations and hence to take correct actions. Contact information of the current transport operators and drivers is needed to resolve the cause of incidents. Especially for multimodal transport, in which there are a lot of logistic providers and their subcontractors involved, this is a challenge. Especially information on changes in plans, such as schedules or transportation routes, is often difficult to receive, making it difficult to assess the validity of exceptions and to be able to contact the driver or the shipper of the consignment. The availability of a detailed transport plan and the procedures for informing on deviations should be part of the contract between customer and security service provider. The information of the logistics operator is also important in those situations where no tracking information is received from the monitoring device. In critical situations the service centre will call the appropriate logistics operator to insure the security of the consignment.

The battery of the monitoring device is a critical factor and should be selected so that it covers the whole supply chain. The lifetime of the battery is determined both by the communication and location frequency and the capacity of the battery. Non-rechargeable batteries have a higher energy density and hence last longer trips. The updating frequency should be optimised depending upon the transport phase: during secure phases (e.g. when the asset is stored in a warehouse) or during periods where communication success is doubtable (e.g. on an ocean vessel), communication frequency could be set low (e.g. once or twice a day). Motion sensors allow optimising battery capacity, by keeping the device asleep when the target is not moving. For efficient recovery, the profile should be adapted over the air so that location and transmission frequency is denser. Also the environmental conditions (temperature), used communication and location options and the availability to access the network and satellite positions affect to the battery lifetime.

Proper handling of the devices is critical. This involves the installation, activation, deactivation and return logistics. Prior to shipment, the battery is fully loaded and the device programmed (updating frequency, geofencing and other alerts).

The device should be attached to the asset and activated. It has to be assured that the device is attached to the correct shipment. Mechanisms have to be taken into use to inform the service centre about the installation (e.g. photograph), the identity of device and consignment and the monitoring device activation. Methods requiring no physical interaction with the device are advisable, e.g. activation through transmission of SMS messages.

The communication medium should be selected towards the need of the supply chain. Cellular communications allow tracking overland shipments. Roaming between different cellular networks should be taken into account in international transport.

The device activation process at the beginning of the trip is critical. It has to be assured that the battery is fully charged and that the device is correctly attached and activated. If the device is sent without proper activation, corrective actions are difficult and the service can't fulfil its promises.

Environmental sensors and location tracking offer both possibilities to identify exceptional and security critical situations. The interpretation of the sensor data is often however cumbersome. For instance, when handling the device during the installation phase accelerations of over 5g easily occur, during transport of large units they can cause damage.

The integrity of containers and trailers can be verified through door and intrusion sensors. Door sensors are most often magnetic or contact. PIR (Passive Infrared sensors) can be implemented as MEMS sensors, and hence be included in monitoring devices. However, the possibility of false alarms is rather high and the sensor view may be obstructed by the load.

Based on the transport phase and mode (road, sea, storage) profiles and rules must be set for alarms and for location and environmental sensors. Guidelines and action plans must be developed for the service centre on how to react on the alarms and warnings dependent on the transport phase. Statistical information received from the shipments can be used to develop heuristics for the guidelines on the reaction to alerts, which make the base for the support centre personnel to manage the security, integrity and efficiency of the consignment deliveries. Identification of the correct transport phase are critical; especially on transitions (e.g. from storage at harbour to road transport) were risks due to the handling are often higher. Data which can assist in identifying transitions for consignments, are location (geofencing), acceleration and environmental (temperature, pressure) data. Satellite location information availability depends on the transport unit (e.g. amount of metal), and alternative methods, such as wireless network based methods improve the availability of location information in critical situations, e.g. for transitions at terminals. During marine and train transport, cellular networks are not always available. In these cases alternative sources for tracking information, such as marine AIS systems and railway traffic tracking systems should be used. This required that the transport mode can be identified and that is assured that the consignment is on the correct transport vehicle (ship, train). When alerts are issued, all data, both sensor and location information, should have the same time stamp, in order to allow to correctly assess the alert.

Competitive factors in monitoring solutions include cost, performance, technology, ease of integration and use, return on investment (ROI), time, and battery life. (Frost&Sullivan, 2012) Monitoring and tracking devices have a cost comparable to mobile phones, and therefore should be returned to the service provider at the end of the transport as much as possible. The profitable security service requires well-functioning return logistics of the monitoring devices. Procedures of the return logistics must be established for different use and customer cases. Agreements on return logistics should be made with the customers and mechanisms developed which make return of the devices easy and cost efficient. Most easy this procedure is when the customer representative is also the consignment receiver. The device case could also function as a return package. When disposable devices are used profitable system may be more difficult to create (Salmela & al., 2010).

CONCLUSION

Multimodal supply chains are complex systems with a multitude of actors, vehicles and transport units involved. Each of the actors is mainly only interested in the part of the supply chain under their responsibility. There is hence no actor responsible for the security of the supply chain. The major risks involved relate both to supply chain issues, such as delays, product damage or deterioration and to actions from outsiders, such as theft. Through the use of monitoring technology the supply chain can be monitored. The main benefits of the technologies are the increase of transparency of the supply chain, reduction of human errors and the possibility to optimise resources. Technology also offers possibility to identify criminal actions at an early stage and to recover goods after theft.

A service concept has been developed for the management of the security through multimodal supply chains. The core of the service is the monitoring of consignments using intelligent devices. Tracking devices monitor location and optionally environmental conditions of the consignment, and transfer these to a service centre. Through comparison of the information received with the trip plan, exceptions can be identified. The service centre performs actions depending on the warning and the transport phase.

The major challenges in the service concept are the identification of exceptions and handling of exceptions. Identification of exceptions demands that an up-to-date transport plan is available to the service centre, so that deviations can be clearly identified. Handling of exceptions requires that exceptions can be validated through contacting of the actor responsible for the transport, e.g. truck driver. Contracts with local police forces are needed in order to be able to recover stolen goods.

The monitoring technology is selected in function of the needs of the supply chain actors and of the properties of the shipment. Key properties in this selection are the transport modes used, the duration of the supply chain, cost of the asset, requirements regarding environmental conditions, and the possibilities to affect in real-time on these parameters. If there are no possibilities to affect, a simple logger is sufficient. Real-time transmission of data allows for immediate actions in case of incidents. Key issues of monitoring equipment are the battery capacity, communication medium and antenna placement, frequency of data transmission. More research is needed on the identification of transitions and identification of exceptional situations, which may require fusion of different sensor values and possibly other data sources. For identifying exceptions, the data of the sensors should be combined with information on the trip plan and external data, such as meteorological, AIS, railway and road traffic information) in order to support the support centre and to improve the level of service.

This project showed clearly the complexity of the supply chain monitoring and security management in multimodal shipments. Management of the security requires the active involvement of all logistic partners, providing and sharing up-to-date planning and transport execution information, so that exceptions can be assessed correctly. Prerequisites for the service are reliable monitoring devices and transparent and secure communication flows.

REFERENCES

4.TS Corporation Oy. (2013). *4TS dSeal device - Ensure the quality and integrity of your deliveries*. Retrieved April 5, 2013, from www.4ts.com

Ballou, R. H. (2004). *Business logistics/supply chain management: planning, organizing, and controlling the supply chain* (5th ed.). Upper Saddle River, NJ: Prentice Hall.

Boukachour, J., Fredouet, C.-H., & Gningue, M. B. (2011). Building an expert-system for maritime container security risk management. *International Journal of Applied Logistics*, *2*(1), 35–56. doi:10.4018/jal.2011010103

Bruneteau, F., Bourhis, O., Cornish, R., & Hallauer, T. (2010). *European location study*. Ptolemus Consulting Group. Retrieved April 5, 2013, from www.ptolemus.com

Burges, D. (2013). *Cargo theft, loss prevention, and supply chain security*. Waltham, MA: Butterworth-Heinemann.

CIPS. (2006). *Supply chain vulnerability*. Retrieved March 29, 2013 from www.cips.org

Craddock, R. J., & Stansfield, E. V. (2005). Sensor fusion for smart containers. In *Proceedings of Signal Processing Solutions for Homeland Security*. IEE. doi:10.1049/ic:20050272

Davydenko, I., Zhang, M., & Tavasszy, L. (2012). Assessment of SMART-CM platform benefits as the result of short transport time. In *Proceedings of e-Freight 2012 e-Freight, International Conference on Paperless Freight Transport Logistics*. Delft, The Netherlands: e-Freight.

Ekwall, D. (2009). The displacement effect in cargo theft. *International Journal of Physical Distribution & Logistics Management*, *39*(1), 47–62. doi:10.1108/09600030910929183

Europol. (2009). *Cargo theft report*. Europol. Retrieved April 5, 2013, from www.europol.europa.eu

FreightWatch International. (2012). *FreightWatch international global threat assessment*. FreightWatch International. Retrieved April 5, 2013, from www.freightwatchintl.com

FreightWatch Security Net. (n.d.). *Geo F2 'pill bottle' tracker*. Retrieved April 5, 2013, from www.freightsecurity.net

Frost & Sullivan. (2012). *Analysis of the real-time location systems (RTLS) market government and healthcare verticals drive growth*. NAE4-11.

Goel, A. (2010). The value of in-transit visibility for supply chains with multiple modes of transport. *International Journal of logistics: research and Applications, 13*(6), 475-492.

Intelliflex. (2013), *The true cause of $17 billion cold chain shrink*. Retrieved March 29, 2013, from www.intelliflex.com

Lahtinen, T. (2010). *Supply chain integrity and efficiency – Creating a new service concept*. Lurea, Finland: Laurea University of Applied Sciences.

McCormack, E., Jensen, M., & Hovde, A. (2010). Evaluating the use of electronic door seals (e-seals) on shipping containers. *International Journal of Applied Logistics*, *1*(4), 13–29. doi:10.4018/jal.2010100102

Meyer-Larsen, N. (2009). Win-win solutions combining security and logistics. In *Proceedings of Second European Conference on ICT for Transport Logistics (ECITL)*. Venice, Italy: ECITL.

Nilsson, F., & Waidringer, J. (2010). Logistics management from a complexity perspective. *The ICFAI Journal of Operations Management*, *3*(2), 59–73.

Papas, N. (2011). Combining EA techniques with bow-tie diagrams to enhance European port security. In *Proceedings of International Conference on Paperless Freight Transport Logistics*. Munich, Germany: IEEE.

Peleg-Gillai, B., Bhat, G., & Sept, L. (2006). *Innovators in supply chain security: Better security drives business value*. Palo Alto, CA: Stanford University.

Permala, A., Rantasila, K., & Pilli-Sihvola, E. (2012). RFID: From closed systems to improving visibility in the manufacturing supply chain. *International Journal of Applied Logistics*, *3*(2), 14–24. doi:10.4018/jal.2012040102

Pilli-Sihvola, E., Rantasila, K., Permala, A., & Huhtaniemi, J. (2012). Solutions for monitoring multimodal supply chain. In *Proceedings of e-Freight 2012. e-Freight, International Conference on Paperless Freight Transport Logistics*. Delft, The Netherlands: e-Freight.

Ritter, L., Barrett, J. M., & Wilson, R. (2007). *Securing total transportation networks: A total security management approach*. New York, NY: McGraw-Hill.

Salmela, H., Toivonen, S., & Scholliers, J. (2010). Enhancing supply chain security with vulnerability management and new technology. *IET Intelligent Transport Systems*, *4*(4), 307–317. doi:10.1049/iet-its.2009.0124

Urciuoli, L. (2010). *Security in physical distribution networks- A survey study of Swedish transport operators*. Lund, Sweden: Lund University.

van den Engel, A. W., & Prummel, E. (2007). *Organised theft of commercial vehicles and their loads in the European Union*. Brussels, Belgium: European Parliament.

Voss, M. D., Whipple, J. M., & Closs, D. J. (2009). The role of strategic security: Internal and external security measures with security performance implications. *Transportation Journal*, *48*(2), 5–23.

Xua, D., Wijesooriyaa, C., Wanga, Y. G., & Beydounb, G. (2011). Outbound logistics exception monitoring: A multi-perspective ontologies' approach with intelligent agents. *Expert Systems with Applications*, *38*(11), 13604–13611.

Yu, Z., Hong Yan, T. C., & Cheng, E. (2001). Benefits of information sharing with supply chain partnerships. *Industrial Management & Data Systems*, *101*(3), 114–121. doi:10.1108/02635570110386625

Chapter 4
RFID:
From Closed Manufacturers' Systems to Supply Chain-Wide Tracking

Antti Permala
VTT Technical Research Centre of Finland, Finland

Karri Rantasila
VTT Technical Research Centre of Finland, Finland

Eetu Pilli-Sihvola
VTT Technical Research Centre of Finland, Finland

Ville Hinkka
VTT Technical Research Centre of Finland, Finland

ABSTRACT

The use of RFID (Radio Frequency Identification) tracking in closed systems is replacing barcode as dominant tracking system in many industrial sectors because RFID enables reading multiple objects simultaneously without visual contact. To enlarge tracking systems to cover global supply chains, all aspects related to RFID, such as radio frequencies, data content, transmission protocols, and message sets, need to be standardised. By collecting, processing, and distributing information efficiently, organisations should be able to improve the efficiency of their transport logistics processes, lower their operational costs, and improve their portfolio of logistics services. This chapter describes the current perspectives, challenges, and benefits of RFID tracking applications in manufacturing industry. The perspectives derived from review of previous research are validated by using case study method.

INTRODUCTION

Due to the complexity of modern logistics networks, the availability of up-to-date information plays a key role in improving efficiency (Yu, Yan, & Cheng, 2001; Nilsson & Waidringer, 2004). Automated identification (Auto-ID) enables tracking and tracing of physical goods in a timely, accurate, and cost-efficient manner.

DOI: 10.4018/978-1-4666-4506-6.ch004

The information gathered using an Auto-ID can be transmitted to an information system in real-time and shared with other parties of the chain. Furthermore, an Auto-ID tracking system represents a major opportunity to improve process control, support inventory management, and develop supply chain as more responsive. Eventually, introduction of Auto-ID based tracking may transform the business processes by enabling re-engineering of the supply chain and removing structural constraints that limit the performance of the chain. (McFarlane & Sheffi, 2003; Wamba & Chatfield, 2009)

The main goal of this paper is to describe the current state of RFID tracking applications in manufacturing supply chains in Europe, as well as present its challenges and benefits based on literature review and empirical evidences.

METHODOLOGY

Research methods, in general, can be divided into two mainstreams; quantitative and qualitative methods (Saunders, Lewis & Thornhill, 2009). Quantitative research is "hard-nosed" and data-driven research that tends to optimize control and generalizability, while "soft" qualitative methods maximize realism (Kotzab, Seuring, Müller & Reiner, 2005). This paper employs case-study method, which is one rather popular form of qualitative research (Yin, 2012). The approach of the paper is analytical as it aims to explain reality by reproducing causal relations (seeking the explaining effects) by finding the current cause (Arbnor & Bjerke, 1997).

In order to create a comprehensive picture of challenges and benefits of RFID tracking systems, a descriptive literature review was conducted by utilizing the most relevant scientific articles and other studies. In empirical part of the paper, the findings of literature review were exposed under verification by conducting a case study of a real-life RFID tracking implementation. A case study method is a widely used technique in studying contemporary phenomena in real-life context (Yin, 2002). Considering the fact that RFID implementations are usually unique and they contain several implementation-specific characteristics, therefore, the authors estimated that the use of case study method would most probably provide the best answers for the research question (McBurnley & White 2009).

As the approach of this study is to verify the findings of literature review with empirical evidence, the main criteria for the case selection is that it should include the majority of those aspects of RFID tracking implementation which arose during the literature review. Therefore, the following three factors favour the selection of RFID tracking implementation in ABB Drivers manufacturing plant in Finland as a case to be studied: (1) The company has a long history with RFID, which allows one to scrutinize the evolution of the system; also, the system was considered to be in mature and stable phase. (2) ABB's RFID tracking implementation is applied to inbound supply chain functions and involves several supply chain partners. (3) ABB is a major player in its industry, and its RFID tracking implementation is conducted in Finland, which is in the frontline of utilising RFID technology.

The information employed in the case study was obtained from two sources. At first, the basic information of implementation was acquired from the case bank of RFID Lab Finland, which is non-profit organization distributing RFID related information. In addition, some press releases of ABB were explored in order to form a solid picture of RFID tracking implementation. Second, the empirical data was acquired by interviewing the Head of Quality and Operational Excellence in ABB Drives Finland. The open questions were derived from the results of the literature review. In practice, the important aspects (i.e. benefits and barriers) identified in previous research were presented to interviewed person, who agreed or rejected the statement, as well as provided additional information for statement in question.

Considering the reliability and validity of employed case study method for the research problem, few issues arise. The level of RFID adoption is still relatively low in many countries, and individual implementation can vary significantly in scope. In addition, RFID can be utilized in several different solutions like personal identification, warehouse management, and product identification. Therefore, this study relies on case study method, which provides deeper understanding of the issues related to research problem than broad scale questionnaires or similar would offer. The reliability of the research has been taken into account by drafting the interview questions based on literature review of earlier research.

AUTOMATED IDENTIFICATION

The two most adapted Auto-ID systems in logistics are barcode identification and RFID. A barcode is an optical machine-readable representation of data relating to the object to which it is attached. RFID is the use of a wireless non-contact system that uses radio-frequency electromagnetic fields to transfer data from a tag attached to an object. RFID technology has the following two advantages over barcode: (1) RFID does not require a line of sight to be read, and therefore dirt, ice or dust on an RFID tag does not prohibit or make reading difficult; and (2) it is possible to read multiple RFID tags simultaneously, while barcodes need to be read individually (Wyld, 2006; Rundh, 2008). These benefits are so fundamental that the use of RFID technology improves the logistics performance of the company more than other automatic identification technologies, particularly when the barcode is used (Johansson and Pålsson, 2009).

RFID Tracking in the Supply Chain

Considering the different actors in the supply chain at a general level, the main phases in which RFID tracking can be employed are identified in Figure 1. By employing an RFID tracking solution, it is possible to automate identification processes that earlier have had to be carried out manually. Improved efficiency and performance is achieved through improved accuracy, throughput time, and decreased personnel.

Current Situation and Challenges in RFID Tracking Implementation

The development in the implementation of RFID tracking has followed a typical progression of an evolving technology. Three phases may be identified in this development: the first phase of enthusiasm was realized in 2004 and 2005. It was followed by a period of disillusionment somewhere between 2006 and 2008. After 2008 RFID has moved into the present situation, where understanding of the technology's relevance and its role in the market or domain is better understood. (Bendavid & Cassivi, 2010)

Regardless of improvements, RFID technology is still relatively marginally adopted by companies in Europe. The level of RFID adoption varies depending on the purpose of use and the size of the enterprise. This is illustrated in Figure 2, which presents the adoption of RFID in EU-27 and case countries for different purposes. As illustrated, both Finland (8%) and Germany (4%) had above EU-27 level shares of enterprises using RFID. The main usage of RFID in supply chain management is related to personnel monitoring or access control (around half of the share), followed by tracking and tracing, which is the second biggest use in all presented areas.

According to a survey conducted in 2006 among Finnish logistics service providers, almost 50 percent of large Finnish manufacturing and trading enterprises that operate internationally expected that they will implement RFID technology in the next five years (Naula, et al., 2006). Wide-scale RFID adoption in the near future was also predicted in the RFID roadmap published in 2006 (Permala, Scholliers, & Granqvist, 2006). At the moment, it seems that adoption of RFID has been slower than predicted in 2006, as demonstrated by the numbers in Figure 2.

Figure 1. Example of applying RFID tracking in the supply chain

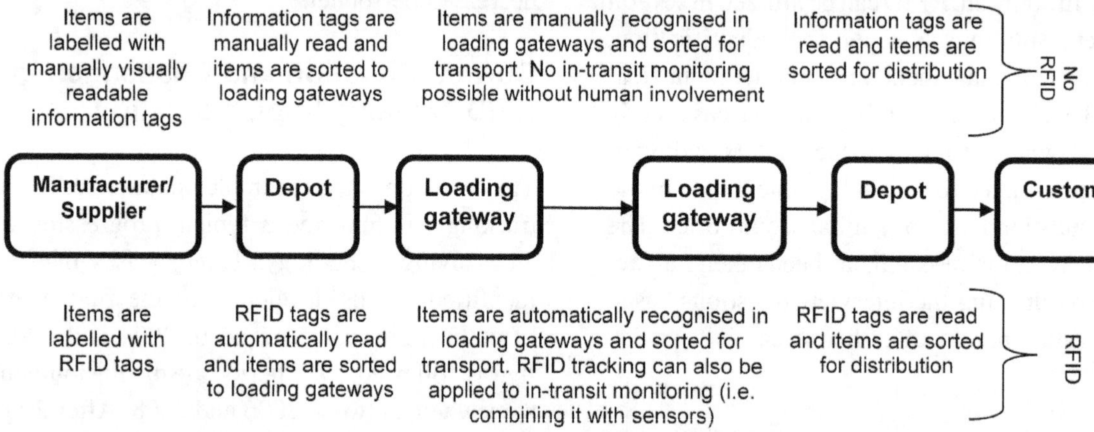

Figure 2. Use of RFID for supply chain management in by purpose (% of enterprises using RFID, multiple answers possible) in EU-27 and Case countries, 2009 (Adapted from Eurostat News Release 12/2010)

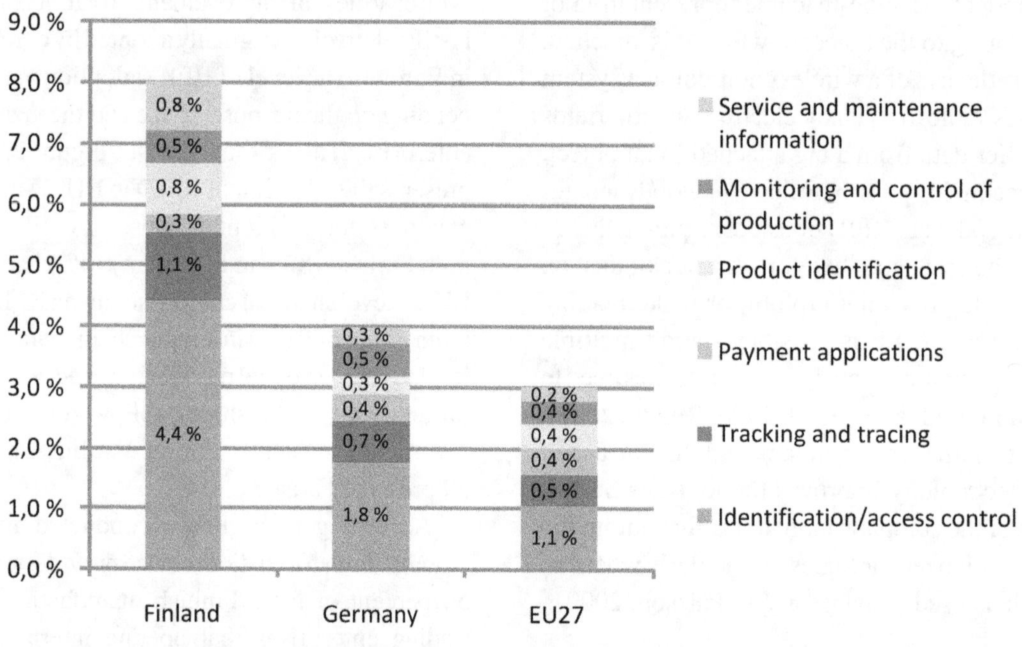

Literature of RFID tracking implementations has identified the following challenges related to the multi-organizational RFID tracking adoption, which had influenced on slowed than estimated adoption of technology: Cost-related issues, technological deficiencies, organizational and inter-organizational problems of adoption (Hinkka, 2012).

The costs can divided as: (1) Cost of new technology (RFID tags, other hardware, software, system integration and data management); (2) the costs of changed operations (tagging costs and the costs of operational automation); (3) implementation costs (support, training and reorganisation); (4) financial problems related to high investments;

and (5) overall uncertainty about the costs versus estimated benefits (Hinkka, 2012).

The significance of technological problems have decreased as the performance of RFID technology has developed rapidly during the recent years, while different manufacturers' products can communicate with each other, and different frequencies used in different continents are not anymore problem. However, there are still some technological problems in the proximity of liquids and metals, where at least more expensive tags are required. (Kapoor et al., 2009) In addition, some authors require more research concerning the security of RFID systems (Zuo, 2010).

The most common organizational problems related to RFID tracking adoption are that company's management is not committed to RFID tracking adoption (Lim and Koh, 2009; Li et al., 2010), or different interest groups inside the company see RFID tracking differently and due to these differences the company cannot form a common understanding about the function of RFID tracking system (Asif and Mandiviwalla 2005).

The biggest inter-organisational problem relates to the fact that if the products are RFID tracked throughout the supply chain, the tag should be attached at the earliest phase possible in the supply chain. However, the common understanding is that the retailers will get the biggest benefits when introducing an RFID tracking system (e.g., Soon and Gutiérrez, 2008; Hinkka et al., 2012). Therefore, product manufacturers may not be willing to pay the largest single cost of the system – the tags – while those downstream in the supply chain would get the biggest benefit (Dutta et al. 2007). Conversely, Whang (2010) also noticed that those downstream on the supply chain may be reluctant to lead the way for tracking adoption, because they may prefer to wait for the RFID tracking decisions of upstream companies before planning their own tracking system, as joining the tracking system when a company upstream attaches RFID tags is cheaper if those downstream can use the existing tags attached by upstream supply chain actors. Whang (2010) names this situation a "free rider problem".

Benefits of RFID Tracking

In supply chain management, RFID is usually deployed for tracking purposes. The literature has identified four main reasons for conducting tracking and building tracking systems: 1) The real-time coordination of material flows and individually tracked items, such as merge-in-transit; 2) the generation of exception notices based on tracking information; 3) the use of tracking in increasing the efficiency of administrative processes such as improved inventory count and goods receipt transactions; 4) the use of tracking information in developing and improving logistics management metrics and analyses (Kärkkäinen et al., 2004). However, the obtainable benefits of RFID tracking depend on the usage level of the technology: on the lower usage level, RFID tracking can automate numerous individual operational processes, for example, by decreasing costs and increasing speed and quality, but RFID tracking can also be used on the higher level by improving managerial processes, for example, by using tracking data for management decisions concerning the whole supply chain (Koh, et al., 2009; Visich et al., 2009).

At a general level, it is difficult to describe the benefits in monetary terms, because the obtainable benefits depend on the surroundings where RFID tracking is used and especially the level of integration of RFID tracking to the information systems of the company. Visich et al., (2009) also pointed out that the information about the existing RFID tracking implementations is confusing, because technology providers usually have signed non-disclosure agreements with their customers, thereby preventing them from discussing these implementations. Lee & Özer (2007) highlight the fact that the best-known papers and reports about the benefits of RFID tracking have been written by technology consultants and other experts representing organizations which gain when RFID technology adoption increases.

However, Visich et al. (2009) has managed to give some estimation about the potential benefits of RFID tracking by searching the reported results of existing RFID tracking applications. According to Visich et al. (2009), in manufacturing, RFID tracking has been reported e.g. to reduce labor costs by 17% and procurement costs by 11%, and reduce cycle time almost 50% and lead time by 27%. In warehouse and distribution centers, time used for different receiving processes have decreased by 10-80%, different shipping processes by 40-80%, and labor needed for inventory management have decreased 14%. In forward supply chain, supply chain response time have decreased 30-50%, supply chain costs have reduced by 10-40%, and supply chain efficiency have increased considerably by enabling doubling the goods processed. (Visich et al., 2009) However, these figures are case-sensitive and they are mostly result of process automation. By integrating RFID tracking as a part of the information system, higher-level benefits of RFID tracking could be even bigger than those reported values.

Standards Related to RFID Tracking

The use of RFID for automated identification in the supply chain requires that each unit or item have a unique number that can be used to identify it. In addition, every supply chain actor should be able to read the identifier and understand its content similarly.

ISO standards basically define the applicable hardware for different purposes. Passive RFID tags operating in the UHF range are the most important for logistics operations because of their relatively long range (2–6 metres) and relatively low cost (approximately 5–10 cents per tag) (Rees, 2004; Bottani et al., 2010). For this frequency range, ISO has developed the 18000-6 standard that allows for three different air interface protocols (18000-6A, B and C).

GS1's EPC is a unique code that is assigned to each tagged object or material and is read when passed through an RFID reader. The RFID technology and associated computer networks use the EPCglobal Network and its EPCIS as an interface to enable EPC-related data to be 'captured' and 'queried' using a defined set of operations and associated EPC-related standards, combined with security mechanisms that satisfy the needs of the parties involved (EPC, 2007). In other words, EPCIS provides a standard interface for storage and access of EPC-related data that can be read and written by the authorized parties involved (Främling et al., 2013).

Originally, the EPCglobal standards for the air interface protocol (originally developed by the Auto-ID Center at MIT) were remarkably different from the ISO standards during the first half-decade of the 2000s. However, the specification and approval of the EPC Generation 2 air protocol in 2006 (RFID Journal, 2006) was a significant breakthrough on the road to RFID interoperability. The EPC Gen. 2 air interface protocol corresponds to the C-mode of 18000-6 ISO standard thus allowing the use of interoperable RFID tags and readers regardless of the chosen approach (ISO or EPCglobal). Even if EPCglobal standards have been estimated to become the dominant data exchange standards, many industries (e.g. metal workshops, paper mills, and printing houses) have decided to use different identifier format such as ID@URI because ID@URI has better scalability than EPCglobal model and suits better for managing product life-cycle information, but it still has the same essential properties as the EPC standards (Främling et al., 2007; Främling et al., 2013).

CASE ABB: RFID TRACKING IN MANUFACTURING PLANT

This chapter presents a manufacturing case related to outbound RFID tracking implementation in ABB's Helsinki manufacturing plant. ABB is a

Swiss-Swedish supplier of power and automation technology with 116,000 employees and revenue of 32 billion USD at the end of 2010. (ABB Annual Report 2010, 2011)

RFID tracking solutions have been part of operations of ABB Drivers in Finland since 2004. The first implementation was related to tracking of transport plywood boxes used in raw material supply. This still existing implementation was based on reading of RFID tagged boxes in the premises of the raw material supplier and ABB's factory that also automatically generates an electronic notification and update to inventory management in ERP. The pilot project was executed in ABB Oy's drives factory in Helsinki in its order supply chain with selected suppliers. (RFID Lab Finland, 2009) Later, a similar kind of inbound RFID tracking solution was implemented in ABB's manufacturing plant located in Estonia. (Head of Quality and Operational Excellence, BU LV Drives, ABB, interview, 14.10.2011)

The first implementation was a part of Tekes (Finnish Funding Agency for Technology and Innovation) ELO (e-Business Logistics) programme and was based on RFID readers, mounted to loading areas to track and record all incoming and outgoing shipments. When moved to the loading dock, empty raw material boxes trigger a material order automatically to the supplier's extranet. Correspondingly, when the supplier fulfils the order, an electronic shipping document into the RFID tag of full boxes is generated. At factory's end the goods are automatically identified and registered in the ERP system via RFID portal, which removes the need of traditional paper shipping documents. As the RFID tracking solution is fully integrated to ERP system, it allows the information exchange between these substantial systems. ("ABB pioneers use", 2005)

The RFID tracking system was further enlarged to cover outbound shipments in 2009 (Wessel, 2009). ABB attaches tags on the pallet or box as soon as a drive is manufactured, and RFID readers at the gate verify if the products are going to the correct vehicle and then transmit the data to the ERP system for updating. The tracking of outbound shipments stops when transport begins. From that point on, it is up to the logistics partners to use and benefit from the data of RFID tags on the boxes. ABB has further expanded the RFID tracking implementation to external warehouses and Estonia manufacturing plants with TekRF (wireless data collection and integration system) solution. Today ABB utilizes EPC Gen. 2 and ISO approach. By the end of 2011 Gen. 2 tags should be implemented to all operations. (Head of Quality and Operational Excellence, BU LV Drives, ABB, interview, 14.10.2011)

The purpose of the interview was to get evidence of experienced benefits and advantages created by RFID tracking. Benefits and challenges of RFID tracking, identified in literature review were presented to interviewee, who was able to provide open answer to respective questions. The questions and answers are provided in Table 1 and Table 2.

Based on interview, it can be stated that most of the benefits identified in previous research have been realized in real world implementation of ABB case. Faster inventory turnover allows the company to improve its efficiency and decrease stock holding costs and the capital tied in inventory. Improved information exchange, on the other hand, reduces the errors and manual work needed to carry out transactions. Since the solution allows real-time tracking, the shipments do not need to be stored to consolidation area, but can now be directly loaded to trailers/containers. Automated identification reduced errors in shipments, as RFID readers in loading area identify the contents of shipment and indicate if there are incorrect items loaded. It is also impossible to close the gate before all the items of shipment are loaded. These functions practically eliminate the possibility of incorrect shipment, which further generates cost savings. In general the transparency of supply chain is increased, lead times are reduced, the quality of goods traffic data is im-

Table 1. Empirical evidences of benefits of RFID tracking adoption in ABB

Benefit identified in previous research	Empirical evidence
Improved inventory management	Faster inventory turnaround Inbound logistics with pull strategy
Added value to customers	Automatic information exchange with logistics service providers with IT-system integrations No direct link to customer
Improved information exchange	All transactions are automated with electronic information exchange
Competitive advantage	Only symbolic value
Cost savings	All investments are covered by the cost savings in less than six months
Enabler of real time tracking and tracing	In warehouses, but not during the transport
Enabler of automated identification	Received products are automatically identified and routed directly to manufacturing process
Improved identification reliability	Somewhat better than with barcode. Gen. 2 tags are more significantly reliable compared to Gen. 1 tags.
Improved security	No influence
More efficient utilization of workforce	Less personnel needed

Table 2. Empirical evidences of challenges of RFID adoption in ABB

Challenges identified in previous research	Empirical evidence
Price of tags	In order to be able to expand RFID tracking implementations, the price should still decrease
Reliability of tags if attached to metal products	No problems detected
Reliability of technology in general	No problems detected
ICT security	No problems detected
Integration to other ICT systems	Customized RFID module in SAP
Problems in measuring return of RFID investments	Measured by savings in labor costs

proved, and more capital is available for investments.

Improved security and competitive advantage were only benefits that could not be directly linked to RFID tracking implementation. Also the ability of RFID tracking to create added value for customers were questioned. However, some added value was produced to other supply chain actors. Finally, the role of RFID tracking as a source of competitive advantage was considered only as symbolic.

According to case study, the main challenges of RFID tracking implementation were related to the price of tags. This could be due to the fact that the company's operating environment requires the use of more expensive tags. All other challenges identified in previous research were overcome by the company. For example, ABB could solve the integration problems by developing their own RFID module for SAP. Measuring the payback time for RFID tracking investments is a commonly agreed problem with implementations. In this case, the interviewee announced that they have conducted comprehensive analysis of the outcome of implementation. In addition to above mentioned problems, interviewee also mentioned that sufficient training is a key for efficient RFID tracking implementations, and RFID technology itself is not a source of challenges.

CONCLUSION

RFID tracking is mainly used in closed systems, but wider applications in global supply chains are still scarce. Both the technology and standardisation have improved significantly during the past 10 years, and based on case study, it seems that the technology itself is reliable enough. The cost of tags, however, decreases companies' willingness to implement supply chain-wide RFID tracking.

The introduction of automatic identification technology requires that each identified unit has a unique number. GS1 codes are widely used, but many sectors, such as machinery industry, still rely on proprietary systems, often based on handwritten labels. Agreement on a common understanding of the numbering system, which can be implemented in RFID tags, is a first prerequisite for inter-company use of RFID tracking. Also standardised processes and used technologies have to be agreed for transfer of information between the partners needing the information.

Tag technology has proceeded tremendously during the past years, and special durable tags have been developed for RFID-unfriendly materials, such as metals and liquids. In cases where tags are attached to a wide set of material (e.g. in a hardware store) proper training of the personnel applying the tags is required, and solutions for tackling RFID-unfriendly materials have to be applied. The need for proper training of personnel was also identified by the case company. It was also indicated that the performance of Gen. 2 tags is significantly better compared to earlier technologies.

In supply chain applications where RFID tracking is used, large players have pushed for the application of the technology in their supply chains. Generally SMEs do not have the power to do this. For efficient tracking and tracing throughout the complete supply chain, each stakeholder should benefit from the technology. Hence, the benefits of the technology should be clearly quantified so that companies will be convinced to invest in RFID tracking. According to case study, it is possible to some extent to quantify the monetary benefits of RFID tracking. One financial indicator to be applied could be return on investments (ROI).

The amount of information exchanged in logistics processes is large, and demand for real-time tracking and tracing services is continuously increasing. That is why more rapid and reliable RFID tracking solutions will gradually replace current barcode-based solutions. Comparing to barcode, RFID technology reduces the manual work required and errors made thus increasing productivity. The examined case also brought out some other concrete benefits. Cost savings were achieved through the automation of many supply chain processes and improved inventory management. Based on the outcome of previous research review, and verified by the empirical case study, it seems evident that RFID tracking can significantly improve processes and has a positive impact on performance in the manufacturing industry, even if it is usually easier for downstream companies to find beneficial RFID tracking applications (see Soon and Gutiérrez, 2008). However, not all the potential benefits of RFID tracking that the literature mentions to be applicable for a manufacturing company could be verified in the case study.

Due to the relatively high price of tags for challenging environment with metals, full RFID integration into supply chain is not yet realized in the case company, even if there are interests among the customers to use RFID tracking. Still, the prerequisites for supply chain integration exist, while the manufacturer actively searches sufficient potential benefits of RFID tracking to justify larger-scale implementations. And when the company finds them, the customers of the case company can use the same RFID tags as "free-riders" (see Whang, 2010), which offers the possibilities for customers of the company to use these tags in their operations leading gradually to integrated supply chain-wide tracking.

REFERENCES

ABB. (2005, April 4). *ABB pioneers use of radio frequency identification technology in material logistics in Finland*. ABB.

ABB Annual Report 2010. (2011). Retrieved October, 14, 2011 from http://www400.abbext.com/2010/ar/servicepages/downloads/files/download.php?file=annual_report_2010.pdf&cat=b

Arbnor, I., & Björn, B. (1997). *Methodology for creating business knowledge* (2nd ed.). Thousand Oaks, CA: Sage Publications.

Asif, Z., & Mandiviwalla, M. (2005). Integrating the supply chain with RFID: A technical and business analysis. *Communications of the Association for Information Systems*, (15): 393–427.

Bendavid, Y., & Cassivi, L. (2010). Bridging the gap between RFID/EPC concepts, technological requirements and supply chain e-business processes. *Journal of Theoretical and Applied Electronic Commerce Research*, *5*(3), 1–16. doi:10.4067/S0718-18762010000300002

Bottani, E., Montanari, R., & Volpi, A. (2010). The impact of RFID and EPC network on the bullwhip effect in the Italian FMCG supply chain. *International Journal of Production Economics*, *124*(2), 426–432. doi:10.1016/j.ijpe.2009.12.005

Dutta, A., Lee, H. L., & Whang, S. (2007). RFID and operations management: Technology, value and incentives. *Production and Operations Management*, *16*(5), 646–655. doi:10.1111/j.1937-5956.2007.tb00286.x

EPC Information Services (EPCIS) Specification, EPCGlobal Ratified Standard. (2007). Retrieved from http://www.gs1.org/gsmp/kc/epcglobal/epcis/epcis_1_0_1-standard-20070921.pdf

Främling, K., Ala-Risku, T., Kärkkäinen, M., & Holmström, J. (2007). Design patterns for managing product life cycle information. *Communications of the ACM*, *50*(6), 75–79. doi:10.1145/1247001.1247009

Främling, K., Parmar, S., Hinkka, V., Tätilä, J., & Rodgers, D. (2013). Assessment of EPCIS standard for interoperable tracking in the supply chain. In *Service orientation in holonic and multi agent manufacturing and robotics*. Berlin: Springer. doi:10.1007/978-3-642-35852-4_8

Hinkka, V. (2012). Challenges for building RFID tracking systems across the whole supply chain. *International Journal for RF Technologies: Research and Applications*, *3*(3), 201–218.

Hinkka, V., Holmström, J., & Främling, K. (2012). RFID tracking in the book supply chain: The transition from postponed to speculative tagging. *International Journal of Logistics: Research and Applications*, *15*(3), 194–214. doi:10.1080/13675567.2012.709834

Johansson, O., & Pålsson, H. (2009). The impact of auto-ID on logistics performance: A benchmarking survey of Swedish manufacturing industries. *Benchmarking: An International Journal*, *16*(4), 504–522. doi:10.1108/14635770910972432

RFID Journal. (2006). *Gen 2 EPC protocol approved as ISO 18000-6C*. Retrieved from http://www.rfidjournal.com/article/view/2481/1/1

Kapoor, G., Zhou, W., & Piramuthu, S. (2009). Challenges associated with RFID tag implementations in supply chains. *European Journal of Information Systems*, *18*(6), 526–533. doi:10.1057/ejis.2009.41

Kärkkäinen, M., Ala-Risku, T., & Främling, K. (2004). Efficient tracking for short-term multi-company networks. *International Journal of Physical Distribution & Logistics Management*, *34*(7), 545–564. doi:10.1108/09600030410552249

Kotzab, H., Seuring, S., Müller, M., & Reiner, G. (2005). *Research methodologies in supply chain management*. Heidelberg, Germany: Springer. doi:10.1007/3-7908-1636-1

RFID Lab Finland. (2009). *Case ABB*. Retrieved from http://www.rfidlab.fi/sites/rfidlab.fi/files/ABB_outbound_RFID_press_FI_090417.pdf

Lee, H., & Özer, Ö. (2007). Unlocking the value of RFID. *Production and Operations Management*, *16*(1), 40–64. doi:10.1111/j.1937-5956.2007.tb00165.x

Li, S., Godon, D., & Visich, J. K. (2010). An exploratory study of RFID implementation in the supply chain. *Management Research Review*, *33*(10), 1005–1015. doi:10.1108/01409171011084003

Lim, S. H., & Koh, C. E. (2009). RFID implementation strategy: Perceived risks and organizational fits. *Industrial Management & Data Systems*, *109*(8), 1017–1036. doi:10.1108/02635570910991274

McBurney, D., & White, T. (2009). *Research methods* (8th ed.). Belmont, CA: Academic Press.

McFarlane, D., & Sheffi, Y. (2003). The impact of automatic identification on supply chain operations. *The International Journal of Logistics Management*, *14*(1), 1–17. doi:10.1108/09574090310806503

Naula, T., Ojala, L., Solakivi, T., Takalokastari, M., Rantanen, M., Kalske, M., ... Stenholm, P. (2006). Finland's state of logistics survey 2006. *Publications of the Ministry of Transport and Communications,* (35).

Nilsson, F., & Waidringer, J. (2004). Logistics management from a complexity perspective. *The ICFAI Journal of Operations Management*, *3*(2), 59–73.

Permala, A., Scholliers, J., & Granqvist, J. (2006). RFID-roadmap for business logistics. AINO Publications, (30).

Rees, R. (2004). *ISO supply chain standards*. Retrieved from http://docbox.etsi.org//erm/open/rfidworkshop/rfid_20%20richard%20rees_bsi.pdf

Roh, J. J., Kunnathur, A., & Tarafdar, M. (2009). Classification of RFID adoption: An expected benefits approach. *Information & Management*, *46*(6), 357–363. doi:10.1016/j.im.2009.07.001

Rundh, B. (2008). Radio frequency identification: Invaluable technology or a new obstacle in the marketing process. *Marketing Intelligence & Planning*, *26*(1), 97–114. doi:10.1108/02634500810847174

Saunders, M., Lewis, P., & Thornhill, A. (2009). *Research methods for business students* (5th ed.). Essex, UK: Academic Press.

Soon, C.-B., & Gutiérrez, J. A. (2008). Effects of the RFID mandate on supply chain management. *Journal of Theoretical and Applied Electronic Commerce Research*, *3*(1), 81–91.

Visich, J. K., Li, S., Khumawala, B. M., & Reyes, P. M. (2009). Empirical evidence of RFID impacts on supply chain performance. *International Journal of Operations & Production Management*, *29*(12), 1290–1315. doi:10.1108/01443570911006009

Wamba, S. F., & Chatfield, A. T. (2009). A contingency model for creating value from RFID supply chain network projects in logistics and manufacturing environments. *European Journal of Information Systems*, *18*(6), 615–636. doi:10.1057/ejis.2009.44

Wessel, R. (2009, November 25). RFID drives up efficiencies at ABB. *RFID Journal.*

Whang, S. (2010). Timing of RFID adoption in a supply chain. *Management Science*, *56*(2), 343–355. doi:10.1287/mnsc.1090.1121

Wyld, D. C. (2006). RFID 101: The next big thing for management. *Management Research News*, *29*(4), 154–173. doi:10.1108/01409170610665022

Yin, R. K. (2002). *Case study research: Design methods* (3rd ed.). Thousand Oaks, CA: Sage Publications.

Yin, R. K. (2012). *Applications of case study research* (3rd ed.). Thousand Oaks, CA: Sage Publications.

Yu, Z., Yan, H., & Cheng, T. C. E. (2001). Benefits of information sharing with supply chain partnerships. *Industrial Management & Data Systems*, *3*(2), 114–121. doi:10.1108/02635570110386625

Zuo, Y. (2010). Survivable RFID systems: Issues, challenges, and techniques. *IEEE Transactions on Systems, Man and Cybernetics. Part C, Applications and Reviews*, *40*(4), 406–418. doi:10.1109/TSMCC.2010.2043949

Chapter 5
Ordering Policy for Imperfect-Quality Deteriorating Items with Initial-Inspection and Allowable Shortage under the Condition of Permissible Delay in Payments

Chandra K. Jaggi
University of Delhi, India

Mandeep Mittal
Amity School of Engineering and Technology, India

ABSTRACT

While developing the inventory model with shortages under permissible delay in payments, it has been observed in the literature that the researchers have not considered the fact that the retailer can earn interest on the revenue generated after fulfilling the outstanding demand as soon as he receives the new consignment at the start of the cycle. Owing to this fact, the present study investigates the impact of interest earned from revenue generated after fulfilling the stock out at the start of the cycle on a single commodity inventory model with shortages for deteriorating item, in which the whole lot goes through an inspection process on arrival before entering into inventory system, under the conditions of permissible delay in payments. After inspection, the non-defective items are retained to fulfill the demand and the defective items are returned to the supplier. The results have been demonstrated with the help of a numerical example using the tools of Matlab7.0.1.

DOI: 10.4018/978-1-4666-4506-6.ch005

1. INTRODUCTION

A very common assumption of the economic order quantity is that all the units produced or purchased are of good quality. But practically it is difficult to produce or purchase items with 100% good quality. Thus, the inspection of an ordered lot becomes indispensable in most of the organizations. The inspected items may be classified into two categories viz., non-defective and defective items. After inspection, the non-defective items are retained to fulfill the demand and the defective items are returned to the supplier. The fraction of defective items is constant and deterministic.

However, when the nature of the item is deteriorating then inspection of the lot becomes more important. The deterioration is now a well-established fact, which may take place in the form of gradual decay, damage or perishability and the process is affected by extraneous and invisible factors. Moreover, the deterioration may also happen due to weather conditions, insufficient and unscientific storage structure etc. This situation occurs particularly in case of foodstuffs, which are damaged due to insects, spoilage, and rodents, where as in other commodities deterioration may occur during normal storage facilities. Ghare and Schrader (1963) were the first to propose an economic order quantity (EOQ) model for items with an exponentially decaying inventory. Covert (1973) introduced variable rate of deterioration in Ghare and Schrader (1963) model. A further improvement was introduced by Shah (1977) considering a model allowing complete backlogging of the unsatisfied demand. A good amount of work has been done by different researchers to explore the effect of deterioration on EOQ model under different circumstances (Dave, 1981; K. L. Mark, 1982; Hollier, 1983; Heng, 1991; Raafat, 1991; Widyadana et al., 2011).

Moreover, while developing a mathematical model in inventory control, it is assumed that the payment will be made to the suppliers for the goods immediately after receiving the consignment. However, in the day-to-day dealing, supplier does allow a certain fixed period to settle the account, during which the supplier charges no interest, but beyond this period interest is charged by the supplier under the terms and conditions agreed upon. Now, in case debt financing, it is often a short-term financing. Thus, interest paid here is nothing but the cost of capital or opportunity cost. Also, short-term loans can be thought of as having been taken from the suppliers on the expiry of the credit period. However, before the account has to be settled, the customer can sell the goods and continues to accumulate revenue and earn interest instead of paying the overdraft that is necessary if the supplier requires settlement of the account after replenishment. Therefore, it makes economic sense for the customer to delay the settlement of the replenishment account up to the last day of the credit period allowed by the supplier. Goyal (1985) presented the model by introducing permissible delay in payments for fixed time period and Aggarwal and Jaggi (1995) extended his work for deteriorating items. Further, Jamal et al. (1997) allowed shortages in the Aggarwal and Jaggi (1995) model, since then, many articles have been appeared under different situations (Raafat 1991; Shah et al. 2000; Goyal and Giri 2001). The primary benefit of taking trade credit is that one can have saving in purchase cost and opportunity cost, which becomes quite relevant for deteriorating items, because in such cases, one has to procure more units then required in the given cycle to account for the deteriorating effect. In particular, when the unit purchase cost is high and decay is continuous, the saving due to delayed payment appears to be more significant.

Lot of work has been published by many authors for finding the economic order quantity with or without shortages for deteriorating items. However, in the literature, it has been found for the inventory models with shortages under permissible delay in payments that the researchers have not incorporated in the interest earned part by the retailer, which he could have also earned on the revenue generated after fulfilling the back order at the start of the new cycle. In this study, an attempt has been made to formulate the retailer ordering policy for deteriorating items with initial-inspection under the condition of permissible delay in payments. Shortages are allowed and are fully backlogged.

2. ASSUMPTIONS AND NOTATIONS

The following assumptions are used in developing the model:

1. The lead time is negligible
2. Shortages are allowed
3. Demand is deterministic at a constant rate and known
4. Interest paid is greater than interest earned
5. Replenishment is instantaneous
6. A constant fraction (θ) of thc on-hand invcntory deteriorate per unit time.

The notations adopted in this chapter are as below:

A Order cost of inventory (dollars per order)
N Number of items received before inspection
R Demand rate (units per unit time)
C Cost price of one item (dollars/unit)
C_1 Holding cost per unit per unit time
C_2 Shortage cost per unit per unit time
S Selling price of one item
I Inspection cost per item
θ Constant fraction of the on-hand inventory deterioration per unit
Np Number of non defective item ($t = 0$)
p fraction of non-defective items ($0 \leq p \leq 1$)
I_e Interest earned
I_p Interest paid, $I_p \geq I_e$
M Permissible delay in settling the accounts
T Inventory cycle length
t_1 Length of the period with positive stock of the item
S_1 Quantity consumed during time t_1
D (T) Deteriorating units per cycle
TVC_1 (S_1, T) Total variable cost per cycle per unit time for case 1, when $0 \leq M \leq t_1 \leq T$
TVC_2 (S_1, T) Total variable cost per cycle per unit time for case 2, when $0 \leq t_1 \leq M \leq T$

3. MATHEMATICAL FORMULATION AND SOLUTION

Let N be the items received at the beginning of the period. Further, It is assumed that after inspection Np be the number of non-defective items at $t = 0$, which are retained to fulfill the demand. The fraction of non-defective items p ($0 \leq p \leq 1$) is known, as it can be easily estimated from the past data.

At the beginning of the cycle, a batch of Np units enters the inventory system, after inspection ($Np - S_1$) units are delivered towards back order leaving a balance of S_1 units as the initial-inventory. With the passage of time, the inventory level gradually decreases mainly due to demands and partly due to deterioration of items up to time t_1 ($0 \leq t_1 \leq T$). Further demand during (t_1, T) are backlogged shown in Figure 1 and Figure 2.

Figure 1. Inventory system for case 1, $M < T_1$ with complete back ordering

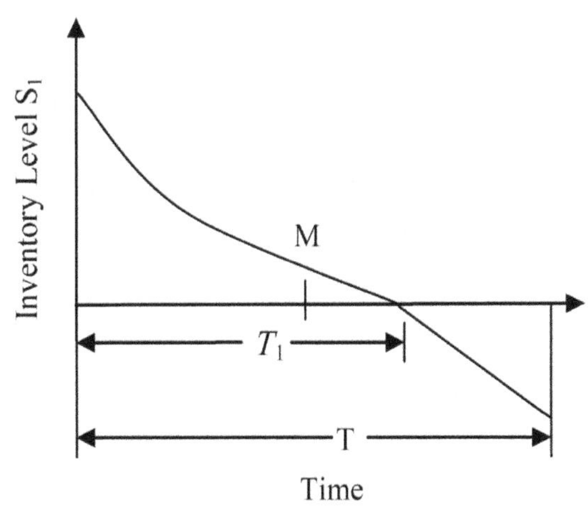

Figure 2. Inventory system for case 2, M > T1 with complete back ordering

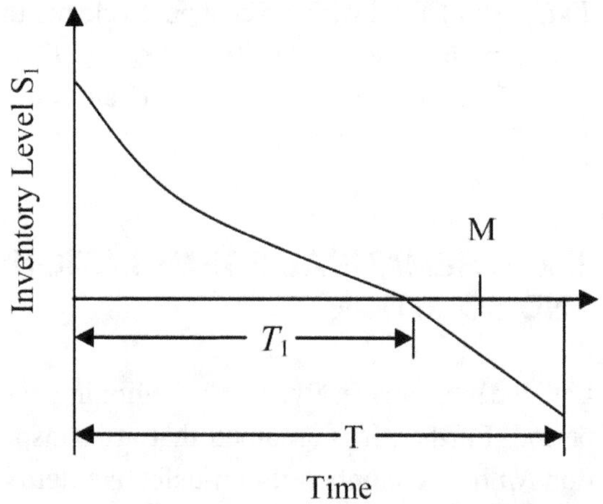

Let $I(t)$ be the inventory level of the system at anytime t ($0 \le t \le T$). The instantaneous states of $I(t)$ over the period $(0, T)$ are given by

$$\frac{dI(t)}{dt} + \theta I(t) = -R \qquad 0 \le t \le t_1 \quad (1)$$

$$\frac{dI(t)}{dt} = -R \qquad t_1 \le t \le T \quad (2)$$

The solution of the above differential equation (boundary condition at $t = 0$, $I(t) = S_1$ = initial inventory) is

$$I(t) = \begin{cases} S_1 e^{-\theta t} - \frac{R}{\theta}[1 - e^{-\theta t}] & 0 \le t \le t_1 \quad (3) \\ R(t_1 - t) & t_1 \le t \le T \quad (4) \end{cases}$$

Since, at $t = t_1$, $I(t_1) = 0$, from Equation (1) and Equation (2), we get

$$S_1 = \frac{R}{\theta}\left(e^{\theta t_1} - 1\right) \qquad (5)$$

i.e.

$$t_1 = \frac{1}{\theta}\log\left(1 + \frac{S_1\theta}{R}\right) \qquad (6)$$

Further, total number of non-defective items is

$$Np = R(T - t_1) + S_1 \qquad (7)$$

or

$$N = \frac{1}{p}\left[R(T - t_1) + S_1\right] \qquad (8)$$

The total demand during t_1 is

$$Rt_1 = \frac{R}{\theta}\log\left(1 + \frac{S_1\theta}{R}\right) \qquad (9)$$

Therefore, number of units that deteriorated $D(T)$, during one cycle is

$$D(T) = \frac{R}{\theta}\left(e^{\theta t_1} - 1\right) - Rt_1 \qquad (10)$$

3.1 Inventory Scenarios

Since, we are considering the permissible delay in payments in this model, where stock out are allowed. Under this situation there will be two cases:

Case 1: $0 \le M \le t_1 \le T$
Case 2: $0 \le t_1 \le M \le T$

Case 1: $0 \le M \le t_1 \le T$

In this case, it is assumed that one can earn interest on revenue generated from the sales up to t_1. Although, he has to settle the account at M, for that, he has to arrange money at some specified rate of interest in order to get his remaining stocks financed for the period M to t_1.

Now, the interest paid per cycle is

$$= CI_p \int_M^{t_1} I(t)\,dt$$
$$= CI_p \left[\frac{S_1}{\theta}(e^{-\theta M} - e^{-\theta t_1}) - \frac{R}{\theta}\left\{ (t_1 - M) + \frac{1}{\theta}(e^{-\theta t_1} - e^{-\theta M}) \right\} \right]$$

where,

$$S_1 = \frac{R}{\theta}\left(e^{\theta t_1} - 1\right)$$
$$= \frac{CI_pR}{\theta^2}(e^{\theta(t_1 - M)} - 1) - \frac{CI_pR}{\theta}(t_1 - M) \quad (11)$$

Interest earned per cycle has got two parts.

Part 1

In first part, one can earn interest till the time period (M),

$$\text{Interest earned} = S\,I_e \int_0^M R\,t\,dt \quad (12)$$

Part 2

Second part includes the revenue generated after fulfilling the back orders (at the start of the cycle) for the time period (M).

$$\text{Interest earned} = SI_eR(T - t_1)\,M \quad (13)$$

∴ Total interest earned in this case

$$S\,I_e \int_0^M R\,t\,dt + S\,I_eR(T - t_1)M \quad (14)$$

Whereas Jamal et al. (1997) has considered the interest earned part as

$$C\,I_e \int_0^{t_1} R\,t\,dt$$

i.e. Jamal et al. (1997) have not taken into account the interest earned from revenue generated after fulfilling the back orders at the start of the cycle, which has got significant impact on their total cost reduction.

$$\text{Inventory Carrying Cost} = C_1 \int_0^{t_1} I(t)\,dt$$
$$= \frac{C_1}{\theta}\left[(1 - e^{-\theta t_1})(S_1 + \frac{R}{\theta}) - R t_1\right] \quad (15)$$
$$= \frac{C_1R}{\theta^2}\left[(e^{\theta t_1} - \theta t_1 - 1)\right]$$

$$\text{Shortage Cost} = C_2 \int_{t_1}^{T} R(t - t_1)\,dt = C_2R(T - t_1)^2/2 \quad (16)$$

$$\text{Inspection Cost} = I\,N \quad (17)$$

The total variable cost per cycle per unit time (TVC_1) is the sum of the order cost, cost of deteriorated units, inspection cost, carrying cost, shortage cost and interest paid, but the interest earned is subtracted from this sum, i.e.

$$TVC_1(S_1, T) = \frac{A}{T} + \frac{CD(T)}{T} + \frac{IN}{T} + \frac{C_1R}{\theta^2 T}\left[(e^{\theta t_1} - \theta t_1 - 1)\right] + \frac{C_2R}{2T}(T - t_1)^2 + \frac{CI_pR}{\theta^2 T}(e^{\theta(t_1 - M)} - 1) - \frac{CI_pR}{\theta T}(t_1 - M) - \frac{SI_eR(T - t_1)M}{T} - SI_e\frac{RM^2}{2T} \quad (18)$$

Now, we discuss some special cases.

Special Case 1

Without inspection and excluding interest earned after fulfilling the back orders (at the start of the cycle) the equation (18) reduces to,

$$TVC_{11}(S_1,T) = \frac{A}{T} + \frac{CD(T)}{T} + \frac{C_1R}{\theta^2 T}\left[(e^{\theta t_1} - \theta t_1 - 1)\right] + \frac{C_2R}{2T}(T - t_1)^2 + \frac{CI_pR}{\theta^2 T}(e^{\theta(t_1 - M)} - 1) - \frac{CI_pR}{\theta T}(t_1 - M) - SI_e\frac{RM^2}{2T} \tag{19}$$

When $S=C$, the above expression reduce to Jamal et al. (1997).

Special Case 2

Without inspection and including interest earned after fulfilling the back orders (at the start of the cycle), the equation (18) reduces to,

$$TVC_{12}(S_1,T) = \frac{A}{T} + \frac{CD(T)}{T} + \frac{C_1R}{\theta^2 T}\left[(e^{\theta t_1} - \theta t_1 - 1)\right] + \frac{C_2R}{2T}(T - t_1)^2 + \frac{CI_pR}{\theta^2 T}(e^{\theta(t_1 - M)} - 1) - \frac{CI_pR}{\theta T}(t_1 - M) - SI_e\frac{RM^2}{2T} - \frac{SI_eR(T - t_1)M}{T} \tag{20}$$

Whereas Jamal et al. (1997) has not taken into account the interest earned from revenue generated after fulfilling the backorders at the start of the cycle on the selling price which will help retailer to reduce his total variable cost.

Special Case 3

With inspection and excluding interest earned after fulfilling the back orders (at the start of the cycle), the equation (18) reduces to,

$$TVC_{13}(S_1,T) = \frac{A}{T} + \frac{CD(T)}{T} + \frac{IN}{T} + \frac{C_1R}{\theta^2 T}\left[(e^{\theta t_1} - \theta t_1 - 1)\right] + \frac{C_2R}{2T}(T - t_1)^2 + \frac{CI_pR}{\theta^2 T}(e^{\theta(t_1 - M)} - 1) - \frac{CI_pR}{\theta T}(t_1 - M) - SI_e\frac{RM^2}{2T} \tag{21}$$

Case 2: $0 \le t_1 \le M \le T$

In this case, one can earn interest on sales revenue up to the permissible period (M), and pay no interest for the units kept in the stock.

Now, interest earned per cycle has got three parts.

Part 1

In first part one can earn interest till the time period (t_1),

$$\text{Interest earned} = SI_e\int_0^{t_1} Rt\,dt \tag{22}$$

Part 2

Second part is having interest earned for the time period ($M - t_1$),

$$\text{Interest earned} = SI_eRt_1(M - t_1) \tag{23}$$

Part 3

Third part includes the revenue generated after fulfilling the back orders (at the start of the cycle) for the time period (M),

$$\text{Interest earned} = SI_eR(T - t_1)M \tag{24}$$

$\therefore$ Total interest earned in this case

$$SI_e\int_0^{t_1} Rt\,dt + SI_eRt_1(M - t_1) + SI_eR(T - t_1)M \tag{25}$$

Whereas Jamal et al. (1997) has considered the interest earned part as

$$CI_e\int_0^{t_1} Rt\,dt + CRt_1I_e(M - t_1)$$

Similar to case 1, here also he has not taken into account the interest earned from revenue generated after fulfilling the back orders at the start of the cycle, which may lead to inaccurate analysis.

$$\text{Carrying Cost} = C_1\int_0^{t_1} I(t)\,dt = \frac{C_1 R}{\theta^2}\left[(e^{\theta t_1} - \theta t_1 - 1)\right] \tag{26}$$

$$\text{Shortage Cost} = C_2\int_{t_1}^{T} R(t - t_1)\,dt = C_2 R(T - t_1)^2 / 2 \tag{27}$$

$$\text{Inspection Cost} = I\,N \tag{28}$$

The total variable cost per cycle per unit time (TVC_2) is the sum of the order cost, cost of deteriorated units, inspection cost, carrying cost and shortage cost but the interest earned is subtracted from this sum, i.e.

$$TVC_2(S_1, T) = \frac{A}{T} + \frac{IN}{T} + \frac{CD(T)}{T} + \frac{C_1 R}{\theta^2 T}\left[(e^{\theta t_1} - \theta t_1 - 1)\right] + \frac{C_2 R}{2T}(T - t_1)^2 - SI_e\frac{R t_1^2}{2T} - \frac{S I_e R t_1 (M - t_1)}{T} - \frac{S I_e R (T - t_1) M}{T} \tag{29}$$

Now, we discuss some special cases.

Special Case 1

Without inspection and excluding interest earned after fulfilling the back orders (at the start of the cycle), the Equation (29) reduces to,

$$TVC_{21}(S_1, T) = \frac{A}{T} + \frac{CD(T)}{T} + \frac{C_1 R}{\theta^2 T}\left[(e^{\theta t_1} - \theta t_1 - 1)\right] + \frac{C_2 R}{T}(T - t_1)^2 / 2 - SI_e\frac{R t_1^2}{2T} - \frac{S I_e R t_1 (M - t_1)}{T} \tag{30}$$

When $S=C$, the above expression reduce to Jamal et al. (1997).

Special Case 2

Without inspection and including interest earned after fulfilling the back orders (at the start of the cycle), the Equation (29) reduces to,

$$TVC_{22}(S_1, T) = \frac{A}{T} + \frac{CD(T)}{T} + \frac{C_1 R}{\theta^2 T}\left[(e^{\theta t_1} - \theta t_1 - 1)\right] + \frac{C_2 R}{T}(T - t_1)^2 / 2 - SI_e\frac{R t_1^2}{2T} - \frac{S I_e R (T - t_1) M}{T} - \frac{S I_e R t_1 (M - t_1)}{T} \tag{31}$$

Whereas Jamal et al. (1997) has not taken into account the interest earned from revenue generated after fulfilling the backorders at the start of the cycle on the selling price which will help retailer to reduce his total variable cost.

Special Case 3

With inspection and excluding interest earned after fulfilling the back orders (at the start of the cycle), the Equation (29) reduces to,

$$TVC_{2.3}(S_1, T) = \frac{A}{T} + \frac{CD(T)}{T} + \frac{IN}{T} + \frac{C_1 R}{\theta^2 T}\left[(e^{\theta t_1} - \theta t_1 - 1)\right] + \frac{C_2 R}{T}(T - t_1)^2 / 2 - SI_e\frac{R t_1^2}{2T} - \frac{S I_e R t_1 (M - t_1)}{T} \tag{32}$$

Total variable cost functions for both the cases are:

$$TVC(S_1, T) = \begin{cases} TVC_1(S_1, T) & 0 \le M \le t_1 \le T \\ TVC_2(S_1, T) & 0 \le t_1 \le M \le T \end{cases}$$

Our objective is to find such values of S_1 and T (for case 1 and case 2) which minimize the total variable cost function (TVC_1 and TVC_2), since this function is differentiable, therefore the general necessary condition for minimization of the function is,

$$\frac{\partial TVC(S_1,T)}{\partial T}=\frac{\partial TVC(S_1,T)}{\partial S_1}=0$$

In Case 1,

$$\frac{\partial TVC_1}{\partial T}=0$$

$$\Rightarrow -\frac{A}{T^2}-\frac{CR}{T^2}\left[\frac{S_1}{CR+\theta S_1}+t_1\right]+\frac{I}{T^2p}(Rt_1-S_1)-\frac{C_1}{T^2\theta}(S_1-Rt_1)-\left[-\frac{SI_eRM^2}{2T^2}\right]-\left[\frac{SI_eRMt_1}{T^2}\right]-\frac{CI_p}{T^2}\left[\left(e^{-\theta M}-\frac{R}{(R+\theta S_1)}\right)\left(\frac{S_1}{\theta}+\frac{R}{\theta^2}\right)-\frac{R}{\theta}(t_1-M)\right]+\frac{C_2R}{2T^2}(T^2-t_1^2)=0 \tag{33}$$

$$\frac{\partial TVC_1}{\partial S_1}=0$$

$$\Rightarrow \frac{C\theta S_1}{T(R+\theta S_1)}+\frac{I\theta S_1}{Tp(R+\theta S_1)}+\frac{C_1S_1}{T(R+\theta S_1)}-\left[-\frac{SI_eRM}{T(R+\theta S_1)}\right]+\frac{CI_p}{T}\left[\frac{e^{-\theta M}R}{\theta(R+\theta S_1)}\right]+\frac{C_2R}{(R+\theta S_1)}\left(\frac{t_1}{T}-1\right)+\frac{SI_eRt_1}{T(R+\theta S_1)}=0 \tag{34}$$

and in Case 2,

$$\frac{\partial TVC_2}{\partial T}=0$$

$$\Rightarrow -\frac{A}{T^2}-\frac{CR}{T^2}\left[\frac{S_1}{CR+\theta S_1}+t_1\right]+\frac{I}{T^2p}(Rt_1-S_1)-\frac{C_1}{T^2\theta}(S_1-Rt_1)-\left[-\frac{SI_eR(M-t_1)t_1}{2T^2}\right]-\left[\frac{SI_eRMt_1}{T^2}\right]-\frac{1}{T}SI_eR\left[\frac{(M-2t_1)}{(R+\theta S_1)}\right]+\frac{C_2R}{2T^2}(T^2-t_1^2)=0 \tag{35}$$

$$\frac{\partial TVC_2}{\partial S_1}=0$$

$$\Rightarrow \frac{C\theta S_1}{T(R+\theta S_1)}+\frac{I\theta S_1}{Tp(R+\theta S_1)}+\frac{C_1S_1}{T(R+\theta S_1)}-\left[-\frac{SI_eRM}{T(R+\theta S_1)}\right]-\frac{SI_eR(M-2t_1)}{T(R+\theta S_1)}-\frac{SI_eRt_1}{T(R+\theta S_1)}+\frac{C_2R}{(R+\theta S_1)}\left[\frac{t_1}{T}-1\right]=0 \tag{36}$$

Solving the above Equations (33), (34), (35) and (36), we get the optimum values of T^* of T and S_1^* of S_1 for TVC_1 and TVC_2.

For sufficient condition, determine the following for both the cases (TVC_1 and TVC_2)

$$\frac{\partial^2 TVC(S_1,T)}{\partial S_1^{\,2}},\ \frac{\partial^2 TVC\ (S_1,T)}{\partial T^2}\ and\ \frac{\partial^2 TVC(S_1,T)}{\partial S_1\partial T}$$

The sufficient condition for minimization, $TVC\ (S_1,\ T)$ is

$$\frac{\partial^2 TVC(S_1,T)}{\partial S_1^{\,2}}>0,\ \frac{\partial^2 TVC(S_1,T)}{\partial T^2}>0$$

$$and\left[\frac{\partial^2 TVC(S_1,T)}{\partial S_1^{\,2}}\right]\left[\frac{\partial^2 TVC(S_1,T)}{\partial T^2}\right]-\left[\frac{\partial^2 TVC(S_1,T)}{\partial S_1\partial T}\right]^2>0$$

Since, these expressions (Appendix A) are complicated, and it is very difficult to prove these conditions mathematically. Alternatively, we have shown convexity by graph for both the cost functions TVC_1 and TVC_2 using plotting tool of Matlab7.0.1. (see Figure 3 and Figure 4).

4. NUMERICAL EXAMPLES

Example 1

Given $p = 0.85$, $\theta = 0.1$, $I_p = 0.15$ /year, $I_e = 0.13$ /year, $R = 1000$ units/year, A=200 dollar/order, $i = 0.12$ /year, $C = 20$ dollars/unit, $I = 0.05$, $C_1 = 2.4$ dollars/unit/yr, $C_2 = 20$ dollars/unit/yr, $S = 24$ dollars/unit and $M = 15$ days.

Figure 3. Optimal total variable cost with respect to S_1 and T for Case 1

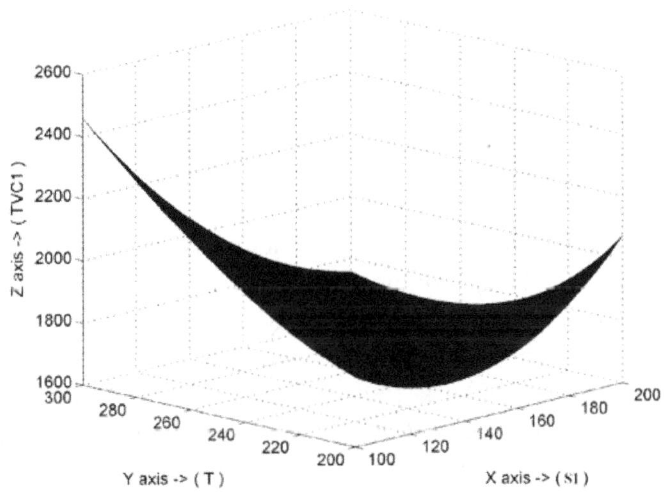

Figure 4. Optimal total variable cost with respect to S_1 and T for Case 2

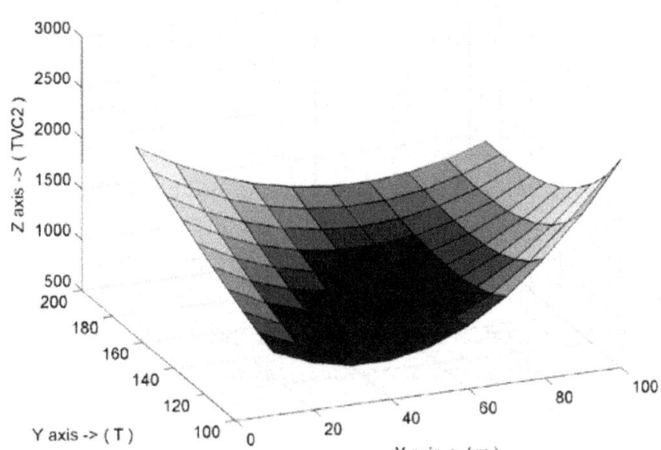

Matlab7.0.1 direct search tool is used to solve the Equation (18), we get following results,

t_1= 72 days T = 99 days

N = 321 units S_1 = 199 units

TVC_1 = \$1407

Special Case 1

Without inspection and excluding interest earned after fulfilling the back orders (at the start of the cycle), by solving Equation (19), we get t_1 = 74 days, T = 99 days, N = 321 units, S_1 = 204 units and TVC_{11} is \$1382.

Special Case 2

Without inspection and including interest earned after fulfilling the back orders (at the start of the cycle), by solving equation (2.1.20), we get t_1 = 72 days, T = 99 days, N = 321 units, S_1 = 199 units and TVC_{12} is \$1348.

While Jamal et al. (1997), T_1= 96 days, T = 117 days, N = 323 units, S_1 = 265 units and total cost function value is \$ 1157.

Special Case 3

With inspection and excluding interest earned after fulfilling the back orders (at the start of the cycle), by solving Equation (21), we get t_1 = 74 days, T = 99 days, N = 321 units, S_1 = 204 units and TVC_{13} is \$1441.

Example 2

Given p=0.85, θ=0.1 I_p = 0.15/year, I_e = 0.13/year, R =1000 units/year, A = 200 dollar/order, i = 0.12/year, C = 120 dollars/unit, I = 0.90, C_1 = 14.40 dollars/unit/yr, C_2 = 10 dollars/unit/yr, S = 138 dollars/unit and M = 30 days.

Matlab7.0.1 direct search tool is used to solve the Equation (29), we get following results, t_1 = 15 days T = 81 days

N = 260 units S_1 = 41 units

TVC_2 = \$1391

Special Case 1

Without inspection and excluding interest earned after fulfilling the back orders (at the start of the cycle), by solving Equation (30), we get t_1 = 24 days, T = 77days, N = 247 units, S_1 = 66 units and TVC_{21} is \$1443.

Special Case 2

Without inspection and including interest earned after fulfilling the back orders (at the start of the cycle), by solving Equation (31), we get t_1 = 15 days, T = 81 days, N = 261 units, S_1 = 41 units and TVC_{22} is \$332. Because interest earned after fulfilling the back order (at the start of the cycle) is very high and reduces to \$1203.

While Jamal et al. (1997), t_1 = 24 days, T= 78 days, N = 212 units, S_1 = 66 units and total cost function value is \$1481.

Special Case 3

With inspection and excluding interest earned after fulfilling the back orders (at the start of the cycle), by solving Equation (32), we get t_1 = 24 days, T = 77 days, N = 247 units, S_1 = 66 units and TVC_{23} is \$2503.

Findings clearly indicate that in both the cases i.e. $M < t_1$ and $M > t_1$, it is beneficial for retailer to introduce inspection and interest earned on the revenue generated after fulfilling the back orders (at the start of the cycle) as he is able to reduce his total cost to that of Jamal et al. (1997).

5 SENSITIVITY ANALYSIS

In order to gain more in sight of the above model the sensitivity analysis has been performed on certain parameters viz. purchase cost (C), Deterioration rate (θ), Shortage cost (C_2) and permissible payment period (M).

Now for fixed C_2, taking θ = {0.1, 0.2}, C = {20, 40, 180, 200} dollars/unit, S = {24, 48, 216, 240} dollars/unit., C_1 = {2.4, 4.8, 21.6, 24} dollars/unit/yr, I = {0.05, 0.1, 0.45, 0.50} and M = {15, 30, 45} days and assuming rest of the data of example 1. No feasible solution marked as "—" in Table 1.

When purchase cost (C) is fixed and taking θ = {0.1, 0.2}, C_2 = {20, 50, 100, 1000} dollars / unit/yr, S = 48 dollars/unit., C_1 = 4.8 dollars/unit/yr, I = 0.1 and M = {15, 30, 45} days and assuming rest of the data of example 1. The solution is given in Table 2.

Further for case 2, keeping fixed C_2 and taking θ = {0.1, 0.2}, C = {120, 150, 180,200} dollars/unit, S = {138, 172.5, 207, 230} dollars/unit, I = {0.90, 1.13, 1.35, 1.50}, C_1 = {14.40, 18, 21.6, 24} dollars/unit/yr and M = {15, 30, 45} days and assuming rest of the data of example 2. No feasible solution marked as "—" in Table 3.

Observations

- It is evident from Table 1 that as M increases, t_1 increases which suggests that the retailer should order more and total average variable cost decreases due to the revenue generated after fulfilling the back order (at the start of the cycle). However, when the deterioration rate (θ) increases, cycle length (T) reduces but the total variable cost increases. This fact encourages the retailer to order more frequently, which helps him to manage the loss due to deterioration effectively.
- However, Table 2 indicates that as C_2 increases, t_1 approaches to T that implies stock out tends to zero, which endorse the fact that if the inventory system is operated at all it is never optimum to have back order.

Table 1. Optimal cycle length and total variable cost with respect to C, θ and M

For case 1: Results for C_2 = 20											
C	θ	0.1					0.2				
	M	t_1	T	S_1	N	TVC_1	t_1	T	S_1	N	TVC_1
20	15	72	99	199	321	1407	61	91	171	295	1539
	30	72	99	199	321	1281	61	91	171	295	1412
	45	72	99	199	321	1154	61	91	171	295	1285
40	15	45	79	124	255	1712	38	74	105	239	1837
	30	45	79	124	255	1459	38	74	104	239	1583
	45	—	—	—	—	—	—	—	—	—	—
180	15	—	—	—	—	—	—	—	—	—	—
	30	—	—	—	—	—	—	—	—	—	—
	45	—	—	—	—	—	—	—	—	—	—
200	15	—	—	—	—	—	—	—	—	—	—
	30	—	—	—	—	—	—	—	—	—	—
	45	—	—	—	—	—	—	—	—	—	—

Table 2. Optimal cycle length and total variable cost with respect to C_2, θ and M

For case 1: Results for $C = 40$											
C_2	θ	0.1					0.2				
	M	t_1	T	S_1	N	TVC_1	t_1	T	S_1	N	TVC_1
20	15	100	122	278	398	1137	79	105	221	343	1337
	30	102	125	284	406	1034	80	107	225	350	1237
	45	105	128	292	419	946	83	110	232	360	1155
50	15	73	86	202	279	1609	58	73	161	238	1907
	30	76	89	209	289	1423	60	76	167	247	1735
	45	80	94	221	305	1276	63	80	177	261	1605
100	15	76	82	210	268	1676	61	69	170	226	2024
	30	79	85	218	278	1495	63	72	177	234	1854
	45	83	90	230	294	1352	67	76	187	248	1731

Table 3. Optimal cycle length and total variable cost with respect to C, θ and M

For case 2: Results for $C_2 = 10$											
C	θ	0.1					0.2				
	M	t_1	T	S_1	N	TVC_2	t_1	T	S_1	N	TVC_2
120	15	—	—	—	—	—	—	—	—	—	—
	30	15	81	41	260	1391	12	79	33	255	1428
	45	16	80	42	259	654	13	76	34	253	691
150	15	—	—	—	—	—	—	—	—	—	—
	30	12	79	33	256	1321	9	78	27	250	1352
	45	13	80	34	255	400	10	77	28	251	431
180	15	—	—	—	—	—	—	—	—	—	—
	30	10	78	28	252	1242	8	77	22	249	1268
	45	11	77	29	251	135	9	76	23	250	162

- Further, Table 3 illustrates that as *M* increases then total variable cost decreases significantly because interest earned after fulfilling the back order increases significantly.

6 CONCLUSION

The present paper investigates the effect of initial-inspection, interest earned on selling price and interest earned after fulfilling the back orders (at the start of the cycle) on the retailer ordering policy. Findings have been very encouraging as retailer is not only able to increase his order quantity but also able to reduce his total variable cost in comparison to Jamal et al. (1997), where inspection cost and interest earned on revenue generated after fulfilling the back order has not been incorporated in the Jamal et al. (1997).

Further, three special cases viz. (i) Without inspection and excluding interest earned after fulfilling the back orders (at the start of the cycle), (ii) Without inspection and including interest earned after fulfilling the back orders (at the start of the cycle) and (iii) With inspection and excluding interest earned after fulfilling the back orders (at the start of the cycle) have been discussed. Results have been validated with the help of numerical examples along with a comprehensive sensitive analysis using Matlab 7.0.1 tools.

7 FUTURE SCOPE

Present paper investigates the impact of *initial-inspection* on retailer's ordering policy for deteriorating items under trade credit. In this study, we have considered *initial-inspection* process, in which the whole lot goes through an inspection process on arrival before entering in to inventory system. After inspection, the effective inventory is the numbers of non-defective items are retained to fulfill the demand and the defective items are returned to the supplier. The fraction of defective items is known and deterministic.

Further, Jaggi and Mittal (2011) developed inventory model which for more realistic situation than the present study, where shortages are not allowed and permissible delay condition is relaxed. In this model, each item goes through *in-process inspection*, and screening rate is assumed to be more than the demand rate. This assumption helps one to meet his demand parallel to the screening process, out of the items which are of perfect quality. Further, the defective items are sold immediately after the screening process is over as a single lot at a discounted price. In this process, the fraction of defective items, α is a random variable with a known probability density function, $f(\alpha)$, which can be calculated from the past data and its expected value is given by

$$E(\alpha) = \int_{a}^{b} \alpha f(\alpha) d\alpha, \quad 0 < a < b < 1$$

Further, it is generally assumed that payment will be made to the supplier for the goods immediately after receiving the consignment. Whereas, in practice, supplier does offers a certain fixed period to the retailer for settling the account. During this period, supplier charges no interest, but beyond this period interest is being charged as has been agreed upon. On the other hand, retailer can earn interest on the revenue generated during this period. Keeping this scenario in mind, Jaggi et al. (2011) extended the Jaggi and Mittal (2011) model for deteriorating items with imperfect quality under permissible delay in payments.

Furthermore, in today's uncertain global economy, due to rising inflation there is a consequent decline in the real value of money, which eventually forces the supplier to provide an attractive trade credit policy to the retailer. Keeping this in view, the model developed by Jaggi et al. (2011) further extended under inflationary conditions by Jaggi et al. (2011a). Results have been demonstrated with the help of a numerical example and sensitivity analysis is also presented to provide managerial insights into practice.

Generally, the inventory models assume that the demand is known and uniform, but this assumption is not always pertinent to many inventory items like basic consumer goods, viz., food items like paddy, wheat, potato, onion, etc. because there may be variation in demand. Many products experience a period of rising demand during the growth phase of their product life cycle. On the other hand, the demand of some products may decline due to the introduction of more attractive products influencing customer's preference. Moreover, the age of inventory has a negative impact on demand due to loss of consumer confidence on the quality of such products and physical loss of the materials. This fact prompted Jaggi et al. (2012) and they developed retailer's inventory model for deteriorating imperfect quality items under inflationary conditions, where demand and price are functions of time. This model is highly beneficial for products like crops (paddy, wheat, potato, onion, ginger, etc.), where demand as well as price varies linearly with time. Moreover, in the present model the screening rate is assumed to be more than the demand rate in order to meet the demand parallel to the screening process by the good quality items. The proposed model optimizes retailer's order quantity by maximizing his total expected profit. A comprehensive sensitivity analysis also has been performed to study the effects of deterioration, inflation and expected number of imperfect quality items on the order quantity and expected total profit.

Recently, Jaggi et al. (2013) developed inventory models with in-process inspection when shortages are allowed and fully backlogged. In this paper, retailer is not only able to meet his current demand but also fulfils his back-orders parallel to the screening process with perfect quality items. The proposed model jointly optimizes retailer's order quantity and shortages by maximizing his expected total profit. A comprehensive sensitivity analysis has also been conducted to explore the effects of key parameters on the optimal results.

REFERENCES

Aggarwal, S. P., & Jaggi, C. K. (1995). Ordering policies of deteriorating items under permissible delay in payments. *The Journal of the Operational Research Society*, *46*(5), 658–662.

Covert, R. P., & Philip, G. C. (1973). An EOQ model for items with weibull distribution deterioration. *AIIE Transactions*, *5*, 323–336. doi:10.1080/05695557308974918

Dave, U., & Patel, L. K. (1981). Policy inventory model for deteriorating items with time proportional demand. *The Journal of the Operational Research Society*, *32*(2), 137–142.

Ghare, P. M., & Schrader, G. F. (1963). A model for exponential decaying inventory. *Journal of Industrial Engineering*, *14*(3), 238–243.

Goyal, S. K. (1985). Economic order quantity under conditions of permissible delay in payments. *The Journal of the Operational Research Society*, *36*(3), 335–338.

Goyal, S. K., & Giri, B. C. (2001). Recent trends in modeling of deteriorating inventory. *European Journal of Operational Research*, *134*(1), 1–16. doi:10.1016/S0377-2217(00)00248-4

Heng, K. J., Labban, J., & Linn, R. J. (1991). An order-level lot-size inventory model for deteriorating items with finite replenishment rate. *Computers & Industrial Engineering*, *20*(2), 187–197. doi:10.1016/0360-8352(91)90024-Z

Hollier, R. H., & Mak, K. L. (1983). Inventory replenishment policies for deteriorating items in a declining market. *International Journal of Production Research*, *21*(4), 813–826. doi:10.1080/00207548308942414

Jaggi, C. K., Goel, S. K., & Mittal, M. (2011a). Economic order quantity model for deteriorating items with imperfect quality and permissible delay on payments. *International Journal of Industrial Engineering Computations*, *2*(2), 237–248. doi:10.5267/j.ijiec.2010.07.003

Jaggi, C. K., Goel, S. K., & Mittal, M. (2013). Credit financing in economic ordering policies for imperfect quality items with allowable shortages. *Applied Mathematics and Computation*, *219*(10), 5268–5282. doi:10.1016/j.amc.2012.11.027

Jaggi, C. K., Khanna, A., & Mittal, M. (2011b). Credit financing for deteriorating imperfect-quality items under inflationary conditions. *International Journal of Services Operations and Informatics*, *6*(4), 292–309. doi:10.1504/IJSOI.2011.045560

Jaggi, C. K., & Mittal, M. (2011). Economic order quantity model for deteriorating items with imperfect quality. *Revista Invetigacion Operacional*, *32*(2), 107–113.

Jaggi, C. K., Mittal, M., & Khanna, A. (2012). Effects of inspection on retailer's ordering policy for deteriorating items with time-dependent demand under inflationary conditions. *International Journal of Systems Science*, 1–9.

Jamal, A. M. M., Sarker, B. R., & Wang, S. (1997). An ordering policy for deteriorating items with allowable shortage and permissible delay in payments. *The Journal of the Operational Research Society*, *48*(8), 826–833.

Mark, K. L. (1982). A production lot-size inventory model for deteriorating items. *Computers & Industrial Engineering*, *6*(2), 309–317.

Raafat, F. (1991). Survey of literature on continuously deteriorating inventory model. *Journal of Operational Research*, *42*, 27–37.

Raafat, F., Wolfe, P. M., & Eldin, H. K. (1991). An inventory model for deteriorating items. *Computers & Industrial Engineering*, *20*(1), 89–94. doi:10.1016/0360-8352(91)90043-6

Shah, N. H., & Shah, Y. K. (2000). Literature survey on inventory model for deteriorating items. *Economic Annals*, *44*, 221–237.

Shah, Y. K. (1977). An order-level lot-size inventory model for deteriorating items. *AIIE Transactions*, *9*(2), 108–112. doi:10.1080/05695557708975129

APPENDIX A

In case 1, Equations (33) and (34) lead to,

$$\frac{\partial^2 TVC_1}{dT^2} = \frac{2A}{T^3} + \frac{2CR\left(e^{\theta t_1} - \theta t_1 - 1\right)}{\theta T^3} + \frac{2IR(Rt_1 - S_1)}{pT^3} + \frac{2C_1R\left(e^{\theta t_1} - \theta t_1 - 1\right)}{\theta^2 T^3} + \frac{C_2Rt_1^2}{T^3} - \frac{2CI_pR\left(-e^{\theta(t_1 - M)} + 1 - \theta t_1 - M\theta\right)}{\theta^2 T^3} - \left(-\frac{2SI_eRMt_1}{T^3}\right) - \frac{SI_eRM^2}{T^3}$$

$$\frac{\partial^2 TVC_1}{dS_1^2} = \frac{CR\theta}{(R+S_1\theta)^2 T} + \frac{IR\theta}{p(R+S_1\theta)^2 T} + \frac{C_1R}{(R+S_1\theta)^2 T} + \frac{C_2R(1+\theta T - \log(R+S_1\theta)/R)}{(R+S_1\theta)^2 T} + \frac{CI_pR}{(R+S_1\theta)^2 T} - \frac{SI_eRM\theta}{(R+S_1\theta)^2 T} - \left(-\frac{SI_eR(-1+\log((R+S_1\theta)/R))}{(R+S_1\theta)^2 T}\right)$$

In case 2, Equations (35) and (36) lead to,

$$\frac{\partial^2 TVC_2}{dT^2} = \frac{2A}{T^3} + \frac{2CR\left(e^{\theta t_1} - \theta t_1 - 1\right)}{\theta T^3} + \frac{2IR(Rt_1 - S_1)}{pT^3} + \frac{2C_1R\left(e^{\theta t_1} - \theta t_1 - 1\right)}{\theta^2 T^3} + \frac{C_2Rt_1^2}{T^3} - \frac{2SI_eRt_1(-M+t_1)}{T^3} - \left(-\frac{2SI_eRMt_1}{T^3}\right) - \frac{SI_eRt_1^2}{T^3}$$

$$\frac{\partial^2 TVC_2}{dS_1^2} = \frac{CR\theta}{(R+S_1\theta)^2 T} + \frac{IR\theta}{p(R+S_1\theta)^2 T} + \frac{C_1R}{(R+S_1\theta)^2 T} + \frac{C_2R(1+\theta T - \log(R+S_1\theta)/R)}{(R+S_1\theta)^2 T} - \frac{SI_eRM\theta}{(R+S_1\theta)^2 T} - \left(-\frac{SI_eR(-1+\log((R+S_1\theta)/R))}{(R+S_1\theta)^2 T}\right) - \left(-\frac{SI_eR(M\theta - 2\log((R+S_1\theta)/R)+2)}{(R+S_1\theta)^2 T}\right)$$

Chapter 6
Innovating Practices in Managing Engineering Design Projects

Fernando Abreu Gonçalves
CEG-IST, Portugal

José Figueiredo
CEG-IST/DEG, University of Lisbon, Portugal

ABSTRACT

We can address innovation from different perspectives. In engineering practices we can look to changes resulting from attempts to discover ways of overcoming difficulties. How can we manage these innovative practices in engineering design projects? Furthering our perspective we use an actor-network way to look at change processes as chains of translations between heterogeneous actors that are enrolled in changes and where patterns of action are inscribed in durable manners. In an actor-network, the chains of inscriptions are stronger if the number of aligned actors is bigger, and this is the case of engineering projects where the cost of change builds up with time. Through the use of some stylized situations, the authors construct a Perturbation Index to obtain numeric values to assess the dynamics of innovations in engineering practices. The aim is that the application of this index to real situations could lead to meaningful descriptions of such innovation processes. Managing innovation in engineering design projects has to do with the management of project scope. The proposal extends scope management from its definition and planning phases through the control of changes along the execution.

DOI: 10.4018/978-1-4666-4506-6.ch006

1. INTRODUCTION

Most references to innovation relate to the development of new products, mainly from the product side characteristics. In this chapter we do not address innovation in these terms, but as practice changes an engineer creatively adopts during engineering design projects.

We adopt Actor-Network Theory (ANT) including the recent developments that go under the denomination of Modes of Existence (Latour, 2013), as a way to understand the change processes along the chain of inscriptions we call translations: "So the phenomenon we are tackling is not inscription per se, but the cascade of ever simplified inscriptions that allow harder facts to be produced at greater cost", Latour (1990), see section 2 of this chapter. The chains of inscriptions are stronger if the number of aligned actors is bigger, and this is exactly the case of a project where the cost of change builds up with the evolving of time (PMBOK, 2008).

We have in mind the management of project scope from its definition and planning through the control of changes and execution. We do it inspired by the Earned Value Management framework and new extensions. These extensions stress the relevance of "time" in projects and add to cost and time performance indicators a schedule performance index. We designed a "perturbation regime index" to account for translation effort. Then we analyze stylized changes "of regime" and conclude that some changes, more than others, can lead to innovative results. This means we gain a wider view about scope, and scope management, being able to observe and challenge good practices changes, something we repute crucial in engineering design (innovative) projects, where requirements and goals permanently drift.

In engineering processes adherence to good practices is positive for replication of things and to assure efficiency, but it can drastically limit innovation. In his day-to-day practice an engineer always has opportunities to innovate, but this always mean challenging good practices and stabilized routines. Good practices sometimes crystallize loosing their edge and hampering innovation opportunities. Innovation in engineering design is not only about the design of new products. We address this dilemma by pointing to a way to challenge innovation in engineering design projects, questioning how engineers commit themselves with innovative practices. Can we mobilize people on their search for innovation? Can we manage the adoption of innovative practices? Is there any possible compromise between good practices and project (design-projects) goals? Can we manage the degree of perturbation in project translations in an advantageous way? To propose a way of research on the answering to these questions is the goal of this chapter.

Project Management standards cover different issues on project management within a roll of *situated* contexts. Maybe it is difficult to generalize but engineering design and technological innovation (R&D) are topics not specifically addressed in the project management different bodies of knowledge. In fact, the uncertainty about the outcomes implies the scope to be dynamic (Pons, 2008) and scope management to be influenced and fertilized by different techniques.

The view of scope as the result of a Work Breakdown Structure (WBS) with its hierarchical formalism may convene impressions of a neat way to go about the scope management of a project. But as all project managers are well aware, these impressions are illusory except for trivial cases. Engineering design is normally characterized by complex interrelated activities and large uncertainties on different aspects, for example about which solution path should be taken, availability of resources, and more. In these circumstances the full scope of a project cannot be anticipated beforehand (Pons and Raine, 2005). When dealing with engineering design projects a main problem is related with evolving uncertainty (Sonnemans et al, 2003). Uncertainty at the formulation level, *what* is the problem and which is the problem, with all the necessary negotiations with key stakeholders in order to formulate the problem well – *problem setting* (Schön, 1983). Uncertainty at the resource level, how can we grab and convince some crucial resources to be part of our project – *enroll and mobilize actors*. And uncertainty at the execution level, how can we estimate efforts to do things that were never done before, and how can we cope with such an amount of different solicitations and demands (strategic nature, operational type, product portfolio)? (Pich et al, 2002). Even from the side of product development, Ford and Chris Coulston (2008), recognize a socio-technical co-evolution of products and markets, emphasizing the need to manage the "fuzzy front end of product development".

The concept of design *drift* can be used to understand the chance of achieving effective design goals. With complexity levels increasing, engineers need to work in ambiguous environments and they need to be flexible over the phases of the project lifecycle. Design drift could traditionally be intended as a negative prospect in project management, when some projects slip away from what the customer actually wanted or simply away from requirements. With a different approach, much inspired by Ciborra (2004), Monteiro & Hepso (2000), Orlikowski (1996) and Hanseth and Braa (2000), we understand design drift as a strategy to take advantage from the fuzzy front end of requirements in engineering design projects. Design drift is a process in which what we obtain at the end of the implementation process can be different from what was intended originally, sometimes with implementation processes moving beyond original system design intentions. In fact, for us, design drift is a way of dealing with positive design risk, a way of being agile facing an environment that is not at all stable.

This is a very general description of innovative practice in engineering design but we argue that scope management can be entailed as an actor-network building a particular "mode of existence", the Project, where things within a time of its own are negotiated with stakeholders, environment, technologies, tools and restrictions. These negotiations are mainly translations, interactions in which actors construct common definitions and meanings and unfold representatives in the pursuit of a collective objective.

After this introduction we briefly explain some relevant Actor-Network Theory (ANT) fundamentals and constructs. Then we address performance in managing projects, using concepts of Earned Value Management (EVM) (Project Management Institute, 2005). Then, acknowledging our social-technical view centered in ANT we develop an approach that looks at scope management in engineering design projects as the management of a *perturbation regime*. This perturbation regime is used as a metaphor to understand allocation of resources, and resource efforts. This is described with a practical illustration where we explore the possibility of managing a fruitful balance between good practices and innovative practices, using a perturbation regime index.

2. REVISITING ANT

This section only intends to state a common ground to some fundamentals of ANT strictly related to what we are dealing with in this chapter and that can be expressed as "circulations", "the social is a certain type of circulation that can travel endlessly" (Latour, 1997).

For a comprehensive understanding of ANT terms we refer to Lancaster University, Centre for Science Studies webpage referenced in reference's list.

The basic process underlying ANT is "translation". An actor translates another and defines-it but in doing so it also redefines itself. The process is reciprocal, with the actors defining one another in interaction. These definitions are "inscriptions" (Akrich, 1992), the activity of actors defining each other. This process of inscription gives rise to actor-networks. Inscriptions can be materialized in technological artifacts. In this process we choose focal actors, for example a project manager. Focal actors initiate the process of translation with different actors such as line managers, team members, clients.

Translation can also be understood as displacement, "to translate is to displace" (Callon, 1986); "Like Michel Serres, I use translation to mean displacement, drift, invention, mediation, the creation of a link that did not exist before and that to some degree modifies two elements or agents."

An Obligatory Point of Passage (OPP) is the first component of a translation. It is the translator, an entity (a thing) in an actors-networks world that intends to speak on behalf of other actors of that world, translating in a way that fits together their interests and behaviors (Gonçalves and Figueiredo, 2010). When one actor speaks on behalf of others, an Actor-Network can be seen as "geography of obligatory points of passage" (Callon, 1986). OPPs become what can be named as "centers of calculation", translating actor's interests and behaviors (matters of concern) into an alignment. The process of alignment can proceed further into "blackboxing, a process that makes the joint production of actors and artifacts entirely opaque", (Latour, 1994). In a project an engineer producing specifications to answer to requirements of the client is an OPP.

Posing the question - "how to act at distance over things that are not familiar with us?" - Latour (1987) answers by defining Immutable Mobiles (IMs): "By conceiving (constructing) means that are able to transform those things as mobile things, while maintaining them immutable, and making them combinable in order to be cumulated". IMs are a kind of go-betweens in the space of OPPs, good enough to find ways to capitalize to a centre. As an example, the specification made by the engineer is an IM.

IMs are said to be immutable but in translations (displacements) there are always transformations and we are particularly interested in the possibility

Figure 1. Program of action

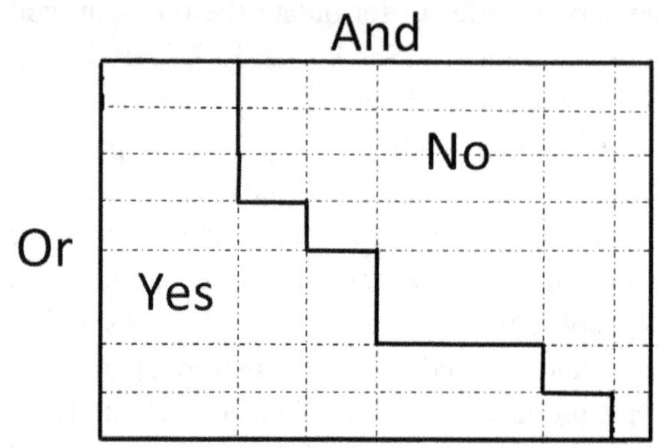

to manage those changes in pursuing innovative practices, (Gonçalves and Figueiredo, 2009, 2010).

A way (Latour *et al.*, 1992) to follow the evolution of networks is the Program of Action (PA). A PA is a semiotic device to describe translations, a two-dimensional chart depicting a succession of proposals (paradigms) along a vertical axis (logical OR axis) where different proposals substitute one after another and a succession of syntagms along a horizontal axis (logical AND axis) where associations among actors align, Figure 1.

We may represent a PA as a flat table, see Table 1 where in the first column we inscribe a focal proposal and in the next columns we inscribe the answers (agreement/disagreement) of the actors involved.

For practical reasons we need to find economic ways to follow the actors. Time, Cost, and Scope are the traditional main actors we follow in projects (in a more demanding scenario we could imagine things like communication, organizational level, and risk, to be also interesting actors to follow). Back to our business of scope management, "time" (duration of a task), is a particular useful actor to follow.

Another way to look at actor-networks is to further explore the idea of materiality above mentioned: translation is "inscribed into durable material" (Law, 1992). Within this view we may turn our attention to Annemarie Mol and John

Law (1994) that contributed with an interpretation addressing displacements in four different types of spaces: fluids – a kind of "Mutable Mobile" object; regions – "Immutable Immobile" object; network – the place for IMs; and fires – "Mutable Immobiles".

More generally we may consider that the four basic addressed "spatialities" are not frozen as if they were some fixed categorizations or "ideal types" of some universal pattern (Boudourides, 2001). We, instead, will retain the image of multiple regimes in a kind of fluid space as a good metaphor for innovative practices in design projects (Law and Mol, 2000) and (Laet and Mol, 2000). This way we could look at translations under the view of fluid circulations - "instead of a fine laminar flow, we will most often get a turbulent flow of whirlpools and rapids. Time becomes reversible instead of irreversible", (Latour, 1993).

Another way to describe an actor-network is by means of some numerical indices. Bruno Latour et al. (1992) proposes some. One of them is the *Reality of the Project*, an evolutionary ratio of the number of associated elements that remain stable in a paradigmatic version in relation to the previous version. The greater this index the more real is the project, that is, the more extended and stronger is the network described. To express this numerically Latour used a formula like this one:

$$R(n) = \frac{\text{number of associated elements that remain stable in a paradigmatic version (n)}}{\text{size of previous version (n-1)}}$$

As we are interested in engineering design projects, to find a way of assessing performance in translations in the network we view innovative practices as a particular *regime* of translation, a kind of fluid where proposals travel amid contexts. It is reasonable to assume that some perturbation will always occur on this circulation. But maybe it will be possible to manage the degree of perturbation in project translations in some profitable ways. To be able to profit from this view of fluid circulation we need to build some dynamics into this picture. Bruno Latour (2013) recently published a book promptly extended in an explorative web site where he addresses an "inquiry into the modes of existence", partly as an involving landscape, or many different involvement landscapes to ANT. Besides actor-networks other domains are considered "modes of existence" with a proper way of being constructed. Latour already defined about fifteen of these modes. In reality we perceive hybrid modes of existence made by crossing over those "pure" modes. We may think of them as networks with a character, as if we colored them with persistent moods. In this new view the network mode of existence maintains its role of following heterogeneous associations but once a relevant domain is found it should be addressed in a way of its own, as if it possessed a particular language, its own way of enunciation.

Operating with modes of existence means to surpass *hiatus,* entering into new spaces, crafting trajectories, and following emerging actors in continuous transformation. In so doing we must be aware that some conditions of *veridiction* (truth) should be applied. The journey is not without risks: there is always the possibility of relaxing to reproduce similar actors in a kind of lineage or, even worst, to retreat to some habitual behavior instead of embarking on a creative path. There is also the risk something really different happens and monsters will emerge.

Framed by this cloud of existences a Project is an actor that unveils, entering existence by a rupture of time, beginning to enroll actors into a new trajectory and keeping them mobilized until a purpose (matter of concern) is reached. The actors enrolled must be all that are relevant for the purpose and only those. They are transformed, in substance and shape, assuring durable artifacts. Along the trajectory there is a continuous concern about ends and means. Calculations are made and sometimes there is a temptation to break rules. One is always aiming to reach as good as possible results.

3. PROJECT PERFORMANCE

Project Management PMI rationale (PMBOK, 2008) uses a Performance Index to measure the performance of the engagement of a resource in a task. Embracing the project as a set of tasks, we adopted a variation, or extension, of the traditional index - PV/AV, where PV is the value of the planned effort and AV is the actual effort exercised to accomplish the task, both in monetary units.

Earned Value Management (EVM) integrates measurements of scope, schedule, and cost relying in a single measure unit for the project performance – cost. Earned value (EV) is defined as the budgeted cost of work performed. That means that EVM systems operate a translation of cost into cost, time into cost and scope into cost that works when integrated into the overall project. This single unit measure system is one of the strongest assets of the EVM system, for the supporters, and the weakest aspect for the detractors, see extensions EVM (PMI, 2005).

Recently and motivated by misleading information provided in very specific project situations (late projects near the conclusion) there was some new developments to the EVM framework (EVM extensions) that converts project performance indicators into time with a schedule performance index – SPIt=PT/AT, where PT is the time planned to obtain EV at Actual time (AT), sometimes referred as Earned Planned Time, (Lipke, 2003), (Rodrigues, 2010).

Discussion is still alive among researchers but this approach has already provided some interesting results, allowing the correction of earlier gaps and we, from our ANT approach, definitely see "time" as the main characteristic of a project.

4. TASK PERTURBATION INDEX

In Project management there are situations where bad performances can be related to bad planning, but there are also bad performances related to bad execution. In real projects both failures of planning and execution occur. In ANT terms we feel the need to extend the network and consider enrolling a new actor, a thing we call good-practice. Good practice should provide a reference against which we can measure alternatives.

In Figure 2 we describe a situation where we have project planning, good-practice, and earned outcome.

In this setting a focal actor P mobilizes G, the target actor, in order to attain project goals. In the sequent chain of translations the outcome attained, E, can be different from what was planned and even G can change in the process. If there is a chance to learn, G may change for the better. In any situation however, P and G are mutually defined and both participate in the performance of the project, translated by E. In fact there are always uncertainties an ambiguity in engineering design, as well as in the evolution of actor-networks.

To deal with a practical example, PT is the planned time for a work package in a Work Breakdown Structure, GT the Good Practice time for that same work package (reference time if supposedly everything goes right).

The schedule performance index SPIt refers to a chain of translations where only PT and AT are addressed. But if we introduce GT, as described, in the Performance Index, the equation can remain mathematically unchanged:

Figure 2. Planned (P), Earned (E), and Good-Practice (G) actors translating in an actor-network

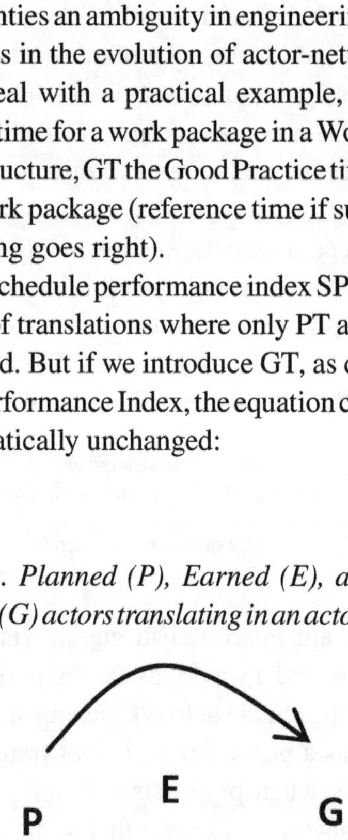

$$SPIt = \frac{PT}{AT} = \frac{PT}{GT} \times \frac{GT}{AT}$$

We use "t "in SPIt to show that we are, inspired by the extensions on EVM, using an index depending on time.

To illustrate the idea of perturbation of the task we may observe that the allocation of most resources is not strictly fixed and their performance is not always stable. We can stress resources to a certain point. For example, if we have a task that normally, in a good practice, is accomplished by 3 $engineer_x day$, maybe it would be possible to "force" it to 2 $engineer_x day$, maintaining quality and time, and reducing cost.

Table 1 is a table that intends to represent a Program of Action for the translation P-G that can be expressed as 'how long it will take Work Package X?'. Good Practice (G) states 3 $engineer_x day$ for the tasks of the work package. The Project Manager is committed to make it in 2 (PT=2) and the outcome is always uncertain.

With this in mind we define a Perturbation regime index, Pr, by taking:

$$Pr = \frac{PT}{GT}$$

Pr can be seen as a measure of the fit between planning (PT) and good practice (GT). Pr will depend mostly on the bargaining power of the project manager face to the functional managers and resources.

Table 1. Program of action of the P-G (E) translation

Work package X		
P	G	E
PT=2	GT=3	
		AT

Figure 3. Pr (Perturbation regime index)

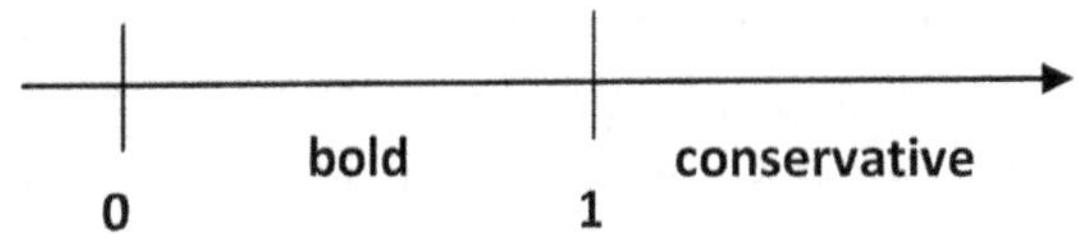

If Pr is less than 1 we face a positive perturbation regime because we are stressing the resources and, perhaps, earning more, if it is higher than 1 we are not taking advantage of the perturbation and should accept good practice durations, Figure 3.

Being bold means that in the project we consider that, due to some circumstances (fighting routine, learning, or others), some task durations can be shorter than usually considered in good practice; being conservative means otherwise, duration would be expected to be longer, so we should stick to good practice. These ideas gain effectiveness if circumstances and tasks are clearly identified, the reason for changing actions can be justified and also if the limits of the regime are perceived.

This conceptual evolution allows us to extend or generalize the meaning of the expressions "effort", "turbulence", or "perturbation" and "stress" used within the resource allocation process of a an engineering design project. This generalization is also aligned with ANT in the way that actors are human and non-human (as we can stress people, we can also stress any technological devices, and even the interpretation of rules and regulations).

5. MANAGING PERTURBATION IN PROJECTS

Our question is now extended. To cope with innovative practices in engineering design project management, how much perturbation would be allowed, while maintaining the process stable, that is, manageable and controllable? Of course these ideas and the meaning of the values of Pr need empirical substantiation from real projects: how much is perturbation; how much perturbation can be tolerated, how can we measure outcomes, etc.

In an approach strictly conceptual we will explore six possible stylized situations defined in terms of the relative positions of PT, GT, and AT, PGE (planned, good practice, earned, as stated in Figure 2) for PV>GV>EV, GPE for GV>PV>EV, and follwing figures.

5.1. PGE: Bold with Worst Outcome

In this first situation we have Figure 4 with the cumulative values in monetary units and showing how PT and GT are to be found in their respective curves from the value of EV.

If we have good reasons to believe in the bold planning then we may ask why it was not possible to obtain a performance better, at least in relation to GT. Can we discern a monster (hybrid) in this setting? Yes, a bad one, planning could be at the origin of the problem assuming an impossible outcome.

5.2. PEG: Bold with Unfavorable Outcome

We have a bold planning and an unfavorable outcome, although better than GT. What went wrong at the execution? Or was PT too bold? If this model allows us to think on what was wrong, the model already proved to be useful. And, as we are obliged to think in actors in the ANT way, we can go far beyond the majority of analysis performed in the management of projects. (see Figure 5)

5.3. EPG: Bold with Favorable Outcome

In this situation there is a bold planning and the outcome is much favorable. Maybe GT was too relaxed. It is a situation in which the good practice could easily be outperformed. We will need to know, or experiment, in which limits. This time the monster is a good one. (see Figure 6)

Figure 4. PT < GT < ET

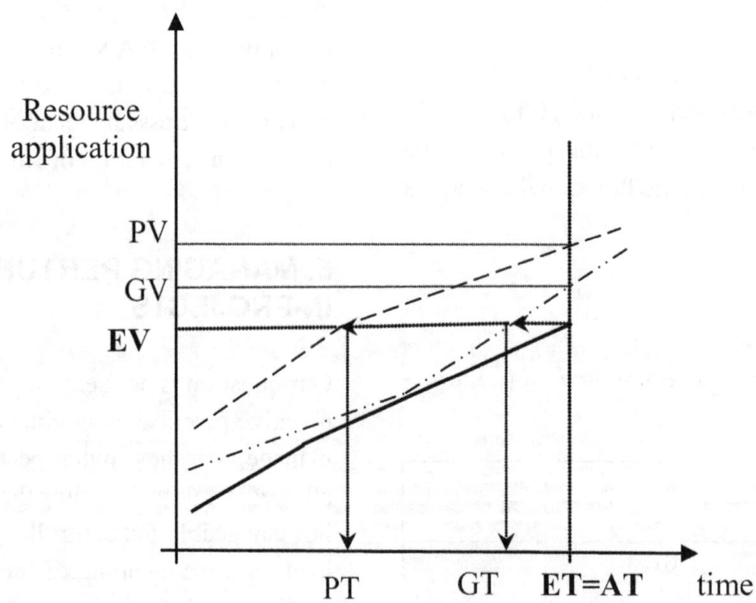

Figure 5. PT < ET < GT

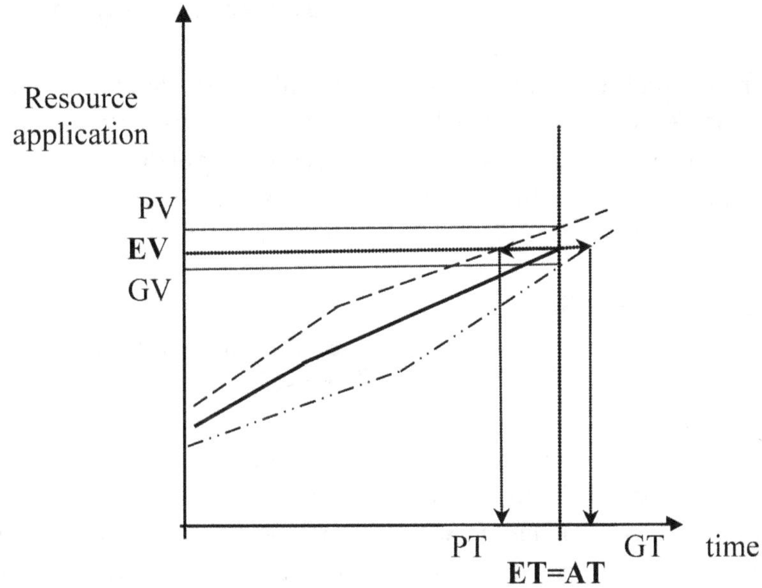

Figure 6. ET < PT < GT

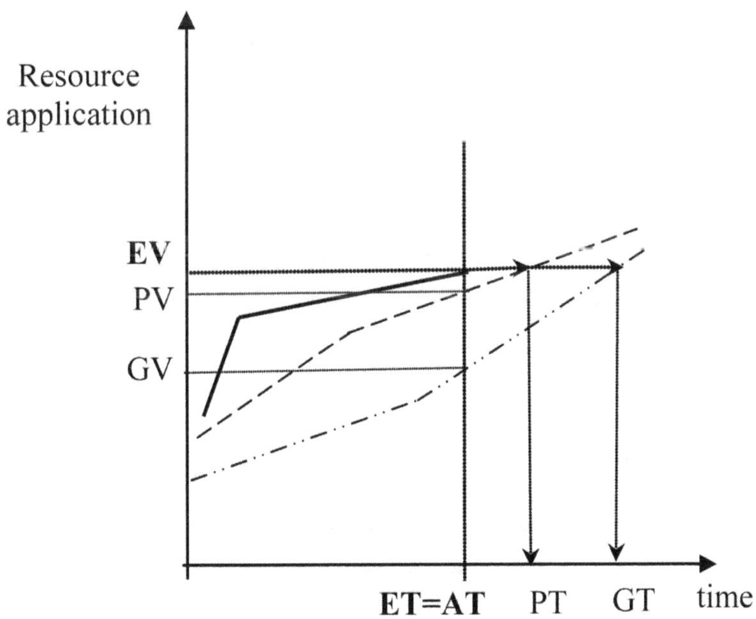

5.4. GPE: Conservative with Worst Outcome

A conservative planning amplified ET, showing a bad outcome. May be too loose a planning impedes good results. Do the circumstances justify this bad outcome? One thing is clear, we should change strategy and rethink our decisions. (see Figure 7)

5.5. GEP: Conservative with Favorable Outcome

Now we have a conservative planning and although GT was not reached, the outcome is better than planned. Maybe it would be helpful to analyze the reasons for the conservative planning. Maybe it would be possible to defy this situation. (see Figure 8)

5.6. EGP: Conservative with Better Favorable Outcome

Finally here we have a very conservative planning and a better favorable outcome. Was this a direct consequence of the conservative planning? Could we have done it even better? The monsters (hybrids) were completely at our side? (see Figure 9)

6. COPING WITH INNOVATIVE PRACTICES

As we are interested in dealing with innovative practices (design) we propose approaches based on looking at how things have been already made (good practices) and admit possible variations. This approach is justified in engineering design settings where there is a need to offer continuous innovations and practice should be included within project settings, see Figure 10. In this diagram we grafted our concept of managing perturbations.

Figure 7. GT < PT < ET

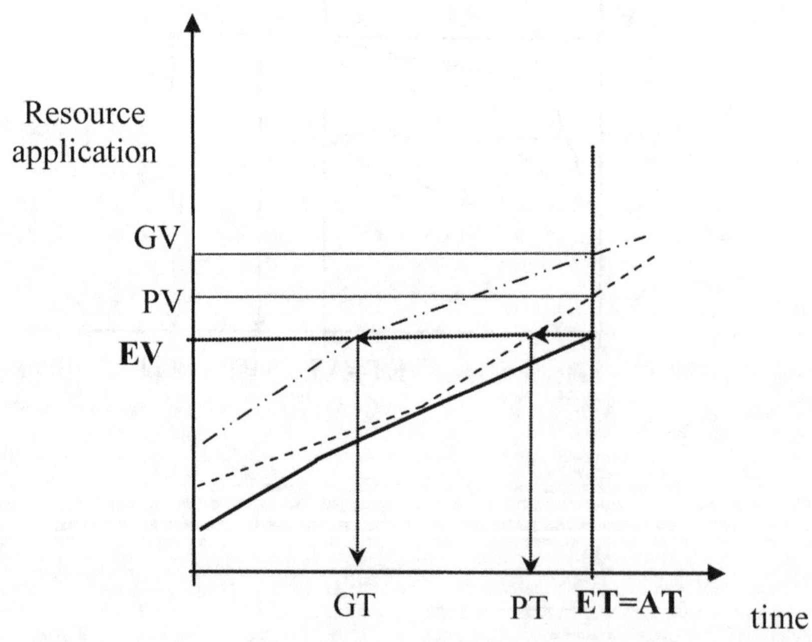

Figure 8. GT < ET < PT

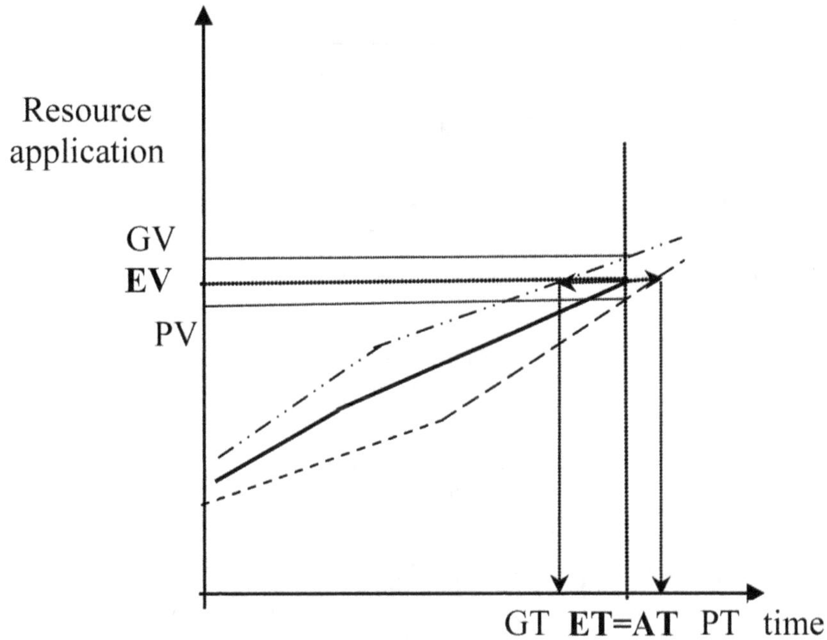

Figure 9. ET < GT < PT

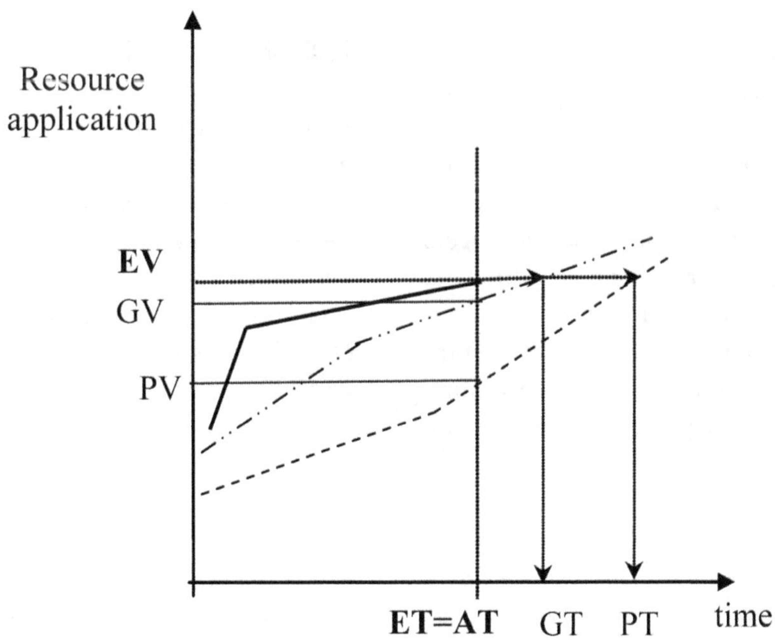

Figure 10. Managing perturbation to cope with Innovative practices (Adapted from Mahmoud-Jouini, 2000)

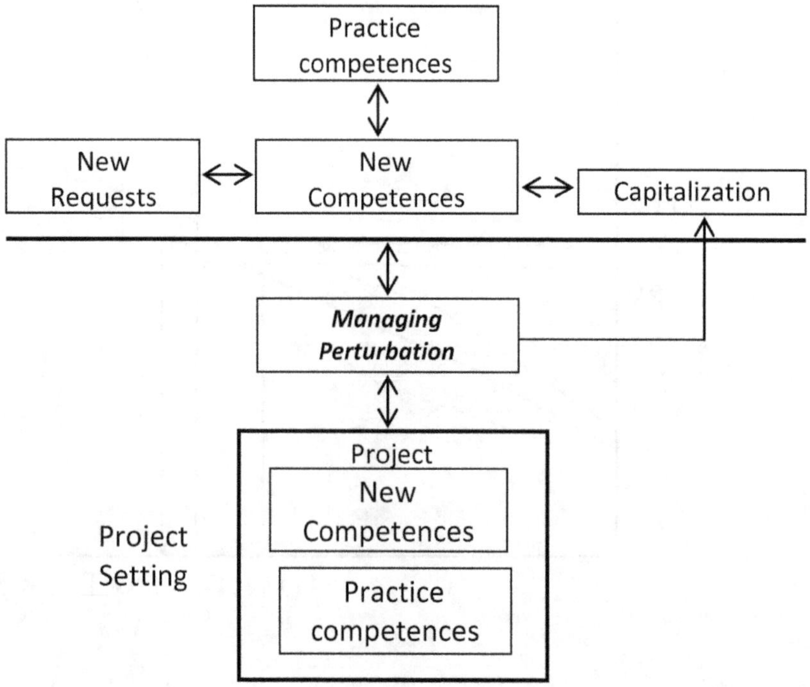

A learning cycle emerges, meaning that "lessons learned" from past project experiences (and from earlier stages of this same project) are incorporated in documentation and then embedded in training programs, discussions with peers, and all kind of negotiations so that project managers learn from experience, from doing, and from the collective learning process of the team and eventually the organization and community. Communication is open in such organizations, leading to a process by which project experiences are "handed down" to next generation project teams (Barkley, 2007).

Our perturbation regime index has with it the burden to collect more data, good practice standards for resource application. However one should bear in mind that these data already exist in functional departments. The only difference is that we propose to import that data *into* project management rationale.

7. CONCLUSION

Resource allocation in engineering design projects may be approached from very different perspectives, not only by scheduling and cost reasons. We must allow for the possibility of resource knowledge to be somewhat fuzzy, as circumstances are often ambiguous and not completely clear. Perhaps the project calls for suppressing parts of some routines on normal practices, or it is mandatory to explore new ones, and you need knowledge for that.

Sometimes you are tempted to cut on features initially addressed but you can't to guarantee adequate project feasibility. There are many different causes for project drifting. The drivers for drifting can be external, or internal. One of the internal ones is surely the one we addressed and is concerned with resource stressing. The way you stress your resources in a project can promote innovation. That is why we set up this model that allows measuring the results of different resource stressing policies. The way we explore these policies is quasi metaphoric, but we tried to scale-it around some extreme, or typical situations.

The use of a number to characterize translations regime enabled us the enumeration of stylized cases that we can analyze further with the lenses of ANT. This analysis integrates Scope Management and Planning Management and even allows some steps towards Risk Management.

Effective project management requires more than good planning, it requires that relevant information be obtained, analyzed, exchanged, and explored in order to decide for reasonably good choices. It requires the ability to question different factors that sometimes are not in the normal roll of a project manager. We think that an ANT approach may uncover new practices that are not addressed in common project management practices. This approach provides an integrative view of the project that can be able to enroll and mobilize different resources. We need to recollect that in ANT terms actors are hybrids and can comprehend, humans, rules, laws, organizations and software. Actors interact in a common language between one another creating links. In fact in ANT terms translation is seen as the process by which a relation is created between two actors.

In engineering design project management, meanings evolve with the project and plans usually drift. This can provide early warnings of pending problems and impact assessments on other activities, which can lead to options for alternate plans and management decisions. Our model also allows to simulate different resource allocation options having a perception of what could occur in the different strategies and controlling the risk of the different option consequences.

Being permanent in our research the inspiration from ANT we usually don't explain the theory in detail, relying on readers having a certain degree of knowledge on the subject. However we decided to lighten the ANT references in favor of more Project Management references hoping to tempt Project Management practitioners to try similar lens on project management approaches.

REFERENCES

Akrich, M. (1992). The description of technical objects. In J. Law (Ed.), *Shaping technology/building society* (pp. 205–224). Cambridge, MA: MIT Press.

Barkley, B. T. (2007). *Project management in new product development*. New York, NY: McGraw-Hill.

Boudourides, M. A. (2001, November 1-3). *Networks, fluids, chaos*. Paper presented at the International Conference Spacing and Timing, Rethinking Globalization & Standardization. Palermo, Italy.

Callon, M. (1986). Some elements of a sociology of translation: Domestication of the scallops and the fishermen of St Brieuc Bay. In J. Law (Ed.), *Power, action and belief: A new sociology of knowledge*. London, UK: Routledge.

Ciborra, C. U. (2004). Digital technologies and the duality of risk. London, UK: CARR, LSE.

Ford, R. M., & Coulston, C. (2008). *Design for electrical and computer engineers: Theory, concepts, and practice*. New York, NY: McGraw-Hill.

Gonçalves, A., & Figueiredo, J. (2009). Organizing competences: Actor-network theory in virtual settings. *International Journal of Networking and Virtual Organisations*, *6*(1), 23–35. doi:10.1504/IJNVO.2009.022481

Gonçalves, A., & Figueiredo, J. (2010a). How to recognize an immutable mobile when you find one: Translations on innovation and design. *International Journal of Actor-Network Theory and Technological Innovation*, *2*(2), 39–53. doi:10.4018/jantti.2010040103

Hanseth, O., & Braa, K. (2000). Who's in control: Designers, managers, or technology? Infrastructure at north hydro. In C. U. Ciborra (Ed.), *From control to drift, the dynamics of corporate information infrastructures*. Oxford, UK: Oxford University Press.

Laet, M., & Mol, A. (2000). The Zimbabwe bush pump: Mechanics of a fluid technology. *Social Studies of Science*, *30*(2), 225–263. doi:10.1177/030631200030002002

Lancaster University. (2000). *Actor network resource: Thematic list*. Retrieved from http://www.lancs.ac.uk/fass/centres/css/ant/ant.htm

Latour, B. (1987). *Science in action: How to follow scientists and engineers through society*. Boston, MA: Harvard University Press.

Latour, B. (1990). Drawing things together. In M. Lynch, & S. Woolgar (Eds.), *Representation in scientific practice* (pp. 19–68). Cambridge, MA: MIT Press.

Latour, B. (1993). *We have never been modern*. Brighton, UK: Harvester Wheatsheaf.

Latour, B. (1994). On technical mediation. *Common Knowledge*, *3*(2), 29–64.

Latour, B. (1997). On recalling ANT. In J. Law, & J. Hassard (Eds.), *Actor-network and after*. Oxford, UK: Blackwell.

Latour, B., Mauguin, P., & Teil, G. (1992). A note on socio-technical graphs. *Social Studies of Science*, *22*(1), 33–57. doi:10.1177/0306312792022001002

Latour, B. (2013). *An Inquiry into Modes of Existence, an Anthropology of the Moderns*. Cambridge, Massachusetts: Harvard University Press.

Law, J. (1992). Notes on the theory of the actor-network: Ordering, strategy and heterogeneity. *Systems Practice*, *5*(4), 379–393. doi:10.1007/BF01059830

Law, J., & Mol, A. (2000). *Situating technoscience: An inquiry into spatialities*. Lancaster, UK: Centre for Science Studies, Lancaster University.

Lipke, W. (2003). *Schedule is different*. The Measurable News.

Mahmoud-Jouini, S. B. (2000). Coment passer d'une demarche réactive à une démarche proactive fondée sur des stratégies d'offres inovantes? In A. Bloch, & D. Manceau (Eds.), *Le cas des grandes entreprises générales de Bâtiment françaises, de l'idée au marché*. France: Vuibert.

Mol, A., & Law, J. (1994). Regions, networks and fluids: Anaemia and social topology. *Social Studies of Science*, *24*, 641–671. doi:10.1177/030631279402400402 PMID:11639423

Monteiro, E., & Hepsø, V. (2000). Infrastructure strategy formation: Seize the day at Statoil. In C. Ciborra (Ed.), *From control to drift: The dynamics of corporate information infrastructure*. Oxford, UK: Oxford University Press.

Orlikowski, W. J. (1996). Improvising organizational transformation over time: A situated change perspective. *Information Systems Research*, *7*(1), 63–92. doi:10.1287/isre.7.1.63

Pich, M. T., Loch, C. H., & De Meyer, A. (2002). On uncertainty, ambiguity, and complexity. *Project Management. Management Science*, *48*(8), 1008–1023. doi:10.1287/mnsc.48.8.1008.163

PMBOK. (2008). *A guide to the project management body of knowledge* (4th ed.). Newtown Square, PA: Project Management Institute.

Pons, D. (2008). Project management for new product development. *Project Management Journal*, *39*(2), 82–97. doi:10.1002/pmj.20052

Pons, D. J., & Raine, J. (2005). Design mechanisms and constrains. *Research in Engineering Design*, *16*, 73–85. doi:10.1007/s00163-005-0008-9

Project Management Institute. (2005). *Practice standard for earned value management* (2nd ed.). Newtown Square, PA.

Rodrigues, A. (2010). Effective measurement of time performance using earned value management – A proposed modified version for SPIU tested across various industries and project types. *PM World Today, 12*(10).

Schön, D. A. (1983). *The Reflective Practitioner: How professionals think in action*. New York: Basic Books. (Reprinted in 1995)

Sonnemans, P. J. M., Geudens, W. H. J., & Brombacher, A. C. (2003). Organizing product releases in time-driven development processes: A probabilistic analysis. *IMA Journal of Management Mathematics, 14*, 337–356. doi:10.1093/imaman/14.4.337

Chapter 7
Essentials of Linear Programming for Managers:
From System of Inequalities to Software Implementation

Hossein Arsham
Johns Hopkins University, USA

M. Bardossy
University of Baltimore, USA

D. K. Sharma
University of Baltimore, USA

ABSTRACT

This chapter provides a critical overview of Linear Programming (LP) from a manager's perspective. The main objective is to provide managers with the essentials of LP as well as cautionary notes and defenses on common modeling issues and software limitations. The authors illustrate the findings by solving a simple LP directly on the original decision variables and constraints space without adding new variables or translating the model to fit a specific solution algorithm. The aims are the unification of diverse set of topics in their natural states in a manner that are easy to understand and providing useful information to the managers. The advances in computing software have brought LP tools to the desktop for a variety of applications to support managerial decision-making. However, it is already recognized that current LP tools, in ample circumstances, do not answer the managerial questions satisfactorily. For instance, there is a costly difference between the mathematical and managerial interpretations of sensitivity analysis. LP software packages provide one-change-at-a-time sensitivity results; the authors develop the largest sensitivity region, which allows for simultaneous dependent and/or independent changes, based on the optimal solution. The procedures are illustrated by numerical examples including LP in standard-form and LP in non standard-form.

DOI: 10.4018/978-1-4666-4506-6.ch007

1. INTRODUCTION

Linear programming (LP) has been a fundamental topic in the development of managerial decision-making. The subject has its origins in the early work of L. B. J. Fourier (1820s) on attempting to solve systems of linear inequalities. However, the wide spread use and acceptance of LP had to wait until the invention of the Standard Simplex Method.

1.1 Standard Simplex Method

Since World War II, LP has been used to solve problems of various dimensions in almost all disciplines. The most popular solution algorithm is the Simplex method that is implemented by most LP software packages.

The Simplex algorithm can be considered as a sub-gradient directional method, jumping from an initial feasible vertex to a neighboring vertex of the feasible region until it arrives at an optimal vertex. The Simplex method requires that the model be expressed in a special format called "Standard Form", that is, some constraints and variables must be transformed. Consequently, when the manager obtains a solution through the Simplex method, she/he is left with the task of interpreting and transforming the Simplex solution back to the original managerial problem. However, this may not be an easy task.

For instance, solving LP problems in which some constraints are in (≥) or (=) form with non-negative right-hand side (RHS) has raised difficulties. With that purpose, the simplex method requires a feasible stating solution. When such starting solution is not readily at-hand, alternative methods are necessary. One version of the Simplex method, known as the two-phase method, introduces an artificial objective function, which is the sum of artificial variables (see Arsham 1997a, 1997b). Another version adds penalty terms, which are the sum of artificial variables with very large, positive coefficients. The latter approach is known as the Big-M method, (see Arsham 2006, 2007). On the other hand, using the Dual Simplex method has its own difficulties. For example, when some coefficients in the objective function are not dual feasible, one must introduce an artificial constraint. Handling equality (=) constraints by the dual simplex method is tedious because of the introduction of two new variables for each equality constraint: one extraneous slack variable and one surplus variable. Also, one may not be able to remove some equality (=) constraints by elimination at the outset, as this may violate the non-negativity condition introduced when constructing the "Standard Form."

Arsham (2013) proposes a tabular interior boundary approach that intents to address these concerns and solves the original managerial LP model, without any need of "Standard Form", artificial variables, artificial constraints, and Big-M. While these variants of the simplex successfully handle an array of constraint forms, they impose a burden and mathematical sophistication on manager that make difficult the success of LP applications.

1.2 The Costly Difference Between the Managerial and Modeler Interpretation

Koltai and Terlaky (2000) state that managerial questions are not answered satisfactorily with the mathematical interpretation of sensitivity analysis. Software packages provide sensitivity results focused on the optimality of a basis and not on the optimality of the values of the decision variables. The implementation of the misunderstood shadow prices and their range of validity may coax managers to take poor decisions with considerable financial losses and strategic consequences. Another source of confusion is the similarity of terms used in operations research (i.e., re-search, as a process), and other related fields. For example, "shadow price" and "opportunity cost" have somewhat different meanings in the LP and Economics literature. The "opportunity cost" of an action in economics can be interpreted as the "shadow price" of that action on the budget.

The LP software packages provide sensitivity results about the optimality of a basis and not about the optimality of the values of the decision variables and the shadow prices that are of interest to the manager. In Section 6 we develop the largest sensitivity region based on the optimal vertex, and allows for simultaneous dependent/independent changes.

1.3 The Confusion between the Model and the Solution Algorithm

Naturally, a modeler, reflecting the modeler's decision problem, creates the LP model. On the other hand, the solution algorithm is designed by, for example, a systems analyst to solve LP problems. Because of the complexity of Simplex method, confusion could creep-in when the modeler's model is developed into a tool using the software solution algorithm. For instance, the algorithm used in QM software and Management Scientists make all variables non-negative. This may create confusion as many LPs may contain negative or unrestricted variables. In this research work, the algorithm adjusts itself to the original model and solves it, instead of changing the model. The decision maker easily understands the result, because our method does not change the structure of the original model. For example, while using Simplex approach the need to transform constraints to standard form changes the manager's LP model to something that is not his and therefore the optimal solution must be transferred back to the manager's problem. This can confuse the managers and lead to costly misinterpretations. One must understand the differences that could creep-in when a model is formulated and when it is solved using a solver's solution algorithm; unfortunately the current state of art of LP software do not differentiate and causes confusions.

There exists a dichotomy between the reality the managers have to deal with, the problems that they face and the type of strategic decisions they need to make, and the methods that modelers have access to solve those problems. Research and technology have brought great advances to the decision-making process. However, more advances are needed to address directly the managers' decision problems.

Figure 1. Manager's world vs. modeler's world

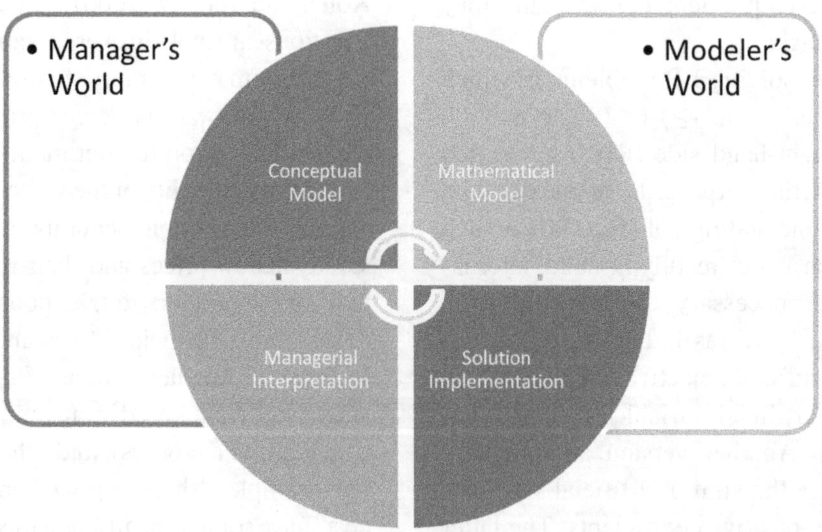

The main issue is that most software packages make assumptions about the structure of the problem, constraints have a certain direction; the decision variables take certain sign-constraints. These assumptions are without loss of generality; however, they do have implications that modify the original model and complicate the interpretation of the solution for implementation. As a result, the conceptual models from the manager's environment have to be translated into mathematical models, gets translated and later into solution models to fit software packages. Afterwards, these solutions must be translated back into the real problem. This translation is often cumbersome for the general user. While translating decision variable might be doable with some effort, interpreting sensitivity analyses from a modified model might be out of reach. In that regard the mathematical model should represent to the conceptual model.

We devoted Sections to show how to build the largest sensitivity region based on the "optimal solution". These sections are independent of all previous sections; in that one needs the optimal solution in any way it is obtained. This development differs significantly from the traditional Simplex's tableau approach that limits itself to the "Preservation of the Basis". No one expects the Manager to understand and interpret the Basis correctly, but it is expected that a Manager should be able to understand and interpret the optimal values of his decision variables and the consequences of any changes to them. The Dual generated from the "Standard Form" does not look like the Manager's dual problem of his primal model.

The Operations Research (OR) processes must be considered from the viewpoint of General Systems Theory. The authors assert that without a whole systems perspective OR can neither be understood nor applied effectively. Typically, OR problems can be broken down into five components viz: (1) the "reality" of the problem, (2) the conceptual model of the problem, (3) the conversion of conceptual model to a mathematical model, (4) solution determination of the mathematical model, (5) managerial interpretation, and (6) the implementation of the solution. The authors believe that in most cases we have under-utilized our knowledge (study) and limited the application of the OR process. There have been extremely few studies and applications of OR that satisfies the systems point of view, Arsham (2012).

The remainder of this chapter is organized in thirteen sections. Almost every section can stand by itself, independent of previous sections. Section 2 introduces linear programming models from model builders and decision makers. Some oddities are included. Section 3 describes the decision maker's environment explicitly in order to enable them to understand the decision problem. Section 4 explores the feasible region by using algebra that is understandable by the decision maker. The algebraic approach to solve linear programs is given in Section 5, which includes computation of slack and surplus variables. The methodology solves LP model in an "as is" state, without any need of "Standard form", artificial variables, artificial constraints, the Big-M, and slack/surplus. Section 6 constructs the largest sensitivity region for the right hand side of the constraints based on the optimal solution. The shadow prices are byproducts of the sensitivity analysis. The dual problem formulation is included in Section 7, which is general and easy to apply. Section 8 considers the general sensitivity region for the cost coefficients of the objective function.

Throughout this chapter we have used LPs of two dimensions to save space but this method can be extended to higher dimensions. Even for two dimension problems we do not use the graphical method because it is limited to two-dimensional problems. Another reason to avoid graphical methods is that in some textbooks the graphical methods for finding the cost sensitivity range, and interpretation of the shadow prices are misleading. We use the algebraic approach throughout this chapter to overcome these limitations. The solution to a non-standard form problem is outlined in Section 9. Software implementation and their limitations are examined in Section 10. The sensitivity analysis of the degenerate optimal vertex to remain optimal is developed in Section 11. Section 12 does sensitivity analysis of multiple optimal solutions. The existence of a curious property; known as More-For-Less, or Less-For-More is covered in Section 13. Concluding remarks are summarized in Section 14.

2. LINEAR PROGRAMMING MODEL

In general terms, a linear programming model is a mathematical approach for determining the set of decisions necessary to achieve the best outcome given a set of resource limitations and requirements represented as a system of linear equations. In business, these can be the outsourcing conditions and quantities to attain the lowest possible total cost, or the production plan to get the maximum profit, or the marketing mix to reach the highest market penetration.

2.1 Linear Programming Model

We define the LP model in the most general way.

Problem P:
Max (or Min) CX
Subject to $AX \leq a$,
$BX \geq b$,
$DX = d$,
$X_i \geq 0, i = 1,..., j$
$X_i \leq 0, i = j+1,..., k$
X_i unrestricted in sign, $i = k+1,..., n$

where matrices A, B, and D have p, q, and r rows, respectively with n columns and vectors c, a, b, and d have appropriate dimensions. Therefore, there are $m = (p + q + r + k)$ constraints and n decision variables. It is assumed that $m \geq n$. Note that the main constraints have been separated into three subgroups. Without loss of generality, we assume that all RHS elements, a, b, and d are non-negative. We do not deal with trivial cases, such as where $A = B = D = 0$ (no constraints), or $a = b = d = 0$ (all boundaries pass through the origin point), or having LP with a single point feasible region. We assume the LP has a no redundant constraints.

2.2 What Can Go Wrong in the Process of Building a Linear Programming (LP) Model?

At the modeling stage potential pitfalls exist which affect any LP application; therefore, the decision-maker and the modeler should be cognizant of modeling phase of LP, Arsham (2008). Thus, potential problems exist which affect any linear programming application. Optimal solution may be infeasible or unbounded, or there may be multiple solutions. Degeneracy may also occur. Figure 2 presents a classification of LP for modeling validation process:

2.3 Dual Problem Construction

Associated with each (primal) LP problem there is a companion problem called its dual problem. Table 1 is a classification of the decision variable constraints is useful and easy to remember in construction of the dual.

General Principles

- If the primal is a maximization problem, then its dual is a minimization problem (and vice versa).
- For each constraint there is a dual decision variable (and vice versa).
- Use the variable type of one problem to find the constraint type of the other problem.
- Use the constraint type of one problem to find the variable type of the other problem.
- The RHS elements of one problem become the objective function coefficients of the other problem (and vice versa).

Figure 2. A classification of LP for modeling validation and sensitivity analysis

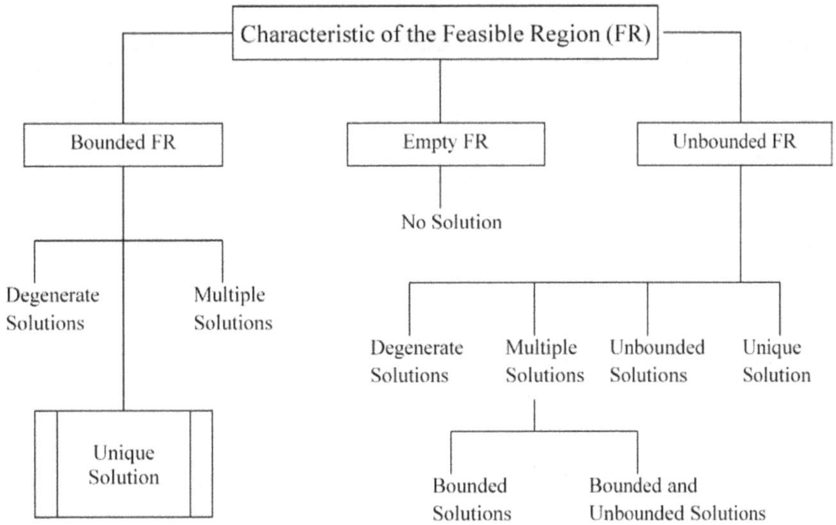

Table 1. A One-to-one correspondence between the primal and the dual LP problems.

	The Dual Problem Construction	
Objective: Max (e.g. Profit) **Constraint types:**		***Objective: Min (e.g. Cost)*** **Constraint types:**
≤ a Sensible constraint		≥ a Sensible constraint
= a Restricted constraint		= a Restricted constraint.
≥ an Unusual constraint		≤ an Unusual constraint.
	Variables types:	
	≥ 0 a Sensible condition	
	unrestricted in sign	
	≤ 0 an Unusual condition	

- The matrix coefficients of the constraints of one problem are the transpose of the matrix coefficients of the constraints for the other problem. That is, rows of the matrix become columns and vice versa.

Applications

One may use duality in a wide variety of applications including:

- It may be more efficient to solve the dual LP problem than the primal Problem.
- The dual solution provides important economical interpretation such as shadow prices.
- If a constraint in the primal is non-binding, i.e., the LHS value is not equal to the RHS value, then the associated variable in the other problem is zero.

Table 2. Possible combinations of primal and dual properties

Primal Problem	Condition Implies	Dual Problem
Feasible; bounded objective	↔	Feasible; bounded objective
Feasible; unbounded objective	→	Infeasible
Infeasible	←	Feasible; unbounded objective
Multiple solutions	↔	Degenerate solution
Degenerate solution	↔	Multiple solutions

- If a decision variable in the primal problem is not zero, then the associated constraint in the dual problem is binding.
- To obtain the sensitivity range of the RHS of primal problem from the sensitivity range of the cost coefficient in the other problem (and vice versa).

Table 2 implies some possible combinations of primal and dual problems.

Problems with LP Packages

Most LP software solvers have difficulties in recognizing the last two cases in the above table namely multiple optimal solutions and degenerate optimal solution. Most LP software do not alert the user about them yet produce sensitivity ranges that are not valid. One must make ensure that the solution is unique, and non-degenerate before analyzing and applying the sensitivity ranges.

2.4 Ensuring the Formulation Is Correct: Odd LP Models

Developing LP models with oddities can assist in the process of ensuring that the formulation is correct. However, almost all LP software solvers have difficulties in recognizing these dark sides of LP and/or giving any suggestions as remedies.

2.4.1 Unbounded Solutions

Identification

If the shadow prices of the primal problem do not satisfy the dual constraint, then the primal problem is unbounded.

Occurrence

An unbound optimal solution means the constraints do not limit the optimal solution and the feasible region is also unbounded, it effectively extends to infinity.

Resolution

In real life, this is not possible. Check the objective function. Possibly, the problem is a minimization problem and maximization was selected instead. Check the formulation of the constraints. Possibly, one or more constraints are missing. Check also the constraints for any incorrect specification in the direction of the inequality, and numerical errors.

2.4.2 Infeasibility

Identification

There is no feasible solution, i.e., there is no feasible vertex. The solution algorithm produces no solution.

Occurrence

An infeasible solution means the constraints are too limiting and have left no feasible region. That is, no solution satisfies all the constraints of the problem.

Resolution

Check the constraints for any incorrect specification in the direction of inequality constraints, and numerical errors. If no error exists, then there are conflicts of interests.

These and other odd situations, including multiple solutions and in particular degeneracy, are rare (except in network models since they often have a redundant constraint, causing degeneracy), and often result at the modeling stage. The decision-maker sets the objective function, while the constraints come from the decision-maker environment, consequently, it is rare to get multiple solutions, i.e. the objective function has the coefficient proportional to those of at least one binding constraint. In these cases, rounding in the coefficient(s) could be the source of the multiple solutions.

3. DECISION-MAKER'S ENVIRONMENT

In the process of conceptualizing and building a mathematical model of the business reality, it is important to identify the most important elements and categorize them into performance measures, uncontrollable inputs, parameters, controllable inputs and strategies.

Understanding the problem requires criteria for grouping entities of the decision model in the same category. Performance measure (or indicators), that is, measuring business performance, should be at the top of the decision-making process and should align with the management goals. The development of effective performance measures is increasingly important for many organizations, even though it remains challenging for organizations in the public sector or not-for-profit organizations, because the performance measures are the drivers for the optimal strategy and determine the direction for the organization.

The decision-making environment elements are further discussed in the context of the following numerical example.

An Illustrative Numerical Example

Consider the following mixed products LP problem of a price-taker small manufacturer producing table and chairs. His objective is to maximize net profit:

$$\begin{aligned} &\text{Maximize } P(\mathbf{X}) = 5X_1 + 3X_2 \\ &\text{Subject to } 2X_1 + X_2 \leq 40, \\ &\qquad X_1 + 2X_2 \leq 50, \\ &\qquad X_1 \geq 0, X_2 \geq 0. \end{aligned}$$

Figure 3 analyzes and depicts his decision environment.

The performance measure is important for the decision maker as a measuring tool for success and identifying the decision problem. Other elements of the decision-maker's environment are classified as follows:

- **Uncontrollable Inputs:** These come from the decision maker's external environment. Uncontrollable inputs often create the problem and constrain the actions.
- **Parameters:** Parameters are the constant elements that do not change during the time horizon of the decision review. These are the factors partially defining the problem. Strategic decisions usually have longer time horizons than both the tactical and the operational decisions, at different levels of organization.
- **Controllable Inputs:** The collection of all possible courses of action the decision maker might take.

Figure 3. Decision-maker's environment in a mixed product LP problem

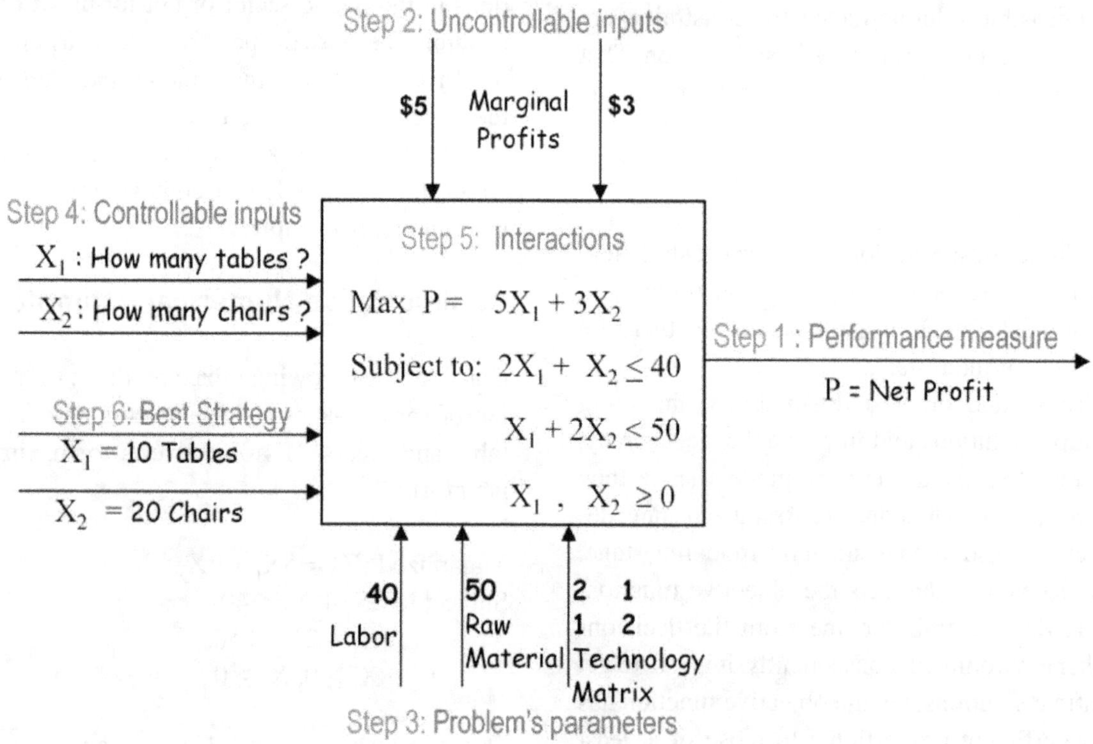

- **Interactions among these Components:** These are logical, mathematical functions representing the cause-and-effect relationships among inputs, parameters, and outcomes.

There are some constraints that are made up of all the parameters. Therefore, they do not need to be treated separately.

- **Actions:** Action is the ultimate decision and is the best course of strategy to achieve the desirable outcome. Decision-making involves the selection of a course of action in the pursuit of one's objective.
- **Controlling the Problem:** Few problems in life, once solved, stay that way. Changing conditions tend to create new problems from the problems that were previously solved. Their solutions create new problems, adding new uncontrollable inputs, and /or new options. One must identify and anticipate these new problems and recognize the cycle of such problems.

4. SOLVING THE SYSTEM OF LINEAR INEQUALITIES

The widely used Simplex solution to an LP problem is strictly based on the theory and solution of a system of linear inequalities (Arsham 2009). The list of constraints in an LP model constitutes a system of linear inequalities that define the feasible region of the problem. The basic solutions to a linear program are the solutions to the systems of equations consisting of constraints at binding positions. Not all basic solutions satisfy all the constraints. Only those that do meet all the constraint restrictions are called the basic feasible solutions. The basic feasible solutions correspond precisely to the vertices of the feasible region.

For illustration purposes we are going to follow the Improved Algebraic Method (IAM) to solve the system of linear inequalities. This method is simple to understand and does not require the formulation of an auxiliary LP problem like it might require the Simplex method. This methodology can also be extended to solve systems of linear inequalities of higher dimensions.

We are interested in finding the vertices of the feasible region of Problem P, expressed as a system of linear equalities and inequalities.

Feasible Region S:

$AX \leq a,$
$BX \geq b,$
$DX = d,$

where some $X_i \geq 0, i = 1,..., j, X_i \leq 0, i = j+1,...,$ k and X_i unrestricted in sign, $i = k+1,..., n$. Matrices A, B, and D as well as vectors a, b, and d have appropriate dimensions.

Therefore, the optimization problem can be expressed as:

Problem P:
Max (or min) $C(\mathbf{X})$
Subject to: $\mathbf{X} \in \mathbf{S}$

4.1 Steps of the Improved Algebraic Method (IAM)

Step 1. Convert all inequalities into equalities including any variable restricted in sign.

Step 2. Calculate the difference between the number of variables (n) and the number of equations (m).

Step 3. Determine the solution to all square systems of equations, i.e., basic solutions. The maximum number of systems of equations to be solved is: m!/[n!(m-n)!].

Step 4. Check feasibility of each basic solution obtained in Step 3 by using all the constraints and identify the basic feasible solutions (BFS).

The coordinates of vertices are the BFSs of the systems of equations obtained by setting some of the constraints at binding (i.e., equality) position. For a bounded feasible region, the number of vertices is at most m!/[n!(m-n)!] where m is the number of constraints and n is the number of variables. Therefore, a BS is obtained by taking any set of n equations then solving them, simultaneously. By plugging this BS in the constraints of other equations, one can check for feasibility of the BS. If it is feasible, then this solution is a BFS that provides the coordinates of a corner point of the feasible region.

4.2 Numerical Example For Solving System of Inequalities

We provide a feasible region problem; FR1 to illustrates the Improved Algebraic Method steps:

Feasible Region FR1:

$12X_1 + 10X_2 \geq 12000$ (1)
$10X_1 + 15X_2 \leq 15000$ (2)
$X_1 \geq 0$ (3)
$X_2 \geq 0$ (4)

Because of constraints 2, 3 and 4, are enclosing feasible region, therefore feasible region is bounded with countable vertices. There will be 4 equations and 2 variables yielding 6 possible combinations. The IAM provides six basic solutions for the system of linear inequalities as shown in Table 3.

There are 3 basic feasible solutions to the system of inequalities namely:

$A= (X_1, X_2) =(1000, 0),$
$B = (X_1, X_2) = (1500,0),$
$C = (X_1, X_2) =(375, 750).$

Table 3. The basic solutions of the numerical example FR1

Binding Constraints	Solution (X_1, X_2)	Feasible? Why?
1 and 2	(375, 750)	Yes, all satisfied
1 and 3	(0, 1200)	No, constraint 2 is not satisfied
1 and 4	(1000, 0)	Yes, all satisfied
2 and 3	(0, 1000)	No, constraint 1 is not satisfied
2 and 4	(1500, 0)	Yes, all satisfied
3 and 4	(0, 0)	No, constraint 1 is not satisfied

4.3 Algebraic Presentation of Feasible Region

Using the scalar parameters λ1, λ2, and λ3 for the 3 vertices, respectively, we obtain the following convex combination of vertices which is a parametric representation of the bounded feasible region:

$X_1 = 1000\lambda1 + 1500\lambda2 + 375\lambda3$

$X_2 = 750\lambda3$

for all parameters λ1, λ2, λ3 such that each is non-negative, and $\lambda1+ \lambda2 + \lambda3 = 1$.

Note that the above representation is valid if the feasible region is bounded. For example the feasible region $X_1 + X_2 \le 1$, $X_1 \ge 1$, $X_2 \ge 0$, is unbounded contains one vertex only.

Moreover if feasible region is bounded it must have more than one vertex. The feasible region of $X_1 + X_2 = 1$, $X_1 - X_2 = 1$, $X_1 \ge 0$, $X_2 \ge 0$, has feasible region with single point (1, 0). In the case of unbounded feasible region one must includes the needed rays, for numerical examples, see Arsham (2008).

Some Useful Comment on Algebraic Solution Method (IAM)

The main purpose of presenting the IAM is for the manager to understand, use, gets useful information, the major software use a refine version of IAM. Therefore, for large problems based on software become apparent. There are special cases when using IAM:

- If there is no feasible vertex, then the problem is infeasible, that is feasible region is empty.
- If the shadow price of the primal does not satisfy the constraints of the dual problem then the problem is unbounded.
- If there are two distinct optimal vertices, then the problem has multiple solutions. In this case the dual has degenerate optimal solution. The reverse is also true.

Notice that these results are not applicable to trivial cases, such as a problem with no vertices, such as: Maximize X1 + X2, subject to: X1+X2 = 1, all variable are unrestricted in sign. For more oddities in LP modeling, visit the website: The Tools for Modeling Validation Process: The Dark Side of LP http://home.ubalt.edu/ntsbarsh/opre640A/partv.htm

5. SOLVING LINEAR PROGRAMS

Consider the following linear program P1:

Problem P1:

Minimize $C(\mathbf{X}) = 18X_1 + 10X_2$

Subject to: $12X_1 + 10X_2 \ge 12000$

$10X_1 + 15X_2 \le 15000$

$X_1, X_2 \ge 0$

Substituting the parametric version of the feasible region into the objective function, we obtain:

$C(\lambda) = 18X_1 + 10X_2 = 18000\lambda1 + 27000\lambda2 + 14250\lambda3$

The optimal solution occurs when $\lambda3 = 1$ and all other λ_i's are set to 0, with a minimum value of 14250. The optimal solution is ($X_1 = 375$, $X_2 = 750$), the vertex corresponding to $\lambda3$.

Let the terms with the largest (smallest) coefficients in $C(\lambda)$ be denoted by λ_L and λ_S respectively. Then the following proposition holds:

Proposition 1: *The maximum (minimum) solution of an LP with a bounded feasible region correspond to the maximization (minimization) of the parametric objective function $C(\lambda)$. Moreover, since $C(\lambda)$ is a linear convex combination, the optimal value of $C(\lambda)$ is obtained by setting λ_L or λ_S equal to 1 and all other $\lambda_i = 0$.*

Proof: Proof follows from applying the IAM to

Max (Min) $C(\lambda)$
Subject to: $\sum \lambda_i = 1$, and all $\lambda_i \geq 0$.

The optimal solution can be found by enumerating all basic feasible solutions by applying the IAM method to this optimization problem. Since this feasible region is bounded it has a finite number of vertices, then this result suggests that optimum is associated with the BFS yielding the largest or smallest objective value, assuming the problem is of maximization or minimization type, respectively. The maximum and minimum vertex of an LP correspond to $\lambda_L = 1$ and $\lambda_S = 1$, respectively. As a result, since $C(\lambda)$ is a (linear) convex combination of its coefficients, the optimal solution of $C(\lambda)$ is obtained by setting λ_L or λ_S equal to 1 and all other $\lambda_i = 0$.

As an application of this proposition, we have the following useful result.

Fundamental Linear Programming Result: *if the optimal solution to an LP is bounded, then it occurs at one of vertices of the feasible region.*

From the above proposition, one might found the tight bound for the objective function, if it exists. For our numerical example we have the following tight bound:

$14250 \leq 18X_1 + 10X_2 \leq 27000$

5.1 Computing Slack and Surplus

Given that the right-hand-side (RHS) of a constraint is non-negative, the slack is the leftover amount of a resource ($\leq$) constraint, and surplus is the access over a requirement ($\geq$) constraint. These quantities represent the absolute values of the difference between the RHS value and the LHS (Left Hand Side) evaluated at an optimal point. Having obtaincd an optimal solution, one can compute the slack and surplus for each constraint at optimality. Equality constraints are always binding with zero slack/surplus.

Since this numerical example is a two-dimensional LP, one expects (at least) to have two binding constraints. The binding constraints at optimality are equations 1 and 2:

$12X_1 + 10X_2 = 12000$
$10X_1 + 15X_2 = 15000$

With surplus and slack value of zero, respectively, i.e., $S_1 = 0$ and $S_2 = 0$ which are the "Reduce Cost" of the dual problem.

6. CONSTRUCTION OF THE LARGEST SENSITIVITY REGIONS BASED ON OPTIMAL SOLUTION NOT "BASIS"

Suppose we obtained the optimal solution by any means, such as graphical (for two dimensional problems), simplex, dual simplex or interior methods. The following two sections construct the largest sensitivity regions based on the given optimal solution.

We develop a novel approach to post-optimality analysis for general LP problems given a unique non-degenerate optimal solution this approach provides a simple framework for the analysis of any single or simultaneous change of right hand side (RHS) or cost coefficients by solving the nominal LP problem with perturbed RHS terms. Assume that the LP is of size (mxn), i.e., it has n decision variables and m constraints including the non-negativity conditions (if any).

6.1 Computation of Shadow Prices

Step 1: Identify the n constraints that are binding at optimal solution. If there is more than n binding constraints, then the problem is degenerate. The degenerate case are dealt with in Section 10.

Step 2: Construct the parametric RHS of the constraints, excluding the non-negativity conditions.

Step 3: Solve the parametric system of equations consisting of the binding constraints. This provides the parametric optimal solution.

Step 4: Construct the simultaneous sensitivity analysis by plugging-in the solution obtained in Step 3, in all other remaining parametric constraints, including the non-negativity conditions (if any).

Step 5: Plug in the parametric optimal solution into the objective function. The coefficient for each parameter is the shadow price for that binding constraint. By definition, the shadow price for a non-binding constraint is always zero.

Following the above steps for our numerical example, one must first solve the following RHS parametric system of equations. By setting all inequalities in binding positions and solving the system of equations, we get the following parametric solution:

$$12X_1 + 10X_2 = 12000 + r_1$$
$$10X_1 + 15X_2 = 15000 + r_2$$

Solving this parametric system of equation for X1 and X2, we have:

$$X_1 = 375 + 3/16r_1 - 1/8r_2$$
$$X_2 = 750 - 1/8r_1 + 3/20r_2$$

The solution can be verified by substitution. For the higher dimension LP, the parametric solution can be obtained by using the JavaScript:

http://www.mirrorservice.org/sites/home.ub-alt.edu/ntsbarsh/Business-stat/otherapplets/PaRHSSyEqu.htm

For large-scale problems one may use symbolic software such as Maple, Matlab, or Mathematica.

The parametric optimal solution shows the impact of changes on the RHS on the optimal decision variables. For example, if the RHS of constraint 1 increases the quantity of X_1 will increase and X_2 decreases. This is often a concept misunderstood or overlooked by managers.

Plugging the parametric solution into objective function, we have:

$$18X_1 + 10X_2 = 14250 + 17/8r_1 - 3/4r_2$$

The shadow prices are the coefficients of the parametric optimal function, i.e., $U_1 = 17/8$ and $U_2 = -3/4$, for the RHS of constraints 1 and 2, respectively. They are the *rate of change* in optimal value with respect to changes in the RHS of each constraint.

6.3 Construction of the RHS Sensitivity Region: Maintaining the Validity of Current Shadow Prices

Notice that the parametric objective function $14250 + 17/8r_1 - 3/4r_2$ is valid when the parametric solution satisfies all other (i.e., non-binding) constraints. In the current numerical example, the other constraints are:

$X_1 \geq 0$ and $X_2 \geq 0$.

These produce the following conditions:

$375 + 3/16r_1 - 1/8r_2 \geq 0$, and $750 - 1/8r_1 + 3/20r_2 \geq 0$.

Notice that this is the largest sensitivity region for the RHS of the binding constraints that allows for simultaneous, dependent/independent changes. This convex region is non-empty since it always contains the origin ($r_1 = 0$, $r_2 = 0$), with a vertex at ($r_1 = -12000$, $r_2 = -15000$) where both right-hand sides binding constraints.

Arsham (1990) gives detail treatments of the most popular sensitivity analysis, including the 100% rule, and the tolerance analysis.

6.4 Sensitivity Range for Each RHS of the Binding Constraints

The above set of inequality sensitivity region can be used to find the ordinary sensitivity range (one change-at a time) for the RHS values of the constraints. The range for the RHS of the first constraint (RHS_1) can be obtained by setting $r_2 = 0$ in the inequalities set (2). This provides $r_1 \leq 6000$ and $r_1 \geq -2000$. Therefore, the allowable increase and decrease in the original value 12000 for RHS_1 are 6000 and 2000, respectively, i.e., the current set of shadow prices remain valid as long as:

$10000 \leq RHS_1 \leq 18000$ and $RHS_2 = 15000$.

Similarly, the range for the RHS of the second constraint (RHS_2) can be obtained by setting $r_1 = 0$ in the inequality set (2). This implies that $r_2 \leq 3000$ and $r_2 \geq -5000$. Therefore the allowable increase and decrease in the original value 15000 for RHS_2 are 3000 and 5000, respectively, i.e., the current set of shadow prices remain valid as long as:

$RHS_1 = 12000$ and $10000 \leq RHS_2 \leq 18000$.

6.5 Sensitivity Range for the Non-Binding Constraints

In the numerical example P1, the non-binding constraints are the non-negativity conditions which are not subject to sensitivity analysis. Consequently, for illustration purposes let us pretend "as if" there is a non-binding constraint $X_1 + 100X_2 \geq 20000$. To find the sensitivity range (one-change-at-a-time), construct the parametric form and plug in the optimal solution. We get

$X_1 + 100X_2 \geq 20000 + r_3$
$375 + 100\,(750) = 75375 \geq 20000 + r_3$
$r_3 \leq 55375$

The amount of increase is 55375 and decreasing amount is unlimited; i.e., ($RHS_3 \leq 75375$).

The following proposition formalizes the shifting of parametric non-binding constraints.

Proposition 2: *For any given point $X^o = (X_1^o, X_2^o, \ldots\ldots, X_n^o)$ the parameter r value for any resource/production constraint is proportional to the ordinary distance between the point X^o and the hyper-plane of the constraint.*

Proof: Proof follows from the fact that the distance from point $X^o = (X_1^o, X_2^o, \ldots\ldots, X_n^o)$ to any non-binding constraint, i.e.

$$a_1 X_1^o + a_2 X_2^o + \ldots.. + a_n X_n^o = b + r$$

is

Absolute $[a_1 X_1^o + a_2 X_2^o + ... + a_n X_n^o - b - r] / (a_1^2 + a_2^2 + ... + a_n^2)^{1/2}$

This reduces to:

Absolute $r / (a_1^2 + a_2^2 + ... + a_n^2)^{1/2}$.

Therefore the parameter r value is proportional to the distance with a constant proportionality $1 / (a_1^2 + a_2^2 + ... + a_n^2)^{1/2}$. This is independent of point X^o. In the above example, X^o is the optimal vertex. This completes the proof.

6.6. Behavior of Changes in the RHS Values of the Optimal Value

To study the directional changes in the optimal value with respect to changes in the RHS (with no redundant constraints present, and all RHS ≥ 0), we distinguish the following two cases:

Case I: Maximization problem

For ≤ constraint: The change is in the same direction. That is, increasing the value of RHS does not decrease the optimal value. It increases or remains the same depending on whether the constraint is a binding or non-binding constraint.

For ≥ constraint: The change is in the reverse direction. That is, increasing the value of RHS does not increase the optimal value. It decreases or remains the same depending on whether the constraint is a binding or non-binding constraint.

For = constraint: The change could be in either direction (see the More-for-less section).

Case II: Minimization problem

For ≥ type constraint: The change is in the reverse direction. That is, increasing the value of RHS does not increase the optimal value (rather, it decreases or has no change depending on whether the constraint is a binding or non-binding constraint).

For ≤ type constraint: The change is in the same direction. That is, increasing the value of RHS does not decrease the optimal value (rather, increases or has no change depending on whether the constraint is a binding or non-binding constraint).

For = constraint: The change could be in either direction (see the More-for-less section).

7. GENERAL SENSITIVITY REGION FOR COST COEFFICIENTS: MAINTAINING THE VALIDITY OF CURRENT OPTIMAL VERTEX NOT "BASIS"

Knowing the unique optimal solution for the dual problem by using any LP solver, one may construct the simultaneous sensitivity analysis for all coefficients of the objective function of the primal problem as follows. Assume that the dual problem has n decision variables and m constraints, including the non-negativity conditions (if any).

7.1 Steps in Finding Sensitivity Region for the Objective Function Coefficients

Step 0: Construct and solve the dual problem.

Step 1: Identify the *n* constraints that are binding at optimal solution (that is, at the shadow prices of the primal problem). If there are more than *n* constraints binding, then the primal problem may have multiple solutions.

Step 2: Construct the parametric RHS of the constraints, excluding the non-negativity conditions.

Step 3: Solve the parametric system of equations consisting of the binding constraints. This provides the parametric optimal solution.

Step 4: Construct simultaneous sensitivity region by plugging in the parametric solu-

Table 4.

Primal Problem P1:	Classifications
Minimize $C(X) = 18X_1 + 10X_2$	Minimization Problem
Subject to: $12X_1 + 10X_2 \geq 12000$	**Sensible Constraint**
$10X_1 + 15X_2 \leq 15000$	**Unusual Constraint**
$X_1 \geq 0$	*Sensible Condition*
$X_2 \geq 0$	*Sensible Condition*
The Dual Problem of Problem P1 is:	
Maximize $12000U_1 + 15000U_2$	Maximization Problem
Subject to: $12U_1 + 10U_2 \leq 18$	*Sensible Constraint*
$10U_1 + 15U_2 \leq 10$	*Sensible Constraint*
$U_1 \geq 0$	**Sensible Condition**
$U_2 \leq 0$	**Unusual Condition**

tion obtained in Step 3, in all non-binding constraints, including the sign conditions such as non-negativity conditions (if any).

We demonstrate the method in our example. The dual problem of our numerical example is constructed in Table 4.

The components of the optimal solution to the dual problem are the shadow prices to the original (primal) LP model, and vise versa. The shadow prices of the primal, i.e., ($U_2 = 17/8$, $U_2 = -3/4$), as calculated earlier (or one can solve it afresh without referring to previous section). The optimal value for the dual $12000(17/8) + 15000(-3/4) = 14250$ that is equal to the optimal value of the primal problem, as always expected. The property that let the primal and the dual optimal values to be equal is known as strong duality, i.e., there is no gape between the dual and primal problem. The slacks of the first two constraints are $12(17/8) + 10(-3/4) - 18 = 0$, $10(17/8) + 15(-3/4) - 10 = 0$, which are the "Reduce Cost" of the primal problem.

To find the ranges for the objective function coefficients, one may use the RHS parametric version of the dual problem;

The parametric presentations of the RHS of the binding constraints are as follows:

$$12U_1 + 10U_2 = 18 + c_1$$
$$10U_1 + 15U_2 = 10 + c_2,$$

Solving these parametric equations we obtain

$$U_1 = 17/8 + 3/16c_1 - 1/8c_2$$
$$U_2 = -3/4 - 1/8c_1 + 3/20c_2$$

Plugging this solution into objective function of the dual problem, we obtain the parametric objective function is $14250 + 375c_1 + 750c_2$ with optimal value of 14250, same as for the optimal value of the primal problem, as expected. Again, this parametric optimal solution is subject to satisfying the unused constraints; namely, $U_1 \geq 0$ and $U_2 \leq 0$. These produce the following largest sensitivity region for the objective function coefficients of the primal problem P1, simultaneously:

$$3/16c_1 - 1/8c_2 \geq -17/8 \text{ and } -1/8c_1 + 3/20c_2 \leq 3/4.$$

Notice that this is the largest sensitivity region for cost coefficients, allowing for simultaneous, dependent/independent changes. This convex region is non-empty because it contains the origin ($c_1 = 0$, $c_2 = 0$), with a vertex at ($c_1 = -18$, $c_2 = -10$) which is the negative of the both cost coefficients.

The ordinary sensitivity range (one change at a time) of the coefficient of first decision variable X_1, currently at 18, can be found by setting $c_2 = 0$ in the above sensitivity region inequality set, yielding $c_1 \geq -6$ and $c_1 \geq -34/3$. Therefore the allowable decrease for the coefficient of X_1 is 6 with unbounded (i.e., Big-M) allowable increase. Similarly, the sensitivity range of the coefficient of second decision variable X_2, currently at 10, can be found by setting $c_1 = 0$ in the above sensitivity region inequality set, yielding $c_2 \leq 5$ and $c_2 \leq 17$. Hence, $c_2 \leq 5$, therefore, the allowable increase for the coefficient of X_1 is 5 with unbounded allowable decrease, (i.e., - Big-M).

Figure 4. Problem P3 feasible region

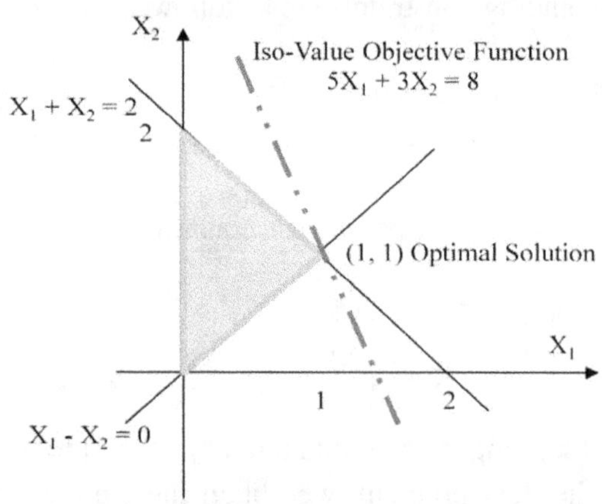

7.2 Finding the Cost Sensitivity Range by the Graphical Method Could Be Misleading

It is a commonly held belief that one can compute the cost sensitivity range by bracketing the slope of the (iso-value) objective function by the slopes of the two lines resulting from the binding constraints. This graphical slope-based method to compute the sensitivity ranges is described in popular textbooks, such as Anderson et al., (2007), Lawrence and Pasternack (2002), and Taylor (2010). However, this approach is only valid when the objective function and the two binding constraints have all slopes with the same sign, that is, all three functions have negative slopes or all three functions have positive slopes. The approach fails if one of the binding constraints has a non-negativity condition or the binding constraints have opposite signs. The following counterexample illustrating the later case.

Problem P3:

Maximize $5X_1 + 3X_2$

Subject to: $X_1 + X_2 \le 2$

$X_1 - X_2 \le 0$

$X_1 \ge 0, X_2 \ge 0$

The optimal solution, $(X_1, X_2) = (1, 1)$, is determined by the two main constraints. The iso-value objective function line and first constraint boundary line have negative slopes, while the other constraint has positive slope. Bracketing the objective function slope between the slopes of the two constraints gives $-1 \le -C_1/C_2 \le 1$, which is an incorrect range. Notice that, not even the current objective function slope -5/3 belongs to this range. The main reason is that in this problem to transition the slope from negative to positive it has to go through being very large and negative, then undefined, and then very large and positive, as opposed to go through zero. Then, the correct ranges are $C_1 \ge 3$, and $-5 \le C_2 \le 5$, respectively, found by equating the ratio of coefficients from the objective function plus and from each of the constraints as follow:

$(5+c_1)/1 = 3/1$ gives $c_1 = -2$

$(5+c_1)/1 = 3/-1$ gives $c_1 = -8$

where c_1 represents the largest possible change in the coefficient. A positive number represents the largest possible increase, while a negative number represents the largest possible decrease.

Therefore, since $c_1 = -2$ is more restrictive than $c_1 = -8$, you may decrease $C_1 = 5$, by 2. There is no lower limit, thus $C_1 \ge 3$.

On the other hand, for C_2 we obtain an upper and lower bound

$(3+c_2)/1 = 5/1$ gives $c_2 = 2$

$(3+c_2)/-1 = 5/1$ gives $c_2 = -8$

Therefore you may decrease $C_2 = 3$, by 8, $-5 \le C_2$, and you may increase $C_2 = 3$, by 2, i.e. $C_2 \le 5$.

7.3 Shadow Prices as the Lagrangian Multipliers

The dual solution provides important economical interpretation such as the marginal values of the RHS elements. The elements of the dual solution are known as the Lagrangian multipliers because they provide (a tight) bound on the optimal value of the primal, and vise versa. For example, considering our numerical example the dual solution can be used to find a lower tight bound for the optimal value, as follow: Multiply each constraint by its corresponding dual solution and then add them up, we get:

$$17/8\ [12X_1 + 10X_2 \geq 12000]$$
$$-3/4\ [10X_1 + 15X_2 \leq 15000]$$

$$18X_1 + 10X_2 \geq 14250$$

Notice that the resultant on the left side is the objective function of the primal problem, and this lower bound is a tight one, since the optimal value is 14250. In other words, the objective function lower bound is 14250, which is the optimal value for the minimization of the objective function $18X_1 + 10X_2$. Moreover, the optimal value for both problems, dual and primal is always the same. This fact is referred to as equilibrium of economical systems, and efficiency in Pareto's sense between the Primal and the Dual Problems. Therefore, there is no duality gap in linear programming.

7.4 Misinterpretation of the Shadow Price

The Shadow Price tells us how much the objective function will change if we change the right-hand side of the corresponding constraint. This is often called the "marginal value", "dual prices" or "dual value", for the constraint. Therefore, the shadow price may not be same as the "Market price, i.e., real price".

For each RHS constraint, the Shadow Price tells us exactly how much the objective function will change if we change the RHS of the corresponding constraint within the limits given in the sensitivity range on the RHS's. Therefore, for each RHS value, the shadow price is the rate of change in the optimal value caused by any allowable increase or decrease in the RHS.

Unfortunately, there are misconceptions regarding the definition of the shadow price. One such misinterpretation is, "In linear programming problems the shadow price of a constraint is the difference between the optimized value of the objective function and the value of the objective function, evaluated at the optional basis, when the right hand side (RHS) of a constraint is increased by one unit." "Shadow Prices: The shadow prices for a Linear Programming problem are the solutions to its dual. The ith shadow price is the change in the objective function resulting from a one-unit increase in the ith coordinate of b. A shadow price is also the amount that an investor would have to pay for one unit of a resource in order to buy out the manufacturer."

Consider the following LP as a counterexample:

Problem P3:

Maximize X_2

Subject to: $X_1 + X_2 \leq 2$

$2.5X_1 + 4X_2 \leq 10$

$X_1 \geq 0, X_2 \geq 0$

This problem attains its optimal solution at (0, 2) with an optimal value of 2. Suppose we wish to compute the shadow price of the first resource that is the RHS of the first constraint. Changing the RHS of the first constraint by increasing it by one unit results in:

Maximize X2

Subject to: $X1 + X2 \leq 3$

$2.5X1 + 4X2 \leq 10$

$X_1 \geq 0, X_2 \geq 0$

The new problem has the optimal solution (0, 2.5) with an optimal value of 2.5.

Therefore, it seems "as-if" the shadow price for this resource is 2.5 - 2 = 0.5. In fact the shadow price for this resource is 1, which can be found by constructing and solving the dual problem.

The reason for this error becomes evident if we notice that the allowable increase to maintain the validity of the shadow price of the first resource is 0.5. The increase by 1 is beyond the allowable change on the first RHS value.

Now suppose we change the same RHS value by, increasing say, by 0.1, which is permissible (compared with its nominal value of 2), then the optimal value for the new problem is 2.1. Therefore the shadow price is (2.1 -2) / 0.1 = 1. We must be careful when calculating shadow prices using this approach the change must be within allowable limits.

If you wish to compute the shadow price of a RHS when its sensitivity range is not available, you may obtain the optimal values for at least two perturbations. If the rate of change for both cases gives you the same values, then this rate is indeed the shadow price. As an example, suppose we perturb the RHS of the first constraint by +0.02 and -0.01. Resolving the problem after these changes using your LP solver, the optimal values are 2.02, and 1.09, respectively. Since the optimal value for the nominal problem (without any perturbation) is equal to 2, the rate of change for the two cases are: (2.02 - 2)/0.02 = 1, and (1.09 - 2)/(-0.01) = 1, respectively. Since these two rates are the same, we conclude that the shadow price for the RHS of the first constraint is indeed equal to 1.

7.5 Managerial Round-Off Errors

You must be careful whenever you round the value of shadow prices. For example, the shadow price of the resource constraint in the above problem is 8/3; therefore, if you wish to buy more of this resource, you should not pay additional price more than \$2.66. Whenever you want to sell any unit of this resource, you should not sell it at an additional price below \$2.67.

A similar error might occur whenever you round the limits on the sensitivity ranges. One must be careful because the upper limit and lower limit must be rounded down and up, respectively.

8. DEALING WITH NON-STANDARD FORM LP PROBLEMS

Creating the Non-Negativity Conditions

By default, most LP software such as Management Scientist (1999), and QM for Windows (2003), assume that all variables are non-negative.

To achieve this requirement, convert any unrestricted variable X_j to two non-negative variables by substituting y - X_j for every X_j. This increases the dimensionality of the problem by only one (introduce one y variable) regardless of how many variables are unrestricted.

If any X_j variable is restricted to be non-positive, substitute - X_j for every X_j. This reduces the complexity of the problem.

Solve the converted problem and then substitute these changes back to get the values for the original variables and optimal value is straightforward. However, to convert sensitivity analyses including the shadow prices are not easy tasks. Consider the following LP problem:

Problem P4:
Maximize $X_1 + 0X_2$
Subject To:

$$X_1 + X_2 \geq 0$$
$$2X_1 + X_2 \leq 2$$
$$X_1 \geq 0$$
$$X_2 \leq 0$$

Following the IAM solution algorithm, optimal solution is at vertex $X_1 = 2$, $X_2 = -2$, which is depicted in Figure 5.

Performing the sensitivity analysis based on optimal vertex to maintaining the optimal vertex, we construct the following parametric RHS of binding constraints:

Figure 5. A typical feasible region for a non-standard form LP

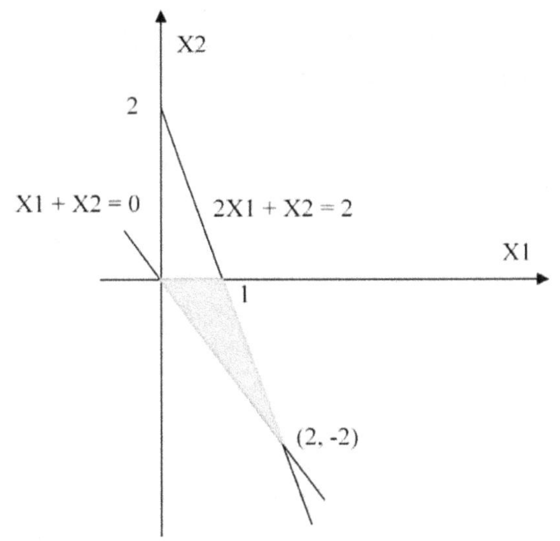

Figure 6. The largest sensitivity region for the RHS of first and second constraints

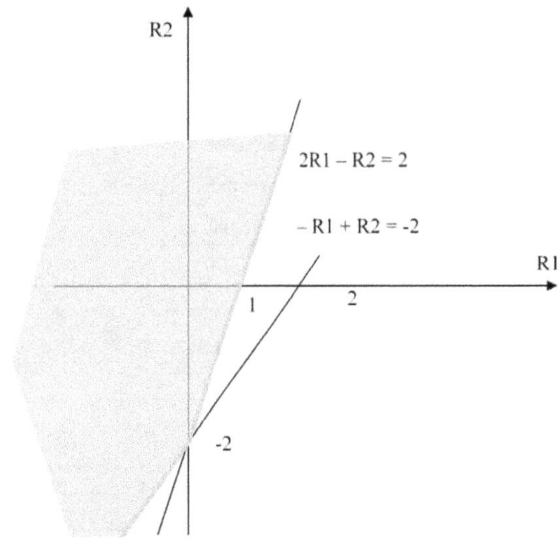

$X_1 + X_2 = 0 + R_1$
$2X_1 + X_2 = 2 + R_2$

The parametric optimal solution is:

$X_1 = 2 - R_1 + R_2$
$X_2 = -2 + 2R_1 - R_2$

We plug the parametric solution into objective function, we have:

$X_1 = 2 - R_1 + R_2$

The shadow prices are the coefficients of the parametric optimal function, i.e., $U_1 = -1$ and $U_2 = 1$, for the RHS of constraints 1 and 2, respectively. They are the *rate of change* in optimal value with respect to changes in the RHS of each constraint.

The shadow prices remain valid as long as this parametric optimal solution satisfy all the other constraints, i.e., $X_1 \geq 0$ and $X_2 \leq 0$. This gives the largest sensitivity region for the RHS.

$2 - R_1 + R_2 \geq 0$
$-2 + 2R_1 - R_2 \leq 0$

The sensitivity region is convex and non-empty, containing the origin (0, 0) and its vertex is at (0, -2), as shown in Figure 4.

Figure 6 presents the largest sensitivity region for the RHS for which the shadow prices remain unchanged, i.e., the optimal dual problem vertex remains optimal.

The ordinary sensitivity range is $R_1 \leq 1$ for the first RHS and $R_2 \geq -2$ for the second, as can be verified from the largest sensitivity region for the RHS depicted in Figure 4.

We now perform sensitivity analysis on the objective function coefficient. The dual problem is constructed in Table 5.

The optimal solution is the shadow prices ($U_1 = -1$, $U_2 = 1$), with optimal value of $0U_1 + 2U_2 = 2$, same as primal, as expected.

$U_1 + 2U_2 = 1 + C_1$
$U_1 + U_2 = 0 + C_2$

The dual parametric solution is:

$U_1 = -1 - C_1 + 2C_2$
$U_2 = 1 + C_1 - C_2$

Table 5.

The Primal Problem:	Classifications
Maximize $X_1 + 0X_2$	Maximization
Subject To:	
$X_1 + X_2 \geq 0$	Unusual
$2X_1 + X_2 \leq 2$	Sensible
$X_1 \geq 0$	Sensible
$X_2 \leq 0$	Unusual
Therefore the dual formulation is:	
The Dual Problem:	
Minimize $0U_1 + 2U_2$	Minimization
Subject To:	
$U_1 + 2U_2 \geq 1$	Sensible
$U_1 + U_2 \leq 0$	Unusual
$U_1 \leq 0$	Unusual
$U_2 \geq 0$	Sensible

The solution must satisfy the other constraints:

$$-1 - C_1 + 2C_2 \leq 0$$
$$1 + C_1 - C_2 \geq 0$$

Figure 7 depicts the largest sensitivity region for the coefficients of objective function, i.e., the optimal solution remains unchanged, i.e., the optimal vertex for the primal problem remains optimal.

Figure 7. The largest sensitivity region for the coefficients of objective function

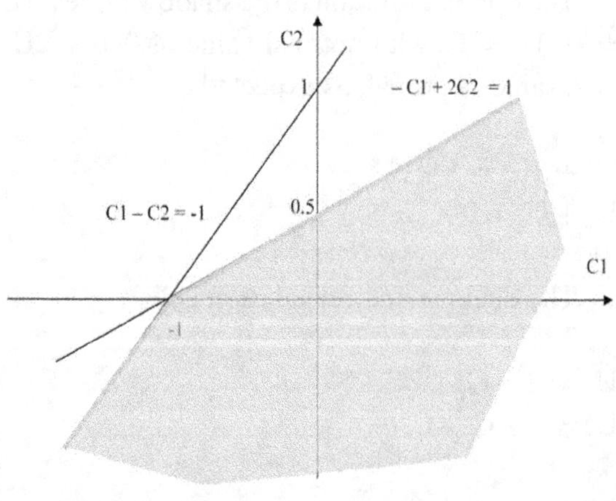

One Change at a Time (i.e., Ordinary Sensitivity Range)

Set $C_2 = 0$, then $C_1 \geq -1$, that is, the allowable decrease is -1, and allowable increase is big-M (i.e., a large nonspecific positive number). Set $C_1 = 0$, Then $C_2 \leq 1/2$, that is, the allowable increase is 0.5, and allowable decrease is -big-M.

The parametric objective function is:

$$0U_1 + 2U_2 = 2 + 2C_1 - 2C_2$$

With shadow prices ($X_1 = 2$, $X_2 = -2$) which is solution to the primal with optimal value of $X_1 + 0X_2 = 2$, as expected.

9. SOFTWARE PACKAGES SHADOW PRICE AND SENSITIVITY ANALYSIS REPORTS

In the previous Sections we solved manually standard form problem P1 and the non-standard form problem P2 and then compared the results with the results obtained by some of the LP packages available in the market.

Unfortunately several of these educational LP software packages give misleading and incomplete results that could increase complexity for the manager. For instance, the widely used LINDO (2003) LP software does not provide any direct warnings about the existence of multiple nor degenerate optimal solutions. Or the other hand WinQSB (2003) might indicates "Note: Alternate Solution Exists!!" when in fact there might be none. As a result, the final report on sensitivity analysis might not be valid in these cases, Lin (2010). Similarly, several other LP software packages have their own limitations listed in Tables 6 and 7.

Table 6. Scope and limitations of some popular software

	Non-Negativity Conditions	Shadow Prices	Optimal Solution to non-standard form problems	Sensitivity Ranges
Lindo (2003)	Imposed on all variables by default, but can be changed to un-restricted	Non-consistent	May need conversion	May need conversion which is not an easy task
WinQSB (2003)	A variable must be either non-negative or unrestricted.	Consistent	May need conversion	Unreliable
QM for Windows (2003)	All variables must be non-negative	Non-consistent	May need conversion	Difficult to Convert
Management Scientist (1999)	All variables must be non-negative	Non-consistent	May need conversion	May need conversion which is not an easy task
Excel Solver (20013)	All variables are non-negative or un-restricted	Consistent	Readily Available	Reliable

Table 7. The standard form numerical example P1 summary results as a base for comparison with popular software

Variables Name	Optimal Solution	Objective Coefficient	Allowable Increase	Allowable Decrease
X1	375.0	18	M	6
X2	750.0	10	5	M
	Shadow	Slack	Allowable	Allowable
Constraints	Price	Surplus	Increase	Decrease
12X1 + 10X2 >= 12000	2.125	0	6000	2000
10X1 + 15X2 <= 15000	- 0.75	0	3000	5000

LINDO produces the output displayed in Listing 1 for the first numerical example P1, which is in standard form.

In this LINDO final report the dual prices are reported to be U_1 = - 2.125, and U_2 = 0.75, while we found earlier that the shadow prices are U_1 = 2.125, and U_2 = - 0.75. Even the term Dual Prices (i.e., Dual solution, are the shadow Prices) is confusing as the values U_1 = - 2.125, and U_2 = 0.75 are not a solution to the dual problem.

This misleading result is not limited to LINDO, unfortunately, at least two other LP software packages namely Quantitative Methods (QM 2003) and Management Scientist (1999) produce misleading results.

9.1 Shadow Prices Might Have Different Sign

When reading and interpreting Shadow Prices from software packages, managers must be especially attentive to the software description of the analysis.

There are no common accepted practices in Sensitivity Reports. For example, LINDO, Excel Solver and GAMS (2013) use different notation in their reports. In general, the terms shadow prices and dual prices are used interchangeable; however, if we compare the reports by LINDO and by hand-computation of numerical example P1,

Listing 1. Input and output of LINDO software package for the first numerical example

```
LINDO INPUT and OUTPUT

Minimize 18X1 +10X2
Subject to 12X1 + 10X2 >12000
10X1 +15X2 < 15000
End

LP OPTIMUM FOUND AT STEP 2
        OBJECTIVE FUNCTION VALUE
      1) 14250.0
VARIABLE        VALUE    REDUCED COST
   X1          375.0        0.0
   X2          750.0        0.0
ROW    SLACK OR SURPLUS    DUAL PRICES
2)          0.0            -2.125
3)          0.0             0.75
NO. ITERATIONS= 2
RANGES IN WHICH THE BASIS IS UNCHANGED:

                          OBJ COEFFICIENT RANGES
      VARIABLE          CURRENT COEF    ALLOWABLE INCREASE  ALLOWABLE DECREASE
         X1                18.0              INFINITY              6.0
         X2                10.0                 5.0              INFINITY

                          RIGHTHAND SIDE RANGES
         ROW            CURRENT RHS     ALLOWABLE INCREASE  ALLOWABLE DECREASE
          2               12000.0             6000.0              2000.0
          3               15000.0             3000.0              5000.0
```

we see that the dual prices from LINDO have the opposite sign of the shadow prices from Solver.

The Lindo (2013) Manual gives the following justification and interpretation:

...improve is a relative term. In a minimization problem, such as our example, interpreting the dual price requires some thought. The dual price ..., is ....-. This means raising the right-hand side of the... constraint by one unit would cause the objective to "improve" by negative 100. That is, it would increase ... In other words, the marginal cost of providing one additional ...in additional salary expense." "Dual prices are sometimes called shadow prices, because they tell you how much you should be willing to pay for additional units of a resource. Going back to the Friday constraint, if we use the shadow price interpretation of the dual price, we can also say that we should be willing to pay .... to have an additional... If ...available for less than .., then we might want to consider this option." "As with reduced costs, dual prices are valid only over a limited range.

Notice that unfortunately Excel Manual (2013) also uses the terms like "unit change", "unit increase" while defining shadow prices.

The dual value for a constraint is nonzero only when the constraint is equal to its bound. This is called a binding constraint, and its value was driven to the bound during the optimization process. Moving the constraint left hand side's value away from the bound will worsen the objective function's value; conversely, "loosening" the bound will improve the objective. The dual value measures the increase in the objective function's value per unit increase in the constraint's bound..... In the case of linear problems, the dual values remain constant over a range

Notice that the change by one unit may not be within the range, see sub-section 7.4.

Just to add to the confusion, other packages such as GAMS report marginal's without any information about their range of validity. (see Listing 2)

Subtle discrepancies like this one are unfortunate. There is a pounding need for consensus on the vocabulary and structure on the output of optimization software packages.

Using Excel LP Solver produces the output shown in Listing 3 for the non-standard form numerical example. (see Table 8)

The results are comparable with our hand computation. However one must be careful in any generalization, since this comparison is done on a specific numerical example.

WinQSB Software produces the following outputs for the non-standard form numerical example. The results with all constraints in explicit form, and then selecting the unrestricted option are shown in Listing 4.

The sensitivity ranges are not correct. They are differed from the results already obtained by hand computations.

9.2 Poor Scaling and Unstable Software Solutions

When modeling a business problem, different conditions might require us to use parameters or inputs having very different order of magnitudes. For example, an LP from manufacturing industry

Listing 2. Excerpt from GAMS output for the first numerical example that is in standard form

```
                  LOWER        LEVEL        UPPER       MARGINAL
EQU constr1      12000.0      12000.0       +INF         2.1250
EQU constr2       -INF        15000.0      15000.0      -0.750
```

Table 8. Non-standard form numerical example P2 summary results

Variables Name	Optimal Solution	Objective Coefficient	Allowable Increase	Allowable Decrease
X1	2	1	M	1
X2	-2	0	0.5	M
Constraints Name	**Shadow Price**	**Slack Surplus**	**Allowable Increase**	**Allowable Decrease**
Constraint 1: X1 + X2 >= 0	-1	0	1	M
Constraint 2: 2X1 + X2 <= 2	1	0	M	2

Listing 3. Excel result with all constraint in explicit form, and then selecting the unrestricted option

Name	Final Value	Reduced Cost	Objective Coefficient	Allowable Increase	Allowable Decrease
X1	2	0	1	1E+30	1
X2	-2	0	0	0.5	1E+30
	Final	Shadow	Constraint	Allowable	Allowable
Name	Value	Price	R.H. Side	Increase	Decrease
Constraint 3: X1 >= 0	2	0	0	2	1E+30
Constraint 4: X2 <= 0	-2	0	0	1E+30	2
Constraint 1: X1 + X2 >= 0	0	-1	0	1	1E+30
Constraint 2: 2X1 + X2 <= 0	2	1	2	1E+30	2

Listing 4. WinQSB Input and Its Combined report

LP Input in Matrix Format

Variable	X1	X2	Direction	R.H.S.
Maximize	1	0		
Constraint 1	1	1	>=	0
Constraint 2	2	1	<=	2
Constraint 3	1	0	>=	0
Constraint 4	0	1	<=	0
Lower Bound	-M	-M		
Upper Bound	M	M		

Variable Type X1 Unrestricted, X2 Unrestricted

Decision Variable	Solution Value	Unit Profit Contribution	Allowable Min. c(j)	Allowable Max. c(j)
X1	2	1	0	1
X2	-2	0	-M	0

Objective Function (Max.) = 2.0

(Note: Alternate Solution Exists!!)

Constraint	Shadow Price	Allowable Min. RHS	Allowable Max. RHS
Constraint 1	-1	-M	1
Constraint 2	1	0	M
Constraint 3	2	-M	2
Constraint 4	2	-2	M

might include dollar amounts in millions and at the same time might include quality based conditions in the range of nanometers (one billionth of a meter). If an LP is modeled using the such units then the solvers such as Lindo and Excel gives us warning about poor scaling but then also gives some updated values. Frontline Solvers (the developers of Excel Solvers) admit to this limitation and the website reports that:

The effects of poor scaling in a large, complex optimization model can be among the most difficult problems to identify and resolve. It can cause Solver to return messages such as 'Solver could not find a feasible solution', 'Solver could not improve the current solution,' or even 'The linearity conditions required by this LP Solver are not satisfied,' with results that are suboptimal or otherwise very different from your expectations. (Frontline Solvers, 2013)

Unfortunately, Solver does not always alert the user about any issues with the problem. The following example clearly is prone to scaling issues with a very large coefficient for X_3 in the third constraint. (Depending on the computer a larger or smaller coefficient yields the issue describe here). However, when the problem is attempted with Solver, Solver sometimes yields an error message as the ones described earlier, but in most cases, it does not give any error message and reports as optimal an infeasible solution.

Numerical Example:

Minimize X_2

Subject to:

$10X_1 - X_2 \leq 0$

$X_1 = 1$

$X_1 - 100000$

$X_3 \leq 0 \ X_3 \leq 1$

X_3 is positive integer

Solver might report as the optimal solution $X_1 = 1$, $X_2 = 10$, and $X_3 = 0$. This solution violates constraint (3). The correct solution is $X_1 = 1$, $X_2 = 10$, and $X_3 = 1$. Therefore, we have numerical instability caused by too large values.

Excel recommends that to avoid such occurrences a model must only be allowed to vary by an order of magnitude of 3 (i.e. 1 to 1000). Excel allows the use of "Use Automatic Scaling" feature for automatically rescaling the values but it cannot rescale the problems "intelligently". Hence if the automatic rescaling feature is used the outcomes could unknowingly become worse rather than better. Automatic rescaling can also slowly down the optimization process significantly (Winston and Albright, 2011). Similar limitations exist for Lindo also (Martin 1999) and may exist for several other solvers as well. However these limitations are largely and conveniently ignored.

One of the possible reasons for such ignorance could be due to the high expectations and the promise of convenience offered by LPs. Using LPs, a manager can get an optimal solution even when the problem includes a huge variety of factors (inputs and constraints) which could be coming from a wide variety of domains. For large and complex LPs, the managers and the coders would want to reduce the complexity associated with LPs and would prefer not to change the units for the sake of convenience. For example when developing LPs for manufacturing sector a manager (and even the coders) might prefer to use dollars values directly in the model instead of converting them to millions (which reduces the order of magnitude by six but then increase the chance of getting optimal dollar values in decimals) and use it with values of quality tolerance in a fraction of millimeter. Such instances are common and widespread. The authors lament that majority of books violates the ideal requirement of order of magnitude in the discussions and leaves the mangers very much vulnerable to being unknowingly dependent on sub-optimal results.

The best approach recommended to deal with this situation is to use consistent scaling at the time of modeling. Managers should be aware of this issue and use consistent units for products and constraints. For example, if working with multiple monetary constraints they all should be scale in hundreds or thousands. Alternative if working with multiple products similar units should be used for all the products.

10. MAINTAINING A PRIMAL DEGENERATE OPTIMAL VERTEX OPTIMAL

Almost all research work in LP literature starts with an assumption of a non-degenerate primal/dual LP. Degeneracy is the case of having multiple vertices at the same point.

The necessary condition for the existence of LP degeneracy: If at optimal point the number of binding constraints is more than n (i.e., the dimension of the problem) then the solution to an LP might be degenerate.

Resolution: Double-check the coefficients of all constraints including the RHS values. There could have been rounding error.

10.1 False Degenerate Optimal Solution

Consider the following n=2 dimensional LP problem:

Numerical Example P2:
Maximize X_2
Subject to:

$$X_1 + X_2 = 5$$
$$-X_1 + X_2 \leq 1$$
$$X_1 \geq 0, X_2 \geq 0$$

This problem has a unique, non-degenerate optimal solution at ($X_1 = 2, X_2 = 3$). However, if one rewrites the equality constraint in the form of two inequalities, the equivalent problem is:

Maximize X2
Subject to:

$$X_1 + X_2 \leq 5$$
$$X_1 + X_2 \geq 5$$
$$-X_1 + X_2 \leq 1$$
$$X_1 \geq 0, X_2 \geq 0$$

Clearly, the optimal solution (2, 3) now makes three constraints binding. Three is more than the dimension n=2 of this problem; therefore one may incorrectly conclude that this solution is a degenerate optimal solution. This simple numerical example serves the purpose of the necessary (but not a sufficient) condition for a degenerate optimal solution.

An occurrence of degeneracy can have a significant impact on the shadow prices and may cause oddities in the output provided by LP solvers. Moreover, the usual sensitivity analyses do not provide complete information in the degenerate case; that is, the information one obtains from most LP packages is a subset of the true sensitivity intervals. There are more effective approaches to cope with this problem; however, they are computationally much more involved, Lin (2010). Almost all software packages do not alert the user that the problem is degenerate.

Numerical Example P4:
Maximize $C(\mathbf{X}) = X_1 + X_2$
Subject to $X_1 \leq 1$

$$X_2 \leq 1$$
$$X_1 + X_2 \leq 2$$
$$X_1 \geq 0$$
$$X_2 \geq 0$$

The optimal solution is $X_1 = 1$, and $X_2 = 1$, at which all three constraints are binding. We are interested in finding out how far the RHS of binding constraints can change while maintaining the degenerate optimal solution. Because of degeneracy, there is dependency among these constraints at the binding position, that is:

$X_1 = 1$
$X_2 = 1$
$X_1 + X_2 = 2$

In order to maintain the degenerate vertex while changing the RHS values, the RHS proportionately must be changed to the coefficients of the decision variables. The resulting parametric RHS system of equations is as follows:

$X_1 = 1 + 1r_1 + 0r_2$
$X_2 = 1 + 0r_1 + 1r_2$
$X_1 + X_2 = 2 + 1r_1 + 1r_2$

The parameters on the RHS allow for the degenerate vertex to move maintaining its structure. Since there are two decision variables, any two of the parametric equations can be used to find the parametric degenerate optimal solution. For example, the first two equations provide ($X_1 = 1 + r_1$, $X_2 = 1 + r_2$), which clearly satisfies the third one. As mentioned earlier, for larger problems one may use JavaScript software: http://home.ubalt.edu/ntsbarsh/Business-stat/otherapplets/PaRHSSyEqu.htm

The perturbed optimal value is $C(X) = X_1 + X_2 = 2 + r_1 + r_2$. In order for this vertex to remain optimal, it must satisfy all other constraints that have not been used. For this numerical example, the non-negativity conditions are the remaining constraints; therefore, we obtain the following conditions for r_1 and r_2:

$\{r_1 \text{ and } r_2 \mid r_1 \geq -1, r_2 \geq -1\}$.

Notice that while the changes satisfy these conditions the degenerate solution remain degenerate, for any change outside of this set ensures the manager that the solution is not any longer degenerate therefore the sensitivity and shadow prices are valid.

11. MAINTAINING THE MULTIPLE OPTIMAL VERTICES OPTIMAL

The necessary condition for the existence of LP multiple solutions: If in the "software final report" the total number of zeros in the Reduced Cost, together with number of zeros in the Shadow Price columns, exceeds the number of constraints, then you might have multiple solutions. Sensitivity analysis is not applicable. That is, the sensitivity analysis based on one optimal solution may not be valid for the others, Lin (2010). Software package like LINDO alert the user if a problem has multiple solutions, while others like Excel LP Solver do not.

Resolution: Check the coefficients in the objective function and the coefficients of the constraints. There could have been rounding error.

Numerical Example P5:

Consider the following LP:

Maximize $C(X) = 6X_1 + 4X_2$
Subject to $X_1 + 2X_2 \leq 16$
$3X_1 + 2X_2 \leq 24$
$X_1 \geq 0$
$X_2 \geq 0$

This LP has two optimal vertices; namely, ($X_1 = 8$, $X_2 = 0$) and ($X_1 = 4$, $X_2 = 6$). Notice that the existence of two solutions means we have innumerable optimal solutions. For example, for the above problem, for all parameter a, $0 \leq a \leq 1$, the following convex combinations are also optimal:

$X_1 = 8a + 4(1 - a) = 4 + 4a,$
$X_2 = 0a + 6(1- a) = 6 - 6a.$

The Dual LP is:

Minimize $16U_1 + 24U_2$
Subject to $U_1 + 3U_2 \geq 6$
$2U_1 + 2U_2 \geq 4$
$U_1 \geq 0$
$U_2 \geq 0$

This dual LP has a degenerate optimal solution ($U_1 = 0$, $U_2 = 2$). In order to maintain the degenerate vertex while changing the RHS, the RHS values must be changed in proportion to the coefficients of the decision variables. The parametric RHS system of equations is:

$U_1 + 3U_2 = 6 + 1c_1 + 3c_2$
$2U_1 + 2U_2 = 4 + 2c_1 + 2c_2$
$U_1 = 0 + 1c_1$

As before, any two of the parametric equations can be used to find the parametric degenerate optimal solution. For example, the first two equations provide $U_1 = c_1$ and $U_2 = 2 + c_2$, which clearly satisfies the third equation. For this vertex to remain optimal, it must satisfy all other constraints that have not been used. For this example, the non-negativity condition, $U_2 \geq 0$, is the only remaining constraint. Therefore, we obtain the following conditions for c_1 and c_2:

$\{c_1, c_2 \mid c_2 \geq -2\}$.

Moreover, the perturbed cost coefficients $(6 + c_1)X_1 + (4 + c_2)X_2$ must be proportional to its parallel constraint $3X_1 + 2X_2 \leq 24$, i.e., $(6 + c_1)/3 = (4 + c_2)/2$. This simplifies to $2c_1 = 3c_2$. Putting together all these conditions, we obtain the largest sensitivity set as follows:

$\{c_1, c_2 \mid c_2 \geq -2, 2c_1 = 3c_2\}$.

Notice that, while the changes satisfy these conditions, the problem has multiple solutions; for any change outside of this set ensures the manager that the problem does not any longer have multiple solutions therefore the sensitivity and shadow prices are valid.

12. THE MORE-FOR-LESS OR LESS-FOR-MORE SITUATIONS

A curious property called the more-for-less (MFL) or less-for-more (LFM) phenomenon is associated with some linear programs (LP). The existing literature has demonstrated the practicality and value of identifying cases where the paradoxical situation exists, Arsham (1996). For example, consider the following production LP.

Numerical Example P6

Maximize $C(X) = X_1 + 3X_2 + 2X_3$
Subject to $X_1 + 2X_2 + X_3 = 4$
$3X_1 + 2X_3 = 9$
$X_1 \geq 0, X_2 \geq 0, X_3 \geq 0$

The optimal solution for this LP occurs at vertex $X_1 = 1, X_2 = 0, X_3 = 3$, with the maximum value of objective function as \$7.

Note that, if the RHS of second constraint is increased to 12, then the new optimal solution gives the optimal value of $C(X) = \$4$; i.e., a decrease in profit while working more hours. This situation arises frequently in LP models and is known as the "more-for-less" paradox where further analysis could bring significant reduction in costs. We apply the primal sensitivity algorithm to solve the RHS perturbed equation constraints with $X_2 = 0$,

$X_1 + X_3 = 4 + r_1$
$3X_1 + 2X_3 = 9 + r_2$

The parametric solution is $X_1 = 1 - 2r_1 + r_2$, $X_2 = 0$, and $X_3 = 3 + 3r_1 - r_2$, with the optimal parametric solution as $C(X) = 7 + 4r_1 - r_2$. Plugging the parametric solution into other non-binding constraints, we obtain

$X_1 = 1 - 2r_1 + r_2 \geq 0$
$X_3 = 3 + 3r_1 - r_2 \geq 0$

Looking at the parametric optimal function, notice that the shadow price of the second constraint, coefficient of r_2, is negative. To find out the best number of hours, one must work to maximize the profit function $7 + 4r_1 - r_2$, by setting $r_1 = 0$ and finding the largest negative value for r_2. Therefore, the constraints reduce to:

$X_1 = 1 + r_2 \geq 0$
$X_3 = 3 - r_2 \geq 0.$

The largest negative value is $r_2 = -1$. This gives the optimal solution of ($X_1 = 0, X_2 = 0, X_3 = 4$) with the optimal value of $C(\mathbf{X}) = 8$. Therefore, the optimal strategy is to work 8 hours instead of 9 hours.

Proposition 3: *A necessary and sufficient condition for the existence of a less-for-more (more-for-less) solution to a maximization (minimization) problem is the existence of an equality constraint with a negative shadow price.*

Theorem 1: *We are in LFM or MFL situation iff at least one of r_i's in the above parametric LP formulation is negative [positive]*

13. CONCLUSION

A business environment is dynamic. A problem solution is valid in a limited time window only and is subject to revision in the next time window. Generally, a constraint set is less subject to change as compared to the objective function. For example, in production and transportation problems, the capacity constraints may remain rather stable over a period of time. On the other hand, profit coefficients of the objective function are inversely related to the price, which may fluctuate, being determined by the market conditions and competition. The proposed method can be used to optimize LP problems with varying objective function. Given a system of linear equalities and/or inequalities, the method provides all vertices of the feasible region. By means of examples, we have illustrated the use of the Improved Algebraic Method to efficiently derive the slack and surplus amounts for the resources of an LP. The parametric representation quickly provides all dual prices to carry out analysis for desirability of obtaining additional resources. The parametric representation also allows one to study variability (i.e, uncertainties) of the coefficients of the objective functions and the right-hand-side of constraints.

We provide a comprehensive managerial coverage of linear programming post-optimality analysis. The collections of presented tools are easy to understand, easy to implement (for small size problems), and provide useful information to the manager. Therefore this collection could prove valuable to involve the manager throughout the decision-making process to understand therefore being implemented.

It is the proposed IAM approach to solve a system of inequalities that provides a bridge between the graphical method and the simplex method. The simplex method is an efficient computer implementation of algebraic methods and almost all the LP software use it.

All the LP tools are covered within the decision variable space; no additional variables such as slack/surplus/artificial variables or constraints are added.

The proposed approach provides useful information for the manager such as shadow prices and expands the sensitivity analysis scope by proving sensitivity regions rather than ranges that LP software provides. Under the proposed methods, the elaborate fundamental LP theorems fall naturally out as by-products. Moreover, it can also be used to fill the gap between the graphical method of solving LP problems and the simplex method.

ACKNOWLEDGMENT

We are most appreciative to the reviewers for their useful comments and suggestions on an earlier version.

REFERENCES

Anderson, D., Sweeney, D., Williams, T., & Martin, R. (2007). *An introduction to management science: Quantitative approaches to decision making*. New York, NY: South-Western College Publisher.

Arsham, H. (1990). Perturbation analysis of general LP models: A unified approach to senstivity parametric, tolerance, and more-for-less analysis. *Mathematical and Computer Modelling*, *13*, 79–102. doi:10.1016/0895-7177(90)90073-V

Arsham, H. (1996). On the more-for-less paradoxical situations in linear programs: A parametric optimization approach. *Journal of Information & Optimization Sciences*, *17*, 485–500. doi:10.1080/02522667.1996.10699297

Arsham, H. (1997a). An artificial-free simplex algorithm for general LP models. *Mathematical and Computer Modelling*, *25*, 107–123. doi:10.1016/S0895-7177(96)00188-4

Arsham, H. (1997b). Initialzation of the simplex algorithm: An artifical-free approach. *SIAM Review*, *39*, 736–744. doi:10.1137/S0036144596304722

Arsham, H. (2006). A big-m free soltuion algorithm for general linear programs. *Internation Journal of Pure and Applied Mathematics*, *32*, 549–564.

Arsham, H. (2007). A computationally stable solution algorithm for large-scalelinear programs. *Applied Mathematics and Computation*, *188*, 1549–1561. doi:10.1016/j.amc.2006.11.031

Arsham, H. (2008). A validation and verification tool for optimization solvers. *Journal of Information & Optimization Sciences*, *29*, 57–80. doi:10.1080/02522667.2008.10699791

Arsham, H. (2009). A simplified algebraic method for system of linear inequalities with LP applications. *Omega: International Journal of Management Science*, *37*, 876–882. doi:10.1016/j.omega.2008.06.001

Arsham, H. (2012). Foundation of linear programming: A managerial perspective from solving system of inequalities to software implementation. *International Journal of Strategic Decision Sciences*, *3*(3), 40–60. doi:10.4018/jsds.2012070104

Arsham, H. (2013). An interior boundary pivotal solution algorithm for linear programs with the optimal solution-based sensitivity region. *International Journal of Mathematics in Operational Research*.

(2013). *Excel: Basic Solve, Frontline Solvers*. New York, NY: Author.

(2013). *GAMS: A User's Guide*. Create Space Independent Publishing Platform.

Koltai, T., & Terlaky, T. (2000). The difference between the managerial and mathematical interpretation of sensitivity analysis results in linear programming. *International Journal of Production Economics*, *65*, 257–274. doi:10.1016/S0925-5273(99)00036-5

Lawerence, J., & Pasternack, B. (2002). *Applied management science: Modeling, spreadsheet analysis, and communication for decision making*. New York, NY: Wiley.

Lin, C. (2010). Computing shadow proces/costs of dengernerate LP problems with reduced simplex tables. *Expert Systems with Applications*, *37*, 5848–5855. doi:10.1016/j.eswa.2010.02.021

LINDO. (2003). *LINDO users's manual*. New York, NY: Author.

(1999). *Management Scientist*. New York, NY: South-Western Educational Publishing.

Martin, K. (1999). *Large scale linear and integer optimization: A unified approach*. New York, NY: Springer. doi:10.1007/978-1-4615-4975-8

(2003). *QM Windows Manual*. New York, NY: Prentice Hall.

Taylor, B. (2009). *Introduction to management science*. New York, NY: Prentice Hall.

(2003). *WinQSB Manual*. New York, NY: Wiley.

Winston, W. W. L., & Albright, S. (2011). *Practical management science*. New York, NY: Cengage Learning.

Chapter 8
Using Holistic Strategy to Govern Today's Extended Enterprise

Kenneth Saban
Duquesne University, USA

ABSTRACT

Governing extended networks is often associated with having the latest collaborative technology and business processes. However, current studies suggest that both are only part of the answer. This chapter argues that a holistic governing strategy needs to be adopted to achieve peak network performance. That strategy requires four conditions: 1) that each organization is ready to work with fellow business partners; 2) that network orchestrators recognize that all business relationships are not created equal and therefore need to be governed differently; 3) that a holistic governing strategy has to be adopted in order to integrate the people, technology, and processes in place; and 4) that people engagement hinges upon six factors (trust, beliefs, communications, culture, reward systems/metrics, and synergy). This chapter is a modification of an earlier work published by IGI Global.

INTRODUCTION

The development of the personal computer and the Internet has dramatically changed the way customers and businesses interact. Armed with more knowledge about the pricing of products, customers are demanding more value-laden solutions. If a business does not meet these demands, customers can quickly search for alternatives at a touch of a button.

In order to compete under these conditions, executives are revamping their business models. The conventional business model was to have all capabilities housed under one roof. While this

DOI: 10.4018/978-1-4666-4506-6.ch008

centrally controlled model worked well in the early-to-mid part of the 20th century, executives are now seeking more flexible structures that resemble networks rather than traditional pyramids.

Miles and Snow (1992) describe the evolution taking place. The earliest model was the Stable Network which was designed to serve a predictable market by linking together independently owned assets along a given product or service value chain. Nike employed this network to quickly introduce new shoe models. The next model was the Internal Network which was developed to allow corporations to gain competitive advantage through shared utilization of scarce assets and know-how. ABB (Asea Brown Bovari) used this model to give it unmatched global synergy in the electrical systems/equipment market. The latest model is the Extended Network or Extended Enterprise which incorporates a large number of partners who - at a moment's notice – can be asked to help deliver a new product or service. A good example of which is Li & Fung Limited who is a global trading group that supplies high-volume, time-sensitive consumer goods. When an order arrives, management selects the best of its 8,300 suppliers located in 40 countries for filling the order at the right quality, time and price (Fung, Fung and Wind 2007).

A Global Outsourcing Study conducted by Deloitte Consulting LLP (2012) underscores the magnitude of the transformation that has and will take place. For example: 76 percent of information technology services were currently outsourced trending to 81 percent; 42 percent of operations were currently outsourced trending to 53 percent; 30 percent of human resource functions were currently outsourced tending to 46 percent; 24 percent of procurement functions were currently outsourced tending to 36 percent; and 11 percent of marketing/sales functions were currently outsourced trending to 24 percent.

While firms like Toyota, Dell Computer and Li & Fung Limited have successfully made this transformation, many firms are still struggling with the process. A study conducted by Dunn and Bradstreet found that 20 to 25 percent of all outsourcing relationships fail within two years and 50 percent fail within five years (McKinsey Quarterly, 2001). Another study conducted by Deloitte consulting LLP (2007) found that "39% of the 300 respondents reported that they had terminated at least one outsourcing contract and transferred it to a different vendor in their careers and...61% reported that they had escalated problems to senior management in their contract's first year, with 15% reporting five or more such escalations." A recent example is Boeing who has experienced a myriad of problems with its 787 Dreamliner jet which is made from parts coming from Canada, Italy, United Kingdom, France, Japan, Korea, Australia and the U.S. Most recently, smoldering batteries – including one fire on the ground – prompted air safety authorities to ground the world's 787s in January 2013.

One reason companies have failed can be linked to the governance structure employed with their extended enterprises. Given this observation, the question this paper seeks to answer is what is the best way to govern extended enterprises? To answer this question, this paper will explore the dynamics of the extended enterprise, the various governing strategies, and the factors that facilitate human collaboration in order to improve network performance.

EXTENDED ENTERPRISE

Davis and Spekman (2004) define the extended enterprise as a "set of collaborating companies, both upstream and downstream, from raw materials to end-use consumption, that work together to bring value to the marketplace." Bititci, Mendibill, Martinez and Albores (2005) view the extended enterprise as "a knowledge based organization which uses the distributed capabilities, competencies and intellectual strengths of its members to gain competitive advantage to maximize the performance of the overall extended enterprise." We define the extended enterprise as a set of collaborating companies who openly share their operational capabilities and intellectual property to generate high-value products that meet and/or exceed customer expectations.

While members in a conventional business model take responsibility for their individual tasks, members of an extended network need to employ a broader perspective. For example, to: recognize the importance of maximizing the value for customers such that each is rewarded based on the individual contribution made; are willing to relinquish total control; accept the fact that an extended enterprise must operate with greater effectiveness and efficiency than as a single firm; and are able to link their individual competencies (people, technology, processes) in such a fashion that speed to market is accelerated, costs are lowered, and new market opportunities are more easily accessed (Davis and Spekman, 2004) .

To achieve this state, the lead organization needs to employ the right form of governance. While conventional business models are normally governed by a rigid control system, extended enterprises require a holistic governing structure that recognizes the rights and obligations of all the stakeholders. A holistic governance structure addresses issues related to asset distribution (people, technology, processes), the roles and performance among the various partners, assigns responsibility for activities and participation in decision-making, ensures equitable treatment of all partners as it relates to sharing both risks and rewards and attempts to build a sense of shared commitment to the broader mission of the enterprise.

Employing the wrong governing structure can dramatically impact the level of inter-firm collaboration, which in turn can impact network performance. A study conducted by the Outsourcing Center (2004) found that the top three causes of outsourcing failure were: unclear expectations before entering the partnership at 23 percent; partners losing interest in the arrangement at 15 percent; and poor governance at 13 percent. These findings underscore the importance of matching the right governance strategy with the right business partnership. The next section will review the governing strategies available and discuss which is most appropriate under what circumstances.

GOVERNING STRATEGIES

In order to successfully manage an extended network, four conditions must be met. The first condition is that all the organizations must be ready to collaborate with the other business partners. Martin, Beimborn, Parikh and Weitzel (2008) argue that an organizational readiness requires three things. First, the business processes that are going to be unilaterally deployed need to be understood by all the stakeholders. It is especially important that the outsourcers thoroughly understand the implications and/or side effects of this reality. Second, all parties need a thorough understanding of the shared information technologies – especially the knowledge and skills required to operate these systems. Third, network orchestrators need to insure that key employees have the experience and/or training to integrate the technology and processes (Provan & Kenis, 2005).

The second condition is that the network orchestrator in the lead organization recognizes that all business relationships are not created equal and therefore may need to be governed differently. Cohen and Roussel (2005) discuss the types of business relationships and the corresponding governing strategy:

1. **Transactional Relationships.** Transactional relationships apply to manufacturer-supplier situations where low risk and low value purchases are made - like cleaning supplies or maintenance repair services. Due to the low criticality of these purchases, the focus is on price and ways to minimize the effort required to acquire these goods. Because

these purchases are very straightforward, there is little need for a high level of collaboration once a relationship is established. Therefore, network orchestrators employ an *arm's-length governing strategy* where neither party is overly concerned with the well-being of the other.

2. **Collaborative Relationships.** The key difference between collaborative and transactional relationships is an awareness of the interdependence and necessity of cooperation. These relationships require that manufacturers and suppliers work much closer due to the criticality of the products being provided - like the hydraulic systems that control the landing gear in an airplane. Business processes and technology are heavily relied upon along with two-way communications. Network orchestrators tend to employ a *co-operative governing strategy* that focuses on trust building, joint efforts, planning and fostering interdependency.
3. **Strategic Relationships.** Due to the high-risk, high-value of critical products being created like fly-by-wire electronic systems, it is mandatory that manufacturers and suppliers operate in a highly collaborative fashion. This requires that information is developed jointly (tacit) rather than being pushed or pulled (explicit). Network orchestrators employ a holistic governing strategy to insure that key individuals have experience working together and have accumulated specialized information and language that allows them to communicate effectively with each other using the technology and processes in place.

What makes strategic relationships so unique to govern is that the business processes employed are spread among all the parties some of which who are not under the direct control of the lead partner(s). Accenture (2012) also found that high performing partnerships also require more than the right business process - it also requires the right technology and right people. For example: 50 percent of high performance businesses see technology within the scope of the external business relationship compared to 25 percent of typical performers; and b) 65 percent of high-performance businesses see the impact of people (client and provider) as compared with only 39 percent of typical performers.

Given the high degree of information exchange in strategic relationships, it is important that a holistic governing strategy be employed in order to integrate the shared people, technology and processes (see Figure 1). This position is supported by Barthelemy and Adsit (2003) who found that overlooking personnel issues is a major stumbling block with Elmuti (2003) who also found that the lack of training/skills can hinder employee participation.

When companies employ the wrong governing strategy – like the arm's length – in an extended enterprise setting, they fail to understand the importance of social impact. Ernst & Whinney (1987) found that companies that did not place a strong emphasis on training performed lower than companies who made employee development a priority. Michigan State University (Bowersox, Droge, Rogers & Wardlow, 1989), Pennsylvania State University (Novack, Langly & Rinehart, 1995), and Fawcett and Magnan (2005) also found that human resource development plays a major role in network perfor-

Figure 1. Components of a holistic governing strategy

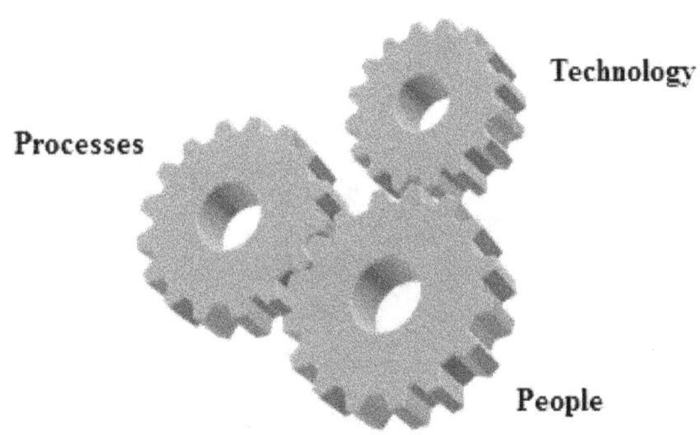

mance. After surveying 358 executives from leading manufacturers and service providers, Gowen and Tallon (2003) concluded that proper human resource management enhances the "value added" of networks by providing more effective resources in terms of better trained employees and enthusiastic employees and managers, which can equate to a more efficient and effective business partnership and one that is hard to emulate. In short, firms that contemplate operating extended enterprises need to recognize two interrelated people issues. First, key employees in each organization must be retained and motivated. Second, that the commitment of employees transferred to the vendor and/or working with the vendor must also be secured as collaboration is often reliant on the transfer of tacit knowledge that resides within the employee. Effective transfer of tacit knowledge normally requires extensive personal contact, periodic face-to-face interaction, and a high degree of trust.

Given that the right governing strategy has been selected, it is also important to insure that the human gear engages the technology and process gears. This is accomplished by dealing with those factors (gear teeth) that facilitate human collaboration which becomes our fourth and final condition. The next section discusses each factor in detail.

HUMAN COLLABORATION FACTORS

One way to think about human collaboration is to consider a large philharmonic orchestra. If each orchestra member were to use a different sheet of music and play independently, the performance would be chaotic. However, when you employ the same sheet of music and add a conductor to organize individual play, real harmony is achieved (McClellan, 2003). Through harmonization, individuals and teams direct their respective talents in such a fashion that they achieve an outcome that can go beyond their own limited vision of what is possible (Gray, 1989). The same can be said about extended networks.

"What factors facilitate human collaboration?" To test the numerous factors that appear in the literature regarding business collaboration, a focus group of 10 regional supply chain executives (representing Fortune 500 corporations as well as consulting organizations) was conducted. The executives were asked to rate the relevance of each factor that most often appeared in the literature based on their day-to-day business experience. This event resulted in the identification of six factors that facilitate human collaboration (see Figure 2).

Figure 2. Components of the Human Collaboration Gear

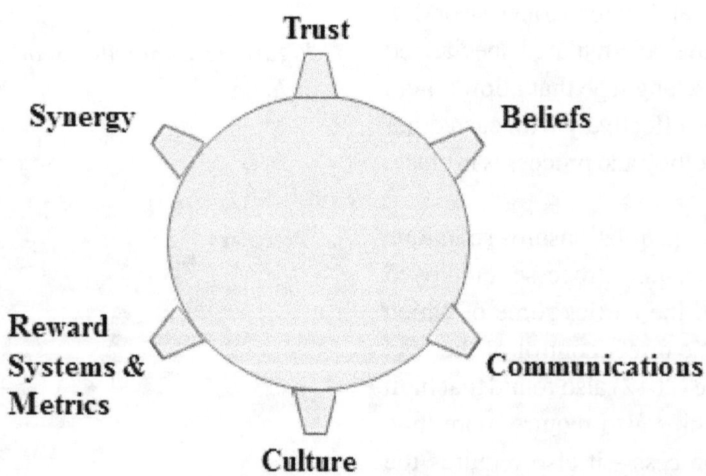

Trust

Grenier & Metes, 1995; Kumar, 2000; Chopra & Meindl, 2001; Jonsson & Zineldin, 2003; Ryssel, Ritter & Gemunden, 2004 agree that building trust among supply chain partners is a critical starting point. However, trust is often the most difficult collaboration factor to achieve. The reason is managers often assume personal relationships will fall into place due to the implementation of today's collaborative technology. However, this is far from the truth. Building trusting relationships where each party has confidence in the other members' capabilities and actions requires more than the right technology and processes, it also requires building personal relationships. As one executive noted: "Supply chain management is one of the most emotional experiences I've ever witnessed." There have been so many methodologies that have developed over the years, people blaming other people for their problems, based on some incident that may or may not have occurred sometime in the past. Once you get everyone together into the same room, you begin to realize the number of false perceptions that exist. People are still very reluctant to let someone else make decisions within their area. It becomes especially tricky when you show people how 'sub-optimizing' their functional area can 'optimize' the entire supply chain." (Handfield & Nichols, 2002).

Trust also requires linking the belief systems of individuals and teams such that each is interested in the other partner's welfare and would not take any action without considering its impact on the other partner (Fukuyama, 1996). Consider this example: a supplier has been brought in by a manufacturer to provide the aluminum wings for an airplane. However, late in the design process, it is discovered that a composite wing will better fit the customer. In addition, the supplier does not have the ability to make composite wings, and has sunk considerable costs into the project. In a trust-based relationship, the supplier has to trust that the manufacturer will treat them fairly whether they decide to stay with the aluminum wing or go with the composite wing. The manufacturer must also trust that each supplier will optimize the enterprise, even when it reduces or eliminates its own role (Goranson, 1999).

Beliefs

Collaboration requires that partners from different organizations not only trust each other, but also believe that the whole is as important as the individual pieces – especially as it relates to meeting the needs of others. A good example is the Kariya Number 1 plant of Aisin Seiki - a major Japanese supplier of brake fluid-proportioning valves to Toyota. Within hours after the plant burned to the ground, Aisin engineers met with their counterparts at Toyota and their other tier one suppliers and passed along their blueprints for valves to any supplier that requested them and distributed whatever undamaged tools, raw materials, and work in process that could be saved. To everyone's surprise, Kyoritsu Sangyo – a small tier two supplier of welding electrodes – delivered 1,000 production-quality values within 85 hours after the accident (Evans & Wolf, 2005).

To be able to work in this fashion, individual/team/organizational beliefs must be aligned. Beliefs are the lenses through which workers see their purpose and responsibilities, set expectations for success and reward, and generally make sense of their place in this environment. Once belief systems are aligned, people are more inclined to share information about one's self as well as process knowledge and domain specific knowledge; for example developing a working knowledge of given protocols to communicate electronically with other team members (Cleland, Bidinda & Chung, 1995). Based on these observations we posit that beliefs and knowledge contribute to individuals, teams and organizations ability to collaborate.

Communications

While the essence of collaboration is the interaction between two parties, it is communication that links individuals and companies. Jonsson and Zineldin (2003) call communications the "factory" of human society, as it is people and not accounting systems, computer terminals or trading agreements that can communicate effectively with each other. While technology plays an increasingly important role in the transfer of information between organizations, it is still an enabler and not the driver of success (Cohen & Roussel, 2005).

Therefore, it is important to understand the type of information that is exchanged in any relationship. Most researchers divide information into two types. Explicit information i.e., facts and figures, is easily codified and can be transmitted across computer networks. Tacit information i.e., know-how, involves knowledge that is complex and difficult to codify. To facilitate both exchanges, Toyota formed its supplier association (*Kyohokia*) which focused on: 1) information exchange between suppliers and Toyota, 2) mutual training and development of suppliers, and 3) establishing socializing events. This has allowed Toyota to develop superior knowledge transfer capabilities which in turn allows all supply chain members to understand each other's goals and coordinate their efforts to achieve common goals while maintaining a satisfactory working relationship (Dyer, 2000).

Culture

By aligning trust, beliefs, knowledge and communications, organizations are able to establish a culture which supports decision-making and work. Schein (1985) defines organizational culture as "a pattern of basic assumptions – invented, discovered, or developed by a given group as it learns to cope with its problems of external adaptation and internal integration – that has worked well enough to be considered valid and, therefore, to be taught to new members as the correct way to perceive, think, and feel in relation to those problems." This definition suggests that culture resides both inside the company and with external partners. While problems with internal integration typically revolve around power/status, leadership and standards, external integration issues normally entail external environments [obtaining a shared understanding of key actors in the environment], mission [developing consensus on who are we, why are we here, and what are our strengths and weaknesses] and correction [developing consensus on how external issues should be solved] (Schultz, 1994). As noted earlier, IBM recognized that if they were going to regain their market dominance, the company had to change their culture that was out-of-sync with its customers (Forrester, 2005). IBM took a number of corrective actions:

- The company communicated its supply chain strategy anchored around four imperatives [Drive focus-flexibility-quality-cost competitiveness, roll-out IBM's core strategic processes across the globe, extend supply chain principles to IBM's labor-based businesses, provide industry leading solutions integration and delivery capability].
- The company appointed some 400 top supply chain executives as "evangelists" to "walk the talk."
- The company secured bottom-line support by changing success metrics.
- The company stayed the course no matter if the strategy changed.

Because each customer order has unique qualities, the possibility exists that management may be confronted with a wide-range of assumptions pertaining to the value of collaboration. By educating front-line managers as to the nuances of creating a collaborative culture, the company will increase the probability that individuals and teams will be more supportive of the network's mission and goals.

Reward Systems and Metrics

To maintain a collaborative culture, it is critical that the proper reward systems and metrics be established as they reinforce certain types of behavior. It is therefore important that manufacturers pay considerable attention to the reward systems that are in place for both their employees as well as the employees of their suppliers (Harrington & Harrington, 1995). The reason for doing so is that these programs not only reward individuals and teams on how they improve corporate performance, but also recognize individuals and teams that deliver high customer value (Forrester, 2005). Therefore, when designing a supply chain reward system, manufacturers need to: a) focus on what is valued, b) demonstrate a clear connection between behavior, the rewards, and metrics, and c) recognize that changes in behavior patterns – such as the adoption of new supply chain technology and processes – may require different types of incentives.

The second requirement is to establish the right metrics. Cohen and Roussell (2005) argue that any effective metrics program must include:

1. Internally focused metrics that address such items as the cost of goods sold, labor rates, and asset turns.
2. Financial and non-financial metrics that address such items as production quality and flexibility, and forecast accuracy.
3. Customer metrics that address such items as on-time delivery, order-fulfillment, and fill rates.
4. Innovation and collaborative metrics that address such items as bringing innovative products to market in record time, developing break-through technology, and solving complex business issues.

Once the metrics are defined, it is important to periodically review one's reward systems and metrics to insure they support the current mission and objectives of the supply chain network.

Synergy

To maintain ongoing collaboration, a number of synergistic activities must take place (Harrington, Hefner & Cox, 1995). For example, making sure that the flow of information between and among partners is maintained. While team interaction within an office is easy to arrange and manage, interaction among virtual teams can be more complex. Therefore, it is important that team building exercises be included in virtual situations coupled with formal exercises like creating charters, mission statements, goals, and operating norms (Kirkman et al., 2002). Another activity is to encourage team members to accept and utilize the dynamics of the team, thus eliminating the traditional view that "I'm right and you're wrong." The Toyota Group's *kyson kyoei* (co-existence and co-prosperity) philosophy underscores the view that what is good for the extended enterprise is good for me, and what is good for me is good for the extended enterprise (Dyer, 2000). The next activity involves internalization. That is, team members must learn to tolerate ambiguity, modify their own views and belief systems, and be receptive to new ways to make group decisions (Goldman, Nagel & Preiss, 1995). The last activity requires that the team implement their plans and strategies from which they will learn, grow and become more synergistic.

This discussion suggests that trust, beliefs, communications, culture, reward systems/metrics, and synergy encourage people to collaborate and therefore employ the technologies and processes in place. These factors encourage employees to adopt the technologies and processes in place which in turn generate better network performance.

Dyer's (2000) work in the automotive field underscores the benefits derived when a collaborative governing strategy is properly applied in an extended enterprise. He compared the performance of Toyota (who employed a collaborative governing strategy) with General Motors, Ford, and Nissan (who tended to employ an arm's length governing strategy) on three levels:

1. **Dedicated Assets.** These are investments made by each partner in order to increase productivity in the network. They include: a) site specialization which is the geographic location of successive product facilities; b) physical asset specialization which is the capital investment made that allows for faster throughput and greater product customization; and c) human specialization which is the specific know-how of the employees assigned the partnership. The results showed: a) Toyota and its suppliers' inventories as a percentage of sales were 50 percent of Nissan and 25 percent of General Motors and Ford; b) Toyota suppliers' had higher capital investments (22 percent) than U.S. suppliers' (15 percent); c) as a result of higher collaboration Toyota was able to develop a new car model in 50 months which was 40 percent faster than U.S. manufacturers.
2. **Knowledge-Sharing.** Strategic relationships require a high level of both tactic (know-how) and explicit (facts) information transfer. Toyota's knowledge-sharing practices resulted in a faster rate of learning with suppliers, which in turn translated into high productivity and quality. The results showed that between 1990-1996, Toyota suppliers: a) reduced their defects (in parts per million) by 84 percent versus 46 percent for the U.S. firms; b) reduced their inventories (as a percent of sales) by 35 percent to only 6 percent for U.S. firms; and c) increased labor productivity (sales per employee) by 36 percent versus 1 percent for U.S. firms.
3. **Inter-firm Trust.** The productivity of an extended enterprise is a function of both production costs and transaction costs. While production costs combine raw material and labor, transaction costs involve all the other costs associated with conducting exchanges between partners. The results showed that: a) Toyota had lower transaction/negotiating costs (21 percent) than General Motors (at 47 percent) and Ford (at 28 percent); b) General Motors procurement costs were six times higher than Toyota; and c) Toyota suppliers shared sensitive information three times more often than General Motors suppliers.

This section has provided an in-depth discussion of six factors that facilitate humans in strategic relationships. Armed with this knowledge, network orchestrators can insure the proper implementation of their holistic governing strategy.

SUMMARY

While the literature on extended enterprises tends to focus on the importance of technology and business processes, less attention has been given to the role of people. This void has in part contributed to the failure of many extended enterprises and subsequent questioning of CEOs as to why their extended enterprises are not delivering at the same level as Toyota, Dell Computer and Li & Fung Limited. To operate at the level of these three companies, four conditions need to be addressed: 1) that each organization in the network is ready to work with fellow business partners; 2) that network orchestrators recognize that all business relationships are not created equal and therefore need to be governed differently; 3) that a holistic governing strategy needs to be employed in order to integrate the people, technology and processes; and 4) that the human gear will best engage the technology and process gears when six human collaboration factors (trust, beliefs, communications, culture, reward systems/metrics, and synergy) are operational.

The work Dyer (2000) completed on the automotive industry underscores the significant benefits that can be derived by matching a holistic governing strategy with a strategic business relationship. In order for this to be accomplished, a different mindset and subsequent framework must be adopted. To validate the applicability of a holistic governing strategy, the four conditions outlined herein will be tested in a mix of industries.

The author hopes that this article will encourage other researchers to explore the competitive advantages gained by using holistic strategy to govern today's extended enterprises.

REFERENCES

Accenture. (2012). *Achieving high performance in BPO*. Retrieved from http://www.accenture.com/us-en/Pages/insight-high-performance-bpo-eight-practices.aspx

Barthelemy, J., & Adsit, D. (2003). The seven deadly sins of outsourcing. *The Academy of Management Executive*, *17*(2), 87–100. doi:10.5465/AME.2003.10025203

Bititci, U., Mendibill, K., Martinez, V., & Albores, P. (2005). Measuring and managing performance in extended enterprises. *International Journal of Operations & Production Management*, *25*(4), 333–353. doi:10.1108/01443570510585534

Bowersox, D. J., Daugherty, P. J., Droge, C. L., Rogers, D. S., & Wardlow, D. L. (1989). *Leading-edge logistics: Competitive positioning for the 90s. Oak Brook*. Council of Logistics Management.

Chopra, S. & Meindl, P. (2001). Supply chain management: Strategy, planning and operation. *Prentice Hall,* USA.

Cleland, D. I., Bidanda, B., & Chung, C. A. (1995). Human issues in technology integration – part 1. *Industrial Management (Des Plaines)*, *37*(Issue 4), 22.

Davis, E. W., & Spekman, R. E. (2004). *The extended enterprise*. USA: Prentice Hall.

Deloitte Consulting, L. L. P. (2007).Why Settle for Less? Deloitte Consulting 2008 Outsourcing Report http://www.deloitte.com/assets/Dcom-Estonia/Local%20Assets/Documents/ce_cons_outs_report_08(1).pdf

Deloitte Consulting, L. L. P. (2012). *2012* Global Outsourcing and Insourcing Survey Executive Summary. http://www.deloitte.com/assets/Dcom-UnitedStates/Local%20Assets/Documents/IMOs/Shared%20Services/us_sdt_2012GlobalOutsourcingandInsourcingSurveyExecutiveSummary_050112.pdf

Dyer, J. H. (2000). *Collaborative advantage*. Oxford University Press.

Elmuti, D. (2003). The Perceived Impact of Outsourcing on Organizational Performance. *American Journal of Business*, *18*(Issue 2), 33–42. doi:10.1108/19355181200300010

Ernst & Whinney. (1987). *Corporate profitability & logistics*. USA: Council of Logistics Management.

Evans, P., & Wolf, B. (2005). Collaboration rules. [July –August]. *Harvard Business Review*, *83*(Issue 7/8), 96. PMID:16028820

Fawcett, S. E., & Magnan, G. M. (2005). Achieving world-class supply chain alignment: Benefits, barriers, and bridges. Center for Advanced Purchasing Studies, http://.www.capsresearch.org.

Forrester (2005). APAC study chain apps spending outlook, http://www.forrester.com.

Fukuyama, F. (1996). *Trust: The social virtues and the creation of prosperity*. New York Free Press.

Fung, V. K., Fung, W. K., & Wind, Y. (2007). *Competing in a flat world*. Wharton School Publishing.

Goldman, S. L., Nagel, R. N., & Preiss, K. (1995). *Agile competitors and virtual organizations*. Van Nostrand Reinhold Publishing.

Goranson, H. T. (1999). *The agile virtual enterprise*. Quorum Books.

Gowen, C. R. III, & Tallon, W. J. (2003). Enhancing Supply Chain Practices through Human Resource Management. *Journal of Management Development*, *22*(Issue 1-2), 32. doi:10.1108/02621710310454842

Gray, B. (1989). Negotiations: Arenas for reconstructing meaning. Working paper, *Pennsylvania State University, Center for Research in Conflict and Negotiation*, USA.

Grenier, R., & Metes, G. (1995). *Going virtual*. USA: Prentice Hall.

Hanfield, Robert B., & Nichols, Jr., Ernst L. (2002). Supply chain redesign, *Prentice Hall*, USA.

Harrington, H. James & Harrington, James S. (1995). Total quality management, McGraw-Hill, USA.

Harrington, H. James, Hefner, Mark B., & Cox, C. Keith (1995). Environmental change plans: Best practices for improvement planning and implementation. Harrington and Harrington's Total Improvement Management, McGraw-Hill, USA.

Kirkman, B. L., Rosen, B., Gibson, C. B., Tesluk, P. E., & McPherson, S. O. (2002). Five challenges to virtual team success: Lessons from Sabre, Inc. *The Academy of Management Executive*, *16*(Issue 3), 67. doi:10.5465/AME.2002.8540322

Kumar, N. (2000). The power of trust in manufacturer-retailer relationships. Managing The Value Chain. *Harvard Business Review*.

Martin, S. F., Beimborn, D., Parikh, M. A., & Weitzel, T. (2008). Organizational Readiness for Business Process Outsourcing: A Model of Determinants and Impact on Outsourcing Success, *Proceedings of the 41st Hawaii International Conference on Systems Sciences*, IEEE, pp. 1-10.

McClellan, M. (2003). *Collaborative manufacturing*. St. Lucie Press.

McKinsey, Q. (2001). Has outsourcing gone too far? https://www.mckinseyquarterly.com/Has_outsourcing_gone_too_far_1126.

Miles, R. E., & Snow, C. C. (1992). Causes of Failure in Network Organizations. *California Management Review*, *34*, 53–72. doi:10.2307/41166703

Novack, R. A., Langly, C. J. Jr, & Rinehart, L. M. (1995). *Creating logistics value: Themes for the future*. USA: Council on Logistics Management.

Outsourcing Center. (2004). What Causes Outsourcing Failures? http://www.outsourcing-center.com/2004-08-what-causes-outsourcing-failures-article-37826.html.

Provan, K. G., & Kenis, P. (2005). Modes of Network Governance: Structure, Management, and Effectiveness, *Public Management Research Association Conference*, September 29 – October 1, University of Southern California, Los Angeles, CA.

Ryssel, R., Ritter, T., & Gemunden, H. G. (2004). The Impact of Information Technology Deployment on Trust, Commitment and Value Creation in Business Relationships. *Journal of Business and Industrial Marketing*, *19*(Issue 3), 197. doi:10.1108/08858620410531333

Schein, E. H. (1985). *Organizational culture and leadership*. Jossey-Baas Publishers.

Schultz, M. (1994). *On studying organizational culture*. Walter de Gruyter Publishers.

Sheffi, Y. (2007). *The resilient enterprise: Overcoming vulnerability for competitive advantage*. The MIT Press.

Shoshanah, C., & Roussel, J. (2005). *Strategic supply chain management*. USA: McGraw-Hill.

Chapter 9
A Stakeholder Perspective of Sustainable Supply Chain Management:
Evidence from a Developing Country

Mohamed Gamal Aboelmaged
Al Ghurair University, Dubai, UAE & Ain Shams University, Egypt

ABSTRACT

Industrialization leaves no doubt that our planet is suffering from global warming, depletion of natural resources, pollution, waste, and other environmental concerns. Consequently, businesses, communities, and governments have been environmentally conscious and shown growing concern for sustainable development, particularly following the establishment of the United Nations Environmental Program (UNEP). This chapter provides further insights into sustainability and supply chain research through adopting a stakeholder perspective to understand drivers and consequences of supply chain sustainability in the United Arab Emirates. The results indicate that pressures enforced by champions and customers are positively related to supply chain sustainability. However, the relationship between government pressures and supply chain sustainability is insignificant. Moreover, the findings provide evidence that the impact of supply chain sustainability on the organizational sustainable performance is significantly positive. Supply managers can learn from these results in developing sustainable initiatives earlier along their supply chain through selecting and evaluating suppliers based on sustainability-related standards. In addition, environmental collaboration with customers and suppliers based on knowledge sharing and application may identify and reduce the total environmental impact.

DOI: 10.4018/978-1-4666-4506-6.ch009

1. INTRODUCTION

Industrialization leaves no doubt that our planet is suffering from global warming, depletion of natural resources, pollution, waste and other environmental concerns. Consequently, communities and governments have been environmentally conscious and shown growing concern for sustainable development, particularly following the establishment of the United Nations Environmental Program (UNEP) in 1972. Since then businesses have been encouraged to rethink how sustainable initiatives should be implemented and linked to products, services and processes. Today's firms are required to impose dramatic changes to their operations and strategies to remain competitive. It is common nowadays to observe banners as 'sustainable operations', 'sustainability for development', 'environmental initiatives', 'go green', or 'eco-designed' are highlighted in firm's documents and websites.

Growing environmental interest has to be a part of the organizational culture to help reengineer the strategies of firms (Madu et al., 2002). Stern (2007) shows that economic consequences of climate change, for example, would cost firms trillions of dollars and early prevention is more economically viable. The same applies to the processes of value creation of a firm which have to maintain environmentally sustainable procurement, production, distribution, use and recycling of products (Hart and Milstein, 2003). Global pressures have also prompted firms to improve their environmental performance (Zhu and Sarkis, 2006). For example, the European Union (EU) implemented Restriction of Hazardous Substances (ROHS) directive that prohibits electrical and electronic equipment containing lead, mercury, cadmium, hexavalent chromium, polybrominated biphenyls (PBB) and polybrominated diphenylethers (PBDE). Linking this directive to Sony's PlayStation game is an instance. Though it's a responsibility of Sony's suppliers, Sony Corporation had to endure much of the consequences when about 1.3 million of PlayStation game consoles were stopped at the Dutch border because of high cadmium levels detected in its cables (Carlton, 2006). Moreover, The EU employed the waste electrical and electronic equipment (WEEE) directive in August 2005 that keeps producers responsible for the costs of the collection, recycling, reuse and recovery of their products at the end of product's usable life in order to reduce its environmental impact. Consequently, leading electrical and electronic firms such as Samsung, Dell, Fujitsu, Toshiba HP, IBM, Motorola, Sony, Panasonic and NEC begin to invest in developing green products and establishing standards associated with using and supplying of hazardous substances aiming at fulfilling environmental directives when exporting to EU countries. This implies that companies are now starting to recognize the role of environmental awareness in improving competitive advantage (Walton et al., 1998), promoting efficiency and synergy among business partners (Rao and Holt, 2005), and creating business value (van Hoek, 1999).

This chapter provides further insights into sustainability research through examining the interaction between sustainability and a business process, namely supply chain management. It also applies stakeholder perspective to understanding the role of stakeholder' pressures in influencing sustainably supply chain which, in turn, impacts sustainable organizational performance in a developing country context. In the next section, we provide a brief synthesis of sustainability from different perspectives including the Triple-Bottom-Line perspective. This synthesis provides the conceptual basis for developing sustainable supply management concept in the third section. Fourth section explores of risks associated with different stages of sustainable supply chain. This is followed by a literature review of sustainable supply chain in developing countries. Section six presents a stakeholder perspective of sustainable supply chain. Finally, empirical evidence from a developing country is delineated, and followed by a discussion and implications for future research.

2. SUSTAINABILITY: THE TRIPLE-BOTTOM-LINE PERSPECTIVE

When the World Commission on Environment and Development (WCED) published "Our Common Future" report in 1987, "sustainability" or "sustainable development" became well-known expressions for practitioners and managers around the world, both developing and industrialized nations. Nevertheless, the terms have been addressed in the literature under different perspectives. The WCED (1987) defined sustainable development as "a development that meets the needs of the present without compromising the ability of future generations to meet their own needs'' (WCED, 1987, p. 8). Hoexter (2006) suggested another concept of sustainability that reflects a balanced and fair exchange between humanity and nature with holistic systems thinking and long time horizon. From a relative perspective, Starik and Rands (1995) considered sustainability as the ability of an individual or collective entity to exist and evolve overtime in such a manner that allows for a relative level of existence and flourishing of other entities. From an ecological perspective, Shrivastava (1995) defined sustainability as the potential for reducing long-term risks associated with resource depletion, fluctuations in energy costs, product liabilities, and pollution and waste management. Other definitions reflect the broad topic of sustainability such as the environmental impact of economic activity in both developing and developed economies (Erlich and Erlich, 1991); meeting human needs (Savitz and Weber, 2006); and maintaining non-renewable resources (Whiteman and Cooper, 2000).

Sustainability is also associated with the fulfillment of the Triple-Bottom-Line (3BL), a concept developed by Elkington (1998, 2004), which simultaneously considers and balances economic, environmental and social goals from a microeconomic perspective. The 3BL is an equal balance among economic development, environmental compatibility, and social equity as pillars for improving quality of life (Carter and Rogers 2008; Goncz et al., 2007; McCue 2010; Sikdar, 2003; Vachon and Mao, 2008). Figure 1 shows a visual representation of 3BL components.

This reflects the perspective of organizational sustainability as a fundamental principle of smart management that works towards incorporating the 3BL into business strategy and daily operations. These sustainable practices are critical ingredients for successful businesses in the 21st century; yet, it is neither simple nor low-cost process to consider.

3. TOWARDS A SUSTAINABLE SUPPLY MANAGEMENT

The essence of sustainability is to maintain environmental, economic and social aspects of firms. From the business perspective, firms play a role throughout the lifecycle of their products and services where sustainability must be an integral component of corporate strategy and supported by management and performance measurement (Epstein, 2008). Milne et al. (2006) conveyed sustainability as a journey of organizational adaption, learning, progress and a movement away from a business-as-usual practices. Among the many approaches that play a significant role in improving environmental impact of any firm is integrating environmental management throughout the supply chain practices (Damali et al., 2008).

The term supply chain has been developed in the late 1980s. Prior to that time, terms like "materials management" and "logistics" were used instead. A supply chain can be defined as an alignment of firms that bring products or services to market (Lambert et al., 1998). It consists of all stages involved, directly or indirectly, in fulfilling a customer request including manufacturer,

Figure 1. The Triple-Bottom-Line (3BL) perspective as an intersection of environmental, social and economic performance

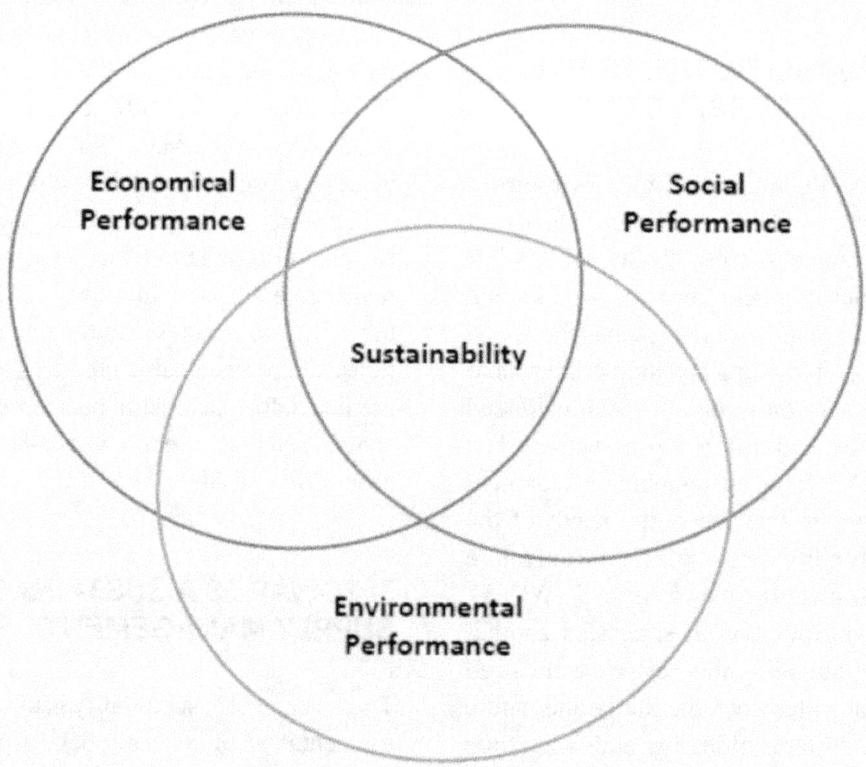

suppliers, transporters, warehouses, retailers, and customers themselves (Chopra and Meindl, 2001). Frankel et al. (2008) conceptualized supply chain as a network of companies from suppliers to end-users, which have the intention of integrating supply/demand via coordinated company efforts. Accordingly, the focus of supply chain management is "The systemic, strategic coordination of the traditional business functions and the tactics across these business functions within a particular company and across businesses within the supply chain, for the purposes of improving the long-term performance of the individual companies and the supply chain as a whole." (Mentzer et al., 2001, p. 18).

Sustainable supply chain is a relatively new area of study and practice, and there is no consensus regarding its definition. Researchers have defined sustainable supply chain from different perspectives and frameworks involving purchasing perspective (Carter and Jennings, 2004; Zsidisin and Siferd, 2001), marketing (Peattie, 2001), operations management (Kleindorfer et al., 2005), lean manufacturing (King and Lenox, 2001; Rothenberg et al., 2001), recycling and remanufacturing (Guide and Van Wassenhove, 2001), logistics (Carter and Jennings, 2002; Tibben-Lembke, 2002), and corporate social responsibility (Amaeshi et al., 2008; Andersen and Skjoett-Larsen, 2009). However, contemporary sustainable supply management literature largely adopts part of or all Triple-Bottom-Line components: economic, environmental, and social performance. Robinson and Wilcox (2008), for example, realized that sustainable supply chain conceptualizes commitment on environment initiatives. Hu and Hsu (2010) defined sustainable supply chain as the management of raw materials, parts/components and processes from suppliers to manufacturers to customers and product take back aiming at improving environmental impacts though lifecycle stages. Vachona and Maoc (2008) used archival data from The Global Competitiveness Report (2004–2005) and the 2005 Environmental Sustainability Index to confirm that supply chain strength is positively linked to the three dimensions of sustainable development (i.e., environmental performance, corporate environmental practices, and social sustainability. Similarly, Carter and Rogers (2008) delineated sustainable supply chain as the strategic, transparent integration and achievement of an organization's social, environmental, and economic goals in the systemic coordination of key inter-organizational business processes. In line with this perspective, Seuring and Müller (2008) defined sustainable supply chain management as the management of material, information and capital flows as well as cooperation among companies along the supply chain while considering economic, environmental and social dimensions. They further stressed that sustainable supply practices should be derived from customer and stakeholder requirements and fulfilled by organizational members. Such integration of sustainability into traditional organizational targets can add value to businesses and offer long-term competitive advantages (Pullman et al., 2009).

Integrating sustainability into supply chain has recently captured the attention of researchers and managers as the impact of sustainability goes beyond the firm's own operations to consider operations of the entire supply chain starting from input suppliers to the consumer including reverse logistics for obsolete products (Fiksel, 2010).

In conclusion, sustainable initiatives can affect every aspect of the supply chain including upstream or inbound activities which relate suppliers to manufacturers, conversion or production processes, downstream or outbound activities that relates manufacturers to customers, and reverse activities that take back products and materials to the manufacturer to be recycled and reused. This conceptualization leads to the integration of economic, environmental, and social performance into supply chain components to yield the following:

- Sustainable logistics;
- Sustainable procurement/purchasing;
- Sustainable product;
- Sustainable operations for manufacturing goods or providing services;
- Sustainable packaging and labeling;
- Sustainable transportation;
- Sustainable reuse and recycling;
- Sustainable sourcing;
- Sustainable buyer-supplier relationship;
- Sustainable inventory management;
- Sustainable materials.

4. SUSTAINABILITY AND SUPPLY CHAIN RISKS

Sustainable supply chain can help firms recognize and manage risks and disturbances associated with procurement, internal supply operations, and product development and stewardship through identifying critical issues on environmental, social and economic levels. Table 1 illustrates environmental, social and economic risks throughout the supply chain process as developed by New Zealand Business Council for Sustainable Development (2003).

5. SUSTAINABLE SUPPLY CHAIN IN DEVELOPING COUNTRIES

Recognizing its importance, a relatively well-developed body of research has investigated aspects of sustainable supply management in western developed contexts. Mustaffa and Potter (2009) indicated that many of supply chain ap-

Table 1. Environmental, social and economic risks of supply chain (Adapted from NZ Business Council for Sustainable Development, 2003)

Impact	Issue	Risk	Example
Procurement Operations			
Environ-mental Impact	Sustainable source of raw materials including all components	Chain of custody is uncertain. May result in consumer concern/NGO activity; Ineligibility for public/private tenders	NGO activity relating to deforestation; UK Gov. procuring doors & desks for Whitehall from endangered species
	Use of chemicals/hazardous substances.	Product safety scare. Precautionary Principle might be applied in certain countries. Product recall	Pesticides in produce. Marks & Spencer ban PVC.
	Animal husbandry: Breeding; Feeding; Food supply	Loss of business because chain of custody cannot be proved	Need to demonstrate animal husbandry.
	Long-term supply	Source drying up will threaten economic viability	Toyota developed hybrid vehicle recognizing diversifying energy supply.
	Short-term supply	Insufficient product to fulfill demand	NZ police not able to source enough LPG vehicles.
	Waste and packaging	Inefficient use of resources. Cost to business	Cost of landfill – levies and charges
Social Impact	Labor standards and practices. Conditions do not conform to ILO standards	Media scrutiny; boycotts; Loss of business, e.g. ineligible for public/private tenders	Overseas labor conditions, e.g. Nike; Gap
	Supplier dependence and viability	Eliminating a supplier from supply base might close their business and cause public outcry	Marks & Spencer's decision to move some of its supply base from UK to overseas.
	Fair pay for suppliers	Sustainability of supply base and potential adverse publicity	Coffee supply; Fair Trade products
Economic Impact	Inventory levels	Negative impact on cash flow	Companies left with seasonal stock due to inefficient procurement
	Local supply or overseas	Buying local may be more expensive and problems with continuous supply.	Sainsbury prefer local produce supply – selling $6bn worth per year
	Escalating cost of supply	Economic viability of product or services	Increase in fuel costs; Energy costs which cannot be passed on to end consumer
Internal Supply Operations			
Environ-mental Impact	Water; Air or soil pollution/ Contamination	Effect on reputation in local and wider community. Cost of remediation and risk of lawsuits	Prestige oil spill; Potential effect of any phosphate/nitrates land pollution to 'clean, green NZ' image.
	Perceived health impacts from local emissions	Effect on reputation in local and wider community	PCBs; Mobile phone masts; Incinerators
	Waste management	Increased costs of disposal to landfill	Potential taxation
Social Impact	Work/Life balance of employees; Unsocial hours	Retention of staff. Additional costs of recruitment. Potentially punitive costs for stress in the workplace or lawsuits	Impact of Occupational Health & Safety Amendment 2003
	Labour standards and practices; Pay & conditions	Staff retention; Strikes; Absenteeism; Claims	Wal-Mart challenged about trade unions; Strikes
Economic Impact	Increased cost of fuel; Energy; Delivery method	Economic viability of product or services	Plant closures on temporary basis because of high power costs
	Increased delivery times due to congestion	Increased resource costs; Service delays to customer; Impact on drivers; Road accidents	New sites with a recycling facility should be away from a residential area to reduce noise and pollution.
	Inefficient operations/Poor productivity/Cost of overheads	Economic viability of product or services; Higher direct and indirect costs; Effect on competitiveness	Companies outsourcing or setting up lower cost operations overseas.

continued on following page

Table 1. Continued

Impact	Issue	Risk	Example
Product Development and Stewardship			
Environmental Impact	Impact and efficiency of product in use	Adverse publicity; Product boycott; Reduction in market share	Major recall of Firestone tyres because of product fault
	Cost or suitability of raw materials increases. Need to find substitutes	Financial impact	Search for alternatives to materials such as PVC.
	End-of-life collection and Disposal	Loss of market share; Non-compliance with potential legislation in home or overseas market	Mobile phones; Car parts; White goods.
Social Impact	Packaging and materials: need to match packaging to product requirements	Increased costs from over packaging; Legislation in some markets relating to Take back systems, e.g. bottles, cans	Returnable levies on bottles in some Countries
	Customer demand	Loss of market share	First mover gains: Shell, Nike
	Increased cost of product/service because of increases in raw materials; Energy; Advertising	Market viability; Loss of market share	Pulp prices effect on the paper industry and can increase the price differential between sustainable and non-sustainable pulp supplies
Economic Impact	Product traceability; bar codes and labels	Costly recall if unable to pinpoint batches of product	Pan Health products recall

plications and practices have occurred within the developed world, with only a very limited range of examples from the developing world available in the literature. Examples of sustainable supply practices in developed contexts may involve US and North America firms (e.g., Min and Galle, 2001; Paulraj, 2011; Vachon and Klassen, 2006), UK public sector (e.g., Hall and Purchase, 2006; Preuss, 2009; Walker and Brammer, 2009), UK food sector (e.g., Rimmington et al., 2006), German manufacturing industry (e.g., Wolf, 2011), civil engineering public procurement in Northern Ireland (Eadie et al., 2011), high-tech medical equipment in the Netherlands (Lindgreen et al., 2008).

In the same vein, there is a misleading impression that Western countries were more sustainable than non-western countries with regard to sustainable supply chain (Kim and Min, 2011). Anbumozhi and Kanda (2005) indicated that most companies in developing countries tend to adopt "end-of-the-pipe" reactive approach for organizational sustainability aiming at reducing the negative environmental impacts rather than implementing a proactive approach to reduce the sources of waste or pollution. Nevertheless, comparatively very little research has recently investigated sustainable supply chain in developing countries. Table 2 highlights recent research findings regarding sustainable supply chain in developing countries.

Table 2 shows that existing research on sustainable supply chain in developing countries tend to be fragmented and theoretical in nature focusing on reviewing sustainable supply initiatives and practices. It is therefore important to unveil different aspects and perspectives of sustainable supply chain in developing countries recognizing that lessons could be learnt and knowledge could be transferred to other countries of a similar profile. Therefore, closer attention to empirical studies on sustainable supply chain in developing countries should be given since the topic is still immature.

Therefore, the purpose of this research is to provide further insights into sustainable supply chain in United Arab Emirates (UAE) from a stakeholder perspective. In other words, it aims

Table 2. Recent research on sustainable supply chain in developing countries

Author	Country	Findings
Rao (2002)	South East Asia	Environmental supply chain practices are taken place within manufacturing sector in South East Asia companies aiming at reducing environmental problems.
Zhu and Sarkis (2004)	China	There is a relationship between specific green supply chain practices and performance (i.e. environmental and economic performance). In addition, quality management and just-in-time (or lean) manufacturing principles may influence the relationship.
Rao and Holt (2005)	South East Asia	Greening the different phases of the supply chain leads to an integrated green supply chain, which ultimately leads to competitiveness and economic performance amongst a sample of companies in South East Asia.
Hus and Hu (2008)	Taiwan	Supplier management, product recycling, organization involvement and life cycle management are critical for implementing green supply chain management in the Taiwanese electrical and electronics industries.
Beatriz et al. (2009)	Brazil	The level of environmental management maturity influences the depth with which Brazilian companies adopt environmental criteria when selecting suppliers. However, Brazilian companies still use traditional criteria to select suppliers, such as quality and cost, and do not adopt environmental requirements in the supplier selection process.
Setthasakko (2009)	Thailand	Lack of system perspective sustainability, absence of top management commitment and cultural diversity, to implementing corporate environmental responsibility are three primary barriers to sustainable supply chain in seafood companies in Thailand.
Ras and Vermeu-len (2009)	South Africa	A qualitative model explaining business performance of South African entrepreneurs in a global supply chain has been developed.
Zuo et al. (2009)	Indonesia	The incorporation of sustainable procurement routes and triple bottom lines criteria in post-disaster construction in Indonesia should be recognized ensuring stakeholder integration and collaboration to reduce the problems in timber procurement.
Mwirigi (2010)	Kenya	Education level of the entrepreneur, exposure to information relating to business and the market dynamics, perceptions of the entrepreneur and actual size of business determine the likelihood of an entrepreneur nurturing strong sustainable supply chain relationships among in Kenyan small firms.
Tencati et al. (2010)	Vietnam	Impact of sustainable sourcing policies created by multinational companies on Vietnamese suppliers is examined. The findings highlight that monetary and non-monetary costs of complying with international standards may turn out to be unsustainable and prohibitive for many Vietnamese enterprises, and especially for small- and medium-sized suppliers.
ElTayeb et al. (2010)	Malaysia	Drivers for green purchasing adoption among EMS 14001 certified companies in Malaysia are examined. The results indicate that Green purchasing is affected by regulations, customer pressures, expected business benefits and firm ownership. The results also suggest that, although Malaysian firms show a high level of social responsibility, it does not constitute a genuine driver for these firms to adopt green purchasing.
Faisal (2010)	Gulf Region	Enablers of sustainable supply chains in Gulf region are examined. Sustainable supply chain adoption requires awareness about sustainable practices like ethical sourcing, green purchasing, environmental purchasing, and logistics social responsibility.
Gupta and Palsule-Desai (2011)	India	Literature on sustainable supply is divided into four broad categories: strategic considerations; functional decisions; regulation and government policies; and integrative models and decision support tools. Environmental initiatives in India and the relevance of sustainability in the context of Indian economy are also discussed.
Jayaratne (2011)	Sri Lanka	Factors that influence sustainable tea supply chain in Sri Lanka are mapped using a theoretical framework.
Zailani et al. (2012)	Malaysia	Environmental purchasing has a positive effect on three categories of outcomes (economic, social and operational), whereas sustainable packaging has a positive effect on environmental, economic and social outcomes. The study indicated that sustainable supply chain management practices have a positive effect on sustainable supply chain performance, particularly from the economic and social perspective.
Mohapatra and Srivastava (2012)	India	Needs of different supply chain stakeholders in rural areas in India are analyzed. The study also provides detailed ROI calculations for all the stakeholders in the value chain and also proposes a road map for implementing the model to improve living conditions in rural areas.
Zhu et al. (2012)	China	The study evaluates the relative roles of innovation and imitation drivers for diffusion of these green supply chain practices. The authors indicates that even though innovation is not insignificant, imitation plays a much larger role for these specific green supply chain practices diffusions amongst Chinese enterprises.
Jabbour et al. (2013)	Brazil	Company size, previous experience with Environmental Management Systems, and the use of hazardous inputs are positively correlated with green supply chain practices adoption in Brazilian companies.

at understanding the role of pressures by firm's stakeholders in influencing sustainable supply chain which, in turn, impacts sustainable organizational performance.

6. A STAKEHOLDER PERSPECTIVE OF SUSTAINABLE SUPPLY CHAIN

Stakeholder models have attracted researchers from different fields since Freeman (1984) published his influential book, *Strategic Management: A Stakeholder Approach*. This perspective suggests that firms should design policies that take into account the pressures from or expectations of any group or individual who can affect or is affected by the achievement of the organization's objectives (Heath, 2006; Stieb, 2009; Walsh, 2005). Therefore, researchers have sought to classify stakeholders to understand how they influence a focal firm and to understand their environment (Mitchell et al., 1997). Stakeholder view has been adopted to investigate corporate sustainability efforts (Clarkson, 1995; Garriga and Mele, 2004; Maon et al., 2009; McWilliams et al., 2006), corporate social responsibility (Jenkins, 2006), and sustainable supply chain management (Ehrgott et al., 2011; Wolf, 2011). Bearing in mind the pressures from and expectation of stakeholders will sustain the future of firm over the long-term. Firms who identify and engage early with customers, employees, government, environmental activists and NGOs stakeholders have the opportunity to act proactively on emerging environmental, social and economic issues in the supply chain and take early leadership in comparison to peers. Also, insights from these stakeholders can help shape relevant sustainable standards of supply chain practices. Therefore, integrating stakeholder's view into sustainable supply strategy will ensure better performance (Narasimhan and Wook, 2001) since it will lead to better performance management (Debrito et al., 2008), improved relationship management (Reuter et al., 2010), effective supply chain partner selection (Jiang, 2009), lower transaction costs (Williamson, 2008) and creating firm resources (Hunt and Davis, 2008). Yet, the application of the stakeholder concept in sustainable supply chain research is still in a comparatively young stage (Ehrgott et al., 2011). Hence, there is a need for examining the specific influence of stakeholder groups on various aspects of supply chain sustainability and the corresponding effects on firm's sustainable performance, particularly in a developing country context.

6.1. The Impact of Sustainability Champions

Champions are individuals within the organization who pioneer new ideas, products, or concepts (Gray and Smeltzer, 1989). Champion pressures are conceptualized as the extent to which employees in the firm pushes for increasing sustainability efforts. Previous research supports the critical role played by organizational champions in adoption and implementation of innovations in an organization (Jeyaraj et al., 2006). Sustainability champions could be at different managerial levels; first, middle or top. Drumwright (1994) found that the primary initial driver of environmental purchasing came from workers who were personally committed to environmental values. Carter et al. (1998) demonstrated a significant relationship between middle as well as top management initiatives and environmental purchasing. By means of case study analysis, Carter and Dresner (2001) explained that "champions" in the organization are an important success factor for sustainable supplier management projects. Similarly, Carter et al. (2007) map the social network within an organization to demonstrate how a manager at the lower ranks effectively championed and drove to fruition a safety related supplier management project. Using data from 244 U.S. and German corporations, Ehrgott et al. (2011) confirmed that the intensity of middle management social pressures is positively related to socially sustainable supplier selection.

It is not just lower-level managers from whom sustainability championing is needed, championing from higher-level managers is also proved to successful implementation of environmental practices. Lambert et al. (1998) submitted that change champions at top managerial levels are important drivers to the implementation of supply chain programs. This line of reasoning is further supported by Pun et al. (2001) and Daily and Bishop (2003) who consider the sustainability championing by top managers as very critical to the success of environment initiatives since it promotes employee empowerment, cultural shift, rewards and incentives systems, training and teamwork concerning sustainability. Moreover, Carter and Jennings (2004) proved that environmental initiatives positively influence purchasing social responsibility. Thus:

Hypothesis 1: The strength of champion pressures is positively related to supply chain sustainability.

6.2. The Impact of Customers

Research has documented customer demands and pressures as a key external driver for supply chain sustainability (Alvarez-Gil et al., 2007; Darnall and Edwards, 2004; Doonan et al., 2005; Lin, 2007; Peng and Lin, 2008; Rao, 2006; Zhu et al., 2008). Lamming and Hampson (1996) indicated that about 75% of US consumers claim that their purchasing decisions are influenced by a company's environmental reputation, and 80% would be willing to pay more for environmentally friendly goods. For example, consumers desire clean products that have been provided through an environmentally sustainable manner (Anbumozhi and Kanda, 2005; Collins et al., 2007; Zhu and Sarkis, 2006). For example, Ford Motor Company requires that all its suppliers comply to the ISO 14001 standard, and other car manufacturers like Toyota, BMW, and Mitsubishi include "supplier activities" in statements of environmental responsibility (ElTayeb et al. 2010). In the same vein, Anbumozhi and Kanda (2005) suggested that pressures and expectations for green products from large companies in developed countries may improve the environmental performance of Asian companies. Also, customers want to understand the conditions under which products have been produced (Locke and Romis, 2007; Locke et al., 2007). Similarly, some customers might choose to boycott the firm's products because of negative environmental impacts (Vachon and Klassen, 2006). Moreover, customer awareness usually is not limited to the knowledge of a firm's product but extends to its conduct and behavior (Brown and Dacin, 1997).

Heightened media coverage and rising legal requirements for customer information make firms' policies more transparent and allow customers to consider the standards to which firms hold their suppliers in their buying decisions (Wolf, 2011). Based on the foregoing argument, the following hypotheses are provided:

Hypothesis 2: The strength of customer pressures is positively related to supply chain sustainability.

6.3. The Impact of Government

With increasing environmental regulations and pressures firms are expected to fulfill socially responsible business practices (Pilkington and Dyerson, 2006). Conformity with these pressures occurs through influence exerted by government agencies as powerful groups that guide the actions of an organization (Rivera, 2004).

Governmental pressures in the form of formal rules, laws, sanctions, and incentives have been found to be the most important drivers for the adoption of sustainability strategies for corporate firms (Gonzalez-Benito and Gonzalez-Benito, 2005; 2008; 2010; Sharma and Vredenburg,

1998). Furthermore, these regulations may cover domestic environmental regulations and policies as well as international environmental agreements that impose direct constraints over firms to make necessary changes in their structure and processes, such as the Kyoto agreement, the Climate Change Treaty and the Montreal Protocol (EIC, 2005; Zhu and Sarkis, 2006; Hall, 2000). Moreover, The EU WEEE directive attempts to tackle the growing quantity of WEEE by making producers responsible for the costs of the collection and recycling of their products at the end of usable life (Gottberg, et al., 2006).

While some organizations try to avoid authoritative rules or legal liabilities related to potential costs, uncertainty, hazardous products, recycling, and expired products (Clemens and Douglas, 2006; Tibben-Lembke, 2002; Zhu and Sarkis, 2006), regulatory institutions may provide inducements for these organizations to behave in a certain way that conforms to the demands of specific agency (Grewal and Dharwadkar, 2002). In addition, parent companies may set rules and standards that oblige their subsidiary companies, in developing countries, to adopt certain sustainable (ElTayeb et al., 2010).

Gupta and Palsule-Desai (2011) explained that government adopt a 'command and control' perspective by mandating levels of environmental taxes (e.g., carbon tax), forcing firms to adopt minimum environmental standards (e.g., mandating a certain percentage of power generation to come from renewable sources), or subsidizing certain technologies and industries (e.g., solar and wind power generation). Consequently, firms are required to comply with these pressures to avoid potential costs, uncertainty, and legal liabilities inherent in existing and anticipated regulations (Clemens and Douglas, 2006). Case study findings from Handfield et al. (1997) suggested that regulations have a positive effect on a firm's environmental initiatives. At a rather specific level, prior research provide valid evidence that government pressures have positive effects on supply chain sustainability including green purchasing in Malaysia (ElTayeb et al., 2010), environmental supply chain practices (Hall, 2000; Walker et al., 2008) and green supply initiatives in South Korea (Lee, 2008). Accordingly, the above arguments lead to the following hypothesis:

Hypothesis 3: The strength of government pressures is positively related to supply chain sustainability.

6.4. The Impact of Sustainable Supply Chain on Sustainable Performance

Realizing sustainable supply chain benefits and outcomes is an important vehicle that sustains implementation of sustainable supply initiatives in a developing context. These outcomes can be realized from three perspectives: environmental, economic, and social.

6.4.1. Environmental Performance

Environmental performance can be defined as the environmental impact that the corporation's activity has on the natural milieu (Sharma and Vredenburg, 1998). This impact could be reflected in operative performance such as materials' consumption, energy management, waste and emission production, and evaluation of real environmental aspects of organizations, as well as the overall organization's environmental management efforts (Papadopoulos and Giama, 2007). Sustainable supply chain paves the way for sustainable performance as a result of environmentally-oriented supply operations that can support the competitiveness of the firm. Prior research links sustainable supply chain practices to better environmental, quality and economic performance (Barney, 1991; Hart, 1995). For example, effective management of sustainable supply chain can promote recycling of raw materials, cut waste and hazardous substances, prevent violating environmental regulations, reduce transaction costs, and enhance energy and water consumption. Generally, empirical researchers have found that firms adopting comprehensive environmental supply chain practices experience better sustainable performance in terms of significantly reduced waste (Melnyk et al., 2003), improved product and process quality (Kleindorfer et al., 2005; Melnyk et al., 2003), and improved recycling and waste reduction (Sroufe, 2003; Rothenberg et al., 2001; Florida, 1996). Appropriate selection of suppliers determines their ability in improving the environmental objectives of a firm (Walton et al., 1998; Min and Galle, 2001; Zhu and Geng, 2001). Furthermore, such collaboration between suppliers and customers in sustainable supply helps firms to develop the environmental prowess of the supply partner (Klassen and Vachon, 2003) such as in case of preventing or mitigating environmental hazards and its associated risks (Linton et al., 2007; Zhu et al., 2007).

6.4.2. Economic Performance

Sustainable supply chain can escort sustainable economic performance in terms of internal cost saving, opening new markets and finding beneficial uses for waste, cutting cost of purchasing materials and energy consumption, reducing the cost of waste treatment and discharge, avoiding a fine in the case of environmental accidents, and increasing profits, sales and market share (Fuentes-Fuentes, et al., 2004; Tsoulfas and Pappis, 2006; Zhu and Sarkis, 2004). For example, cases from Malaysian context show that sustainable supply can insure economic returns such as cost savings, marketing opportunities and financial returns to justify the cost associated with compliance to a formal environmental standard (Anbumozhi and Kanda, 2005, ElTayeb, 2010, Zulkifli and Amran, 2006). Zhu and Sarkis (2004) found that the existence of environmental supply management programs led to both positive and negative economic performance in Chinese manufacturing enterprises. However, Carter et al. (2000) found that environmental supply management is significantly related to both net income and cost of goods sold after controlling for firm size, leverage and primary earnings per share. Also, Melnyk et al. (2003) found that firms adopting formal environmental supply management systems significantly reduced overall costs.

6.4.3. Social Performance

From a social perspective, much previous research has focused on sustainable supply chain because its relation to social performance. Sustainable performance should recognize, value and promote the capability of people with appropriate human policies and practices for equity, development and well-being (Daily and Huang, 2001). Sustainable supply chain practices involve reciprocal collaboration among supply partners in terms of knowledge resources and capabilities which in turn can improve total sustainable performance (Geffen and Rothenberg, 2000; Klassen and Vachon, 2003). Marshall et al. (2005) found that a concern for vineyard employee welfare was linked to the reduction of toxic spray applications and other potentially damaging environmental practices. Worker participation and training have been positively related to environmental improvement in supply operations (Florida, 1996; Johnson et al., 2006; Rothenberg et al., 2001). Additionally, collaborative management of supplier relationships in sustainable supply can help to design new socially oriented products and improve operational efficiencies (Damali et al. 2008; Wittmann et al., 2009; Lao et al., 2010; Pagell et al., 2010).

Based on the foregoing arguments, we expect sustainable supply chain practices to positively influence sustainable performance, leading to the following hypothesis:

Hypothesis 4: Supply chain sustainability is positively related to performance sustainability.

Figure 2. Research hypotheses

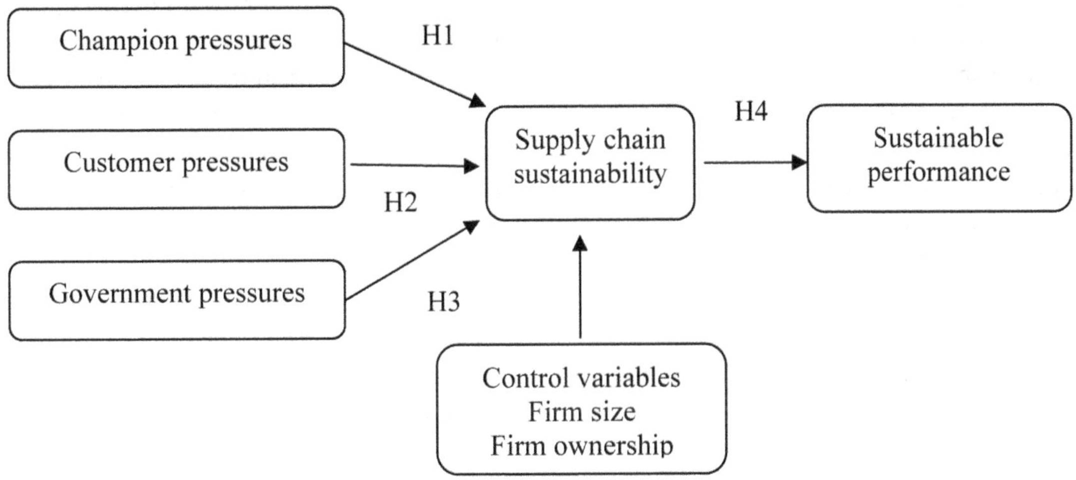

6.4.4. Control Variables: Size and Ownership

To ensure robustness of the results, three control variables were considered in the analysis. Firm size and ownership type may influence the sustainable practices of the firm. Also, firms in different industries may behave differently pertaining to sustainable practices. The selection of these variables is based on previous studies that found a significant effect of these variables on green initiatives. For instance, Bowen (2002) and Carter and Jennings (2004) argued that larger firms are more committed to corporate social responsibility and sustainable performance because they have more resources and they are more visible to the society. In addition, Branzei and Vertinsky (2003) indicated a positive influence of firm size on 'eco-sustainability' orientation in Chain and Japan. In addition, the pressure on corporations to improve their environmental performances comes from globalization rather than localization (Sarkis and Tamarkin, 2005). ElTayeb et al. (2010) revealed that multinational firms are more oriented to green practices than local firms in Malaysia. This seems reasonable since international firms are exposed to international environmental agreements as well as domestic environmental regulations which both have a great and immediate effect on sustainable supply practices. Figure 2 illustrates the hypothesized relationships.

7. THE EMPIRICAL STUDY

7.1. Data Collection

A self-administered questionnaire was used to target top managers with particular emphasis on purchasing/supply managers and officers in the UAE firms. Because of the difficulties facing researchers in collecting data in Arab countries, a personal delivery and pick-up system seems to be the most appropriate method of collecting questionnaires (Elbanna, 2010; Hunt and At-Twaijri, 1996). Immediately after the questionnaire distribution, a letter soliciting internal promotion of the study was emailed to all corresponding managers in participating organizations.

Respondents were asked to complete the questionnaires individually and return them to designated department secretaries, from whom the questionnaires were collected.

A total of 400 questionnaires were distributed to randomly selected organizations which were drawn from economic and commercial directories published by chambers of industry and commerce in seven UAE emirates including Abu Dhabi, Dubai, Sharjah, Ajman, Ras al-Khaimah, Fujairah and Umm al-Quwain. A total of 118 completed questionnaires were returned. A total of 6 questionnaires were discarded because of incomplete data, leaving a balance of 112 usable questionnaires for this study with a response rate of 28 per cent. The majority of respondents were in operations or supply specialization (41 per cent) with occupational experience over ten years (79 per cent). Most of the respondent firms were from service sector (72.6 per cent), small/medium in size (SMEs) (58.5 per cent) and with local ownership (61.4 percent).

7.2. Measures

All the measurement items in the study were adapted from prior research. The three constructs describing pressures from customers (CP), the government (GP), and management employees (MP) were operationalized based on the work of Carter and Jennings (2004) and Ehrogott et al. (2011). Pressures from suppliers (SP) are measured by five items adapted from Vachon and Klassen (2006). Following past research, sustainability of supply chain (SS) is conceptualized as a second-order factor including fifteen items that tap supplier selection, environmental collaboration and supplier evaluation adopted from Zhu et al. (2008) and Zhu and Sarkis (2004). Supplier selection includes five items that reflect the extent to which the firm selects their potential suppliers based on their environmental competence and environmental performance. Environmental collaboration has six items that measure the extent to which firms cooperate with their suppliers to develop environmental strategies and provide them with resources to support their environmental goals. Finally, supplier evaluation dimension uses four items to capture the extent to which firms regularly monitor suppliers' internal operations, goods and performance.

Sustainable organizational performance is operationalized to cover the economic, environmental and social dimensions of sustainable performance based on the work of Carter and Rogers (2008) and Paulraj (2011). Economic performance dimension measures cost, return on investments and earnings indicators. Environmental performance dimension covers improvements in air emission, waste, consumption of hazardous materials, environmental accidents and energy savings. Social performance covers improvements in social welfare and betterment, community health and safety, risks to the general public and occupational health and safety of employees. For all measures, a seven-point Likert-type scale (ranging from 1 = strongly disagree to 7 = strongly agree) was used.

Control variables were included and operationalized using ordinal scales to measure potential effects. The number of employees was included as a dummy variable with firms having < 500 employees coded as 0 and firms having > 500 coded as 1. Firm ownership was measured as a dummy variable with local firms coded as 0 while non-local firms (multinational, JV, etc.) coded as 1.

7.3. Reliability and Validity

To ensure validity and reliability, several steps were taken to refine and rigorously pre-test the questionnaire. The questionnaire was pre-tested by three experts in environmental sciences and five experts in supply chain. The comments collected from these experts provided a basis for revisions of the construct measures and modifications of the wording and item sequence. All of our multiple-item constructs achieved Cronbach alpha's of 0.73 or higher (0.73 - 0.92 was obtained), which is higher than the suggested cut-point alpha value of 0.70 (Hair et al., 2006), indicating strong internal consistency. Moreover, Harman's one-factor test

Table 3. Correlation matrix and summary statistics

	Variable	Mean	SD	1	2	3	4	5	6
1	Firm size	0.415	0.581						
2	Firm ownership	0.381	0.623	- 0.153					
3	Champion pressures	4.243	0.784	0.087	0.404				
4	Customer pressures	3.676	0.659	0.133	0.252	0.058			
5	Government pressures	3.752	0.921	0.215	0.379	0.145	0.177		
6	Supply chain sustainability	3.819	0.834	0.412	0.553	0.356	0.263	0.159	
7	Sustainable performance	4.137	1.026	0.370	0.348	0.158	0.220	0.172	0.501

Note: Correlations with absolute value greater than 0.19 are significant at $p<0.05$, and those greater than 0.28 are significant at $p < 0.01$

was used to address the issue of common method variance. Significant common method variance could result if one general factor accounts for the majority of covariance in the variables (Podsakoff et al., 2003). A principal components factor analysis on the questionnaire measurement items yields all factors with eigenvalues >1.0 that account for 71.16% of the total variance, and the first factor accounts for 26.14% of the variance. Since a single factor does not emerge and one general factor does not account for most of the variance, common method bias is unlikely to be a serious problem in this research (Podsakoff et al., 2003).

7.4. Research Results

This study attempts to understand the role of stakeholder pressures by government, customers and organizational champions in determining the sustainability of supply chain practices as well as its impact on sustainable organizational performance. Table 2 shows correlations, means, and standard deviations for all variables. The correlations range from 0.058 to 0.553; with no pair of measures exceeding the value of 0.60, suggesting no need for concern with respect to sever multicollinearity problems among research variables (Hair et al., 2006). The correlations also indicate a preliminary support for the significant relationships among key variables.

Table 3 shows the results of regression analysis predicting supply chain sustainability. Overall, it suggests that the variables are largely related in the theoretically predicted manner. Model I in Table 3 provide the base models that include the two control variables where size and ownership type have significant effect on supply chain sustainability. Model II captures the direct effects of pressures by government, customers and organizational champions on supply chain sustainability, The two models are both significant, at $p < 0.01$ for model I and $p < 0.001$ for Model II which explain an additional 41% of variance over what the control variables alone explain ($R^2 = 0.23$ and 0.64, respectively).

Based on this model, the study revealed many interesting results. For Hypothesis 1, the coefficients of pressures by organization champions are positive and significant for supply chain sustainability ($\beta = 0.31, p < 0.001$). Pressures by customers are also emerged as a significant variable influencing supply chain sustainability ($\beta = 0.19$, $p < 0.01$), providing support for hypothesis 2. On the contrary, the coefficients of government pressures are insignificant for supply chain sustainability ($\beta = 0.11$, na). Therefore, hypothesis 3 is not supported. Next, the study examines how supply chain sustainability affects sustainable organizational performance. Model III in Table 4 is the base model that includes the two control variables where firm ownership significantly

Table 4. Results of regression analysis predicting supply chain sustainability

Dependent Variable	Supply Chain Sustainability	
	Model I	Model II
Control variables		
Firm size	0.154*	0.043
Firm ownership	0.238**	0.091
Independent variables		
Champion pressures		0.305***
Customer pressures		0.186**
Government pressures		0.112
Model statistics		
R^2	0.236	0.649
Adjusted R^2	0.214	0.635
F	13.796***	41.351***

* *p* <0.05; ** *p* <0.01; *** *p* <0.001

Table 5. Results of regression analysis predicting sustainable performance

Dependent Variable	Sustainability Performance	
	Model III	Model IV
Control variables		
Firm size	0.051	0.013
Firm ownership	0.48***	0.221**
Independent variable		
Supply chain sustainability		0.583***
Model statistics		
R^2	0.276	0.461
Adjusted R^2	0.258	0.447
F	15.036***	28.911***

* *p* <0.05; ** *p* <0.01; *** *p* <0.001

influences sustainable performance with no significant effect of firm size. Model IV in Table 4 presents the results with inclusion of the sustainable supply chain factor in the model which is significant ($\beta = 0.58, p < 0.001$) and yields a R^2 of 0.46. The results suggest that supply chain sustainability is a significant determinant of sustainable organizational performance, providing support for hypothesis 4.

8. DISCUSSION AND IMPLICATIONS FOR FUTURE RESEARCH

The purpose of this paper is to provide further insights into drivers and consequences of sustainable supply chain in a developing country from a stakeholder perspective. Specifically, it aims at empirically examining the impact of stakeholder pressures on sustainable supply chain practices as well as the impact of these practices on sustainable performance including environmental, economic and social dimensions. The results indicate that pressures by organizational champions and customers are positively related to supply chain sustainability. However, the relationship between government pressures and supply chain sustainability is insignificant. Moreover, the findings provide evidence that the impact of supply chain sustainability on sustainable performance is significantly positive. The findings contribute to theoretical development in several ways.

First, the results of regression analysis provide support for the notion that some stakeholder pressures over the firm can lead to a higher sustainability level of its supply chain. This is in line with stakeholder research that focuses on their critical impact on sustainable practices and performance (e.g., Debrito et al., 2008; Reuter et al., 2010; Jiang, 2009; Williamson, 2008). Also, the present findings imply that some stakeholders would have a higher significant impact than others with regard to sustainable practices and firms, therefore, should first understand and react to particular needs of the most influential stakeholders.

Second, the findings reveal strong evidence of the critical role played by customers in influencing sustainable supply chain initiatives in UAE firms. This is consistent with previous research that relates environmental pressures by customers to firms' environmental supply management practices. For example, Rao (2002) addressed that collaborative relationships between firms and their customers can contribute to high levels of advanced environmental supply chain management practice of the firms. Lee (2008) addressed that buyer environmental requirements and support were positively linked to their suppliers' willingness to participate in green supply chain initiatives in South Korea. However, Zhu et al (2008) did not find these pressures from customers to be significant in the context of green supply practices in an emerging Chinese market. Although firms in developing countries may relatively be less exposed to the customer enforcement on environmental issues, customers' awareness and expectations for environmental practices are growing over time and pushing for sustainable organizational reaction (Gilmore and Pine, 2007; Reynolds and Yuthas, 2008). Hence, customers' environmental requests can directly affect firm's supply behavior through turning their attention to environmental issues. In the same vein, Hall (2000) predicted that a wave encouraging the greening of a supply chain is likely to be triggered by final customers who reflect market pressures onto their firms. Supply chain managers in UAE firms need to bear in mind customer expectations and its impact on sustainable supply decisions. Since they are dealing directly with customers, a strategic coordination with the firm's sales and marketing managers ought to be beneficial to establish customer-centric performance in supply management.

Third, the findings provide support for the role of organizational champions in influencing supply chain sustainability. This is supported by a number of researchers who have looked at environmental pressures and support by employees in different context (e.g., Carter et al., 1998; Zhu et al., 2008). This confirms the predictions of stakeholder theory that organizational members can use their power to influence the firm's supplier management practices (Ehrgott et al., 2011). From the person–organization perspective, Behery (2009) suggested that person–organization and person–job fit affect organizational commitment through the psychological contract in UAE firms. Within this perspective, the congruence among individual, job and organizational values pertaining to sustainability and environmental concerns in UAE firms may boost the commitment of organizational members toward environmental practices and shape their willingness to exert considerable environmental efforts on the organization's behalf.

Moreover, the positive relationship between organizational champions and supply chain sustainability will play a role in diminishing the barrier to sustainable practices. Thus, many firms in UAE may identify with these findings through creating supportive environment that encourage their employees, particularly supply officers and managers, to express their intrinsic motivation to advance supply chain sustainability concerning supplier selection, evaluation and environmental collaboration dimensions. The other issue that may be arising in this situation is that most of local UAE firms are in early adoption stages of sustainable supply chain practices. Thus, linking consumers and environmental organizational champions through training and education programs or other sharing information mechanism will facilitate collaborative development concerning dimensions of supply chain sustainability (e.g., assessing environmental performance of suppliers and their products) and foster its implementation.

Fourth, pressures from government have not emerged in this study as a significant driver for sustainable supply chain. This result suggests that formal rules, laws, sanctions, and incentives do not motivate sustainable supply chain practices as strongly as hypothesized. This finding contradicts prior research that considers governmental pressures as the most important driver for the adoption of sustainable supply chain initiatives (ElTayeb et al., 2010; Hall, 2000; Lee, 2008; Walker et al., 2008). However, findings of this research were in accordance with the results by other studies such as Ehrgott et al. (2011), Bowen et al. (2001), Zhu et al. (2007; 2008). For Example, Zhu et al (2008) who did not find these pressures from government to be significant in the context of green supply practices in an emerging Chinese market. Further, Carter and Jennings (2004) found no impact of government regulation on purchasing social responsibility. Rather, Ehrgott et al. (2011) found no significant relationship between pressures from government and socially sustainable supplier selection in US and German firms. A plausible explanation for the insignificant effect of government pressures in our study is that these pressures may encourage reactive environmental policies and practices that aim at avoiding governmental sanctions rather than supporting firms' proactive sustainable initiatives. This explanation is in line with Mardi (1992) and Dean and Brown (1995) who found that governmental regulations can create operating barriers and increases costs for businesses. Buysse and Verbeke (2003) arrived at a similar result when they segment the European firms into environmentally leading, environmentally proactive, and environmentally reactive. They find a positive governmental effect only for reactive companies.

Fifth, the positive relationship between supply chain sustainability and sustainable performance is very promising; however, not all studies have shown this relationship to be significant. For example, Levy (1995) found little relationship between green supply chain and sustainable performance. Moreover, there has been some trepidation of confirming that environmental performance can be improved through applying environmental practices such as ISO 14001 (King et al., 2005). Bowen et al. (2001) suggested that financial performance is clearly not being reaped in short-term profitability and sales performance. Despite these findings, previous research confirms the role of supply chain practices in improving environmental performance (Melnyk et al., 2003; Sroufe, 2003; Linton et al., 2007; Zhu et al., 2007), economic performance (Alvarez Gil et al., 2001; Carter et al., 2000; Zhu and Sarkis, 2004) and social performance (Daily and Huang, 2001; Marshall et al., 2005). For example, Alvarez Gil et al. (2001) indicated that environmental management such as GSCM has a positive relationship with an organization's economic performance. There is also an evidence to suggest that sustainable supply chain can prepare enterprises for superior longer-term performance through improved management of environmental risks and the development of capabilities for continuous environmental improvement (Zhu et al., 2010).

Supply managers can learn from these results in developing sustainable initiatives earlier along their supply chain through selecting and evaluating suppliers based on sustainability-related standards. Also, managers should consider environmental collaboration with customers and suppliers to identify and reduce the total environmental impact. Such collaboration may involve knowledge sharing and application through exchanging physical as well as intellectual resources that maintain tools and skills. Finally, this research provides support for the significant impact of firm size and ownership type on supply chain sustainability. This is in harmony with previous empirical studies that consider larger firms to be more committed to sustainable practices than smaller firms (e.g., Branzei and Vertinsky, 2003; Bowen, 2002; Carter and Jennings, 2004). The findings are also consistent with ElTayeb et al. (2010) and Sarkis and Tamarkin (2005) who suggest that international firms are more exposed to sustainable practices than local firms. Managers in larger firms and multinational firms tend to have sufficient resources to invest in sustainable supply practices as they are more exposed to environmental pressures.

Like any other research, this study has some limitations. First, the self-reported data used in this study may suffer the problems of common method variance. Though this study uses the Harman one-factor test to verify that common method variance is not a significant problem, the issue may still exist and accordingly is addressed as a potential limitation. Thus, future research can benefit from asking more than one respondent in each firm using objective measures of research variables.

Second, this study did not examine the direct effects of stakeholder pressures on sustainable performance and the potential mediating effect of supply chain sustainability. Further research may address these relationships using a structural equation model to explore the direct and indirect effects between the variables in the study simultaneously. Third, this study only concerns the effects of some stakeholders on supply chain sustainability. Other internal organizational capabilities such as knowledge management capabilities, strategic orientation of supply function and innovation performance may potentially affect supply chain sustainability and/or sustainable performance. Hence, future research may work on examining the impact of internal capabilities on sustainable practices. Fourth, investigating UAE firms may limit the generalizability of the findings to other contexts due to potential cultural effects. Nevertheless, the study provides researchers with rich data that can guide future empirical research to generalize or modify the concepts in different cultural contexts using different methods such as longitudinal design providing further insights into the dynamic nature of research variables.

To conclude, pressures by organizational champions and customers in a developing country context are positively related to supply chain sustainability. However, the pressures by government do not show significant effect on supply chain sustainability. Moreover, the impact of supply chain sustainability on sustainable performance is significantly positive.

REFERENCES

Alvarez-Gil, M., Berrone, P., Husillos, F. J., & Lado, N. (2007). Reverse logistics, stakeholders' influence, organizational slack, and managers' posture. *Journal of Business Research*, *60*(5), 463–473. doi:10.1016/j.jbusres.2006.12.004

Amaeshi, K., Osuji, O., & Nnodim, P. (2008). Corporate social responsibility in supply chains of global brands: A boundaryless responsibility? Clarifications, exceptions and implications. *Journal of Business Ethics*, *81*(1), 223–234. doi:10.1007/s10551-007-9490-5

Anbumozhi, V., & Kanda, Y. (2005). Greening the production and supply chains in Asia: Is there a role for voluntarily initiatives? *IGES Kansai Research Center Discussion Paper*, KRC-2005 (6). Retrieved from www.iges.or.jp

Andersen, M., & Skjoett-Larsen, T. (2009). Corporate social responsibility in global supply chains. *Supply Chain Management, 14*(2), 75–86. doi:10.1108/13598540910941948

Barney, J. (1991). Firm resources and sustained competitive advantage. *Journal of Management, 17*, 99–120. doi:10.1177/014920639101700108

Beatriz, A., Jabbour, C., & Jabbour, J. (2009). Are supplier selection criteria going green? Case studies of companies in Brazil. *Industrial Management & Data Systems, 109*, 477–495. doi:10.1108/02635570910948623

Behery, M. (2009). Person/organization job-fitting and affective commitment to the organization: Perspectives from the UAE. *Cross Cultural Management: An International Journal, 6*(2), 179–196. doi:10.1108/13527600910953928

Bowen, F. E. (2002). Does size matter? Organizational slack and visibility as alternative explanations for environmental responsiveness. *Business & Society, 41*(1), 118–124. doi:10.1177/0007650302041001007

Bowen, F. E., Cousins, P. D., Lamming, R. C., & Faruk, A. C. (2001). The role of supply management capabilities in green supply. *Production and Operations Management, 10*(2), 174–189. doi:10.1111/j.1937-5956.2001.tb00077.x

Branzei, O., & Vertinsky, I. (2003). Eco-sustainability orientation in China and Japan: Differences between proactive and reactive firms. In *Research in corporate sustainability*. Northampton, MA: Edward Elgar Publishing.

Brown, T. J., & Dacin. (1997). The company and the product: Corporate associations and consumer product responses. *Journal of Marketing, 61*(1), 68–84. doi:10.2307/1252190

Buysse, K., & Verbeke, A. (2003). Proactive environmental strategies: A stakeholder management perspective. *Strategic Management Journal, 24*(5), 453–470. doi:10.1002/smj.299

Carlton, J. (2006, June 29). EU's environmental hurdles for electronics, rules to require mitigation of toxic materials that are common in many products. *Wall Street Journal*, p. B5.

Carter, C., Ellram, & Tate, W. (2007). The use of social network analysis in logistics research. *Journal of Business Logistics, 28*(1), 137–170. doi:10.1002/j.2158-1592.2007.tb00235.x

Carter, C., & Dresner, M. (2001). Purchasing's role in environmental: Cross-functional development of grounded theory. *Journal of Supply Chain Management, 37*, 12–27. doi:10.1111/j.1745-493X.2001.tb00102.x

Carter, C., & Jennings, M. (2004). The role of purchasing in corporate social responsibility: A structural equation analysis. *Journal of Business Logistics, 25*(1), 145–186. doi:10.1002/j.2158-1592.2004.tb00173.x

Carter, C., Kale, R., & Grimm, C. (2000). Environmental purchasing and firm performance: An empirical investigation. *Transportation Research Part E, Logistics and Transportation Review, 36*(3), 219–228. doi:10.1016/S1366-5545(99)00034-4

Carter, C., & Rogers, D. (2008). A framework of sustainable supply chain management: Moving toward new theory. *International Journal of Physical Distribution & Logistics Management, 38*(5), 360–387. doi:10.1108/09600030810882816

Carter, C. R., Ellram, & Ready, K. (1998). Environmental purchasing: Benchmarking our German counterparts. *Journal of Supply Chain Management, 34*(4), 28–38.

Chopra, S., & Meindl, P. (2001). *Supply chain management: Strategy, planning, and operations*. Upper Saddle River, NJ: Prentice-Hall, Inc.

Clarkson, M. (1995). A stakeholder framework for analyzing and evaluating corporate social performance. *Academy of Management Review*, *20*(1), 92–117.

Clemens, B., & Douglas, T. J. (2006). Does coercion drive firms to adopt 'voluntary' green initiatives? Relationships among coercion, superior firm resources, and voluntary green initiatives. *Journal of Business Research*, *59*, 483–491. doi:10.1016/j.jbusres.2005.09.016

Collins, C. M., Steg, & Koning. (2007). Customers' values, beliefs on sustainable corporate performance, and buying behavior. *Psychology and Marketing*, *24*(6), 555–577. doi:10.1002/mar.20173

Daily, B., & Bishop, J. (2003). TQM workforce factors and employee involvement: The pivotal role of teamwork. *Journal of Managerial Issues*, *15*, 393–417.

Daily, B., & Huang, S. (2001). Achieving sustainability through attention to human resource factors in environmental management. *International Journal of Operations & Production Management*, *21*(12), 1539–1552. doi:10.1108/01443570110410892

Damali, N., Jolley, G., & Handfield, R. (2008). Environmental management systems and green supply chain management: Complements for sustainability? *Business Strategy and the Environment*, *17*(1), 30–45. doi:10.1002/bse.557

Darnall, N., & Edwards, D., Jr. (2004). *Predicting the cost of environmental system adoption*. Paper presented at Academy of Management National Meetings. New Orleans, LA.

Dean, T., & Brown, R. (1995). Pollution regulation as a barrier to new firm entry: Initial evidence and implications for future research. *Academy of Management Journal*, *38*, 288–303. doi:10.2307/256737

Debrito, M., Carbone, V., & Blanquart, C. (2008). Towards a sustainable fashion retail supply chain in Europe: Organisation and performance. *International Journal of Production Economics*, *114*(2), 534–553. doi:10.1016/j.ijpe.2007.06.012

Doonan, J., Lanoie, P., & Laplante, B. (2005). Analysis determinants of environmental performance in the Canadian pulp & paper industry: an assessment from inside the industry. *Ecological Economics*, *55*, 73–84. doi:10.1016/j.ecolecon.2004.10.017

Drumwright, M. (1994). Socially responsible organizational buying: Environmental concern as a noneconomic buying criterion. *Journal of Marketing*, *58*(3), 1–19. doi:10.2307/1252307

Ehrgott, M., Reimann, F., Kaufmann, L., & Carter, C. (2011). Social sustainability in selecting emerging economy suppliers. *Journal of Business Ethics*, *98*, 99–119. doi:10.1007/s10551-010-0537-7

Elbanna, S. (2010). Strategic planning in United Arab Emirates. *International Journal of Commerce and Management*, *20*(1), 26–40. doi:10.1108/10569211011025934

ElTayeb, T., Zailani, S., & Jayaraman, K. (2010). The examination on the drivers for green purchasing adoption among EMS 14001 certified companies in Malaysia. *Journal of Manufacturing Technology Management*, *21*(2), 206–225. doi:10.1108/17410381011014378

Epstein, M. (2008). *Making sustainability work: Best practices in managing and measuring social and environmental impacts*. Sheffield, UK: Greenleaf.

Erlich, P. R., & Erlich, A. H. (1991). *The population explosion*. New York: Touchstone.

Faisal, M. (2010). Sustainable supply chains: A study of interaction among the enablers. *Business Process Management Journal*, *16*(3), 508–529. doi:10.1108/14637151011049476

Fiksel, J. (2010). Evaluating supply chain sustainability. *Chemical Engineering Progress*, *106*(5), 28.

Florida, R. (1996). Lean and green: The move to environmentally conscious manufacturing. *California Management Review*, *39*, 80–105. doi:10.2307/41165877

Freeman, E. (1984). *Strategic management: A stakeholder approach*. Boston: Pitman.

Garriga, E., & Mele. (2004). Corporate social responsibility theories: Mapping the territory. *Journal of Business Ethics*, *53*(1/2), 51–71. doi:10.1023/B:BUSI.0000039399.90587.34

Geffen, C., & Rothenberg, S. (2000). Suppliers and environmental innovation: The automotive paint process. *International Journal of Operations & Production Management*, *20*, 166–186. doi:10.1108/01443570010304242

Gilmore, J. H., & Pine. (2007). *Authenticity: What consumers really want*. Cambridge, MA: Harvard Business School Press.

Go'ncz, E., Skirke, U., Kleizen, H., & Barber, M. (2007). Increasing the rate of sustainable change: A call for a redefinition of the concept and the model for its implementation. *Journal of Cleaner Production*, *15*(6), 525–537. doi:10.1016/j.jclepro.2006.05.018

Gonza'lez-Benito, J., & Gonza'lez-Benito. (2005). An analysis of the relationship between environmental motivations and ISO14001 certification. *British Journal of Management*, *16*(2), 133–148. doi:10.1111/j.1467-8551.2005.00436.x

Gonza'lez-Benito, J., & Gonza'lez-Benito. (2008). Operations management practices linked to the adoption of ISO 14001: An empirical analysis of Spanish manufacturers. *International Journal of Production Economics*, *113*(1), 60–73. doi:10.1016/j.ijpe.2007.02.051

Gonza'lez-Benito, J., & Gonza'lez-Benito. (2010). A study of determinant factors of stakeholder environmental pressure perceived by industrial companies. *Business Strategy and the Environment*, *19*(3), 164–181.

Gray, E. R., & Smeltzer. (1989). *Management: The competitive edge*. London: Collier MacMillan Publishers.

Guide, V., & VanWassenhove, L. N. (2001). Managing product returns for remanufacturing. *Production and Operations Management*, *10*(2), 142–155. doi:10.1111/j.1937-5956.2001.tb00075.x

Gupta, S., & Palsule-Desai, O. (2011). Sustainable supply chain management: review and research opportunities. *IIMB Management Review*, *23*, 234–245. doi:10.1016/j.iimb.2011.09.002

Hair, J. F. Jr, Black, W. C., Babin, B. J., Anderson, R. E., & Tatham, R. L. (2006). *Multivariate data analysis* (6th ed.). Upper Saddle River, NJ: Pearson Prentice Hall.

Hall, J. (2000). Environmental supply chain dynamics. *Journal of Cleaner Production*, *8*(6), 455–471. doi:10.1016/S0959-6526(00)00013-5

Hall, M., & Purchase, D. (2006). Building or bodging? Attitudes to sustainability in UK public sector housing construction development. *Sustainable Development*, *14*, 205–218. doi:10.1002/sd.265

Handfield, R., Wallon, S., Seegers, L., & Melnyk, S. (1997). Green'Value Chain Practices in the Furniture Industry. *Journal of Operations Management*, *15*, 293–315. doi:10.1016/S0272-6963(97)00004-1

Hart, S. (1995). A natural-resource-based view of the firm. *Academy of Management Review*, *20*(4), 986–1014.

Hart, S., & Milstein, M. (2003). Creating sustainable value. *The Academy of Management Executive*, *17*(2), 56–69. doi:10.5465/AME.2003.10025194

Heath, J. (2006). Business ethics without stakeholders. *Business Ethics Quarterly*, *16*(3), 533–557. doi:10.5840/beq200616448

Hoexter, M. (2006). *The concept of sustainability: Internal diversity and points of conflicts*. Retrieved from http://greenthoughts.us

Hsu, C., & Hu, A. (2008). Green supply chain management in the electronic industry. *International Journal of Environmental Science and Technology*, *5*(2), 205–216. doi:10.1007/BF03326014

Hunt, D., & At-Twaiijri, M. (1996). Values and the Saudi managers: An empirical investigation. *Journal of Management Development*, *15*(5), 48–56. doi:10.1108/02621719610117259

Hunt, S., & Davis, D. (2008). Grounding supply chain management in resource-advantage theory. *Journal of Supply Chain Management*, *44*(1), 10 21. doi:10.1111/j.1745-493X.2008.00042.x

Jabbour, A., Jabbour, C., & Govindan, K. et al. (2013). Factors affecting the adoption of green supply chain management practices in Brazil: Empirical evidence. *The International Journal of Environmental Studies*. PMID:24179250

Jayaratne, P. (2011). Sustainable supply and supply chain mapping - Sri Lankan tea supply chain. *HDR Student Conference, Paper 21*. University of Wollongong, NZ. Retrieved from http://ro.uow.edu.au/sbshdr/2011/papers/21

Jeyaraj, A., Rottman, J., & Lacity, M. (2006). A review of the predictors, linkages, and biases in IT innovation adoption research. *Journal of Information Technology*, *21*, 1–23. doi:10.1057/palgrave.jit.2000056

Jiang, B. (2009). The effects of interorganizational governance on supplier's compliance with SCC: An empirical examination of compliant and non-compliant suppliers. *Journal of Operations Management*, *27*(4), 267–280. doi:10.1016/j.jom.2008.09.005

Johnson, P.R., Leenders, & Fearon. (2006). Supply's growing status and influence: A sixteen-year perspective. *The Journal of Supply Chain Management*, *13*, 33–43. doi:10.1111/j.1745-493X.2006.00010.x

Kim, L., & Min, H. (2011). Measuring supply chain efficiency from a green perspective. *Management Research Review*, *34*(11), 1169–1189. doi:10.1108/01409171111178738

King, A., & Lenox, M. (2001). Lean and green? An empirical examination of the relationship between lean production and environmental performance. *Production and Operations Management*, *10*(3), 244–256. doi:10.1111/j.1937-5956.2001.tb00373.x

Klassen, R., & Vachon. (2003). Collaboration and evaluation in the supply chain: The impact on plant-level environmental investment. *Production and Operations Management*, *12*(3), 336–352. doi:10.1111/j.1937-5956.2003.tb00207.x

Kleindorfer, P. R., Singhal, K., & Van Wassenhove, L. N. (2005). Sustainable operations management. *Production and Operations Management*, *14*(4), 482–492. doi:10.1111/j.1937-5956.2005.tb00235.x

Lambert, D., Stock, J., & Ellram, L. (1998). *Fundamentals of logistics management*. Boston, MA: Irwin/McGraw-Hill.

Lambert, D. M., & Cooper, M., & Pagh. (1998). Supply chain management: Implementation issues and research opportunities. *International Journal of Logistics Management*, *9*(2), 1–20. doi:10.1108/09574099810805807

Lamming, R., & Hampson, J. (1996). The environment as a supply chain management issue. *British Journal of Management*, *7*, 45–62. doi:10.1111/j.1467-8551.1996.tb00147.x

Lao, Y, Hong, & Rao. (2010). Supply management, supply flexibility, and performance outcomes: An empirical investigation of manufacturing firms. *Journal of Supply Chain Management*, *46*(3), 6–22. doi:10.1111/j.1745-493X.2010.03195.x

Lee, S.-Y. (2008). Drivers for the participation of small and medium-sized suppliers in green supply chain initiatives. *Supply Chain Management: An International Journal*, *13*(3), 185–198. doi:10.1108/13598540810871235

Lin, C. Y. (2007). Adoption of green supply chain practices in Taiwan's logistics industry. *Journal of International Management Studies*, *2*(2), 90–98.

Lindgreen, A., Antioco, M., Harness, D., & Sloot, R. (2008). Purchasing and marketing of social and environmental sustainability for high-tech medical equipment. *Journal of Business Ethics*, *85*, 445–462. doi:10.1007/s10551-008-9740-1

Linton, J. D., Klassen, & Jayaraman. (2007). Sustainable supply chains: An introduction. *Journal of Operations Management*, *25*(6), 1075–1082. doi:10.1016/j.jom.2007.01.012

Locke, R., & Romis. (2007). Improving work conditions in a global supply chain. *MIT Sloan Management Review*, *48*(2), 54–62.

Locke, R. M., Fei, & Brause. (2007). Does monitoring improve labor standards? Lessons from NIKE. *Industrial & Labor Relations Review*, *61*(1), 3–31.

Madu, C., Kuei, C., & Madu, I. (2002). A hierarchic metric approach for integration of green issues in manufacturing: A paper recycling application. *Journal of Environmental Management*, *64*, 261–272. doi:10.1006/jema.2001.0498 PMID:12040959

Maon, F., Lindgreen, & Swaen. (2009). Designing and implementing corporate social responsibility: An integrative framework grounded in theory and practice. *Journal of Business Ethics*, *87*, 71–89. doi:10.1007/s10551-008-9804-2

Mardi, A. G. (1992). Environmentally friendly development: Can the private sector succeed where others have failed? *The Columbia Journal of World Business*, *27*(3&4), 194–200.

Marshall, R. S., Cordano, M., & Silverman, M. (2005). Exploring individual and institutional drivers of proactive environmentalism in the US wine industry. *Business Strategy and the Environment*, *14*, 92–109. doi:10.1002/bse.433

McCue, D. (2010). Sustainability and the triple bottom line. *World Trade*, *23*(10), 24.

McWilliams, A., Siegel, & Wright. (2006). Corporate social responsibility, international perspectives. *The Journal of Business Strategy*, *23*(1), 1–12.

Melnyk, S. A., Sroufe, R. P., & Calantone, R. (2003). Assessing the impact of environmental management systems on corporate and environmental performance. *Journal of Operations Management*, *21*(2), 329–351. doi:10.1016/S0272-6963(02)00109-2

Mentzer, J., DeWitt, W., Keebler, J., Min, S., Nix, N., Smith, C., & Zacharia, Z. (2001). Defining supply chain management. *Journal of Business Logistics*, *22*(2), 1–25. doi:10.1002/j.2158-1592.2001.tb00001.x

Milne, M., Kearins, K., & Walton, S. (2006). Creating adventures in wonderland: The journey metaphor and environmental sustainability. *Organization*, *13*, 801–839. doi:10.1177/1350508406068506

Min, H., & Galle. (1997). Green purchasing strategies: Trends and implications. *International Journal of Purchasing and Materials Management*, *33*(3), 10–17. doi:10.1111/j.1745-493X.1997.tb00026.x

Min, H., & Galle, W. P. (2001). Green purchasing practices of US firms. *International Journal of Operations & Production Management*, *21*(9), 1222–1238. doi:10.1108/EUM0000000005923

Mitchell, R. K., Agle, B. R., & Wood, D. J. (1997). Toward a theory of stakeholder dentification and salience: Defining the principle of who and what really counts. *Academy of Management Review*, *22*(4), 853.

Mohapatra, S., & Srivastava, N. (2012). Sustaining competitive advantage in social entrepreneurship - A case study. *International Journal of Logistics Economics and Globalisation*, *4*, 197–220. doi:10.1504/IJLEG.2012.050207

Mustaffa, N., & Potter, A. (2009). Healthcare supply chain management in Malaysia: A case study. *Supply Chain Management: An International Journal*, *14*(3), 234–243. doi:10.1108/13598540910954575

Mwirigi, F. (2010). The challenge of building sustainable supply chain relationships among small firms in developing economies: The case of Kenya. *International Review of Business Research Papers*, *6*(4), 189–201.

Narasimhan, R., & Wook. (2001). Information system utilization strategy for supply chain integration. *Journal of Business Logistics*, *22*(2), 51–75. doi:10.1002/j.2158-1592.2001.tb00003.x

NZ Business Council for Sustainable Development Report. (2003). *Business guide to a sustainable supply chain*. Retrieved from http:// www.nzbcsd.org.nz/supplychain

Oppenheim, J. Bonini, Bielak, Kehm, & Lacy. (2007). *Shaping the new rules of competition: UN global compact participant mirror*. McKinsey & Company. Retrieved from http://tinyurl.com/69shws

Pagell, M., Wu, & Wasserman. (2010). Thinking differently about purchasing portfolios: An assessment of sustainable sourcing. *Journal of Supply Chain Management*, *46*(1), 57–73. doi:10.1111/j.1745-493X.2009.03186.x

Paulraj, A. (2011). Understanding the relationships between internal resources and capabilities, sustainable supply management and organizational sustainability. *Journal of Supply Chain Management*, *47*(1), 19–37. doi:10.1111/j.1745-493X.2010.03212.x

Peattie, K. (2001). Towards sustainability: The third age of green marketing. *Marketing Review*, *2*(2), 129–146. doi:10.1362/1469347012569869

Peng, Y. S., & Lin, S. S. (2008). Local responsiveness pressure, subsidiary resources, green management adoption and subsidiary's performance: Evidence from Taiwanese manufactures. *Journal of Business Ethics*, *79*(1), 199–212. doi:10.1007/s10551-007-9382-8

Podsakoff, P. M., MacKenzie, S. B., Lee, J., & Podsakoff, N. P. (2003). Common method biases in behavioral research: A critical review of the literature and recommended remedies. *The Journal of Applied Psychology*, *88*(5), 879–903. doi:10.1037/0021-9010.88.5.879 PMID:14516251

Preuss, L. (2009). Addressing sustainable development through public procurement. *Supply Chain Management: An International Journal*, *14*(3), 213–223. doi:10.1108/13598540910954557

Pullman, M., Maloni, M., & Carter, C. (2009). Food for thought: Social versus environmental sustainability practices and performance outcomes. *J. Supply Chain Manag.*, *45*, 38–54. doi:10.1111/j.1745-493X.2009.03175.x

Pun, F., Chin, S., & Gill, R. (2001). Determinants of employee involvement practices in manufacturing enterprises. *Total Quality Management*, *12*(1), 95–109. doi:10.1080/09544120020010129

Rao, P. (2002). Greening the supply chain a new initiative in south East Asia. *International Journal of Operations & Production Management*, *22*, 632–655. doi:10.1108/01443570210427668

Rao, P. (2006). Greening of suppliers/in-bound logistics in the South East Asian context. In *Greening the supply chain*. London: Springer. doi:10.1007/1-84628-299-3_11

Rao, P., & Holt, D. (2005). Do green supply chains lead to competitiveness and economic performance? *International Journal of Operations & Production Management*, *25*(9), 898–916. doi:10.1108/01443570510613956

Ras, P., & Vermeulen, W. (2009). Sustainable production and the performance of South African entrepreneurs in a global supply chain: The case of South African table grape producers. *Sustainable Development*, *17*(5), 325–341. doi:10.1002/sd.427

Reuter, C., Foerstl, Hartmann, & Blome. (2010). Sustainable global supplier management: The role of dynamic capabilities in achieving competitive advantage. *Journal of Supply Chain Management*, *26*(2), 45–63. doi:10.1111/j.1745-493X.2010.03189.x

Reynolds, M. A., & Yuthas. (2008). Moral discourse and corporate social responsibility reporting. *Journal of Business Ethics*, *78*, 47–64. doi:10.1007/s10551-006-9316-x

Rimmington, M., Smith, J. C., & Hawkins, R. (2006). Corporate social responsibility and sustainable food procurement. *British Food Journal*, *108*(10/11), 824–837. doi:10.1108/00070700610702082

Robinson, D. R., & Wilcox, S. (2008). The greening of the supply chain. *Logistics Management*, *47*(10), 67–72.

Rothenberg, S., Pil, F. K., & Maxwell, J. (2001). Lean, green, and the quest for superior performance. *Production and Operations Management*, *10*(3), 228–243. doi:10.1111/j.1937-5956.2001.tb00372.x

Sarkis, J., & Tamarkin, M. (2005). Real options analysis for green trading: the case of greenhouse gases. *Eng. Econ.*, *50*, 273–294. doi:10.1080/00137910500227208

Savitz, A. W., & Weber, K. (2006). *The triple bottom line*. San Francisco: Jossey-Bass.

Setthasakko, W. (2009). Barriers to implementing corporate environmental responsibility in Thailand: A qualitative approach. *The International Journal of Organizational Analysis*, *17*(3), 169–183. doi:10.1108/19348830910974905

Seuring, S., & Müller, M. (2008). From a literature review to a conceptual framework for sustainable supply chain management. *Journal of Cleaner Production*, *16*(15), 1699–1710. doi:10.1016/j.jclepro.2008.04.020

Sharma, S., & Vredenburg. (1998). Proactive corporate environmental strategy and the development of competitively valuable organizational capabilities. *Strategic Management Journal*, *19*(8), 729–753. doi:10.1002/(SICI)1097-0266(199808)19:8<729::AID-SMJ967>3.0.CO;2-4

Shrivastava, P. (1995). The role of corporations in achieving ecological sustainability. *Academy of Management Review*, *20*(4), 936–960.

Sikdar, S. K. (2003). Sustainable development and sustainability metrics. *AIChE Journal. American Institute of Chemical Engineers*, *49*(8), 1928–1932. doi:10.1002/aic.690490802

Sroufe, R. (2003). Effects of environmental management systems on environmental management practices and operations. *Production and Operations Management*, *12*(3), 416–431. doi:10.1111/j.1937-5956.2003.tb00212.x

Starik, M., & Rands. (1995). Weaving an integrated web: Multilevel and multisystem perspectives of ecologically sustainable organizations. *Academy of Management Review*, *20*(4), 908–935.

Stern, N. (2007). *The Stern review*. Cambridge, MA: Cambridge University Press.

Stieb, J. (2009). Assessing Freeman's stakeholder theory. *Journal of Business Ethics*, *87*, 401–414. doi:10.1007/s10551-008-9928-4

Tencati, A., Russo, A., & Quaglia, V. (2010). Sustainability along the global supply chain: The case of Vietnam. *Social Responsibility Journal*, *6*(1), 91–107. doi:10.1108/17471111011024577

Tibben-Lembke, R. S. (2002). Life after death: reverse logistics and the product life cycle. *International Journal of Physical Distribution & Logistics Management*, *32*(3), 223–244. doi:10.1108/09600030210426548

Vachon, S., & Klassen. (2006). Extending green practices across the supply chain: The impact of upstream and downstream integration. *International Journal of Operations & Production Management*, *26*(7), 795–821. doi:10.1108/01443570610672248

Vachon, S., & Mao, Z. (2008). Linking supply chain strength to sustainable development: A country-level analysis. *Journal of Cleaner Production*, *16*(15), 1552–1560. doi:10.1016/j.jclepro.2008.04.012

van Hoek, R. (1999). From reversed logistics to green supply chains. *Supply Chain Management: An International Journal*, *4*(3), 129–135. doi:10.1108/13598549910279576

Walker, H., & Brammer, S. (2009). Sustainable procurement in the United Kingdom public sector. *Supply Chain Management: An International Journal*, *14*(2), 128–137. doi:10.1108/13598540910941993

Walker, H., Sisto, L. D., & McBain, D. (2008). Drivers and barriers to environmental supply chain management: Lessons from the public and private sectors. *Journal of Purchasing and Supply Management*, *14*(1), 69–85. doi:10.1016/j.pursup.2008.01.007

Walsh, J. P. (2005). Review essay: Taking stock of stakeholder management. *Academy of Management Review*, *30*(2), 426–438. doi:10.5465/AMR.2005.16387898

Walton, S. V., Handfield, R. B., & Melnyk, S. A. (1998). The green supply chain: integrating suppliers into environmental management processes. *International Journal of Purchasing & Materials Management*, *34*(2), 2–11. doi:10.1111/j.1745-493X.1998.tb00042.x

WCED (World Commission on Environment and Development). (1987). *Our common future*. Oxford, UK: Oxford University Press.

Whiteman, G., & Cooper, W. H. (2000). Ecological embeddedness. *Academy of Management Journal*, *43*(6), 1265–1282. doi:10.2307/1556349

Williamson, O. (2008). Outsourcing: Transaction cost economics and supply chain management. *Journal of Supply Chain Management*, *44*(1), 5–16. doi:10.1111/j.1745-493X.2008.00051.x

Wittmann, C.M., Hunt, & Arnett. (2009). Explaining alliance success: Competences, resources, relational factors, and resource-advantage theory. *Industrial Marketing Management*, *38*, 743–756. doi:10.1016/j.indmarman.2008.02.007

Wolf, J. (2011). Sustainable supply chain management integration: A qualitative analysis of the German manufacturing industry. *Journal of Business Ethics*, *102*(2), 221–235. doi:10.1007/s10551-011-0806-0

Wynstra, J. Y. F. (2006). *Inkoop, leveranciers en innovatie: Van VOC tot space shuttle*. Erasmus Research Institute of Management. Retrieved from http://hdl.handle.net/1765/7439

Zailani, S., Jeyaraman, K., Vengadasan, G., & Premkumar, R. (2012). Sustainable supply chain management (SSCM) in Malaysia: A survey. *International Journal of Production Economics*, *140*(1), 330–340. doi:10.1016/j.ijpe.2012.02.008

Zhu, Q., Sarkis, & Lai, K. (2008). Confirmation of a measurement model for green supply chain management practices implementation. *International Journal of Production Economics*, *111*(2), 261–273. doi:10.1016/j.ijpe.2006.11.029

Zhu, Q., & Geng, Y. (2001). Integrating environmental issues into supplier selection and management: A study of large and medium-sized state-owned enterprises in China. *Greener Management International*, *35*, 27–40.

Zhu, Q., Geng, Y., Fujita, T., & Hashimoto, S. (2010). Green supply chain management in leading manufacturers: Case studies in Japanese large companies. *Management Research Review*, *33*(4), 380–392. doi:10.1108/01409171011030471

Zhu, Q., & Sarkis, J. (2006). An inter-sectoral comparison of green supply chain management in China: Drivers and practices. *Journal of Cleaner Production*, *14*, 472–486. doi:10.1016/j.jclepro.2005.01.003

Zhu, Q., Sarkis, J., & Lai, K. (2007). Initiatives and outcomes of green supply chain management implementation by Chinese manufacturers. *Journal of Environmental Management*, *25*(6), 179–189. doi:10.1016/j.jenvman.2006.09.003 PMID:17084502

Zhu, Q., Tian, Y., & Sarkis, J. (2012). Diffusion of selected green supply chain management practices: An assessment of Chinese enterprises. *Production Planning and Control*, *23*, 837–862. doi:10.1080/09537287.2011.642188

Zhu, Q. H., & Sarkis, J. (2004). Relationships between operational practices and performance among early adopters of green supply chain management practices in Chinese manufacturing enterprises. *Journal of Operations Management*, *22*, 265–289. doi:10.1016/j.jom.2004.01.005

Zsidisin, G. A., & Siferd, S. P. (2001). Environmental purchasing: A framework for theory development. *European Journal of Purchasing and Supply Management*, *7*, 61–73. doi:10.1016/S0969-7012(00)00007-1

Zuo, K., Potangaroa, R., Wilkinson, S., & Rotimi, J. (2009). A project management prospective in achieving a sustainable supply chain for timber procurement in Banda Aceh, Indonesia. *International Journal of Managing Projects in Business*, *2*(3), 237–251. doi:10.1108/17538370910971045

Chapter 10
Key Success Drivers for Large High-Technology Projects:
Prediction and Practice

Phil Crosby
CSIRO, Australia

ABSTRACT

Success in project management, and particularly in large, high-technology projects, continues to test the resources of organisations and their sponsors. This chapter revisits the conclusions of an earlier meta-study (Crosby, 2012a) that examined a large number of published case investigations and research efforts relating to the success and failure of projects. In that study, the success factors for general, and high-technology, projects were grouped and ranked as strategic success drivers for use prescriptively by project practitioners and approvers, and the principal drivers were examined closely to reveal any less obvious characteristics influencing project success. This chapter takes the original findings of ranked success drivers and investigates how these align with the experiences of three large contemporary high-technology projects. The conclusions show that, while the original set of drivers remains valid as predictors of project success, the ranking is likely to vary, even between projects that are technically and structurally similar. Two additional success factors are added as a result of the present study.

DOI: 10.4018/978-1-4666-4506-6.ch010

INTRODUCTION

Despite present global financial constraints, the world is currently in an era of great innovation realised through many large science/engineering projects (e.g. the Large Hadron Collider, the ITER Fusion Reactor, and the Square Kilometre Array). These breakthrough projects, by their nature, have inherent high-risk, yet their international scale and huge cost implications demand that success measures are achieved and project performance is maximised. Many modern studies (Jugdev & Müller, 2005; Williams et al., 2009; ICCPM, 2011; Crosby, 2012c) show that traditional (i.e. PMBoK style) project management and execution techniques alone are insufficient to meet the demands placed on complex high-tech mega-projects. A fuller understanding of success drivers, professionally applied in the early stages, is required to lift project performance.

My original meta-study (Crosby, 2012a) of project success factors (based on analyses of 2820 cases) concluded that sound project management control and execution systems, and a clear project definition and goal set, are by far the two most important drivers of project success. Competent information management systems ranked third in importance followed by 15 other significant factors empirically shown to improve project outcomes (See Table 1). Many of these success drivers are wholly or partly newly ranked 'soft' drivers, indicating the importance of leadership, motivation, expectations, and team engagement. The study findings encouraged further research to benchmark the performance of contemporary large engineering and science projects against the reported strategic drivers to further our knowledge of factors for project success.

This paper takes the original findings of ranked success drivers (beyond the fundamental project management artifacts e.g. project plans, WBS, etc) and investigates how well these align with the recent experience of three large high-technology (high-tech) projects. The cases are purposely selected from the astronomy world, each being an example of a new breed of large and expensive array-style radio telescopes.

Table 1. Success drivers ranked by relative importance (from Crosby, 2012a)

#	Success Driver	High-Tech Project Ranking
1	Project management (PM) control & execution systems in place, with robust policies, planning, procedures, document control, audit, etc	23.87
2	Clear project definition, requirements, goals, objectives, scope, and project mission; sound business case	19.53
3	Mature project communication, information systems; effective public relations management	11.18
4	(Top) management (or sponsor) support with sustained commitment, appropriately engaged	8.96
5	Project baseline, budget estimates accuracy, schedule and project phasing, effective project performance (reviews) and measurement	8.96
6	Leadership skills, PM experience & stability; motivating & socially capable PM	5.79
7	Agreed realistic customer / user expectations; frequent customer contact	3.37
8	PM/Organisational understanding & competence in project management	3.37
9	Adequate resourcing of the project	2.37
10	Aligned perceptions of project goals & success - management and team; sense of urgency instilled	2.37
11	Effective stakeholder engagement / partnership (e.g. client, contractors, etc)	2.37
12	Organisational responsibilities assigned to right-sized capable team	1.64
13	Mature, effective project management change control process; effective deviations handling & configuration control	1.64
14	Understanding & continuous management of risk; visibility of risk register	1.13
15	Project Manager & PM systems matched to project complexity, and culturally aligned	1.13
16	Effective means of learning from experience and continuous improvement environment	0.78
17	Full understanding, and early engagement, of host government environment and institutional requirements	0.78
18	Right-sized systems engineering; managing and procuring in right sized project 'chunks'	0.78

A precise definition of a "large" project (also referred to as a mega or giant project) is not offered, except that these endeavours are characterised by multi-million or even billion dollar budgets, time-frames usually measured in years, significant infrastructure requirement, and attract a high level of public or political attention. Flyvbjerg (2008, p. 1) sets an important programmatical context when he states:

mega also implies the size of the task involved in developing, planning, and managing projects of this magnitude. The risks are substantial [and] cost overruns of 50% are common.

LITERATURE OVERVIEW

The original research (Crosby, 2012a) was underpinned by material published from the mid 1970s to 2010, an era broadly covering the professionalisation of project management, and a full cycle of global economic activity.

The literature review in that paper referenced the most prolific authors relating to project management (PM) success including works by Morris & Hough (1986); Turner (2004); Müller & Turner (2007 & 2010); Cooke-Davies (2001 & 2002); Dvir et al. (1998 & 2003); Pinto & Slevin (1989); Williams (2004 & 2008); Shenhar et al. (2001 & 2007); de Wit (1988); Belassi & Tukel (1996); Murphy et al. (1974); and Winter et al. (2006). These and other cited authors contribute useful research around the triple constraints of scope, time and cost, (commonly termed the 'iron' or 'golden triangle') and begin to expand these factors to include quality, risk, sustainability and commercial success. Shenhar & Wideman's (1996) "success dimensions" include customer use, customer satisfaction, market share creation, and new technologies/product lines.

An important initiative adding to the literature was the UK Government funded research activity called Rethinking Project Management (RPM) (Maylor, 2006). This network of project management (PM) professionals and subject authors followed a research program framed to question mainstream ideas. Crosby (2012a) cites several papers from that publication including Winter et al. (2006) who offer a useful summary of findings and commend new research directions beyond the rational and intellectual foundations that underpin the triple constraints paradigm. Of particular note is the need for increased recognition of human issues and exogenous factors as potent success drivers.

Crosby (2012a) found much material concerning PM performance published within the practice genre, including professional magazines (e.g. NASA's regular *ASK* publication). Material from the PM professional organisations (e.g. the Project Management Institute's PMBOK (PMBOK® Guide, 2008) were found to be useful practice guides, though aimed more at project structuring and execution than research analysis (Morris et al., 2006).

Recent work concerning the identification and application of success drivers for large projects has continued to develop understanding by focusing in three areas: complexity, resilience, and project review and learning:

- **Complexity:** A task force report from The International Centre for Complex Project Management (ICCPM) includes detailed and profound recommendations concerning improved management of project complexity, and argues strongly that traditional PM approaches to complex projects are counter-productive (ICCPM, 2011). Ernst (2011) in his Asia Pacific report posits that true innovation not only requires complexity, but adds to it through capability dimensions. Jackson (2011) promotes a holistic approach to complex PM through application of soft systems methodologies. Remington's recent book (Remington, 2011) shows how complex projects lie on a spectrum between merely complicated, and chaos, and builds on earlier work showing that complex projects may be better understood when viewed as complex adaptive systems perhaps more easily grasped by practitioners (Remington & Pollack, 2008);

- **Resilience:** Much discussed as a concept, there is little research material on resilience building in high-tech projects. Crosby (2012b) perhaps offers the most comprehensive recent study of project resilience, concluding that project success is not random, and that early adoption of certain approaches, activities, and launch conditions will position start-up projects for success and resilience. Nine resilience factors are revealed, grouped into three 'attitudinal' factors, and six 'conditioning' factors. Denyer & Kutsch, as cited in PMI, (2011) highlight the project-strengthening effect of managers that strategise and "imagineer" risk scenarios early. Jani (2010) adds the human trait of "self efficacy" as enabling individuals to face setbacks and be more willing to exert efforts to carry on and achieve their goals. Looking at risk in the defence project arena, Thomson et al. (2011, p. 14) assert that *"even the best-managed projects will suffer setbacks or even failures from time to time"*, and question whether robustness really is improved when more detailed information is available. They conclude that sometimes gathering more information leads to worse outcomes;
- **Review and learning:** Harris (2009) upholds the main purpose of project reviews as events to evaluate outcomes against aims and success criteria, and supports new thinking by identifying reviews as opportunities to improve project performance through organisational learning. Contemporary casework is largely documented in PM practice literature where informative narratives examine project perturbations (and sometimes crises), and how these are reviewed and tackled. Again, NASA's *ASK* magazine series provides excellent examples including and beyond the space arena (e.g. programmatic recovery of the National Ignition Facility program – NASA, 2011). Early stage project review and recovery may include procurement failure as detailed by Rendleman & Faulconer (2011). Modern Government project guides are devoting increased attention to reviews and lessons-learned (DoE, 2011; OGC, 2002). Relevant to the cases in this paper, Fisher et al. (2010, p. 8) look at large instruments for astronomy projects, concluding that: *"Cost and schedule overruns and de-scoping...must be taken very seriously and the lessons thoroughly understood before attempting new large endeavours."* The topic is comprehensively summed up as an early stage strategic conversation within a collection of essays edited by Williams et al. (2009).

RESEARCH METHOD AND DATA SOURCE

Before describing the collection, reduction, and analysis of the data, the underlying value of casework is briefly considered.

Case study research protagonist Yin (2009) joins Flyvbjerg (2006) in arguing that casework reveals 'context-dependent' knowledge that encourages learning maturity. Flyvbjerg goes on to explain the richness of information in the personal case-narrative, (even in single cases) and its ability to describe realities which are hard to reveal or define in scientific parlance. Ika (2009, p. 15) supports in-depth interviews as allowing: *"project actors to tell their professional life stories or talk about success factors"* and adds the rhetorical question, *"Are not words, by their very nature, infinitely richer than numbers?"*

As reported by Crosby (2012c), the methodologies of literature research, complemented and validated through case-based investigations - though by its nature always incomplete - enables the extraction of recurring topics offering new insights when dealing with large, high-tech, project success.

The questions and responses that form the conversation and ultimate narrative that informs casework research require careful construction and interpretation – a process known as discourse analysis. With an emphasis on reflexivity, discourse analysis inevitably introduces the researcher as a component of the discourse, though as a constructive effect (Phillips & Hardy, 2002). A key point is that the researcher or analyst should not attempt to interpret individual statements or texts, but consider summations of remarks and bodies of texts to derive a broader reality. As Phillips and Hardy (2002, p. 9) put it: *"Content analysis...can be used to connect textual content to broader discursive contexts."*

The case data presented here relates to three contemporary high-tech radio-telescope projects. While different in terms of scale and location, they are each a national or international endeavour, comprising a networked array of innovative receptors coupled to a high-performance computing (HPC) facility, and supported by infrastructure. All three projects were reviewed at the stage of post phase 1 commissioning; essentially an end-end system and signal path test. Table 2 describes and compares key project features.

Data from each case project were sourced by different means, although collated for analysis using a single method described below. The projects/sources were:

- **ALMA:**
  - **Primary:** Face to face interview with (immediate past) Deputy Director of the Joint ALMA Observatory (April 2013);
  - **Secondary:** Project documentation in the public arena;
- **ASKAP:**
  - **Primary:** Internal records from periodic project reviews conducted by CSIRO Division of Astronomy and Space Science (June 2012 - Feb 2013);
  - **Secondary:** Observations from ASKAP project management and team meetings;
- **LOFAR:**
  - **Primary:** Notes and final report from the LOFAR Project Post-Project

Table 2. Profile of case study projects

Project Acronym & Web Link	Sponsor Organisation	Description	Approx Budget
ALMA http://www.almaobservatory.org/en/home	ESO – NAOJ – NRAO Collaboration	Radio telescope array of 66 dishes located on the Andes in northern Chile.	US1.4 billion
ASKAP http://www.atnf.csiro.au/projects/askap/	The Commonwealth Scientific and Industrial Organisation (CSIRO), Australia	Radio telescope array of 36 dishes located in the mid-West of Western Australia. ASKAP is a precursor for the SKA project currently under construction.	A$170 million
LOFAR http://www.lofar.org/	ASTRON, The Netherlands	Radio telescope consisting of thousands of omni-directional dipole antennas. Centred in North Holland, it commenced operations in 2010, and is continually being expanded.	€166 million

and Lessons-Learned Review (April 2011);
   - **Secondary:** ASTRON newsletters 2010 – 2012.

The data were collected against the topics and issues listed in a questionnaire developed by the author and successfully used during a post-project review of the LOFAR project (Crosby & Boonstra, 2011). The actual topics were derived from the ranked programmatic success drivers identified in my earlier study (see Table 1). Some text was modified for easier interpretation by practitioners, and driver #5 was ungrouped (into topics *vii, ix,* and *x*). Four supplementary topics were integrated covering; organisational structure and governance, collaborations and culture, task forces, and procurement, plus two new topics that emerged from the LOFAR review – defined requirements, and the capability to meet technical challenges.

The key data sought was qualitative, in the sense that the interviews and document analyses were targeted to elicit clear and substantiated answers to two principle questions posed to each project:

1. Which of the topics identify as key factors for project success?
2. Which of the topics showed diminished performance, or could have been managed much better?

To aid the process, each topic was complemented by examples of issues to consider, thus encouraging depth to the conversation/investigation. No limits were specified around how many topics could be selected as positive (✓) or underperforming (x) factors; the task being simply to identify (without ranking) those having the highest impact. Where no strong influence on success or failure emerged, the topic box was left blank. The results of this information gathering and assessment process are shown in the three right-hand columns of Table 3.

KEY FINDINGS

The extraction of key success factors from the appraisal in Table 3 reveals some interesting and informative results:

- The process using the 20 topic questionnaire elicited no further topics of note, even when prompted by the author;
- The first topic (top level commitment) was unanimously recorded as a key success driver, upholding the original research that ranked it a close 2nd out of 18 factors;
- Two other drivers were ranked unanimously positive; effective engagement with the socio-political environment, and the capability to meet the project's technical challenges. This latter factor did not appear in my original list of success drivers due to (a) that meta-study being wholly concerned with programmatic topics, and (b) an underestimation of the positive contribution that a 'common-challenge' can make to large, high-tech project execution.

CONCLUSION

Given that none of the interviews, group discussions, or document review activities brought forth any additional programmatic topics, it is hypothesised that the listing of 18 success driver types in the original study (Crosby, 2012a), supplemented with the additional topics of *"defined requirements"* and *"technical challenge capability"* is comprehensive and theoretically complete.

The present study finds the addition of this latter topic as important because it draws in the vital impetus to succeed that is stimulated by the technical demands of the project goal(s). While this was clear in all projects, it stood out in the ALMA project where technical advances were tested across a number of domains.

Table 3. Key success drivers from ALAMA, ASKAP, and LOFAR projects

Topic	Examples of Issues to Consider	ALMA	ASKAP	LOFAR
i. Top level commitment and support	Support from collaborating parties Interest and support of the project at high level Help from project sponsors when necessary Top level interference in project execution	✓	✓	✓
ii. Project Mission, goals, objectives, scope, shared purpose, urgency	Clarity of the project goals Objectives communicated to all teams The pace of project execution Management of the project scope	✓		
iii. Project management systems (PMS) and controls, discipline, tools, applications.	Quality of project tools and applications Effectiveness of project tools and applications Match of tools and PMS to the project tasks, complexity, and organisation culture How well did people use the tools and PMS?	✓	x	x
iv. Top level project management competency, leadership	Match of management style to the project Skill in building trust, motivation, and confidence Ability to deal with all stakeholders Managed problems and situations effectively	✓	✓	
v. Organisation structure, entity, governance, responsibilities & authorities (R&A), matrix?, number & size of teams, diversity	Match of organisation structure to the project needs R&A in relation to the key positions The size and make-up of teams Interaction between work groups	x		
vi. Collaborations, cultures, standards, obligations	Effective working of project collaborations Matching of work standards and behaviour Trust between collaborators to deliver promises			x
vii. Budget & schedule, realism, viability	Fact based budgets and schedule Experience based budgets and schedule Confidence in budget/schedule forecasts		x	
viii. Resources – human, financial, physical	Sufficient people with correct skills Sufficient resources to sustain project costs Promised supplies/support items			x
ix. Project planning & execution, baselines, lifecycles, phasing, monitoring	Accuracy and detail of project plans Correctness of project phasing & schedule Realism of time and expense budget Adequacy of project supervision and monitoring			
x. Project reviews, feedback, project and organisational learning	Effectiveness of reviews (Concept, PDR, CDR etc.) Follow-up and action form reviews Team and organisational lessons learned		✓	
xi. Change management, configuration management	Effectiveness of project change management Capability of PMS to deal with changes Maintenance of project configuration data Communication of project changes to teams	✓		
xii. Risk management, contingency, Mission assurance	Preparation of risk management plans and ongoing review and assessments Preparations for unknown unknowns Communication of risk management plans Adequacy of project contingency (reserves) Application of risk contingency (reserves)		x	

continued on following page

Table 3. Continued

Topic	Examples of Issues to Consider	ALMA	ASKAP	LOFAR
xiii. Project communications and information management	Effective management of project technical information Effective control of PMS information Ambiguity and confusion within the project Spurious negative communications		✓	x
xiv. Task Forces, project emergencies	Readiness to form and use task forces (tiger teams) Capability of task forces to address problems			
xv. Procurement & contractor management	Effectiveness of procurement function to support the project Processes to award contracts Value for money obtained for the project	x	✓	✓
xvi. Stakeholder engagement, suppliers, users, expectations	Good relations with suppliers/contractors Strategic engagement to support the project Satisfaction of user expectations Collaborative partner engagement Funding agencies requirements met	✓		
xvii. Socio-political environment, early engagement. Physical, infrastructure, technical, economic, social, political.	Legislation to enable the project Host site issues Economic benefits achieved Political issues Industrial spin-offs Intellectual property Social and community matters	✓	✓	✓
xviii. Project complexity, dealing with – technical, environment, operational	Understanding of project complexity Management of technical complexity Adequacy of planning and resources to meet complex project problems Operational complexities foreseen			
(new) xix. Technical challenges	Research and development capability Technical expertise in the project teams System engineering, and project interfaces Solving of technical problems during project execution Maintenance of technical discipline Industry capability	✓	✓	✓
(new) xx. Defined requirements	Requirements engineering process Non-vague and unambiguousness Developed collaboratively Defined early enough Responsibilities clear			✓

The value of top level commitment from project supporters is strongly endorsed by all three projects (confirming the original conclusion), as is the requirement for minimal meddling by sponsors.

Managing the wider socio-economic environment was similarly identified by each project as pivotal to success. This factor was also in the original study (#17), and is endorsed by a recent report of major astronomy projects that warns of changing political agendas and agency priorities, budget pressures, as well as unanticipated events. It recommends to: *"start with a more realistic sense of [sponsor], budgetary and policy environments... so that [projects] can be more resilient"* (Fellows & Alexander, 2010, p. 3).

The sample of three projects is too small to confidently assert further alignment with the original study (beyond agreement with the topic list). Remaining findings are distributed over the topics in Table 3. However the results do *infer* agreed importance of project management competency, and attention to competent procurement and contractor management. Moreover, selection, application, and use of project management systems clearly challenged two of the three cases.

Accepting the limitations of data validity, I reason that the 'smearing' of other findings across the project success drivers strongly implies that large high-tech projects each have their own project character, and thus their own management challenges. Certainly, the narratives from project members, and data from associated reports, give emphasis to distinct unexpected perturbations. Each project survived at least one major threat to successful completion, confirming the value of post-project review, analysis, and learning processes (Driver #5 in the original study).

Dr. Phil Crosby works at CSIRO's Astronomy and Space Science Division, Australia, where he manages strategic business planning, industry engagement, and is a major projects specialist. Between 2009 and 2011 he was seconded to the SKA Program Development Office in the UK as Domain Specialist for the SKA industry engagement strategy. Philip trained with ICL and BT, before operating his own medical and industrial electronics firm. Then followed 12 years with NATA in technical management standards; leading major field assessments including reviews of Antarctic science impacts, and ANSTO's radio-isotope production facility. During 2005, Philip worked in Boeing Australia managing Industrial Participation. In addition to technical engineering qualifications, he holds a BA (Business Administration), is a graduate of the Australian Institute of Company Directors, and holds a PhD in mega-science project management. He is a member of the International Centre for Project Management, and an Associate Member of Engineers Australia.*

ACKNOWLEDGMENT

The author thanks The LOFAR team at ASTRON (the Netherlands), Dr. Lewis Ball (ex-ALMA), and CSIRO staff for their views on the ASKAP project. The author acknowledges the Editor-in-Chief of this publication, and the anonymous reviewers for their helpful comments in finalising this paper.

REFERENCES

Belassi, W., & Tukel, O. I. (1996). A new framework for determining critical success/failure factors in projects. *International Journal of Project Management, 14*(3), 141–151. doi:10.1016/0263-7863(95)00064-X

Cooke-Davies, T. (2002). The 'real' success factors on projects. *International Journal of Project Management, 20*(3), 185–190. doi:10.1016/S0263-7863(01)00067-9

Cooke-Davies, T. J. (2001). *Towards improved project management practice.* (Unpublished PhD Thesis). Leeds Metropolitan University, Leeds, UK.

Crosby, P. (2012a). Key success drivers: Meta-study findings applicable to large high-technology projects. *International Journal of Information Technology Project Management, 3*(2), 1–20. doi:10.4018/jitpm.2012040101

Crosby, P. (2012b). Building resilience in large high-technology projects: Front end conditioning for success. *International Journal of Information Technology Project Management, 3*(4), 21–40. doi:10.4018/jitpm.2012100102

Crosby, P. (2012c). *Predictive indicators of success in science & engineering projects - Application to the SKA initiative* (Unpublished PhD thesis). Curtin University, Canberra, Australia.

Crosby, P., & Boonstra, A.-J. (2011). *LOFAR project – Post project and lessons learned review (internal report)*. ASTRON.

de Wit, A. (1988). Measurement of project success. *International Journal of Project Management*, *6*(3), 164–170. doi:10.1016/0263-7863(88)90043-9

DoE. (2011). Cost estimating guide. DOE G 413.3-21, Current Directives 400 series. Washington, DC: U.S. Department of Energy.

Dvir, D., Lipovetsky, S., Shemhar, A., & Tishler, A. (1998). In search of project classification: A non-universal approach to project success factors. *Research Policy*, *27*, 915–935. doi:10.1016/S0048-7333(98)00085-7

Dvir, D., Raz, T., & Shenhar, A. J. (2003). An empirical analysis of the relationship between project planning and project success. *International Journal of Project Management*, *21*(2), 89–95. doi:10.1016/S0263-7863(02)00012-1

Ernst, D. (2011, May 11). China's innovation policy is a wake-up call for America. *Asia Pacific Issues*, *100*, 2.

Fellows, J. D., & Alexander, J. K. (Eds.). (2010). *Decadal science strategy surveys: Report of a workshop*. Retrieved from http://www.nap.edu/catalog/11894.html

Fisher, J. R. (2010). *Large instrument development for radio astronomy: Astro2010 technology development white paper.* Observatory Technical Council, National Radio Astronomy Observatory.

Flyvbjerg, B. (2006). *Five misunderstandings about case-study research.* Sage Publications. Retrieved from http://qix.sagepub.com/content/12/2/219

Flyvbjerg, B. (2008). *What is a mega-project?* Retrieved from http://flyvbjerg.plan.aau.dk/whatisamegaproject.php

Guide, P. M. B. O. K. (2008). A guide to the project management body of knowledge. The Project Management Institute, 4.

Harris, E. (2009). *Strategic project risk appraisal and management*. London: Gower Publishing Limited.

ICCPM. (2011). *Complex project management: Global perspectives and the strategic agenda to 2025 (task force report)*. International Centre for Complex Project Management.

Ika, L. A. (2009). Project success as a topic in project management journals. *Project Management Journal*, *40*(4), 6–19. doi:10.1002/pmj.20137

Jackson, M. C. (2011). *Systems thinking: Creative holism for managers*. London: John Wiley & Sons.

Jani, A. (2010). Escalation of commitment in troubled IT projects: Influence of project risk factors and self-efficacy on the perception of risk and the commitment to a failing project. *International Journal of Project Management*, *29*(7), 934–945. doi:10.1016/j.ijproman.2010.08.004

Jugdev, K., & Müller, R. A. (2005). Retrospective look at our evolving understanding of project success. *Project Management Journal*, *36*(4), 19–31.

Maylor, H. (2006). Guest editorial - Special issue on rethinking project management. *International Journal of Project Management*, *24*, 635–637. doi:10.1016/j.ijproman.2006.09.013

Morris, P. W. G., Crawford, L., Hodgson, D., Shepherd, M. M., & Thomas, J. (2006). Exploring the role of formal bodies of knowledge in defining a profession – The case of project management. *International Journal of Project Management*, *24*, 710–721. doi:10.1016/j.ijproman.2006.09.012

Morris, P. W. G., & Hough, H. H. (1986). *Preconditions of success and failure in major projects.* Oxford, UK: The Oxford Centre for Management Studies.

Müller, R., & Turner, R. (2007). The influence of project managers on project success criteria and project success by type of project. *European Management Journal, 25*(4), 289–309. doi:10.1016/j.emj.2007.06.003

Müller, R., & Turner, R. (2010). Leadership competency profiles of successful project managers. *International Journal of Project Management, 28*, 437–448. doi:10.1016/j.ijproman.2009.09.003

Murphy, D. C., Baker, B. N., & Fisher, D. (1974). *Determinants of project success.* Boston: School of Management, Boston College.

NASA. (2011). ASK magazine: The NASA source for project management and engineering excellence. *NASA, 41.*

OGC. (2002). *Contract management guidelines: principles for service contracts.* London: Office of Government Commerce, HMSO.

Phillips, N., & Hardy, C. (2002). *Discourse analysis: Investigating processes of social construction.* Thousand Oaks, CA: Sage Publications.

Pinto, J. K., & Slevin, D. P. (1989). Critical success factors in R&D projects. *Research Technology Management, 32*, 31.

PMI. (2011). PM network magazine: Too big to handle - Megaprojects and meeting the triple constraints. *Project Management Institute, 25*(3), 22–23.

Remington, K. (2011). *Leading complex projects.* London: Gower Publishing Ltd.

Remington, K., & Pollack, J. (2008). Are there special tools for complex projects? In *World congress project management to run.* IPMA.

Rendleman. J. D., & Faulconer, J. W. (2011). Escaping the space acquisition death spiral, three part series. *High Frontier Journal, 7*(4).

Shenhar, A. J., & Dvir, D. (2007). *Reinventing project management: The diamond approach to successful growth and innovation.* Boston: Harvard Business School Press.

Shenhar, A. J., Levy, O., & Maltz, A. C. (2001). Project success: A multi-dimensional strategic concept. *Long Range Planning, 34*, 699–725. doi:10.1016/S0024-6301(01)00097-8

Shenhar, A. J., & Wideman, R. M. (1996). *Improving PM: Linking success criteria to project type.* Project Management Institute, Symposium Creating Canadian Advantage through Project Management.

Thomson, M., Davies, A., Jenkins, C. (2011). Three views of risk: Selecting and acquiring military equipment. *ASPI Special Report, 42.*

Turner, J. R. (2004). Five conditions for project success. *International Journal of Project Management, 22*, 349–350. doi:10.1016/j.ijproman.2004.02.001

Williams, T. (2004). Identifying the hard lessons from projects – Easily. *International Journal of Project Management, 22*, 273–279. doi:10.1016/j.ijproman.2003.11.001

Williams, T. (2008). How do organisations learn lessons from projects – And do they? *IEEE Transactions on Engineering Management, 55*(2). doi:10.1109/TEM.2007.912920

Williams, T., Samset, K., & Sunnevåg, K. J. (Eds.). (2009). *Making essential choices with scant information: Front end decision making in major projects*. London: Palgrave Macmillan. doi:10.1057/9780230236837

Winter, M., Smith, C., Morris, P., & Cicmil, S. (2006). Directions for future research in project management: The main findings of a UK government-funded research network. *International Journal of Project Management*, *24*, 638–649. doi:10.1016/j.ijproman.2006.08.009

Yin, R. K. (2009). *Case study research: Design and methods*. Thousand Oaks, CA: Sage Publications.

Chapter 11
Analysis, Critique, and Proposed Revision of Crew Resource Management for Cockpit Automation:
The R-MPM and ERICA

Ronald John Lofaro
Embry-Riddle Aeronautical University, USA

ABSTRACT

It is well over 30 years since the first (then called) Cockpit Resource Management (CRM) training, now called crew resource management was introduced. It is a shibboleth, a sacred cow as it were, despite many issues, concerns, and changes over the years. Some 21 years ago, 1992, an Air Transport Association (ATA)/Federal Aviation Association (FAA)-Sponsored Workshop was convened in an attempt to deal with some specific CRM issues. Yet the issues and needs as articulated in that workshop, and some newer ones, remain. Thus, this chapter is 21 years overdue, leading to the questions: Why now and is it still relevant? As said, some needs, issues, and concerns remain. The relevancy is that both a critique of civil aviation CRM on many levels and a comparison with current USAF, USCG, and USN CRM are presented. The proposed skeletal template for the long-overdue revision of civil aviation CRM, the R-MPM is shown. Next, a new model for an intelligent cockpit automated decision aid/advisory system, Event Response Integrated Decision Advisories (ERICA), is shown. ERICA came about from 2009-2012 work in automated decision-making tools for the cockpit and the realization that the Revised Mission Performance Model (R-MPM) and ERICA were interrelated.

DOI: 10.4018/978-1-4666-4506-6.ch011

FORWARD

While this Chapter may seem specific to US civil and military operations and training, it is doubly generalizable. First, much of foreign pilot training is modeled on the US and many pilots flying for other flag carriers were trained in US. Secondly, what we will present has generic aspects and can be modified to fit planes and types of operations/missions not in the US military and/or US flag carriers milieu.

INTRODUCTION

This section will give a brief overview of why and how this Chapter came to be written. First, came a request by IGI Global to take a recently published article, up-date and expand on it. The strictures were to rewrite 20% of text; do a full update of data and appropriate update of conclusions; and do an exploration of future research potential. These were, as they say, a walk in the park; a serendipity as many of these tasks had already been done. In point of fact, more than 20% of the article has been re-written and there has been added approximately 40% of new material and, future research needs have been delineated along with the consequent potential of such research. With that said, let us look at the sequence of writing the article and this Chapter. Much of what is written in this Introduction will be expanded on and explicated in the Chapter itself.

Civilian CRM, by the mid-1990's, had continually encountered such problems that the "Big 3" (Delta, American and United Airlines) US flag air carriers almost completely revamped and renamed their CRM programs; basically focusing on human factors (McKenna, J.T., 1996). These sea changes received little publicity; we shall later look at some reasons why. Currently, CRM is now mandated in one Federal Aviation Regulations (FAR) for aircrew training using an optional training methodology while all the US military services have developed their own CRM, initially modeled on the civilian paradigm but significantly changed over the years. These changes reflect an emphasis on the skills and abilities needed for performance and mission success. The FAA's CRM model had been little changed.

Since 1993, Captain Smith and the author have worked together on developing, publishing and presenting a paradigm for flightcrew risk identification/management for operational decision making (ODM), the arena is most needed in terms of crew training, evaluation, performance... safety. At this point in time, much of Kevin Smith and the author's work on risk identification and management in flight are both in print and have been the subject of international participant Workshops. However, the author had a particular interest in the identification of CRM issues and flaws; for example, the author would often explicate and discuss these issues with military Instructor Pilots (IPs) from all services, in aviation human factors and aviation psychology classes that he taught from 2004 through 2011. It became clear that it was incumbent on him to provide both a cogent critique of the CRM model and, provide the CRM overhaul, based on a new model; this became the revised mission performance model (R-MPM). The original MPM being a product of a 1992 ATA-FAA Workshop wherein an attempt to develop a systematic method to integrate and evaluate CRM's role in flight control and safety. This Chapter completes a full circle and returns to what is acknowledged should have been done, circa 1992/3. Finally, as will be presented, there is new work on automated decision advisories/cockpit displays, based on decision-making algorithms, for the flightcrew. This is where the future of aviation safety seems to reside.

BACKGROUND

CRM training was first developed in the late 1970's/early 1980's after a series of disastrous and fatal air-carrier accidents; accidents where perfectly functioning planes crashed. beginning with the Eastern Airlines L-1011 in 1972 that crashed on an approach to Miami, killing 104 persons and culminating, so to speak, with the 1978 United Air Lines DC-8 which ran out of fuel, after circling for an hour, approaching Portland, OR. That pilot repeatedly ignored verbal input from crew members that they were to the point of not having enough fuel on board to reach their destination. Indeed and sadly, this was true and the plane crashed. Fortunately very few (10 out of 189) people perished. But, that accident that was seized on as paradigmatic; the pilot's attitude and actions were somehow seen as endemic to airline Captains. Nothing could be further from the truth...this type of Captain were anomalies. Perhaps 1% to 2% of airline Captains acted in such a manner; after all who wants to die? What that pilot did was egregious, but also not typical. Yet, why was this accident allowed to become a cornerstone for CRM? Why indeed?

The American flying public wanted assurances that something was being done to counteract the root causes of those highly-publicized 1970-1980's accidents. The air-carriers were extremely concerned about the impact of negative publicity on their bottom line, profit. The Federal Aviation Administration (FAA) regulates aviation in the United States (U.S.). The federal regulations covering aviation are all found in Combined Federal Regulations (CFR), Title 14, Aeronautics and Space; these are commonly referred to as the FARs. These are divided into parts and each part has a descriptive title and is a specific and detailed regulation. A major example is CFR 14, Part 121 is the FAR titled Operating Requirement: Domestic Flight and Supplemental Operations, usually referred to simply as Part 121. Remember that, back then, the FAA's stated mission was to ensure flight safety AND promote airline travel; difficult if not oxymoronic.

To reassure the flying public, the FAA approached the NASA Ames Research Center to help with this problem of eroding public confidence. NASA Ames in conjunction with some in academia developed a new type of training, then called cockpit resource management (CRM) training. The developers were grounded in social psychology and a related discipline, I/O (organizational and industrial psychology). So, it was natural that these CRM developers focused on 2 factors, group dynamics and communication. The United Air Lines Portland accident was indeed a model of terrible group dynamics and ignored communications; ergo, it was trumpeted as the rationale for the new training as if all the controlled flight into terrain (CFIT) accidents... and others...were caused by interaction and communication shortfalls.

And, the FAA wanted the air-carriers to implement this new flightcrew training, i.e. CRMT. Rather than append existing FARs to make CRM mandatory, the FAA chose another path. The FAA in order to make CRM training costs palatable to air-carriers, offered to waive some hours of pilot mandatory (yearly) recurrent training in lieu of CRM training. As training is a "big buck" item for air-carriers, this allowed the carriers to "save" (not spend additional funds), yet give CRMT. Recently, an alternate explanation has been given by a reputable source: United had been running their C-L-R training (i.e., their version of CRM) under a FAA waiver, which was expiring. United came to the FAA and said the training was costing money and was not mandated. The FAA, to ensure United continued with their C-L-R/CRM training granted this new waiver: all line Captains undergo, every 6 months, a line check for proficiency. The FAA responded that they would waive this requirement if the Captains took C-L-R/CRM instead. This waiver was to apply to all carriers IF they instituted a CRM training effort. Thus, under either explanation, which are somewhat similar, the FAA ensured the CRM training would, to a great extent be given. This also had the effect of silencing some critics who, understandably, wanted new, human factors (HF)-oriented training to counter the rash of accidents. The FAA also contracted out some/all of the early CRM training development.

However, there still was a sticking point; jeopardy. Somewhat simply put: as a further inducement to carriers to give CRMT, the FAA and Air Line Pilots Association (ALPA) agreed that CRM would be "no jeopardy" training. That means that there could be no "downs" (failure) on a check ride or LOFT that could be attributed solely to a CRM error/deficiency. This still seems to be the case.

To continue with a brief history of CRM: CRMT in the beginning, was a stand-alone, often off-site, ground training course for flight crews operating under FAR Part 121, appendix H. The stated purpose of CRM was to reduce/eliminate CFIT accidents by improving interaction and team performance. While a number of airlines implemented this voluntary program, the development of a rationale and metrics to directly measure the where, why and how of CRM's impact on flightcrew performance was not and, still has not been seriously attempted since 1992.

In 1986, a "one over the world" conference was held in San Francisco, CA. It was hosted by NASA and the USAF Military Airlift Command (MAC) and was entitled Cockpit Resource Management Training. (Orlady, H. & Foushee, H.C., Eds, 1986)

The avowed purpose was that, since CRM training was being done in a variety of venues and it was 6 years since the kick-off CRM(T) conference had been held, this new conference would allow reporting of past efforts and more importantly, identify issues and propose future pathways to deal with these issues. The author was a participant in one of the 6 Conference Working Groups, built around four major questions; one of which was is CRM training effective or not and how do we find this out? In short, how do we assess/evaluate CRM training?

Some keys findings, 27 years ago by that Conference, were that 1.There is a need for pilot development in decision-making. 2. While the need for objective, measurable CRM skills assessment was identified, it was agreed that there was no system in place to do this, although anecdotal evidence seemed to show a need for CRM and the study of the effectiveness of CRM training programs. However, a disturbing statement in the 1985 Conference Proceedings was that "... there was a need for [emphases added] *selling* CRM on a supportive basis to the industry and [emphases added] *invoking standards at the time would hinder that process.*" (Orlady, H. & Foushee, H.C., Eds, p.232)

Now, 27 years later, aviation does not have a template for evaluation of the need for which CRM skills as well as when, where and why, are needed in the operational arena in flight control maneuvers for safety of flight... and, how vital are they to safety? One fall-back of the CRM establishment for all these years has been that CRM skills are global concepts and cannot be evaluated; this is still promulgated in the latest FAA CRM advisory circular of 2004 (FAA CRM AC 120-51E). The other fall-back can be seen in this statement from a United States government accountability office (GAO) report: "However, FAA stated that science has not yet developed valid, reliable criteria for measuring CRM performance" (GAO/RCED-98-7, *Human Factors: FAA Guidance and Oversight of Pilot Crew Resource Management training Can Be Improved, 1997,* p.15).

A REALISTIC LOOK BACK IN TIME

Over the years, CRMT had often been a contentious issue with flightcrews at air carriers. In response, CRM has been modified to have line pilots actually do the instruction and not rely on outside facilitators with a canned course of lectures and materials. This was in response to severe criticism by pilots that many outside instructors did not have the aviation knowledge needed to answer pilot's queries. Also, many air carriers established CRM training units within their training centers; the CRM instructional team became two pilots; many air carriers developed some of their own CRM training materials. It is to be noted that the basic CRMT structure and goals still remained relatively unchanged. Yet, in spite of changes and modifications, it all fell apart, quietly and without fanfare, in 1996. At that time, the "Big Three" American flag carriers made sea changes in CRM. American Airlines, in July of 1996, set aside much of CRM as they were doing it. Their reason was that their flight crews has valid objections to, and concerns about, CRM: "CRM as too often viewed as a number of interpersonal issues that simply do not define the problems that we face in aviation." "CRM training will most likely always be defined and suffer in terms of the first generation of courses…which were seen as" touchy feely", "getting along", and "managing human relations or resolving personality conflicts" rather than dealing with truly important concerns (Ewell and Chidester, 1996). American's new focus is now on preparing flight crews for the daily challenges of normal and normal operations encountered flying the line. Delta Airlines, in the same timeframe, revamped their "CRM for New Captains" course and now calls it "In Command." As with American, Delta is emphasizing leadership, responsibility and performance. So, in 1996/7, we saw these two of the Big 3 air carriers eschewing over-emphasis on communication. United's CRM program was and is, called C-L-R, where the C is for Command and the L is Leadership, so it seems that United had already gone past the interpersonal in C-L-R and on to crew performance issues. Yet, even United changed aspects of their C-L-R in 1997 to re-emphasize the pilot's command and leadership functions. (McKenna, J.T., 1996).

ANOTHER PATH OPENS AND IS CLOSED

Thus, 27 years ago, some important questions and concerns were raised about CRM/T and a certain amount of skepticism existed in the airline community. As one response, the U.S. Air Transport Association (ATA) in 1990 formed a joint air-carrier/FAA/academe Working Group to deal with integrated CRM, other CRM issues and LOFT; both the author and Captain Smith were in that group. This Working Group recognized that the CRM skills must be integrated with a corresponding set of technical skills (flight control skills) in an interactive matrix in order to fully evaluate overall crew proficiency. A set of flight crew CRM markers ("CRM behavioral markers") with behaviorally-anchored rating scales was developed by the Dr. Robert Helmreich at the University of Texas. Yet, there was no corresponding set of flight control skills, with analogous behavioral markers. This lack was resolved in 1991 when Captain Kevin Smith (United Airlines) and Jan DeMuth (FAA Flight Standards) developed an initial set of performance markers for the Technical/Flight Control skills. Both the CRM and the Technical sets of markers were used in the next step of the actual integration of CRM and flight control skills. Captain Smith took the flight control skills and added the CRM skills as then in use to form a template as an basis for CRM Integration; this was called the Mission Performance Model (MPM; please see Appendix A).

Issues raised at the 1985 NASA/USAF MAC conference as well as respond to some of the recommendations of the ATA/FAA Working Group using a new concept, integrated CRM. The integration, and assessment, of Crew Resource Management (CRM) and flight control skills received considerable attention—and, a fair share of concern and skepticism in spite of the fact that such an "Integrated CRM" approach would serve as training tool, both in line-oriented flight training (LOFT) design and in specifying when/where and even if, the CRM/flight control skill linkages existed.

An approach to Integrated CRM, along with both human factors and flight control/technical skill evaluation scales, was partially developed during that ATA/FAA-hosted workshop, using the MPM. The author was the designer and facilitator of this Workshop. Six training Captains from Northwest Airlines, Delta Airlines, United Airlines, to include Captain Smith, the Chief Pilot for Boeing, 2 FAA researchers and 3 others were the participants. (Lofaro, R.J., 1992a)

Upon completion of the 1992 Integrated CRM Workshop, a new set of problems arose. The Integrated CRM concept and the mission performance model (MPM) were well received by the major air carrier CRM/AQP training pilots as well as Boeing's then-chief pilot attending the workshop. There were tentative plans for a follow-on workshop. However, a high-ranking FAA human factors official's view that integrated CRM was transparently, almost intuitively clear to all pilots, ergo FAA will not back further efforts, was coupled with a CRM "establishment" that was not open to taking CRM to either another level or in new directions. These factors indicated that any effort at another Workshop or any further work on Integrated CRM whereby the identification of where and which of the CRM skills were critical to safety of flight...or, indeed, to just to fly the mission itself... were dead ended. Thus, in 2013, twenty-seven (27) years after the above cited Workshop, aviation still does not have a template for evaluation of the when, where and why CRM skills are needed in the operational arena of flight control maneuvers for safety of flight. Put another way: how do CRM skills function in the ops arena... and, how vital are they to safety?

The author would be remiss if a clear distinction is not made here between what we call Integrated CRM and what is now referred to by the same name in the current FAA Advisory Circular (AC) on CRM (FAA AC 120-54E; 2004). This document speaks of integrating CRM training into an advanced qualification program (AQP) for flightcrew training...more later on AQP. The AC takes the stance that, within AQP, you can do "appropriate integration of CRM and technical training." However, its view, almost a polar opposite of our integrated CRM concept, goes on to say that CRM is composed of global activities which do not fit neatly into a hierarchical set of technical activities and, *de rigueur,* cannot be evaluated in conjunction with, if at all, flightcrew technical skills as CRM skills are generalized, not particular to any situation. Thus, this latest FAA on CRM warps the term "Integrated CRM" by defining it as simply putting CRM into flight training. Nothing on when/where/how; no integrated evaluation schema and, in point of fact, no evaluation at all.

The only new concept in this latest, and possibly last FAA CRM AC, are fatigue/stress and their impact. Since fatigue in flying has been studied since WWI, it seems that it has taken the FAA a fairly long timc to include it in their latest AC on CRM.

CONCLUSION

In retrospect, too much may have been hoped for and/or expected from CRMT; CRM may have been over-hyped to the flying public. CRMT is a process, not an outcome and CRMT is not the silver bullet for ending CTIT accidents as we all initially had hoped. And, after 30 plus years, why are there no CRM "'best practices" ? Not the best way to teach CRM, but the best CRM practices for use in the operational/flight venue. It is posited that the CRM developers' resistance to such evaluation was partly founded on not having a real grasp of the transport mission, the pilot's role and function in that mission and, partly on a commitment to a social psychology model. Another reason: development of the methodology for integrating and evaluating CRM and flight control/technical skills was attempted only once, and while partially successful, was never followed through on and placed in the dustbin by the FAA/CRM establishment.

Similarly, it is posited that, from the beginning, CRM used a questionable paradigm and took an incorrect path; this resulted in what we will term "flaws." Herein, the unaddressed and/or uncorrected flaws in initial and current CRM will be delineated . At this point, the purpose of this Chapter has been made very clear. It is possible, nay, probable that this Chapter will receive some broadsides, opprobrium and more from the well-entrenched CRM establishment. Along with that, there are those who will say that, after thirty plus years, what has changed or why now with such an article? Inertia and cognitive dissonance being what they are, these are obstacles to really examining CRMT. Thomas Kuhn has dealt extensively with the difficulties of paradigm change in the sciences (Kuhn, 1962, 2002). One finds in the *Chronicles of Narnia, Book Seven*, that Aslan has a description which can apply to many persons wedded to a paradigm: "They will not let us help them. ...Their prison is only in their own minds. Yet they are in that prison and so afraid ...that they cannot be taken out." The reader is asked to consider the arguments and reasoning with, at least, an open mind. Remember that this Chapter not only critiques CRM, but shows what are the beginnings of a viable and operational alternative.

CRM FLAWS

Time and Goals

CRM was developed at a time when flight crews consisted of 3, 4 and sometimes more persons; viz. DC-8; B-747 early models; B-707; B-727; L-1011; et al. It seemed natural to look at group dynamics and interpersonal communication. Time has long passed this social psychology model by since modern jet liners have 2 person crews, sitting a few feet away from each other and have had this configuration for many years. A modern paradigm must have as its goals two things which pilots have, again and again identified as core: decision-making in the context of risk management and workload management...with some emphases on the automation, instrumentation and information now available to such automation/instrumentation.

What We All "Know?"

Let us begin with the "common wisdom" that pilots made good decisions easily and almost naturally, aided by (some) increase in experience. The facile assumption that additional experience will teach pilots to make better decisions has proven to be a dangerous fallacy. Experience can be a nasty teacher, often giving the test before, or without, giving the lessons and materials needed for the test. Experience can also reinforce poor decisions and behaviors that seemingly "worked" in the past (the "not your day to die" phenomenon).

Communication

Inspection of early and even later CRM shows social psychology concepts were the core of CRMT. As said, CRM was developed using a social psychology paradigm with emphases in communication and group dynamics. However, both decision-making and workload management, recognized by pilots as their key functions, get only a passing mention.. Please be aware that although mention is made of decision-making/DM and later in CRM evolution, situational awareness/SA, there are no specifics given as to what/when/how...even why...to incorporate these vital elements into what is now known as crew resource management training. Additionally, SA and DM are sub-summed as parts of/under the rubric of Communication.

Communication can be a focus at some times/ in many ways. But, nowhere in CRMT do we see communication instantiated (Universal Instantiation or even as an Existential Instantiation) in the symbolic logic meaning, that is to say we do not see communication distributed in/over aspects of the other CRMT clusters/categories as they are termed. In universal instantiation, the rationale says that there is a property of something such that, if there is an occurrence, it must have that property. In other words, one reasons from generalized quantified premises by considering an "instance" to which that generalized statement applies. If all accidents have communication problems/issues as a property, than ANY accident must have communication as a problem/issue This plainly is not true of communication.

Where in traditional CRM do we see specific examples/instances of communication across what were termed clusters/categories? We only see communication within its own Cluster. And, even then we only see actual communications in the pre-flight briefings; thesc are really SOP and flight-plan driven, not CRM-driven. [Note: as will be seen in the R-MPM, communication is "instantiated/distributed]. Re-stated: Traditional (and current) CRM, with its emphases on communications and team dynamics, can be one enabler of good decisions. Communication may/ may not be part of decision making; it is NOT decision- making.

Psychological Underpinnings and Assumptions

The psychological underpinnings were, and still are, based on some doubtful assumptions, to wit: CRM training, (which was originally off-site and stand-alone; unusual for flight crew training) was an attempt to ascertain and modify crew attitudes. Returning to the fact that the original CRM was developed by social psychologists, with the emphases on group dynamics and communication: the goal of CRMT was to change attitudes about communication, group dynamics, et al. The belief was that positive attitudinal changes gained from the CRMT course would bring about behavioral changes in the cockpit

The latest FAA CRM AC still shows that these aims have not changed... "This [CRM] topic includes internal and external influences on interpersonal communications. External factors include communication barriers such as rank, age, gender, and organizational culture, including the identification of inadequate SOPs. Internal factors include speaking skills, listening skills... conflict resolution techniques, and the use of appropriate assertiveness and advocacy." (Federal Aviation Administration AC 120-51E: Crew Resource Management; 2004. p.10)

First, there was no effort, indeed no way, to correlate CRMT attitudinal change with changes in cockpit behavior. Secondly, there is evidence, cited below, that a significant number of CRMT participants had no real attitude change or developed negative attitudes as a result of/after their training. Add in this finding; any positive attitudinal change engendered by the course can last about a year or less. More importantly, the assumption that attitude change from CRMT necessarily leads to behavioral change in the cockpit is more than just questionable.

[Aside: We know how to change attitudes and behavior (at least short-term); research into cults and the Korean and 'Nam wars have taught us these methods. They are: the use of isolation coupled with real physical torture; sometimes the torture or the isolation, in and of themselves, will be sufficient. The third ingredient that can be added to either/both isolation and torture are drugs to weaken the will, produce dream/fugue states wherein the mental processes are more susceptible to manipulation and the line between reality and fantasy is blurred. All of the above need to be done repeatedly and on a long-term basis to be effective. None of the above are used in/after the typical initial CRM training.]

Evaluation/Measurement/ Assessment

CRM should have a T; it is training. It is standard practice to evaluate training when completed, the first stage being a test to see how much of the content was even learned. Passing strange, this does not happen in CRMT. In fact, the United States government accountability office (GAO) report had this recommendation to the FAA:"We believe that until FAA establishes a process for CRM curriculum development that includes an assessment of the extent to which pilots have mastered that curriculum, it will not be possible to measure CRM's performance in contributing to aviation safety." (GAO/RCED-98-7, 1997, p.16). It is difficult, if not impossible to measure the impact of training when one does not know how much of the training content was even learned. In point of fact, the only real evaluation of CRMT was that of any attitudinal change from before to after taking the CRM course (see below).

The relationship between CRMT and safety, which was and is the prime rationale offered for teaching CRM, has really not been proven In fact, the same GAO report cited has this statement: "However, FAA stated that science has not yet developed valid, reliable criteria for measuring CRM performance." (GAO/RCED-98-7,1997, p.17) At this point, we can almost hear many stating CRMT has been researched a myriad of times. Two points here: The results of most of the CRM research as to safety is not truly outcome driven, often includes anecdotes and, has this major flaw: the fact is that over the past 30 plus years, the accident rate (and total number of accidents) are both extraordinarily low and have been/are at a plateau. This results in such low numbers of accidents in a statistically-based study renders results moot.

Much of the testing has used what is incorrectly referred to as a CRM evaluation, the crew management attitude questionnaire, CMAQ. This instrument appears to be an amalgam of 4 industrial/organizational (I/O) tests, modified for aviation. The results of taking the CMAQ are used to show whether or not the CRM training made any difference in flightcrew attitudes towards team functioning, distributed decision making, communication along the lines of advocacy, conflict resolution, and the like: the social psychology model. It seems that the CMAQ developers defined what constituted positive flightcrew attitudes, in line with their emphases on group dynamics and communication.

Thus, CRMT, being an attempt at attitude change, was evaluated with a specially developed attitude test, to be given before and after the training in an effort to see what/if attitudes changed

However, a boomerang, or negative, effect was found, one in which 20% of those flightcrews having finished their CRMT course were seen as having better attitudes; 60% were basically unchanged and the boomerang? 20% had more negative attitudes towards CRM. These findings did come from the developers and initial proponents. (Helmreich, R. L.; Chidester, T. R.; Foushee, H. Clayton; Gregorich, Steve; Wilhelm, J.A.,1990 and Hackman, 1995)

Even more important was another finding that, after a year had passed from the initial CRMT, attitudes have reverted to then pre-CRMT levels. (Weiner, Kanki, Helmreich, Eds.,1995) These findings come from the CRM developers and initial proponents; they should have been, and was, cause for concern then... and still should be.

It is interesting that very similar results were found by 2 psychologists then working for the United States Air Force (USAF). These 2 researchers worked with 100 USAF pilots, using two personality tests plus the CMAQ . They found the same 20%-60%-20% split; they chose to call the 20% (of what were referred to on the civilian side as boomerang effect) as "the wrong stuff." The top 20% they chose to call "the right stuff," possibly influenced by a movie of the same name. (Retzlaff & Gibertini, 1987). As all their subjects were USAF pilots, it goes without saying they no longer work for USAF. However, it could and, possibly should be a research project to determine if this 20%-60%-20% breakout is true across all pilot communities. If so, then questions about as to pilot selection, self-selection, safety and more arise.

[Aside: There is a European system to evaluate the non-technical (CRM) flight crew skills: NOTECHS, standing for non-technical system for assessing pilot's CRM skills (Flin, Martin *et al.*, 2003; reprinted 2005). This system has many good points, the chief two being that both decision making and situational awareness are separated out for training and assessment. An interesting point is that NOTECHS was [emphases added] *specifically not designed* to be a "...tool for introducing psychological jargon into the [pilot's] evaluation." (Flin, Martin, et al. 2005, p.151). All this having been said, NOTECHS does not have a methodology for integrating CRM and flight control/technical skills for evaluation and, as with CRM evaluation in the USA, the evaluation of NOTECHS skills in check ride cannot be grounds for a "failure," unless there is a corresponding technical consequence which compromises safety of flight (Flin, Martin, *et al.* 2005, p.148). However, as said, NOTECHS does not have a methodology for integrating CRM and flight control/technical skills for evaluation, which makes a CRM failure assessment somewhat arbitrary.]

SUMMARY

It is at least doubtful that the hoped for positive attitudinal changes occurred and/or persisted in many of the flightcrews taking CRMT; in point of fact, some negative attitudes developed. Even more doubtful was that attitudinal change, in any sense, was permanent and necessarily resulted in permanent behavioral changes.

All flightcrew training must have this goal: performance enhancement which will lead to safety enhancement. However, with CRMT, there was no methodology by which to measure the crew's performance because of two factors; the first being that no template for linkage of CRM and technical flight performance actions/skills existed. Again, the GAO report cited said: "For example, FAA recommends that airlines train crews in 'workload management and situational awareness.' For this training, FAA suggests such topics as 'preparation/planning/vigilance' and 'workload distribution/distraction avoidance.' For those airlines that choose to integrate these topics with technical flying skills, FAA does not explain how the airlines are to do so." (GAO/RCED-98-7, 1997, p. 13). Add in the CRM establishment's insistence that CRM skills/knowledge were of a global nature and not amenable any evaluation approach that tried to break these out for evaluation. This still seems to be the official CRM stance as evidenced in the latest FAA AC CRM.

STARTING ANEW

This renewal began during and after the 1992 Integrated CRM Workshop (Lofaro,1992a). There it was clear that aeronautical decision-making was greatly different than decision-making on the ground and that a new paradigm was necessary, one that both articulated the differences and had a new set of decision-making (DM) techniques specific to what pilots/crew encounter. To that end, another FAA-sponsored Workshop was convened in Denver, CO later in 1992 to explore what was then called aeronautical decision-making. This Workshop was attended by experts in DM from academia, industry and aviation; there was a resultant FAA report: (Lofaro & Adams, R.J. & Adams, C.N., 1992b). The two Workshops indicted a path different than that 1990's CRM was taking...as will be shown.

Defining the Primary Roles of the Pilot/Flightcrew

In 1993, (Lofaro & Smith, 1993), a concept entitled "Pilot as Risk Manager," was proposed and it was further proposed that this is the quintessential captain's activity. Decision-making is the primary tool to be used by the pilot/crew to their primary functions: risk management by risk identification and risk reduction. This primary function of the pilot as a risk manager has been emphasized, over the past 20 years, multiple times (Lofaro & Smith, 2003, 2001, 2000, 1999, 1993).

The Operational Decision Making Paradigm (ODM)

A recent issue of AW&ST (Flattau, J., 2011) has highlighted the need for pilot decision-making in training and even gone further; there is a call for the skills to deal with the unexpected and for a sea-change in the current aviation training model... to wit: the skills that need to be trained..."include the capability to make quick decisions as well as analyze what is possible and what crew and aircraft are unable to do...we need to re-define the whole training perspective"...because, in point of fact," things that have never happened happen all the time-and these are the things that today's pilot's are the least trained to deal with."

Lofaro and Smith recognized the need for the capability to make such one-of-a-kind, time-compressed decisions over 20 years ago. To that end was developed an operational decision-making model(ODM) to be used in flight. The ODM paradigm deals with risk: its identification, level and management as the basis for making decisions in the operational environment; decisions based on a precise and accurate understanding of risk (Lofaro & Smith or Smith & Lofaro: 2001, 2000, 1999)

By risk is meant specific risk or danger to the aircraft, passengers and crew. If the level of risk rises, effective risk reduction strategies and maneuvers need to be employed by the pilot/crew to keep that risk within manageable limits. If decisive action is not taken as risk rises, the risk will continue to rise to a point where no crew action/response may be effective as to the plane's trajectory and safety.

ODM is the process that provides the flightcrew a method for risk management by decision making...DM done by the pilot/flightcrew, often under time pressure and often with little or no margin for error. Take note that, before any action (or inaction), there is a decision...or missed decision point.

ODM involves integrating situational awareness (SA) with the recognition and assessment of those factors that are critical to safe flight in order to identify and respond to the risk level.

[Aside: Since approximately the early-mid 2000's, the concept of risk management as an integral part of safety seems finally to have emerged in aviation, most notably beginning in military aviation where it is termed organizational risk management (ORM). ORM is somewhat static in that the risk assessment is done prior to launch and somewhat constrained in that, if the pre-launch risk level is too high, permission to change plans or to go-no, must come from higher up the chain of command. This having been said, ORM is a very useful tool.]

Components of the ODM Model

The ODM model and any resultant training program begins with the concept that all operational decisions have the same decision-making structure. Integrated throughout the model is the pilot's role as risk manager. The other components are:

The Operational Envelope (Mission Space): The operational envelope portion of the ODM and the mission performance model (MPM) are "the keys to the cockpit door." They allow for entry to and understanding of the pilot's mental model and the displays needed to ensure safety.

Situational Knowledge/Risk Location (Location within the Ops Envelope); the Critical Mission

Impact Areas and the 72 Critical Mission Factors which comprise the Impact Areas/Risk Location. There is a recently developed adjunct to the critical mission factors. It is an analogy to the USN/USMC ORM in that should any of these factors/conditions be present before launch, there should be no launch...until/unless the factors /conditions change or a new flight plan is devised. (these are called Show Stoppers).

The Rising Risk Continuum/ Risk Location

And, the undergirding: Cumulative effect (a concept whose use is embedded throughout all the above components). *Finally, it must be strongly noted that all of these components are so intertwined that any separation or sequencing of them is artificial, since it will become apparent that to speak of any one component brings in the others. Actually, there is no sequence per se. Rather, there is a constant interplay and branching across the components; perhaps a continuously rotating spiral best describes the process the pilot is involved in.* Thus, optimal ODM is achieved and, with it, optimal risk management and successful mission completion. (More complete treatments of the ODM paradigm can be found in Lofaro & Smith, 2001 and Lofaro & Smith, 2011c.)

LINE ORIENTED FLIGHT TRAINING (LOFT) AND CRM

There are four basic types of LOFT defined by the FAA; a more complete exposition can be found in the following FAA Advisory Circulars: AC 120-35C: Line Operation Simulations (9/6/91); AC 120-40B: Airplane Simulator Qualification (7/29/91); AC 120-45A: Airplane Flight Training Device Qualification (2/5/92).

The Air Transport Association of America (ATA), through many committees and working groups, and the FAA have tried to find the best ways to design, develop and implement LOFT and a particular type of LOFT, pilot/crew line-oriented evaluation (LOE) in a flight simulator (FS). This collaboration led to FARs and FAA Advisory Circulars (AC). The end-result is that LOFT, which is done entirely in a FS, has become over the past 20 years what could be called the crown jewel of air carrier pilot/crew training and evaluation. LOFT has thus become more and more recognized as the indispensable training, practice and evaluation tool for flight crew.

LOFT is now the pre-eminent tool, as well as test situation, for training and evaluating civil captains/crews.

Over the years, the need for a process by which LOFTs can be designed and crew performance enhanced as well as evaluated became abundantly clear. The author did develop a dual process for such LOFT design. A complete presentation with the dual methods for developing LOFT scenarios is available (Lofaro & Smith, 2008)

This process used the Mission Performance Model (MPM) as developed by Captain Kevin Smith, then of United Air Lines, with some input from Jan DeMuth of the FAA. The MPM came from the recognition that the CRM skills must be integrated with a corresponding set of technical skills (flight control skills) in an interactive matrix in order to fully evaluate overall crew proficiency. Further, such an "Integrated CRM" approach would serve a as training tool – and, in LOFT design and, in specifying where the CRM/flight control skill linkages existed and, to actually design the LOFT (two variants as above). Remember, if you will, that CRM was to be an integral component of LOFT and LOFT is now the primary tool for both type rating and crew performance evaluation. Nowhere is found, to the best of my knowledge, any LOFT design template that is aimed at evaluation of CRM skills as they impact flight control and trajectory.

However, the MPM used in the initial work on LOFT design thus needed to be and was greatly changed; the result is the initial version of the R-

MPM, as presented later in this Chapter. Future work will be needed to continue its develop and evaluate its use-value in LOFT design. It would seem that it would be effective in both the design and performance evaluation aspects of LOFT. But, this is a topic for future R&D, using flight simulators

THE MILITARY AND CRM: USCG, USAF, AND USN

There is neither time nor space to go through a complete history of how the US military developed their own CRM programs, beginning with using the civilian aviation model and, as the years passed, modifying it for their own mission and needs. Therefore, we will sketch out a exceedingly brief history before we see how much their 2012 CRM programs differ and/or are the same as the FAA's latest CRM AC.

United States Coast Guard (USCG)

Before we look at the US Navy and the Air Force, let us look at the US Coast Guard. They are, as to aviation assets, a small service with since 9/11, a whole set of new duties in the US security arena. Yet, being a small service serves them well as they can design, change, implement new training programs quickly and easily. The USCG recently changed their CRM course. As Ron Lofaro has taught many aviation psychology courses to USCG Instructor Pilots over the past 8 years, he had access to some of these new CRM training materials.

The USCG CRM core principles are, in rank order:

1. Situational Awareness (SA)
2. Communications
3. Leadership and Decision Making
4. Assertive Behavior
5. Operational Risk Management/ORM (a principle that USN and USMC use and we have mentioned before).

Since the USCG only flies multi-crew planes and helicopters, assertive behavior is of import as the pilot must be made aware of such things as sea state; whether rescue diver can be/is deployed; whether the sling to pick up persons from water or roof tops as with Hurricane Katrina has been deployed and has a person in it.

That having been said: situational awareness and decision making are 2 of first 3 core principles. Again, of import is that stress and fatigue are specified as negative factors in SA and DM; USCG calls them CRM roadblocks. We now look at the core principles of USAF and USN CRM.

United States Air Force (USAF)

The USAF was the military leader in CRM. The first Air Force command to initiate a military version of CRM training for aircrews was the Military Airlift Command (MAC) in the mid-1980s. Almost immediately, other major commands began implementing their own mission-specific versions of CRM training for their aircrews. The Military Airlift Command, as has been noted, co-hosted the 1985 CRM conference in San Francisco. In 1992, the Air Force realized that the mission differences between commands and the differences between civilian airline pilots and their CRM and military pilots needed to be addressed and conducted its first Air Force-wide CRM conference.

The USAF steering committee was established at this conference to standardize requirements for developing, implementing, and managing all USAF CRM training programs for aircrews. This guidance was published as Air Force Instruction (AFI) 36-2243, *Cockpit/Crew Resource Management Program*, in 1994, and later changed to AFI 11-290. At this time, the guiding document for

USAF CRM is still AFI 11-290 but had a key Supplement was added in 2005; see below.

The USAF (formerly named) Strategic Air Command fielded a very comprehensive CRM training program under contract to Hernandez Engineering. This course, in turn, was adapted for training USAF fighter crewmembers stationed in Europe. The USAF Inspection and Safety Center initiated the development of a prototype course focused on single-seat fighters in 1990. It was intended to improve inter-cockpit as well as intra-cockpit Decision Making. This program was a modified and enhanced version of the US Navy ACT program. The US Naval Safety Center, working with MAC, had developed militarized CRM programs, labeling them Aircrew Coordination Training (ACT) and, developed by CAE-Link. Units of the US Army have applied this same course to utility and attack helicopters. The USAF and USN training commands also integrated these materials into their respective undergraduate pilot and navigator training. Thus, both the USAF and USN initially had civilian contractor-developed CRM as developed from the civilian CRM model. However, it must be strongly stated that current USAF, USN and USCG CRM really have changed since their original development and implementation, mainly becoming more performance and mission success oriented.

USAF/USN Current CRM: Core Principles

In 2005, the USAF issued an important Supplement (Supplement 1 to AFI 11-290) to their CRM training program of 1999; revised 11 April, 2001 (Air Force Instruction 11-290: Cockpit/Crew Resource Management Training Program). This supplement contains the CRM core curriculum components, their definitions, and has the corresponding names of the USN 7 critical CRM components in parentheses next to each USAF core principle, as seen below.

In rank order, the USAF core components and USAF definitions (emphases added) are:

Core CRM Curriculum. CRM knowledge and skill objectives will be tailored to fit the unique characteristics of each primary mission. Navy Aircrew Coordination Training (ACT) equivalent terms are in parenthesis.

1. **Situational Awareness. (Situational Awareness)** Includes knowledge and skill objectives for preventing the loss of situational awareness, skills for recognizing the loss of situational awareness, and techniques for recovering from the loss of situational awareness.
2. **Crew Coordination/Flight Integrity. (Leadership and Assertiveness)** Knowledge and skill objectives covering the impact on aircrew performance of command authority, leadership, responsibility, assertiveness, conflict resolution, hazardous attitudes, behavioral styles, legitimate avenues of dissent, and team-building.
3. **Communication. (Communication)** Includes knowledge of common errors, cultural influences, and barriers (rank, age, experience and position). Skills will encompass listening, feedback, precision and efficiency of communication with all members and agencies (i.e., Crewmembers, Wingmen, Weather, ATC, Intelligence, etc.).
4. **Risk Management/Decision Making. (Decision Making)** Includes risk assessment, the risk management process, tools, breakdowns in judgment and *flight* discipline, problem-solving, evaluation of hazards, and control measures.
5. **Task Management (Adaptability/ Flexibility).** Includes establishing

priorities, overload, underload, complacency, management of automation, available resources, checklist discipline, and standard operating procedures.

6. **Mission Planning/Debrief (Mission Analysis).** Includes pre-mission analysis and planning, briefing, ongoing mission evaluation, and post mission debrief.

Again, SA is number one and again, and neither SA nor DM are part of communication. This is important as we shall soon see. Please also note that while USAF and USN agree on the core components, their ranking is slightly different and the USN uses its own definitions for their 7 critical CRM components. Also note that compared to USCG, DM and leadership are ranked 2 and 4; very informative and very telling. USAF also brings in risk management as part of DM which we have posited is true since 1993.

The USN 7 critical CRM factors, in rank-order:

1. Situational Awareness
2. Decision Making
3. Assertiveness
4. Mission Analysis
5. Communications
6. Leadership
7. Adaptability/Flexibility

Again, we see SA and DM as priority number's 1 and 2.

In sum: over the years, the US military services have come to realize that SA (not communication) is one key and DM, especially DM as a part of/linked to risk management, is another. Further, both SA and DM are separate components, and are *not* sub-groups or sub-summed under any other rubric. We also note that all the US military services cited have, to some degree, put CRM in as a component of flight training and all have evaluation flightcrew check rides on CRM which can be "busted," i.e. failed.

THE FAA'S LATEST VERSION OF CRM

Let us turn to the aforementioned latest (2004) FAA advisory circular on CRM. *Caveat*: Please be aware that there are authors' critiques/question on many of the CRM components; these are always in parentheses and in italics.

A. **Communications Processes and Decision Behavior.** This topic includes internal and external influences on interpersonal communications. External factors include communication barriers such as rank, age, gender, and organizational culture, including the identification of inadequate SOPs. Internal factors include speaking skills, listening skills and decision making skills, conflict resolution techniques, and the use of appropriate assertiveness and advocacy. The importance of clear and unambiguous communication must be stressed in all training activities involving pilots, flight attendants, and aircraft dispatchers. The greater one's concern in flight-related matters, the greater is the need for clear communication. More specific subtopics include the following:

   1. **Briefings.** Training in addressing both operational and interpersonal issues, and training in establishing and maintaining open communications. A captain's briefings should reaffirm established SOPs and should address the most threatening safety and security situations.

      a. **Safety.** A captain's briefing should address emergencies that might require an airplane evacuation (e.g., cabin fire or engine fire) and should highlight the functions of flightcrew and flight attendants during an evacuation. A captain's briefing should stress to flight attendants the importance of identifying able-bodied passengers and briefing them, in turn. Passengers in exit rows are particularly important resources, and flight attendants should brief them on what to do during an evacuation. *(All SOP's)*

b. **Security.** A captain's briefing should address general security topics, especially hijack, and any known or suspected specific threat pertaining to the flight. Flight attendants should identify able-bodied passengers, including exit row seat occupants, and may enroll them as resources who might be called upon to help contain a disruption caused by a passenger(s). *(All these activities are also SOP's)*

2. **Inquiry/Advocacy/Assertion.** Training in the potential benefits of crewmembers advocating the course of action that they feel is best, even though it may involve conflict with others.
3. **Crew Self-Critique (Decisions and Actions).** Illustrating the value of review, feedback, and critique focusing on the process and the people involved. One of the best techniques for reinforcing effective human factors practices is careful debriefing of activities, highlighting the processes that were followed. Additionally, it is essential that each crewmember be able to recognize good and bad communications, and effective and ineffective team behavior. *(How?)*
4. **Conflict Resolution.** Demonstrating effective techniques of resolving disagreements among crewmembers in interpreting information or in proposing courses of action. Demonstrating effective techniques for maintaining open communication while dealing with conflict. *(Group dynamics...social psychology paradigm)*
5. **Communications and Decision making.** Demonstrating effective techniques of seeking and evaluating information. Showing the influence of biases and other cognitive factors on decision quality. There are benefits in providing crews with operational models of this group decision process. Crews may refer to these models to make good choices in situations when information is incomplete or contradictory. *(DM is still part of Communications; what models are referred to and how is this to be done?)*

B. **Team Building and Maintenance.** This topic includes interpersonal relationships and practices. Effective leadership/followership and interpersonal relationships are key concepts to be stressed. Curricula can also include recognizing and dealing with diverse personalities and operating styles. Subtopics include*: (Group dynamics again)*

1. **Leadership/Followership/Concern for Task.** Showing the benefits of the practice of effective leadership through coordinating activities and maintaining proper balance between respecting authority and practicing assertiveness. Staying centered on the goals of safe and efficient operations.
2. **Interpersonal Relationships/Group Climate.** Demonstrating the usefulness of showing sensitivity to other crewmembers' personalities and styles. Emphasizing the value of maintaining

a friendly, relaxed, and supportive yet task-oriented tone in the cockpit and aircraft cabin. The importance of recognizing symptoms of fatigue and stress, and taking appropriate action. *(Group dynamics yet again)*

3. **Workload Management and Situation Awareness.** Stressing the importance of maintaining awareness of the operational environment and anticipating contingencies. Instruction may address practices (e.g., vigilance, planning and time management, prioritizing tasks, and avoiding distractions) that result in higher levels of situation awareness. The following operational practices may be included:
   a. **Preparation/Planning/ Vigilance.** Issues include methods to improve monitoring and accomplishing required tasks, asking for and responding to new information, and preparing in advance for required activities.
   b. **Workload Distribution/ Distraction Avoidance.** Issues involve proper allocation of tasks to individuals, avoidance of work overloads in self and in others, prioritization of tasks during periods of high workload, and preventing nonessential factors from distracting attention from adherence to SOPs, particularly those relating to critical tasks. (*Here one sees SA and Workload Management grouped together, but under team building as the next-to-last components listed. Again, the group dynamics/social psychology paradigm. There is a new component, added since 1998...see below)*
4. **Individual Factors/Stress Reduction.** Training in this area may include describing and demonstrating individual characteristics that can influence crew effectiveness. Research has shown that many crewmembers are unfamiliar with the negative effects of stress and fatigue on individual cognitive functions and team performance. Training may include a review of scientific evidence on fatigue and stress and their effects on performance. The content may include specific effects of fatigue and stress in potential emergency situations. The effects of personal and interpersonal problems and the increased importance of effective interpersonal communications under stressful conditions may also be addressed. Training may also include familiarization with various countermeasures for coping with stressors. Additional curriculum topics may include examination of personality and motivation characteristics, self-assessment of personal style, and identifying cognitive factors that influence perception and decision making to determine if CRM training is achieving its goals. *(Where does the flightcrew get the requisite psychological knowledge/training/experience to perform these "examinations" and "self-assessments"?)*

Finally, as we have pointed out earlier, the roles of stress and fatigue in aviation have been studied since WW I. It is heartening to see FAA has finally recognized these factors in degraded performance. As one last example: while the military assesses flightcrew CRM performances where failure is an option, we here show the FAA CRM AC's words: For optimal assessment, data on crewmembers' [emphases added] *attitudes and behavior* should be collected before CRM indoctrination and again at intervals after the last component of CRM training, to determine both initial and enduring effects of the program. The goal should be to obtain an accurate picture of the organization's significant corporate personality traits before formal adoption of CRM training, and to continue to monitor those traits after implementation. Hence, the goal of the FAA CRM is not performance evaluation but organizational culture/climate assessment.

Comparison of FAA CRM and Military CRM

When the military and FAA/Civil aviation CRM components are compared, it is clear that the civil aviation CRMT, as exemplified by FAA CRM AC is very different. The FAA AC seems little different that earlier ones dating back to 1992 while the military has gone on to SA, DM and risk management as keys. It seems clear that the FAA still advocates a social psychology model while the military holds to an operational focus on mission success and safety.

At this point, little more needs to be said.

THE R-MPM AND ERICA

Over 23 years of working with Kevin Smith, we have often taken a step back and thought: well and good, we finally have done all we can; what are working on is complete and nothing is left to do. In each and every case, we were wrong. We continued to expand our work, finding out that what we had done simply and seamlessly led to expansion, to new directions, to realizations that we were not "finished" and there was more to work to be done. It happened again in 2012.

In trying to descriptively capture what it termed Event Response Integrated Decision Advisories (ERICA) for the pilot/crew, Captain Smith was looking back at some of our earlier articles and presentations. It occurred to Captain Smith (epiphany of sorts) that we had not seen the reality, possibilities, use-value of our prediction models as presented at our 2009 and 2011 ISAP Workshops; very likely a result of being too close to what we were doing and always, it seemed, re-working, extending and modifying our seminal work in operational decision making. Captain Smith went on to first conceptualize, then develop ERICA and its components, resulting in a provisional patent for this intellectual property/invention. ERICA is seen as an automated part of an existing cockpit display system, based on the ODM with 2 new functionalities: PMG and MPE.

An epiphany of sorts happened to the author as he was finalizing two Journal articles on CRM flaws and a proposed template to replace CRM: the Revised Mission Performance Model (R-MPM). He saw that parts of ERICA can be embedded in the R-MPM and realized that the R-MPM and ERICA were interrelated.

Re-Opening the Closed Path

It had long been clear that the next step, in this age of automation, was to develop an automated ODM. However, circa 2009 an attempt was made to show the components and architecture of an automated display, an in-flight risk management display then termed Automated Risk Management Display to include configuration, display placement, and operational algorithms. The ODM was the core and development sequencing along with protocols and procedures for its usage was delineated. The display would be integrated into the lower engine indicator and crew alerting system (EICAS) of the cockpit display system and be connected to a similar display utilized by the Dispatcher. To that end we published 2 interrelated articles (Lofaro & Smith, 2011a and Lofaro & Smith, 2011b).

However, the ultimate goal and vision is an Risk Identification/ Risk Management Decision Aid; an automated cockpit display(s) for pilot/crew advisories/guidance to include data on Terrain Critical Trajectory Management Flight

Guidance; Takeoff Alternate (divert) Flight and Procedural Guidance; Destination Alternate (divert) Flight and Procedural Guidance; Enroute Divert Flight and Procedural Guidance; ETOPS Divert Flight and Procedural Guidance; Organized Track Escape Maneuver Guidance; Critical Flight Maneuver Trajectory Optimization and Oceanic Flight Optimization Decision Aids. The above data, which are contained in, or can be developed from, the modern airliner's systems, needs to be made available to the flightcrew. Lofty goals but doable..at least was believed doable.

RATIONALE

Why opt for an automated decision aid/advisory system? First and foremost: there were almost insane decisions/non-decision errors made by pilots/crew; forcing quest for developing systems that are as idiot (perhaps the word error should be used in place of idiot)-proof as possible, while keeping the Captain in his rightful role and responsibility. Secondly, up to now, cockpit design approaches have used essentially the "shotgun" method where increasing amounts of information are "dumped" into the cockpit environment, believing incorrectly that more information is always better. This design approach has resulted in too much irrelevant information, difficult to understand information, or no meaningful information actually being provided to the flightcrew.

There are real prices to pay for these data difficulties: "plan continuation bias" and workload overload. In times of stress, the decision maker will choose from among all the information available only those facts which support a preconceived solution or position..."tunnel vision," i.e., functional fixation and decision bias. As workload increases to overload, the initial response is to prioritize, then shed tasks to reduce workload. What is here proposed is not more information transmitted to the cockpit, but "better" information; better meaning on-going, timely information that incorporates risk identification, evaluation and management. And, the incorporation of this information with on-board information to be used in an automated system for decision-making aids/advisories and flight control guidance to the pilot/crew.

ERICA

NOTE: What follows is taken from Kevin Smith's work on conceptualizing and developing ERICA to the point of being granted a provisional patent. It is the author's best and, hopefully adequate, efforts to present Captain Smith's work on ERICA.

ERICA represents a particular set of functions in the general area of mission adaptive cockpit displays for air transport aircraft. ERICA contains three unique, automated sub-systems that perform before and then engage throughout the course of a mission. ERICA is specifically designed to address operational problems that require accurate decisions and decisive action by providing the flight crews with meaningful operational data. The function of ERICA's automated systems and cockpit displays is to provide flight crews with the right information at the right time to make the right decision(s). The flight crew remains pro-active in response to all mission critical events (risk) and operational decision making is optimized. This has been aviation's most elusive goal.

ERICA Components

ODM: Operational Decision Making
PMG: Precision Maneuver Guidance.
MPE: Mission Performance Evaluator

ODM

This is essentially an onboard risk management decision aid. To incorporate this decision aiding function, ODM performs the following. 1). ODM collects all pertinent risk information; and 2). Displays this information to the flight crew as distinctive levels of risk. ODM is initialized prior to departure and operates continuously for the duration of the flight.

PMG

The Precision Maneuver Guidance feature—PMG—provides the crew with essential critical flight maneuver guidance when urgent response is needed to avoid catastrophe. This critical flight maneuver performance package is triggered by available activation devices (i.e. Microburst Alert) and contains all known escape, recovery, avoidance, and reject maneuvers. PMG only activates during specific triggering events.

MPE

The Mission Performance Evaluator—MPE—evaluates the overall performance of the mission with respect to not only significant parametric criteria, but to "best practices" as well. Importantly, a look-ahead function, the core attribute of MPE, aids the crew in maintaining a pro-active posture throughout the course of the mission. One key feature of the MPE are anchor points; comparison of a plane's position, airspeed, configuration et al to pre-specified parameters at various in-flight points, with a focus on final approach. MPE operates continuously for the duration of the flight.

We now look at the R-MPM and, will show where components of ERICA are being placed. This is a work in progress.

REVISED MISSION PERFORMANCE MODEL (R-MPM)

Remember that the MPM was developed prior to and, used at, the 1992 ATA/FAA Workshop. However, it was part of what was termed Integrated CRM, which was neither used by the FAA/industry nor followed up on. While it was realized that CRM skills must be integrated with a corresponding set of technical skills (flight control skills) in an interactive matrix in order to fully evaluate overall crew performance and proficiency, no further work was done in the arena either.

[Aside: At the 1992 Workshop cited above, an additional component was added to the CRM side of the MPM, based upon comprehensive mission studies conducted by the ATA/AQP working group. This component was called Operational Integrity. It was removed from the FAA CRM ACs at some point and by some authority. As a first step, it is included in the R-MPM.]

Over the years, the author never stopped looking at why CRM needed to be revised and what a revision needed to capture. Any CRM/MPM revision must also be able to accomplish what had long been called for: act as a training tool, a tool for LOFT design and, for flightcrew performance evaluation in a LOFT as well as in any other evaluation flight. In doing the revision of the MPM, he quickly became aware that what was termed the CRM skills side of the MPM contained much of what he had been critical of for some years. Moreover, the goal was to specify what performances were needed; performances in the cognitive realm. Thus, while keeping the basic title of the flight control/technical skills portion of the revised paradigm being worked on, the old CRM skills were re-titled Cognitive Performances. Next, the needed cognitive performances were almost completely changed, as well as modifying some aspects of the original flight control portion (now re-titled Flight Trajectory Performances). The result is a revised mission performance model; R-MPM.

R-MPM, at this stage, is a skeletal outline/template of a proposed flightcrew performance model, built for performance evaluation and LOFT design; a template which is flexible and can be easily modified for different aircraft types, mission types, operating conditions (winter ops; carrier ops, et al) by inserting new boundary/relevant boundary conditions; mission objectives/constraints; aircraft type parameters. And to return to what was mentioned earlier, can be a design base for automated intelligent cockpit design.

The R-MPM is the outcome of seeking to replace a flawed CRM template with one which could be used in both training, design (LOFT) and performance evaluation. A now-recognized connection with ERICA also occurred.

THE R-MPM

Note: following group theory, groups and sub-groups are used as headers. All sub-group listings are or will be broken out into elements. The elements will be used to develop be our analog to the old CRM behavioral markers...we will call them: "performance indicators." It is hoped to have some civil and military pilots provide assistance in this further decomposition into elements. (see Figure 1.)

Cognitive Performances

Group1: DM

Sub-groups:

1. ODM elements: a. model
   b. Role of ODM in ERICA: risk ident/management
   c. in-flight communication
2. SA in DM
   elements: a. in-flight communication

Group 2: Workload Management

Sub-groups:

1. Pre-flight briefing
   elements: a. Pre-flight workload assignments
2. In-flight workload
   elements: a. in-flight communication
3. Workload management in risk management
   elements: a. Recognition of overload
   b. Re-prioritize and re-assign tasks
   c. in-flight communication
4. Captain's authority (FAR 91.3) wrt # 4. above

Group 3: Ops Integrity...*caveat*: civil and military may well differ on the sub-groups/elements and their import BUT again, this is a generalizable template that can be modified as needed.

Sub-groups:

1. Specification of mission concerns (Pre-flight briefing)
   elements: a. Conscious avoidance of undue risk: ERICA/ODM; Show Stoppers; critical mission factors
   b. Use of ERICA in an automated risk guidance display
2. Schedule reliability
   elements: a. departure time
   b. arrival timc
   c. delay awareness
3. Regulatory compliance
   elements: a. parameter/deviation awareness: ERICA/MPE
   b. limits/thresholds
   c. in-flight communication

Flight Trajectory Performances

Group 1: Flight Maneuvers and Attitude Control

Sub-groups:

1. Instrument Interpretation
   elements: a. ERICA advisories and actions needed
2. Instrument Selection
3. Lateral Navigation
4. Vertical navigation
5. Altitude Control
6. Velocity Vector

Figure 1. The R-MPM diagram

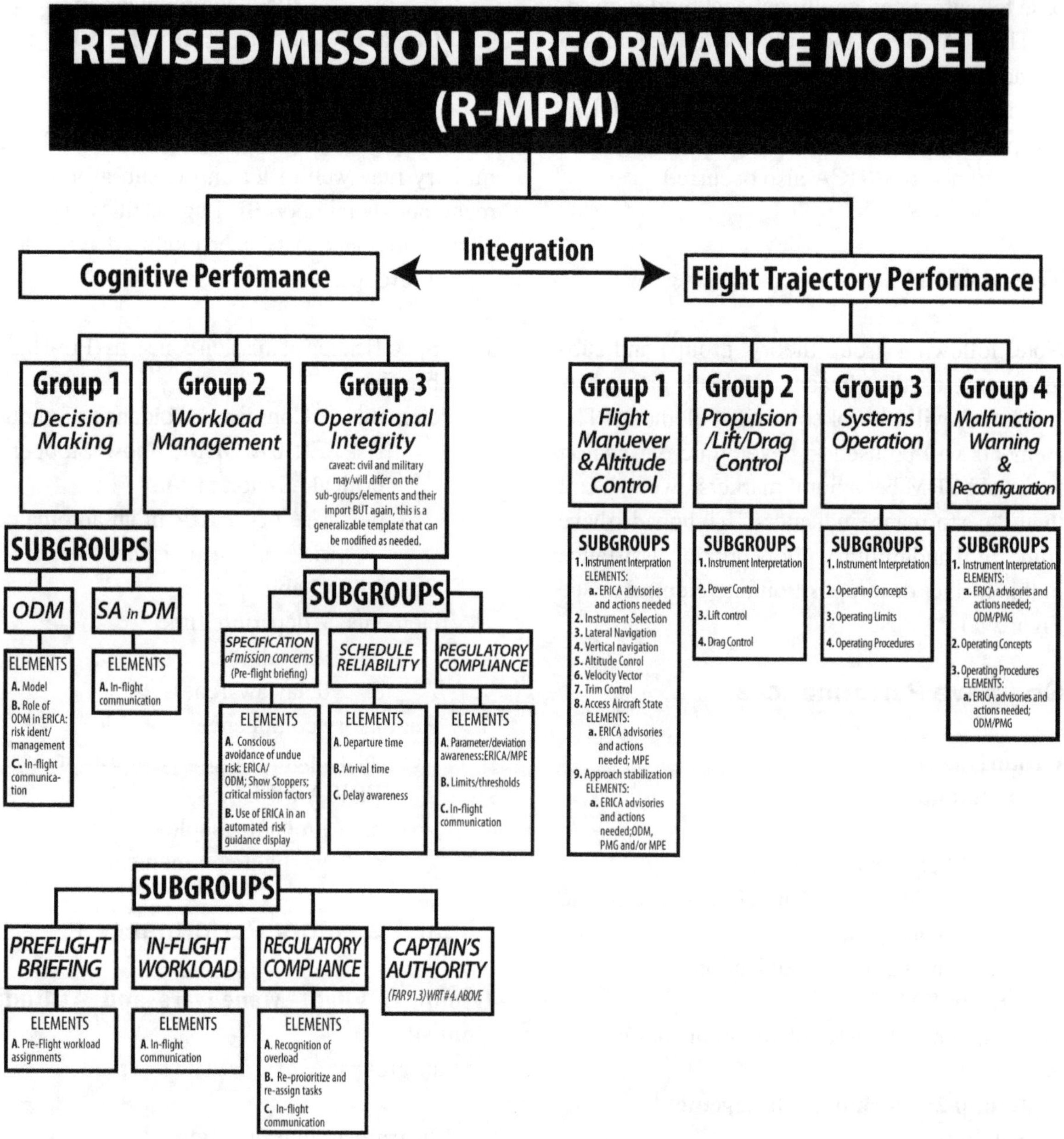

7. Trim Control
8. Access Aircraft State
   elements: a. ERICA advisories and actions needed; MPE
9. Approach stabilization
   elements: a. ERICA advisories and actions needed; ODM, PMG and/or MPE

Group 2: Propulsion/Lift/Drag Control
Sub-groups:

1. Instrument Interpretation
2. Power Control
3. Lift control
4. Drag Control

Group 3: Systems Operations

Sub-groups:

1.Instrument Interpretation

2. Operating Concepts
3. Operating Limits
4. Operating Procedures

Group 4: Malfunction Warning and Reconfiguration

Sub-Groups:

1.Instrument Interpretation

elements: a. ERICA advisories and actions needed; ODM/PMG

2. Operating Concepts
3. Operating Procedures

elements: a. ERICA advisories and actions needed; ODM/PMG

SUMMARY, CONCLUSIONS, FUTURE R&D

A Completed Trip?

It seems that these is a closure to a full circle from 1990 to 2013. For clarity, some sort of recapitulation for the reader is needed; one which shows the sequence of events and work. To that end: the circle began with an ATA/FAA Working Group in 1990. The Working Group was looking at integrating CRM into evaluation as well as LOFT design. Next was a 1992 ATA/FAA-hosted Workshop on Integrated CRM and more. After the Workshop, integrated CRM and the MPM were not followed-up on by the FAA and the CRM community. But, at that same Workshop, it seemed clear that, while CRM was not the answer to CFIT, aeronautical decision-making well could be. This was followed by a second FAA/Academia/Industry Workshop on aeronautical decision-making.

At this point, a the development of model for decision-making in the operational arena was initiated; a model to be used to instruct and to aid pilots in real time in what was called operational decision making (ODM). This paradigm defined the primary role and function of the pilot/crew: risk identification and management. From 1993 on, the ODM paradigm was refined, expanded and published on. Usage of ODM and the MPM in LOFT scenario design and for flightcrew performance evaluation in a LOFT was explored.

Circa 2008, automating the ODM and incorporating this into existing cockpit displays took a lead role. A prototype template for the design of such an automated ODM display as well as an R&D protocol for building an automated ODM system...which identified safety of flight threats, could be up-dated at any point in the flight (either from the ground/Dispatch or by crew) and evaluated the risk level to aid the decisions that needed to be made...was worked on.

Next, it was realized that the full potential of automating ODM was yet to be explored; Captain Kevin Smith went forward, working with an aeronautical engineer /patent attorney and others. He then laid the groundwork for ERICA by working out the complexities, diagrams, schemata, flow charts and interrelationships of a tri-partite, automated decision aid/advisories, system: ERICA

The author went back to an area that he had neglected as to formally writing on it: the revision of CRM. In developing and publishing analyses/critiques of CRM, he realized that the MPM (if it was to serve as a new model) then also needed revision. The reason being that many of the CRM components of the MPM were exactly what were being criticized and in dire need of not revision, but replacement. When finished, this new model was termed the R-MPM. The R-MPM is seen as a template for LOFT design/development and flightcrew training and evaluation.

[Aside: this of course means that the LOFT design/development process done circa 2007/8 was in need of revision as it would use the R-MPM.]

Finally, while working on the R-MPM, it was realized that ERICA's sub-systems played in many of the R-MPM components; identifying all these relationships and connections are a task to be completed.

At this point, our joint journey is over, as it were, yet there remain many avenues for possible modification and refinement of these works.

Conclusions

ERICA and the R-MPM have tremendous potential for aircraft safety. It incorporates 20 years of work on operational decision making The outcome is an interrelated template for training, flightcrew performance evaluation, LOFT, mission planning, accident prevention...and, even though it was not mentioned in the Chapter, accident analyses. In point of fact, ODM has been successfully used for such analyses in some of our publications and, at an internationally-attended Workshop. The ODM and ERICA-embedded components of the R-MPM can be a template for such accident analyses. ERICA may prove to be a watershed in the thought, theories and operational aspects of safety.

The above may seem to be a small set of conclusions. However, they are germane to and, vital for, aviation safety.

Future R&D

- Validation/modification of the R-MPM components; the same for the embedded ERICA components.
- Revision of the formerly MPM-based LOFT design model; now to be R-MPM-based.
- Flightcrew performance evaluation methodology using R-MPM for LOFT and check-rides.
- Validation of the new LOFT design in flight simulators; same for the evaluative methods/criteria.
- Studies that show the ERICA + R-MPM can yield valuable accident analyses data; this can be done by re-analyzing prior accidents plus new ones.
- Study showing where and how to embed ERICA and R-MPM into flight training... and, unlike CRM, showing what was learned in the training and how it transfers to the cockpit/FS.
- Building a prototype for ERICA; testing it, building fully functional ERICA for tests in a FS (breadboard, brassboard and beyond)

NOTE

The views and opinions expressed are solely those of the author. They do not necessarily represent the positions or policies of any private, public or governmental organizations. However, it must be noted that the author worked on various government/industry committees that dealt with crew resource management (CRM), line-oriented flight training (LOFT) scenario use and development; took several CRM courses both military and air-carrier and, had contractual oversight for various CRM projects: in short, possesses firsthand knowledge about much of what is presented herein.

ACKNOWLEDGMENT

Many Sections in this article are one resultant of a 23 year association and collaboration by the author with Captain Kevin M. Smith (USNR, ret.),United Air Lines (ret). It must be said that this Chapter reflects and encompasses a good deal of our past collaborations and, consequently owes a good deal to those collaborations. In particular, ERICA, which Kevin Smith has a provisional patent for, is almost entirely his concept and the fruit of his efforts. It has been my privilege and honor to have worked with, written with and presented with Captain Smith. We now have chosen to go down different paths.

REFERENCES

Code of Federal Regulations (CFR). (n.d.). *Title 14, Aeronautics and Space; Part 121; Operating Requirement: Domestic Flight and Supplemental Operations*. (This is a regulation that is periodically revised.)

Ewell, C. D., & Chidester, T. (1996). American Airlines Converts CRM in Favor of Human Factors and Safety Training. *The Flightdeck, July/August*. Flight Department, American Airlines, DFW Airport.

Federal Aviation Administration (1991). *AC 120-40B: Airplane Simulator Qualification* (7/29/91).

Federal Aviation Administration (2004). *AC 120-35C: Line Operational Simulations: Line Oriented Flight Training, Special Purpose Operational Training, Line Operational Evaluation* (09/27/04).

Federal Aviation Administration (2004). *AC 120-51E: Crew Resource Management* (01/22/04).

Federal Aviation Administration AC (1992). *120-45A: Airplane Flight Training Device Qualification* (2/5/92).

Flauttau, J. (2011). Regaining control. *Aviation Week & Space Technology*, (38): 87–88.

Flin, R., Martin, L., et al. (2003, 2005). Development of the NOTECHS (non-technical skills) System for Assessing Pilots' CRM Skills. In Contemporary Issues in Human Factors and Aviation Safety (Harris, Don and Muir, Helen C., (Eds). (pp.133-154). Burlington, VT: Ashgate Publishing Co

Government Accountability Office (GAO). (1997). Human Factors: FAA Guidance and Oversight of Pilot Crew Resource Management training Can Be Improved. In GAO/RCED-98-7. GAO, Washington DC.

Hackman, J. R. (1995). Teams, Leaders, Organizations. In Crew Resource Management; Weiner, E; Kanki, B; Helmreich, R (Eds) (p.52). Academic Press (now part of Elsevier), Maryland Heights, MO.

Helmreich, R. L., Chidester, T. R., & Foushee, H. Clayton; Gregorich, Steve; Wilhelm, J.A.(1990). How Effective is Cockpit Resource Management Training? in Flight Safety Foundation's Flight Safety Digest, May 1990, 1-16, Alexandria, VA

Helmreich, R. L., & Foushee, H. C. (1995). Why CRM? In Crew Resource Management; Weiner, E; Kanki, B; Helmreich, R (Eds) (p.38). Academic Press (now part of Elsevier) 1995, Maryland Heights, MO.

Kuhn, T. S. (1962). *The Structure of Scientific Revolutions*. Chicago: University of Chicago Press.

Kuhn, T. S. (2000). *The Road Since Structure: Philosophical Essays, 1970-1993*. Chicago: University of Chicago Press.

Lofaro, R. J. (1992a). Workshop on Integrated Crew Resource Management (CRM). *In DOT/FAA/RD-95/5*. Springfield, VA: National Technical Information Service.

Lofaro, R.J., Adams, R.J., and C.N., (1992b). Workshop on Aeronautical Decision-Making (ADM) In *DOT/FAA/RD-92/14* (Vol. I, II). Springfield, VA: National Technical Information Service.

Lofaro, R. J., & Smith, K. M. (1993). The Role of LOFT in CRM Integration. In *Proceedings the Seventh International Symposium of Aviation Psychologists*. Columbus, OH: ISAP.

Lofaro, R. J., & Smith, K. M. (1998). Rising Risk? Rising Safety? The Millennium and Air Travel. *Transportation Law Journal, 2*(25).

Lofaro, R. J., & Smith, K. M. (1999). Operational Decision-Making (ODM) and Risk Management (RM): Rising Risk, the Critical Mission Factors and Training. In *Proceedings of Tenth International Symposium on Aviation Psychology*. Columbus, OH: ISAP.

Lofaro, R. J., & Smith, K. M. (2011a) Similar Accidents: Lessons Learned and Countermeasures. In *Proceedings of the 16th International Symposium on Aviation Psychology*. Dayton, OH.

Lofaro, R. J., & Smith, K. M. (2011b). Collaborative decision-making between cockpit and dispatch for safety. *Technology and Innovation, 3*(12), 225–231.

Lofaro, R. J., & Smith, K. M. (2011c). The Aviation Operational Environment: Integrating a Decision-Making Paradigm, Flight Simulator Training and an Automated Cockpit Display for Aviation Safety. In E. Abu-Taeih, A. El-Sheikk, & M. Jafari (Eds.), *Technology Engineering and Management in Aviation: Advancements and Discoveries* (pp. 241–282). Hershey, PA: IGI Global. doi:10.4018/978-1-60960-887-3.ch015

McKenna, J. T. (1996). Carriers Hone CRM Programs. *Aviation Week & Space Technology*, (145): 146–147.

Orlady, H., & Foushee, H. C. (Eds.). (1986). Cockpit Resource Management Training. *NASA Conference Publication, 2455*, ntrs.nasa.gov

Retzlaff, P. D., & Gibertini, M. (1987). Air Force Pilot Personality [Psychology Press/ Taylor and Francis Group, Philadelphia, PA]. *Hard Data on the Right Stuff in Multivariate Behavioural Research, 22*, 383–389. doi:10.1207/ s15327906mbr2204_1

Smith, K. M., & Lofaro, R. J. (2001). A Paradigm for Developing Operational Decision-making (ODM). In *Proceedings of 2001 SAE World Aviation Congress (WAC) Conference*. Seattle, WA: SAE WAC

Smith, K. M., & Lofaro, R. J. (2003). The Finalized Paradigm for Operational Decision-Making (ODM) Paradigm: Components and Placement. In *Proceedings of the 12th International Symposium on Aviation Psychology*. Dayton, OH: ISAP.

U.S. Air Force: Air Force Instruction (AFI) (2005). 11-290: Cockpit/Crew Resource Management Training Program: Supplement 1 (03/22/2005).

KEY TERMS AND DEFINITIONS

Crew Resource Management (CRM): The application of team management concepts in the flight deck environment. It was initially known as cockpit resource management, but now, CRM programs can also include cabin crews, maintenance personnel, and others.

Cockpit Displays: All the cathode ray tube (CRT) and liquid crystal display (LCD) display screens, as well as dials and gauges in the instrument panel of an aircraft. Together, these provide pilot and crew with the information needed for a safe flight.

Decision Making: The process of the identification of alternatives for a course of action. this is followed by the selection of one of the alternatives and resultant action.

Federal Aviation Administration: (FAA) is the national aviation authority of the United States of America. An agency of the United States Department of Transportation, it has authority to regulate and oversee all aspects of civil aviation in the U.S.

FAA Advisory Circular (AC): A series of documents issued by the FAA which function as an air-carrier road-map, a "how to" for specific programs or addressing specific issues.

Group Dynamics: the interactions that influence the attitudes and behavior of people when they are grouped with others through either choice or other circumstances.

LOFT: An acronym for line-oriented flight training. It is given to flight crews in an FAA-rated level D flight simulator. It is a realistic, pre-programmed scenario of a flight and is both practice and assessment of flight crew performance; it can now be used to obtain type-ratings.

Paradigm: One that serves as a pattern or model. More broadly: A set of assumptions, concepts, values, and practices that constitutes a way of viewing reality.

Situational Awareness: is the perception of environmental elements with respect to time and/or space, the comprehension of their meaning, and the projection of their status after some variable has changed; it involves perception of the environment critical to decision-makers in complex, dynamic areas such as aviation. Situational awareness (SA) involves being aware of what is happening in the vicinity, in order to understand how information, events, and one's own actions will impact goals and objectives, both immediately and in the near future. Thus, situational awareness is especially important in work domains where the information flow can be quite high and poor decisions may lead to serious consequences (e.g., piloting an airplane).

Social Psychology: The study of how people's thoughts, feelings, and behaviors are influenced by the actual, imagined, or implied presence of others; a discipline that studies social influence, social perception and social interaction.

Type-Rating: An allowance to fly a certain aircraft type that requires additional training beyond the scope of initial license and aircraft class training. What aircraft require a type rating is decided on, in the USA, by the FAA.

APPENDIX A: THE MISSION PERFORMANCE MODEL (1991/2)

Figure 2.

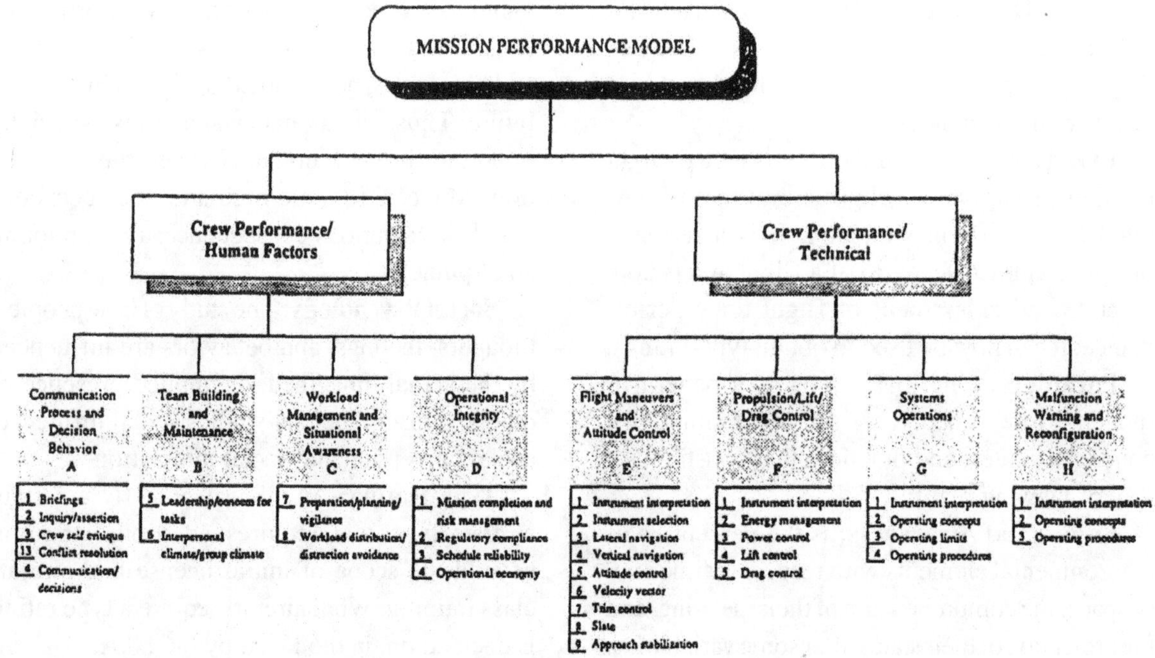

Chapter 12
Beyond the Paradox of Service Industrialization:
Approaches to Design Meaningful Services

Jesús Alcoba González
La Salle Campus Madrid, Spain

ABSTRACT

Highly industrialized services add value to a company, but at the same time they can destroy it, should an almost identical proposal be made to two different clients who are seeking different experiences. The analysis of human subjectivity shows that the interaction with services goes beyond its mere use, as it is related to human beings' search for meaning, and it can potentially become a part of its users' biography and identity. The present chapter compiles some of the models that can contribute to getting over the paradox of service industrialization and which are here divided into two types: those that allow for a deep knowledge of the customer and those that are based upon designing experiences as a value proposition. The present chapter also suggests a research agenda that aims to get over the paradox of service industrialization.

THE PARADOX OF SERVICE INDUSTRIALIZATION

If we consider a continuum between one pole in which the service is wholly industrialized and another where it is completely craft work, it seems clear that if a company is willing to capture the same value along the continuum, in the pole where the service is purely craft work, the cost for the customer will be high. However, this is also the point in which the maximum level of personalization takes place, due to a high interaction between the company and the customer, and also the point

DOI: 10.4018/978-1-4666-4506-6.ch012

in which few customers can acquire the service. On the other hand, if the level of industrialization is high, the service is produced at a very low price and the company captures the same value or even more, being able to sell it to the customer at a low price. Consequently, many more customers can make use of this service, which in turn fosters the potential increase of the company's profits in a considerable way. In most cases, however, the service offered is identical for all the customers, which has a very clear impact on lower levels of personalization.

It is here that the genesis of the paradox of service industrialization is to be found: the interesting issue for discussion is that in this scenario the industrialized service is, by definition, replicable and therefore similar to itself, being thus little personalized: broadly speaking, all the customers have the same experience with the same company. Moreover, due to consultancy processes and benchmarking, competitive advantage is moved from one company to another, which provokes that in the end many companies from the same sector end up offering essentially the same service to millions of customers that perceive the world in a clearly different way and, as a result, try to have different experiences. Just a glance at some services, such as flights, communications or restaurants, shows that, in essence, all the services offered are basically the same.

Even in the certain cases in which service delivery was exclusively offered by one of the companies of a sector at the beginning, after some years all of them have come to offer basically the same as a result of consulting processes and benchmarking. What happens then is that the whole sector essentially provides the same service with slight alterations. It can be deduced from Porter's classic ideas (1996) that if the service is the same (and if it is not the case of a market niche), the companies will be forced to compete in cost since they cannot compete in differentiation. The tourism sector is a good example of this phenomenon: if holidays are reduced to the mere addition of a flight and a hotel, it is clear that the customer would look for the travel agency offering the best price.

It is easy to find some side effects of an excess of industrialization: let's consider, for instance, the way in which economy class in planes frequently has an almost identical way of providing the catering and entertaining services. The passengers cannot eat when they want, buy when they want and cannot, in some cases, watch a film when they want either, but have to adapt to the rhythm imposed by the cabin crew. In the hotel industry there is a similar phenomenon: millions of passengers come and leave hotels at different times because their flights land and depart at different times. However, the hotel industry insists, quite surprisingly, on the concept of room occupancy from noon till noon the following day. The effect of industrialization can also be seen in call centers: in many cases, the customer is forced to repeat their problem incessantly to different operators as they each cover a very small area in the value chain; there is nobody at the other end who can understand the customer's situation and problems globally. This frequently leads to the common situation where a customer who has placed a complaint is startled when some days later, another department from the same company, unaware of an upset customer, offers him a new product. The most extreme case of call centers are the phone calls that many people get at ungodly hours offering them services they have neither requested nor need. Almost thirty years after Levitt (1976) proposed that the services could benefit from industrialization in terms of efficiency and productivity, critical voices started to raise on the industrialization of sales calls. They stress that in many cases they left no room for initiative and that what was before a conversation between two people, has become a dialogue between one person and a machine simulating a person (Roman, 2004).

Schlesinger and Heskett (1991) described long ago the negative impact of a wrong application of service industrialization, which results in what they call the cycle of failure: the service is designed so that in the contact point with the customer, work is repetitive, boring and it requires minimum training. This almost inevitably implies high staff rotation and customer dissatisfaction. Perhaps the most significant aspect of this approach, which is also a reason that may explain its failure, is that it considers people as mere parts in an assembly line. What is more, intelligence is considered to be in the company and not in the worker, s/he needs to be told exactly what to do (leaving him/her little or no margin at all for decision-making) and the worker is absolutely dispensable in this contact point.

Leon and Davies (2008) have described another paradox in the service sector when considering the advance in industrialization in technology and traditional services companies. From their point of view, whereas the first ones base their activity upon the efficient organization of a series of capabilities that technological systems have, traditional service companies as entertainment, the hotel industry or travelling, mainly try to create an experience for the customer, at the same time replicable and customizable. According to these authors, due to this difference of perspective, industrialization in companies of traditional services has been more successful than in high technology companies. Indeed, it is very likely that support forums have rapidly proliferated on the Internet as a defense mechanism for customers when facing an excessive technology industrialization in which the supplier companies fail to provide enough information and support on the use of the service. These authors conclude that the key to successful service industrialization is to take a user-centric approach. Nonetheless, the interesting point of this statement, which every company would share without a doubt, is not *what* but *how*. That is to say, how to create an experience for each customer while at the same time maintaining a sufficient level of industrialization that creates enough value for the company.

HUMAN SUBJECTIVITY AND THE CREATION OF MEANING

The paradox of service industrialization takes place when the service is industrialized so that it is replicable, predictable and can also be provided with a high quality level and at a low production cost. At this juncture, there is a risk of destroying what lies at the very essence of service delivery: clients seeking diverse experiences because they themselves are different and they perceive the world in different ways. In short, industrialization theoretically increases service value, but as service value is based on subjectivity, industrialization can also destroy it.

This fact should lead us to bear in mind the phenomenon of subjectivity to understand its role in human beings, for which we could take into account the available research carried out in Psychology and Neuroscience. Perhaps the most significant fact in this field is that human beings do not perceive reality as it is. For example, human beings do not perceive colors as they really are just because we cannot detect frequencies beyond violet nor below red color. Likewise, human beings cannot perceive the whole sound specter either. Therefore, unlike what could be assumed, very little exterior information reaches the brain. For example, regarding visual information, in contrast to the infinite quantity of information surrounding us, the retina only receives ten thousand million bits per second. However, only ten thousand out of these reach the cerebral cortex, and that which constitutes conscious perception is hardly up to a hundred (Raichle, 2010).

Furthermore, it must be noted that it is not only that the human brain perceives the exterior in a limited way, but it also reads reality in a way which is not objective. This can be seen in the development of concept maps, which are visual representations of concepts and of the relations existing among them (Novak & Cañas, 2006). Naturally, the relevance that each person gives to each concept and to the relations established among the concepts is different. In these models of the world, language plays a key role since it contributes to cognitive development by providing the setting of analogies and relationships and representations that allow abstraction (Gentner, 2003). Therefore, it has been pointed out that there is a difference between the logical and the psychological meaning of concepts. The latter refers to each person's idiosyncratic and individual experience (Ausubel, Novak, & Hanesian, 2000). As an example of to what extent human beings employ subjective versions of language, there is a remarkable study which aimed to find out the different conceptualizations that different people had of the expression "having had sex", which proved a surprisingly disagreement among participants (S. A Sanders et al., 2010.) Taking for granted that sex is an important and unequivocal coordinate for human beings, since it has belonged to their experience for millions of years, even before it could be defined as such, this fact is an obvious indicator of the clear subjectivity with which human beings approach reality.

Another relevant fact in subjectivity studies is that, besides neither perceiving nor registering reality faithfully, human beings cannot remember things as they happened either. Studies on the reminiscence effect have proved that the period of years from which most events and activities are remembered is between 10 and 30, and that these events and activities are judged as the most important or the best from the cultural point of view and also for their impact on the public life and on the individual (Rubin & Rahhal, 1998). This triggers the perception that time speeds up as the individual grows up (Draaisma, 2009). On the other hand, according to a recent review, a significant percentage of participants in different studies (between 18% and 37%) was implanted false memories about events such as an animal attack, and accident, or surgery (Lilienfeld, Lynn, Ruscio, & Beyerstein, 2010).

There is enough evidence to affirm that these variations in the register of what happens are connected with human beings' need to construct meaning. Thus, the elaborations carried out by the mind and based on experience help to construct identity in the shape of an autobiographical narrative that needs to be meaningful for the person but also for those in contact with them (McAdams, 2001). The search for meaning in human beings (Steger & Kashdan, 2007) may seem solely restricted to philosophical or transcendental aspects if meaning is to be considered as a link with something bigger (Peterson, Peterson, & Seligman, 2005), but the truth is that it is also related to everyday issues, like labor supply (Ariely, Kamenica, & Prelec, 2008). Human beings need an understanding of the world that helps them know how to act, and meaning indeed provides an environment that lets them contextualize what is valued, what is believed, what is accepted or what is desired (Diller, Shedroff, & Rhea, 2008).

In order to understand why human beings need to make meaning out of experience, the question that should be asked is what the ultimate function of this meaning is. Based on Mountcastle's hypothesis (1978), it has been stated that the ability to predict is the main function of the cerebral cortex and the basis of intelligence (Hawkins & Blakeslee, 2004). This statement is compatible with the fact that the human brain gets relatively little information about the reality that surrounds him, since what this fact seems to suggest is that the brain is constantly making internal predictions to plan ahead for whatever happens outside (Raichle, 2010). And it is also compatible with the current tendency to consider memory as a constructive process in which information is stored so that it can be retrieved to make simulations of the future (Schacter & Addis, 2007). In fact, the processes that allow us to imagine or simulate the future depend on the same neural processes involved in episodic memory (Schacter, Addis, & Buckner, 2007).

This can explain human beings' search for meaning and the happiness or unhappiness related to this search. As research shows, the presence of meaning in life has a positive correlation with happiness and positive feelings, and a negative correlation with depression and negative feelings. On the contrary, the absence of meaning, and therefore the search for it (except for people who have already achieved it in a substantial say), shows the opposite relation pattern (Park, Park, & Peterson, 2010). For all this, people are constantly trying to create meaning in their lives. As the course of events develops, the human mind makes a considerable effort to make sense out of different news, people, objects, products and services with which it interacts, trying to connect them in a meaningful way with a previously created representational and relational matrix.

In sum, the mental model which supports human beings' activities does not happen to meet reality; it is subjective and it is mainly meant to make meaning in order to make predictions for the future. As Taleb (2008) has argued, predicting events instead of narrating them is the true evidence of our ability to understand the world. Our hypothesis is that the ability to predict is a mechanism at the service of the survival of human beings. Therefore, the activities of searching and constructing meaning that human beings carry out in contact with the exterior are vital activities. And in this way, the products and services with which s/he interacts are not just objects or events, since they acquire or lose value as they become or not part of his/her autobiography and identity.

Lemley (2009) has argued that the way in which tribes would worship their idols in the past has found a contemporary parallel in the way in which customers organize themselves around brands. Consequently, it is not unimportant for a customer to choose between one product or service or another, since it is not only a matter of creating an emotional bond with a brand, but also this bond is now part of the way in which this person describes his/her existence. This explains why in some cases the feeling of belonging to a brand is so explicitly stated, as can be observed, for instance, in expressions such as "I am a PC" or "I am a Mac" (Ariely & Berns, 2012).

When a person gets in contact with a service s/he constructs a particular way of describing it that may or may not be linked to his/her life plot. There is meaning when an event connects to a person's biographical narrative. People will be likely to use services that naturally adapt to these parameters, and they will not be likely to use services that do not adjust or that are contrary to them. We can also presume that the meaning that people give to the different events of everyday life may be related to what they each consider as the meaning of life.

For instance, those people whose hobby is related to nature will be likely to live experiences through the use of products and services that are coherent with that concept. Consequently, they will tend to purchase suitable outdoor vehicles, such as mountain bicycles or Jeep vehicles, and they will preferably choose suitable clothing for outdoor activities. Quite often, the colors of this clothing are found in nature, such as green or khaki, and accessories that usually match this clothing relate to camping activities, such as torches and compasses. On the other hand, and according to the fact that user value is created not only by the use or possession of a product but also by the complete opposite (Boztepe, 2007), these people will prefer not to use distinctly polluting vehicles, uncomfortable clothing or objects that denote artificiality. It is quite possible that, for these people, enjoying outdoor activities is linked to wider ideas such as ecology or sustainability.

This process resembles the idea widely spread by constructivist authors that every time the human mind finds a new concept there is an interaction between the matrix of previous knowledge and the incoming knowledge, producing an integrating effort usually called learning. In a similar way, it is possible to model the interaction between customers and the services they are finding. Thus, as the new service is substantially related to the previous autobiography of the individual, meaning is created and the service is accepted by the user, described with his/her own language and finally integrated into his/her life plot, connecting with other related products and services. On the contrary, if the experience does not find a fixing point in the consumer's identity, quite probably it will be rejected. Obviously, there is a whole range of partial integration hints between both ends. Although there is no evidence yet to affirm it, it can be assumed that on many occasions the existence or lack of meaning can be a much more determining factor regarding the use of a service than the service utility and quality. This would explain why there are customers who continue buying and using products and services in spite of being out-fashioned or defective, in some cases even justifying such failures in surprising ways. Similarly, there are people who will never use determinate value propositions, no matter their utility and high quality, simply because they have nothing to do with them.

In brief, probably the keys that mark the difference between successful and unsuccessful services for each person are found further away from issues related to simple quality assurance and even to the mere assumption that there is an emotional bond between people and the services they use. It is indeed much more than that because people create life bonds with services searching for meaning in them.

BEYOND THE PARADOX OF SERVICE INDUSTRIALIZATION

The most significant effect of the paradox of service industrialization is that, as it adds value to the company by creating a replicable and predictable service, and which is thus identical to itself, value can be destroyed in the case of the customer, since each client looks for a different experience as part of a different biography. Obviously, a solution to this problem is customization, but it must be noted that the most important problem takes place before the creation of the service and not after, that is, in its design. Because, indeed, once the service that customers consider valuable has been created, adapting the service, depending on its costs and complexities, so that other different customers can use it, seems a rather simple task. What is more, it is possible to design it so that the customer himself/herself personalizes it with a self-service format. But in principle this does not add qualified information about how to design services so that they become part of the customer's biographical identity. Thus, the most important question is how to decode the way in which customers read reality and construct their identity, so that a service adapted to them can be designed.

Lately, both research and practice have made valuable reflections to get over the paradox of service industrialization. Even at the risk of finding the classification excessively simplistic, it can be said that there are two types of initiative, which are related to the two poles between which the interaction takes place: on the one hand, those related to the deep knowledge of the customer, and on the other hand, those focused on designing value propositions. Clearly, this classification is only useful to arrange the different contributions, since in practice both polarities regard the other and feedback mutually.

KNOWING THE CUSTOMER

The ultimate goal of knowing the customer is to enable the design of a value proposition which is really close to his/her necessities, so that a solid and lasting relation can be built. This idea is latent in the concept of customer intimacy that Treacy and Wiersema (1993) described already two decades ago as one of the strategies to lead the market.

Indeed, intimacy with the customer, closeness to his/her necessities, is one of the important keys to construct value.

The problem is to know how the customer assesses the service, and, in order to find out this answer, the most extended approach for a long time has been the use of questionnaires such as SERVQUAL (Parasuraman, Zeithaml, & Berry, 1988). However, there has been an important evolution from this starting point. Possibly, the first evolution, still close to the questionnaire format, is the appearance of the Net Promoter Score (Reichheld, 2003). What we get with it is a figure that tells us whether the users would recommend the services of a company or not, so that we can draw comparisons, either with other temporal moments or with other companies from the same sector. Perhaps the most important evolution of this index is that it aims to exceed the approach of traditional questionnaires which customers usually find tedious to fill in, and which only collect a limited series of issues. The Net Promoter Score focuses on the aspect which possibly summarizes them all, and this is whether the customer feels so satisfied with the experience that s/he would recommend it to people s/he knows. Nevertheless, what the score does not indicate is the reason *why* a customer says one thing or the other, and, therefore, while it is an index that can help to understand what the relationship between the customer and the service is, it is not useful to design the service. Maybe what both approaches have in common, and precisely the reason why they do not provide useful information for the design, is because both are based upon the service quality assessment, while they should focus on assessing the quality of the experience. Quality of experience differs from the traditional approach of quality assurance in that experiences are subjective and dependent on the context (Jain, 2004). The assessment of quality of experience should then try to capture subjectivity and, therefore, instead of measuring, it should attempt to describe, given that potentially, experiences are not better or worse than others, but some are closer to specific customers depending on their biographies and identities. It has been argued that using objective and quantitative data is necessary when evaluating the quality of experience and, possibly, a combination of objective and subjective measurements may be required (Brooks & Hestnes, 2010). But probably the greatest difficulty lies precisely in this last end. In this regard, we may need to think that traditional numerical evaluations are nonetheless an artificial recreation of what the human mind registers in its autobiographical memory, which are mainly images or sensations and of course aspects that can be translated into words but not easily into numbers. In this line, the study by Nurkka, Kujala and Kemppainen (2009) is an attempt to approach the qualitative description of an experience by using incomplete sentences.

As has been stated earlier, in the model of the world that each person elaborates, language plays an important role (Gentner, 2003). This means that if one has to describe an experience rather than measure it to compare it with others, one of the best approaches consists of analyzing what the customers of a given service say or write when they use it. In this line, there is a turning point in which there is a move from sending questionnaires to the customers to just analyzing what they spontaneously say. The advantages of this approach are evident: firstly, the information the customer provides is not closed by a series of options previously biased by the researcher. Secondly, the customer gives his/her opinion voluntarily, and in doing so, s/he does not feel the external pressures that usually come with filling in the questionnaire, which usually have to do with finishing it early, or with thinking more about the prize the brand has offered to those who accomplish it than about the answers themselves. Thirdly, the customer expresses his/her opinions exactly as they mentally appear, that is, s/he does not have to translate them into numerical artifices as for instance Likert scales do. Social Networks, forums and blogs have nowadays big amounts of information that customers give voluntarily about the services they use. Unfortunately, until recently there was no tool to analyze these opinions, but there is more and more research on this issue every day. Liu (2010) has approached the study of sentiment analysis and subjectivity, covering the following aspects: the problem of formalizing the study of sentiments as a scientific problem, the classification of sentiments and subjectivity, finding the features or attributes according to which opinions are given, the study of comparisons in sentiment analysis, the search and recovery of opinions, and, finally, the detection of spam opinions and the assessment of the usefulness or quality of the opinions. In a more specific way, Pang and Lee (2008) have described the fundamental challenges posed by the construction of an opinion search engine. Firstly, the kind of documents that are relevant to detect opinions has to be established, and then the parts of the documents that contain opinions must be selected. The next step is to establish what the prevailing feeling expressed in the opinions is and the specific opinions about certain aspects or features about the given topic. Finally, the system must give the information in a synthesized way, for which it can include all or some of the following actions: vote additions, selective opinion highlighting, representation of agree and disagree points, community or opinion holder identification, and identification of the different levels of authority within opinion holders.

A different approach to know what a customer thinks or feels when s/he relates to a service is a group of techniques globally known as neuromarketing, which consists of capturing the register of brain activity when a person is subjected to a stimulus with a marketing research focus. The methods usually employed are: electroencephalography (EEG), magneto encephalography (MEG) and functional magnetic resonance imaging (fMRI), although this latest technique is perhaps the most popular one. fMRI is based upon the measuring of BOLD (Blood Oxygen Level Dependant), which is the level of oxygen detected in a specific area of the brain. The basis of studies with fMRI lies in considering that the increase of the BOLD level points to brain activity, although this technique does not measure directly the electromagnetic signs generated by neurons (Morin, 2011).

This approach is related to two main objectives (Ariely & Berns, 2012): firstly, they are based on the belief that people cannot articulate correctly their product or service preferences, and thus, these techniques can disclose hidden information about their wishes. Secondly, it is believed that this direct detection of what people want can help in the design process, detecting precociously what the customers prefer to dismiss whatever is not appropriate before the product is finished.

The novelty of these techniques and their seemingly promise to make marketing research gain momentum is triggering the appearance of studies which aim at determining exactly what they are about and what implication their use can have (Fisher, Chin, & Klitzman, 2010). On the other hand, critical voices are being heard, claiming that finding out precise information about the customer's behavior with these media is not possible, given that the cognitive processes associated with buying decisions base on multiple factors and cannot be reduced to one single activation area (Ariely & Berns, 2012). In this sense, it must be noted that the research with fMRI is subject to a series of limitations (Shermer, 2012): firstly, the subjects have to enter a machine, which is a non-natural atmosphere. Secondly, what these equipments show are indirect measures of brain activity. Thirdly, the colored images which are usually displayed can constitute an exaggeration beyond the real activity of the brain. Fourthly, brain images are statistic compilations. And, finally, brain areas are activated for varied reasons. In this sense, a great part of the recommended caution as far as research with fMRI is concerned relates to what has been called reverse inference: when researchers detect an activity in a specific area of the brain, they often assume that a certain cognitive activity has taken place. This way of reasoning has been called reverse inference because it goes in the opposite direction it usually goes (when eliciting a particular cognitive process, a particular brain area will be activated). Reverse inference could be valid if the particular area activated exclusively before a certain stimulus or task, but if it activates as the result of a long series of cognitive processes, then the firmness of this affirmation is rather weak (Polldrack, 2006).

DESIGNING A VALUE PROPOSITION

The ground-breaking work by Pine and Gilmore (1998) described a progression in the creation of value which goes from extracting commodities to making goods, then to the delivery of services and finally to staging experiences. Experiences are memorable, personal, and are mainly related to the world of the senses. For these authors, experiences are located within a framework defined by two polarities: absorption-immersion and passive-active participation. Since that first study which focused the attention of the services sector on experiences, both research and practice have been progressing in their definition and design. According to Hassenzahl and Tractinsky (2006), an experience is the consequence of the interaction among the user's inner state, the features of the designed system s/he interacts with and the context in which the interaction takes place. According to Diller, Shedroff & Rhea (2008), an experience is the conscious feeling that something changes within the human being (whether it may be the context, their body, mind or spirit). It is something that happens every day, although some experiences may be superficial while others are deeper.

It is becoming more and common that customers do not just demand commodities and services, but they are also looking for experiences. The reason why this is so lies in the information presented previously about the way in which human beings perceive reality and incorporate it into their life plot. As the now classic study by Csikszentmihalyi and Rochberg-Halton (1981) aptly describes, even domestic objects have a meaning that exceeds their use; in the same way that music equipments help to shape mood or that a bicycle may mean freedom for a child. Accounting for these reasons, the design of objects has gone on to have a greater importance, as the fact that people relate to them at an emotional level has been confirmed. Norman (2004) believes that there are three levels in the design of objects: visceral design has to do with the features found in the environment which surrounds us and is related to physical attributes, such as appearance, sound or the feeling they provoke. For behavioral design, aspect is not important, and what really matters is its use. The four components which affect this kind of design are function, comprehension, usability and physical feeling. Finally, reflective design has to do with the message, culture and the meaning of a product or its use, and its value primarily exists in the mind of the user.

It is easy to understand that if these meanings take place with simple objects, their consideration acquires more relevance in the world of services because of two main reasons: the first one is that in many cases the provision of a service involves objects. The second reason is that the value of services generates an interaction with the user in a dynamic way, and therefore, the control that must be exercised over the meanings it evokes on the user must be higher.

Possibly, the greatest advance regarding experiences as a value proposition has taken place when research and practice have generated a corpus of knowledge adequate to allow their design. One of the frameworks which approaches the management of the customer's different experiences has been described by Schmitt (2007), who believes that these find support in five main sources: sense (sensory experiences), feel (experiences that trigger feelings, moods and emotions on the part of the customers), think (experiences of a cognitive or creative kind), act (bodily experiences, life styles and interaction with other people), and relate (experiences which connect the individual to his/her social and cultural context). For Shedroff (2009), an experience has five dimensions: significance (meaning, status, emotion, price, function), breadth (product, services, brand, nomenclature, channel/environment, promotion, price), intensity (reflex, habit, engagement), duration (initiation, immersion, conclusion, continuation), triggers (senses –taste, sight, sound, smell, touch-, cognitive –concepts, symbols-), and interaction (static, passive, active, interactive).

As in the case of objects, and maybe more significantly, experiences connect with human beings through the meanings they evoke, which can vary: accomplishment, beauty, creation, community, duty, enlightenment, freedom, harmony, justice, oneness, redemption, security, truth, validation and wonder (Diller et al., 2008).

The design of experiences aims at getting in touch with the customer in a way that goes beyond the mere economic transaction in order to go on to become part of his/her set of vital meanings, of his/her biography and identity. This current tendency to reach the customer from his/her most intimate and emotional side possibly has its greatest exponent in the concept of lovemark (Roberts, 2005), which argues that the way in which companies can stand out in the market and get the customers' attention is by creating an emotional bond with them based on love and respect in order to get a loyalty beyond reason in which mystery, sensuality and intimacy are essential elements.

Nevertheless, what happens is that, in order to design experiences or create brands which wake up such intense emotions again, an approach that allows for knowing what the customer is looking for as part of his/her narrative autobiography is needed, an approach that moves beyond the usefulness of the objects or the services with which the customer interacts. Previously, the main approaches that aim at finding this knowledge of the customers have been collected, but there is another way to design value propositions adapted to their needs, and which does not necessarily imply asking, researching what they think or analyzing the responses of their brains, as it implies the customers' implication in the very design.

This approach understands that the traditional design method, which focused on the user, cannot satisfy the complexities of the world nowadays, and that the essential function of the design does not consist of creating products, but experiences, emotions, interaction, sustainability, service or transformation (Sanders & Stappers, 2008). From this perspective, value is not uniquely created by companies, but co-created through the effort of the different market agents, among which customers are necessarily found, and thus are considered as value creators instead of value destroyers (Vargo, Maglio, & Akaka, 2008). From the perspective of Prahalad and Ramaswamy (2004), the key is to create a context of dialogue in which customers can create personalized experiences. In such conversational setting, the assessment of risks and benefits, transparency and access to information are pivotal. In sum, the essential issue in the co-creation of experiences is that it is not about designing *for* the customer, but *with* the customer.

A RESEARCH AGENDA

The paradox of service industrialization is a contradictory phenomenon in which, while customers are increasingly looking for services that fit more closely their necessities, the pressure toward value generation through industrialization makes them more identical to themselves. There is a turning point in the craft-industrialized continuum where value starts being destroyed in the case of customers as a consequence of the company's necessity to capture value through industrialization. Value is destroyed because the customer sees no vital sense in a service which has nothing to do with his/her life plot.

The reason why this is so, as has been discussed, lies in the particular way in which human beings perceive, register and remember reality, which is loaded with the particular idiosyncrasy of each customer, and whose background is related to seeing the necessity to make sense, which is possibly connected to the prediction of events and to the survival of the human race. In this manner, each individual's subjectivity makes some customers find meaning in particular services while they reject others, observing the opposite phenomenon in other services. Therefore, it is necessary to approach each of these customers in a much deeper way, decoding the way in which they read reality in order to design services which are really adapted to their necessities.

In this realm, valuable initiatives, such as the creation of ultra-short questionnaires, language analysis, the use of brain analysis techniques, emotional design and design of experiences or co-creation, add valuable contributions to design value propositions which can actually become part of the customers' life. And Service Science can contribute to the systematization of knowledge in relation with this phenomenon. We have proposed elsewhere (Alcoba, 2011) a series of research topics, which by no means is exhaustive but provides a broad perspective. Here we provide an abridged and revised version of this list. The agenda is classified in four areas: information systems, first contact with client, service-client interaction and measuring:

- **Information systems:**
  - How to increase the amount of available information and how it is interconnected in customers' databases so that every department can get access to the whole information about every client;
  - How to present those large amounts of information in a way that can be rapidly understood by computers and humans;
  - How to extract information from client databases so that the systems can assess specific actions for them, in order to customize and enrich the service experience;
  - How to analyze the language that customers use when describing their experience with a service in order to draw conclusions from it;
  - How to store customers' perception of the experience (positive and negative) and how to use that information;
  - How to capture and store subjectivity in databases;
- **First contact with a client:**
  - How to gather information about client expectations;
  - How to benefit from the analysis of the first moment in which a customer comes into contact with the service either virtually or personally;
  - How to classify customers who come into contact with a service from the first moment (classifying better to provide a more adapted service);
  - How to increase interpersonal adaptive behavior and service offering adaptation to the staff in contact with the customers (Bettencourt & Gwinner, 1996), by devising specific ways to deal with the client depending on his/her profile;
- **Service-client interaction:**
  - How to improve research on customers' perception of the experience by introducing qualitative research techniques (Glaser & Strauss, 1967);
  - How to increase clients' self disclosure of information;
  - How to benefit from service improvement ideas given by clients, whether spontaneously or as part of a structured action;
  - How to interpret the behavior of a client in a specific moment according to the past history of that customer with the service, and also how to predict

    - the future behavior of the customer based on that history;
    - How to interpret a client's emotional state by analyzing natural language automatically (written or spoken);
- **Measuring:**
    - How to measure immersion, participation or customization of a service;
    - How to calculate the degree in which these factors should exist for the customer to perceive a meaningful experience;
    - How to measure the impact of outsourcing, offshoring and lean manufacturing in customization, and also the impact of customization in service production cost;
    - How to measure the influence of using automated systems in customers' perception of service value;
    - How to measure the number of contacts or information which a service needs in order to gather enough information from clients to provide the proper experience;
    - How to calculate the semantic mean of customers' opinions on a service (Alcoba, 2012);
    - How to deeply and accurately measure clients' satisfaction with the experience.

As economy becomes more and more complex and more basic needs are satisfied, it seems more evident that human beings need to find experiences which complete their biographies and, consequently, their identities, following Maslow's classic motivation outline (1943). Therefore, the essential point upon which this article bases is that human subjectivity and each individual's need to access a different world of experiences will be, with all probability, more and more frequent as countries reach an adequate level of development. Thus, the growth potential for research and value generation is immense.

REFERENCES

Alcoba, J. (2011). The Paradox of Service Industrialization. In J. Morin, J. Ralyté, & M. Snene (Eds.), Exploring Services Science. Geneva: Springer.

Alcoba, J. (2012). The Paradox of Service Industrialization and the Creation of Meaning. International Journal of Service Science, Management, Engineering, and Technology, 3(2), 50–62.

Ariely, D., & Berns, G. S. (2012). Neuromarketing: the hope and hype of neuroimaging in business. Nature Reviews Neuroscience, 11(4), 284–292.

Ariely, D., Kamenica, E., & Prelec, D. (2008). Man's search for meaning: The case of Legos. Journal of Economic Behavior & Organization, 67, 671–677.

Ausubel, D. P., Novak, J. D., & Hanesian, H. (2000). Psicología Educativa. Un punto de vista cognoscitivo. Mexico: Trillas.

Bettencourt, L. A., & Gwinner, K. (1996). Customization of the service experience: the role of the frontline employee. International Journal of Service Industry, 7(2), 3–15.

Boztepe, S. (2007). User Value: Competing Theories and Models. International Journal of Design, 1(2), 55–63.

Brooks, P., & Hestnes, B. (2010). User Measures of Quality of Experience: Why Being Objective and Quantitative is Important. IEEE Network, March/April, 8–13.

Csikszentmihalyi, M., & Rochberg-Halton, E. (1981). The meaning of things. Domestic symbols and the self. New York: Cambridge University Press.

Diller, S., Shedroff, N., & Rhea, D. (2008). Making Meaning: How Successful Businesses Deliver Meaningful Customer Experiences. Berkeley: New Riders.

Draaisma, D. (2009). Por qué "el tiempo vuela" cuando nos hacemos mayores. Madrid: Alianza Editorial.

Fisher, C. E., Chin, L., & Klitzman, R. (2010). Defining neuromarketing: practices and professional challenges. Harvard review of psychiatry, 18(4), 230–237.

Gentner, D. (2003). Why we're so smart. In D. Gentner & S. Goldin-Meadow (Eds.), Language in mind: Advances in the study of language and thought. Cambridge: MIT Press.

Glaser, B. G., & Strauss, A. L. (1967). The Discovery of Grounded Theory. Strategies for Qualitative Research. Chicago: Aldine Publishing Company.

Hassenzahl, M., & Tractinsky, N. (2006). User experience - a research agenda. Behaviour & Information Technology, 25(2), 91–97.

Hawkins, J., & Blakeslee, S. (2004). On intelligence. New York: Times Books. Henry Holt and Company.

Jain, R. (2004). Quality of Experience. IEEE Multimedia, 11(1), 95–96.

Lemley, D. (2009). The road to authentic brand is littered with design. In T. Lockwood (Ed.), Design Thinking: Integrating Innovation, Customer Experience, and Brand Value. New York: Allworth Press.

Leon, N., & Davies, A. (2008). Managed service paradox. IBM Systems Journal, 47(1).

Levitt, T. (1976). The industrialization of service. Harvard Business Review, September/October, 63–74.

Lilienfeld, S. O., Lynn, S. J., Ruscio, J., & Beyerstein, B. L. (2010). 50 Great Myths of Popular Psychology. Shattering Widespread Misconceptions about Human Behavior. Sussex: Wiley-Blackwell.

Liu, B. (2010). Sentiment Analysis and Subjectivity. In N. Indurkhya & F. J. Damerau (Eds.), Handbook of Natural Language Processing. London: Chapman & Hall/CRC.

Maslow, A. H. (1943). A Theory of Human Motivation. Psychological Review, 50, 370–396.

McAdams, D. P. (2001). The psychology of life stories. Review of General Psychology, 5(2), 100–122.

Morin, C. (2011). Neuromarketing: The New Science of Consumer Behavior. Society, 48(2), 131–135.

Mountcastle, V. B. (1978). An organizing principle for cerebral function: the unit module and the distributed system. In V. B. Mountcastle & G. M. Edelman (Eds.), The mindful brain. Cortical organization and the group-selective theory of higher brain function. Massachusetts: The MIT Press.

Norman, D. A. (2004). Emotional Design. Why we love (or hate) everyday things. Cambridge: Basic Books.

Novak, J. D., & Cañas, A. J. (2006). The theory underlying concept maps and how to construct them. Retrieved from http://cmap.ihmc.us/Publications/ResearchPapers/TheoryUnderlyingConceptMaps.pdf

Nurkka, P., Kujala, S., & Kemppainen, K. (2009). Capturing users' perceptions of valuable experience and meaning. Journal of Engineering Design, 20(5), 449–465.

Pang, B., & Lee, L. (2008). Opinion mining and sentiment analysis. Foundations and Trends in Information Retrieval, 2(1-2), 1–135.

Parasuraman, A., Zeithaml, V. A., & Berry, L. B. (1988). SERVQUAL: A Multiple-Item Scale of Measuring Consumer Perceptions of Service Quality. Journal of Retailing, 64(1), 12–40.

Park, N., Park, M., & Peterson, C. (2010). When is the Search for Meaning Related to Life Satisfaction? Applied Psychology: health and well-being, 2(1), 1–13.

Peterson, C., Peterson, N., & Seligman, M. E. P. (2005). Orientations to hapiness and life satisfaction: the full life versus the empty life. Journal of Happiness Studies, 6, 25–41.

Pine, J., & Gilmore, J. H. (1998). Welcome to the Experience Economy. Harvard Business Review, Julio-Agosto, 97–105.

Polldrack, R. A. (2006). Can cognitive processes be inferred from neuroimaging data? TRENDS in Cognitive Sciences, 10(2).

Porter, M. E. (1996). What is strategy? Harvard Business Review, Noviembre-Diciembre, 61–78.

Prahalad, C. K., & Ramaswamy, V. (2004). Co-creation experiences: the next practice in value creation. Journal of Interactive Marketing, 18(3), 5–14.

Raichle, M. E. (2010). La Red Neuronal por Defecto. Investigacion y Ciencia, (Mayo), 20–26.

Reichheld, F. F. (2003). The One Number You Need to Grow. Harvard Business Review, December 2003, 46–54.

Roberts, K. (2005). Lovemarks. Barcelona: Urano.

Roman, M. (2004). The industrialization of service: a personal journey. Ivey Business Journal, September/October, 1–7.

Rubin, D. C., & Rahhal, T. A. (1998). Things learned in early adulthood are remembered best. Memory & Cognition, 26(1), 3–19.

Sanders, E. B.-N., & Stappers, P. J. (2008). Co-creation and the new landscapes of design. CoDesign, 4(1), 5–18.

Sanders, S. A., Hill, B. J., Yarber, W. L., Graham, C. A., Crosby, R. A., & Milhausen, R. R. (2010). Misclassification bias: diversity in conceptualizations about having 'had sex'. Sexual Health, 7, 31–34.

Schacter, D. L., & Addis, R. L. (2007). The ghosts of past and future. Nature, 445(4 January), 27.

Schacter, D. L., Addis, R. L., & Buckner, R. L. (2007). Remembering the past to imagine the future: the prospective brain. Nature Reviews. Neuroscience, 8(8), 657–661.

Schlesinger, L. A., & Heskett, J. L. (1991). Breaking the Cycle of Failure in Services. Sloan Management Review, 32(3), 17–28.

Schmitt, B. (2007). Experiential Marketing. Barcelona: Deusto.

Shedroff, N. (2009). Experience design 1.1. Experience Design Books.

Shermer, M. (2012). El necesario escepticismo sobre los escáneres cerebrales. Cuadernos Mente y Cerebro, 1, 90–96.

Steger, M. F., & Kashdan, T. B. (2007). Search for meaning in life. Retrieved from http://sage-ereference.com/socialpsychology/Article_n469.html

Taleb, N. N. (2008). El cisne negro. El impacto de lo altamente improbable. Barcelona: Paidós.

Treacy, M., & Wiersema, F. (1993). Customer Intimacy and Other Value Disciplines. Harvard Business Review, Enero-Febrero, 84–93.

Vargo, S. L., Maglio, P., & Akaka, M. A. (2008). On value and value co-creation: A service systems and service logic perspective. European Management Journal, 26, 145–152.

Chapter 13
Adoption and Utilization of ICT in the Chinese Third-Party Logistics Industry

Kwok Hung Lau
Royal Melbourne Institute of Technology University, Australia

Haibo Huang
Royal Melbourne Institute of Technology University, Australia

ABSTRACT

This chapter reports the findings of a questionnaire survey on the adoption and utilization of information and communication technology (ICT) – an important tool to enhance efficiency and responsiveness in modern-day supply chains – in the Chinese third-party logistics (3PL) industry. As a global manufacturing base, China has a burgeoning 3PL industry serving a large number of domestic firms as well as multinational corporations. With more and more organizations outsourcing their logistics function to 3PL firms, the latter have to make use of ICT for efficient communication with clients and coordination of activities in order to work hand-in-hand with customers to meet their day-to-day logistics needs. To investigate the status quo of ICT utilization in China, a questionnaire survey was conducted in 2009 to investigate the level of ICT adoption by the 3PL firms. For comparison, the findings of the Annual Third-Party Logistics Studies from 2007 to 2013 in this regard were also analyzed. The questionnaire survey results reveal that ICT is being widely adopted in the Chinese 3PL industry suggesting a high level of awareness of its significance and benefits to both the service providers and the clients. The ICT used ranges from low-tech telephone, facsimile, Internet access, to more advanced and sophisticated radio frequency identification technology and enterprise resource planning system. While small 3PL companies are using less expensive ICT at a tactical level mainly to cut costs and reduce errors in the day-to-day logistics operations, medium-sized and large firms have evolved to make use of more expensive ICT for planning and strategic purposes, such as business control, customer integration, and service differentiation. Transportation management system is regarded as the most important IT for business to most 3PL firms surveyed whereas the significance of other systems for warehouse management, or-

DOI: 10.4018/978-1-4666-4506-6.ch013

der management, and inventory management, etc. varies depending on the services provided and the resources available. A follow-up review of the recent literature suggests that ICT adoption in the 3PL industry of China has not significantly increased since the 2009 survey but the general awareness of the importance of ICT capabilities is growing.

INTRODUCTION

Efficiency and responsiveness can be regarded as the primary goals in supply chain management (Chopra & Meindl, 2012). To achieve these two objectives, effective communication and real-time information sharing are the keys. Information and communication technology (ICT) therefore plays a vital role in managing modern-day complex supply chains. It permits effective coordination of day-to-day replenishment and distribution activities hence minimizes delay and waste. It also provides visibility to enable tracing and tracking of inventory, thereby allows better control and greater flexibility in matching supply and demand. In addition, the technology enables internal as well as external integration. It helps promote collaboration among business partners through collective decision making using real-time data and information stored and shared on a common platform. Together with automation employing data capture technologies such as barcode, global positioning, and radio frequency identification, ICT can significantly reduce labor, errors, and response time across the entire supply chain. This will not only lower total supply chain costs (i.e., higher efficiency) but also enhance overall customer service (i.e., improved responsiveness). In fact, significance and benefits of ICT in the logistics industry has long been recognized in many studies such as Bowersox and Daugherty (1995), Bowersox et al. (1989), Bowersox et al. (1999), Edwards et al. (2001), and Kathuria et al. (1999). Benefits of ICT adoption and utilization in other industries such as the healthcare industry have also been explored and acknowledged (Topacan et al., 2010).

ICT for the Third-Party Logistics Industry

Globalization and outsourcing have impacted significantly on modern-day supply chain management. The former has resulted in extended supply chains with global configurations and multiple entry and exit points. The latter has given rise to the prosperity of integrated logistics service providers who look after all the logistics activities in a supply chain for various customers. Information technology plays a significant role in these changes (Gunasekaran & Ngai, 2004). To facilitate communication, coordination, and collaboration, ICT is often employed to link up business partners in a supply chain. Electronic Data Interchange (EDI) that operates on private value-added networks (VAN) has long been used to automate business transactions between large organizations (Angeles, 2000). Fast expansion of the Internet and rapid advancement of Web technologies in the last couple of decades have extended the use of EDI from large to small and medium-sized enterprises (SMEs). With greater functionalities and most importantly no requirement of expensive VAN to operate, Web-based EDI (or WEDI) and applications developed using eXtensible Markup Language (XML) protocol have become more affordable alternatives to the traditional EDI allowing SMEs to collaborate and directly compete with big companies in business. Similar experience is also witnessed in the development of other information technology (IT) systems such as Enterprise Resource Planning (ERP) that are used for both internal and external integration. New technologies such as Web Services, which allows information systems to be assembled from components downloadable from the Internet, permit organizations to easily integrate with business partners that run their own different internal systems (Fensel & Bussler, 2002; Hagel, 2002; Vidgen et al., 2004; Wu, 2004). The emerging Radio Frequency IDentification (RFID) technology for real-time tracking and efficient order assortment of fast moving products enables seamless integration between supply chain stages (Deloitte, 2003). Working together with other systems, it also facilitates full automation of routine logistics processes and may one day replace the widely used barcode technology completely (Vijayaraman & Osyk, 2006).

With more and more organizations outsourcing their logistics activities to third-party logistics (3PL) firms, ICT has become an important asset to 3PL firms because they must integrate their systems with those of their clients in order to take over the logistics function effectively. As a bridge connecting clients and other members in the supply chain, ICT in 3PL firms plays a critical role in synchronizing and coordinating complex supply chain activities across logistics users and their customers (Lai et al., 2006). However, the high investment cost of some sophisticated IT systems, such as ERP, and the lack of IT expertise have dampened the adoption rate of ICT by 3PL firms (Sum et al., 2001). This issue is particularly critical for small logistics firms, which are usually limited in resources. With further technological advancement and reduction in utilization cost, however, there should be ample opportunities for 3PL service providers to further adopt IT to improve operational efficiency and competitiveness (Bhatnagar et al., 1999).

Objectives of the Study

While there are abundant studies on the use of ICT in logistics and supply chain management, many of them have only looked at the general benefits (Bowersox & Daugherty, 1995; Introna, 1991), strategy formulation (Carr, 2003; Kim et al., 2008), applications (Davenport & Brooks, 2004; White et al., 2005), and impacts on competitive advantage and firm performance (Lai et al., 2006; Wu et al., 2006). In comparison, research on ICT adoption and utilization by 3PL firms particularly in developing countries is relatively limited. Studies by Lai et al. (2005) and Lin and Ho (2009) on RFID adoption are among the few exceptions. Nevertheless, these studies tend to focus on a particular technology instead of investigating the situation from a holistic perspective. To fill the gap in the literature, this study attempts to investigate ICT adoption and utilization in the 3PL industry using China as an example and explore the various reasons for using the technology. The findings may shed light on the situation in other developing countries comparing to that of developed countries. The following research questions are posed:

RQ1: What is the level of ICT adoption and utilization in the 3PL industry of China?
RQ2: What are the reasons for the adoption as perceived by the 3PL firms?
RQ3: Are there any differences in the usage of ICT across the 3PL industry?
RQ4: How is the situation in China different from that of the global 3PL industry?

To answer RQ1-RQ3, a questionnaire survey was conducted to find out the level of ICT adoption and utilization in the 3PL industry of China. Views of the surveyed 3PL firms on why they used ICT and how important ICT was to their business were also sought. As a comparison, the current utilization of ICT by 3PL firms from a global perspective (i.e., RQ4) was investigated through a desktop research using the survey data of the Annual Third-Party Logistics Studies from 2007 to 2013 (Langley Jr., 2007, 2008, 2009, 2010, 2012, 2013).

LITERATURE REVIEW

Third-Party Logistics and IT Capabilities of Logistics Service Providers

Third-party logistics is a big industry. The findings from the 2013 Third-Party Logistics Study (or 2013 Annual 3PL Study for short) show that expenditures on outsourcing as percentages of total logistics expenditures in the four regions under study – North America, Europe, Asia-Pacific and Latin America – are 33%, 55%, 31% and 39% respectively (Langley Jr., 2013). Driven by the ever-increasing global competition and rising customer expectations, more and more companies are outsourcing their activities to 3PL firms so that they can concentrate on their core competencies (Vaidyanathan, 2005). Many multinational corporations (MNC) seeking integrated logistics services tend to form alliances with large 3PL firms to simplify management and improve service quality (Lau & Ma, 2008). Table 1 shows the range of outsourced logistics services in all the four regions surveyed as revealed in the 2010, 2012, and 2013 Annual 3PL Studies. It can be seen that transportation (both domestic and international), warehousing, forwarding, and custom brokerage, which demand frequent communication among different parties in the supply chain and intense information sharing and processing, are among the top most outsourced logistics activities.

Table 1. Commonly outsourced logistics services

Outsourced Logistics Service	2010	2012	2013
Domestic transportation	83%	78%	76%
International transportation	75	71	71
Warehousing	74	62	63
Freight forwarding	53	57	53
Customs clearance and brokerage	58	48	52
Reverse logistics (defective, repair, return)	35	27	26
Cross-docking	38	26	25
Product labelling, packaging, assembly, kitting	36	24	25
Transportation planning and management	31	23	22
Inventory management		21	19
Freight bill auditing and payment	28	17	18
Order management and fulfilment	16	14	16
Information technology (IT) services	20	15	13
Service parts logistics	Not asked	14	12
Customer service	13	11	10
Supply chain consultancy services provided by 3PLs	18	11	10
Fleet management	15	10	8
Lead logistics provider (LLP) / 4PL services	13	9	8
Sustainability/Green Supply Chain-Related Services	Not asked	4	6

Source: 2010-2013 Third-Party Logistics Studies (Langley Jr., 2010, 2012, 2013)

In order to provide efficient and responsive service to clients, 3PL firms have to invest in ICT to improve communication, enhance visibility, and integrate with clients' information systems. Table 2 shows that transportation management (execution and planning), electronic data interchange, warehouse/distribution center management, visibility, Web portals, and bar coding are the six most outsourced IT-based services reported by shippers in the 2013 Annual 3PL Study. The percentages have increased significantly in recent years indicating that shippers are relying more and more on 3PL firms in this regard. In fact, IT is viewed as "central to the overall performance and effectiveness of 3PL-customer relationship" (Langley Jr., 2009, p. 13).

While there continues to be significant opportunities for outsourcing IT-based services, insufficient information capabilities is one of the major continuing problems with 3PL providers as reported by customers in the annual 3PL surveys (Table 3). Other continuing IT-related problems with 3PL providers include errors caused by excessive manual steps within business processes (i.e., insufficient process automation) and lack of business process integration across regions and supply chain services (i.e., inadequate system integration). Among the major reasons for non-

Table 2. Current use of 3PL-provided technology

IT-Based Service	2009	2012	2013
Transportation management (execution)	69%	75%	72%
Electronic data interchange (order, advanced shipment notices, invoicing))	22	66	68
Transportation management (planning)	34	68	67
Warehouse/Distribution center management	59	67	64
Visibility (order, shipment, inventory, etc.)	19	63	60
Web portals for booking, order checking, inventory management, and billing	21	55	59
Bar coding	18	47	50
Transportation sourcing	35	41	45
Global trade management tools (customs processing and import/export document management)	40	37	43
Customer order management	5	42	41
Collaboration tools (SharePoint, Lotus Notes, video conferencing, etc.)	Not asked	25	32
Supply chain planning	4	31	30
Network modelling and optimization	13	25	30
Supply chain event management	13	31	26
Advanced analytics and data mining tools	Not asked	19	26
RFID	11	21	24
Yard management	38	16	17

Source: 2009-2013 Annual Third-Party Logistics Studies (Langley Jr., 2009, 2012, 2013)

Table 3. Continuing problems with 3PL providers as reported by customers

Problem with 3PL Provider	2007	2008	2009
Service level commitments not realized	46%	51%	46%
Lack of continuous, ongoing improvements and achievements in offerings	40	42	46
Information technology capabilities not sufficient	35	39	43
Cost reductions not realized	35	36	36
Lack of project management skills	33	35	35
Unsatisfactory transition during implementation stage	30	30	31
Ineffective management of key performance indicators (KPIs)	30	34	Not asked
Errors caused by excessive manual steps within business processes	29	28	Not asked
Lack of consultative/knowledge-based skills	26	25	Not asked
Lack of global capabilities	19	27	30
Lack of business process integration across regions and supply chain services	20	20	26
Benefits not achieved in a timely manner	Not asked	Not asked	25
Promising premium services that they do not have	Not asked	Not asked	25
Inability to form meaningful and trusting relationships	13	17	12
Poor post-merger integration of acquired companies	13	13	Not asked
No problems	13	12	Not asked

Source: 2007-2009 Annual Third-Party Logistics Studies (Langley Jr., 2007, 2008, 2009)

users electing not to use 3PLs, "too difficult to integrate our IT systems with the 3PL's systems" ranks high from No. 10 in 2009 to No. 4 in 2012 although it falls back to No. 7 in 2013 (Langley, Jr., 2013). The Annual 3PL Studies from 2002 to 2012 also consistently find a big expectancy-performance gap between shipper expectations of 3PL IT-based services and the actual experiences with these services (Figure 1). While more and more shippers feel that IT is a necessary element of 3PL expertise, many of them do not satisfy with their 3PLprobider's IT capabilities. This so-called "IT Gap" seems to narrow a bit and stabilize in recent years probably due to increased investment in ICT by the 3PL industry (Langley Jr., 2013, p. 21). However, there is still a 40% gap that needs to be filled. Therefore, the need for 3PL firms to further invest in ICT in order to meet the requirements of their clients has become more and more critical.

Theories of Supply Chain Management and IT Investment

In explaining inter-organizational phenomena in supply chain management and outsourcing, many theories have been used which include network theory (NT) and resource-based view (RBV). NT contends that performance of a firm depends not only on how efficiently it cooperates with its direct

Figure 1. Continuing gap between customer expectations and experiences relating to 3PL IT capabilities

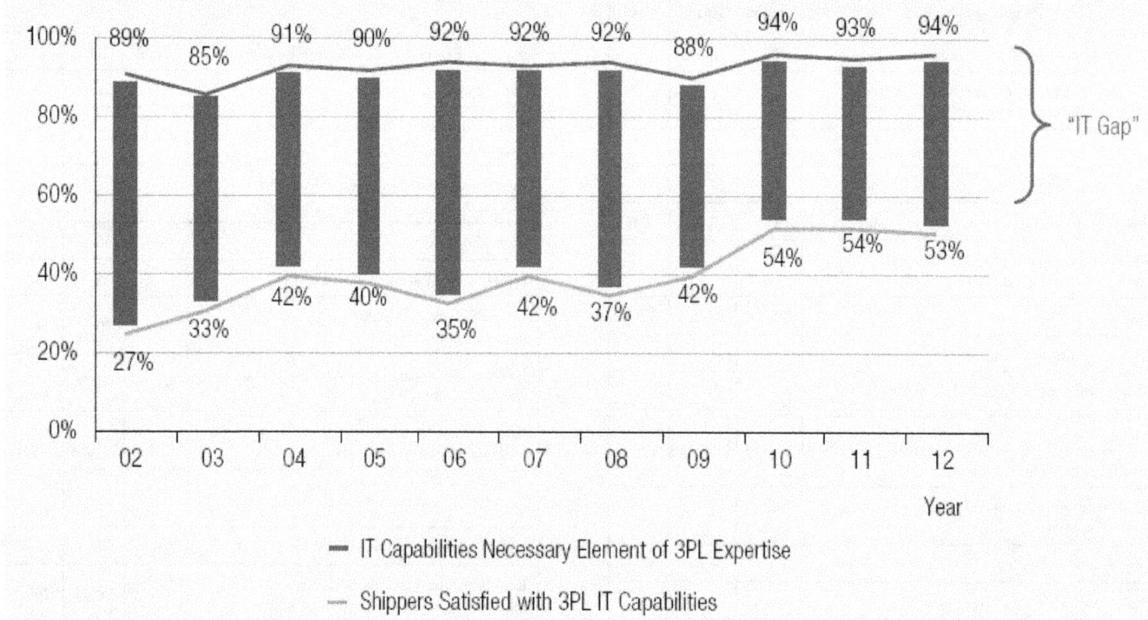

Source: 2013 Annual Third-Party Logistics Study (Langley Jr., 2013, Figure 6, p. 11)

partners but also how well these partners cooperate with their own business partners (Halldórssan et al., 2007; Jarillo, 1988). Interactions between firms enable the development of new resources as cooperative relationships permit the combination of resources from separate firms to achieve greater advantage than through individual effort (Haakansson & Ford, 2002). NT emphasizes on trust developed through positive long-term cooperative relations among organizations and mutual adaptation of routines and systems through exchange processes. From this perspective, supply chain management is seen as reciprocated interactions between organizations. Through direct communication and information sharing, firms in a supply chain work closely together to customize operation to meet individual customer requirement. Links between firms are developed through exchange of goods and information and mutual adaptation in personal, technical, logistics and other elements (Johanson & Mattsson, 1987).

In contrast, RBV sees a firm as a bundle of resources and assets and emphasizes the use of rare, valuable, in-imitable and un-substitutable resources to gain sustainable competitive advantage (Barney, 1991; Conner, 1991; Grant, 1991; Wernerfelt, 1984). Resources include tangible physical assets, intangible information and knowledge, and capabilities that are built upon effective use of tangible and intangible resources, repeated and refined practices, and accumulated experiences. Capabilities are more difficult to imitate or substitute and therefore add greater value to the firm. The theory helps explain a firm's investment in building capabilities that are difficult for its competitors to imitate or mitigate in a short period of time so as to attain long-term sustainable competitive advantage (Collis 1994; Reed and DeFillippi, 1990).

Both NT and RBV can explain why it is important for 3PL firms to invest in ICT. From a NT perspective, IT facilitates the development of positive cooperative relations between the 3PL firm and its customers as well as the business partners of the customers. The tight cooperation allows organizations involved to achieve greater advantage in managing the entire supply chain than through individual effort. From a RBV perspective, IT as a resource can be converted into a unique capability and ultimately distinctive competence through proper investment and development for a 3PL firm to gain sustainable competitive advantage over its competitors.

IT Advancement and Alignment

Using RBV, Wu et al. (2006, p. 494) contend that "IT can facilitate the development of higher-order organizational capabilities, namely supply chain capabilities, which are firm specific and hard to duplicate across organizations". It is the information advantage achieved through the adoption of sophisticated technologies and the synergistic benefits achieved through an integrated system that provide a firm with sustainable competitive advantage (Bharadwaj, 2000).

According to Wu et al. (2006), supply chain capabilities encompass four dimensions: information exchange (which refers to a firm's ability to share knowledge with its supply chain partners), coordination (which refers to a firm's ability to coordinate transaction-related activities with supply chain partners), inter-firm activity integration (which refers to the extent to which a firm coordinates its strategic channel activities such as planning and forecasting with its supply chain partners), and supply chain responsiveness (which refers to the extent to which channel members respond cooperatively to environmental changes). Without investing in the appropriate ICT, a 3PL firm will not be able to acquire these capabilities in order to operate and manage the logistics and supply chain activities of its clients effectively.

Wu et al. (2006) also propose that the adoption of ICT should focus on proper advancement and alignment. IT advancement refers to the extent to which a firm adopts the most sophisticated technology. For a 3PL service provider, proactive adoption and implementation of advanced IT to find solutions for customers ahead of competitors is crucial in gaining a competitive edge. IT alignment refers to the extent a firm's IT is compatible with that of its channel partners. For a 3PL firm, it is important that the IT systems it uses can integrate with those of its clients and their supply chain partners to enable effective and efficient information sharing and coordination. The IT capability gap as revealed in the Annual 3PL Studies clearly reinforces the significance of these two considerations. Study by Posey and Bari (2009) also suggests that the degree to which supply chain members' information systems are compatible with each other plays a significant role in information-processing capabilities. Adapted from the strategic alignment model developed by Henderson and Venkatraman (1991), Figure 2 depicts how IT advancement and alignment, both internal and external, can be achieved through strategic fit and functional integration. While the capability building process as advocated in RBV accounts for the internal IT advancement, external IT alignment is made possible through the exchange and mutual adaptation processes as emphasized in NT.

Logistics Industry in China

Since her accession to the World Trade Organization, China has experienced a tremendous growth in the logistics industry with an average annual growth of 23%, which is much higher than the GDP growth rate of 9.5% (Research and Markets, 2009). However, as shown in Figure 3, her logistics costs as a percentage of GDP (average 18%) is also the highest in comparison with that of developed countries such as Japan (11%), US (8%), and EU (7%) (KPMG, 2008). Although the percentage drops gradually from 18.4 in 2007 to 17.8 in 2011, it is still exorbitantly high. The total logistics cost recorded 8,400 billion yuan in 2011, up by 18.5% year on year (Li & Fung Research Centre, 2012). Total logistics value, which

Figure 2. IT advancement and alignment between 3PL firm and its customer

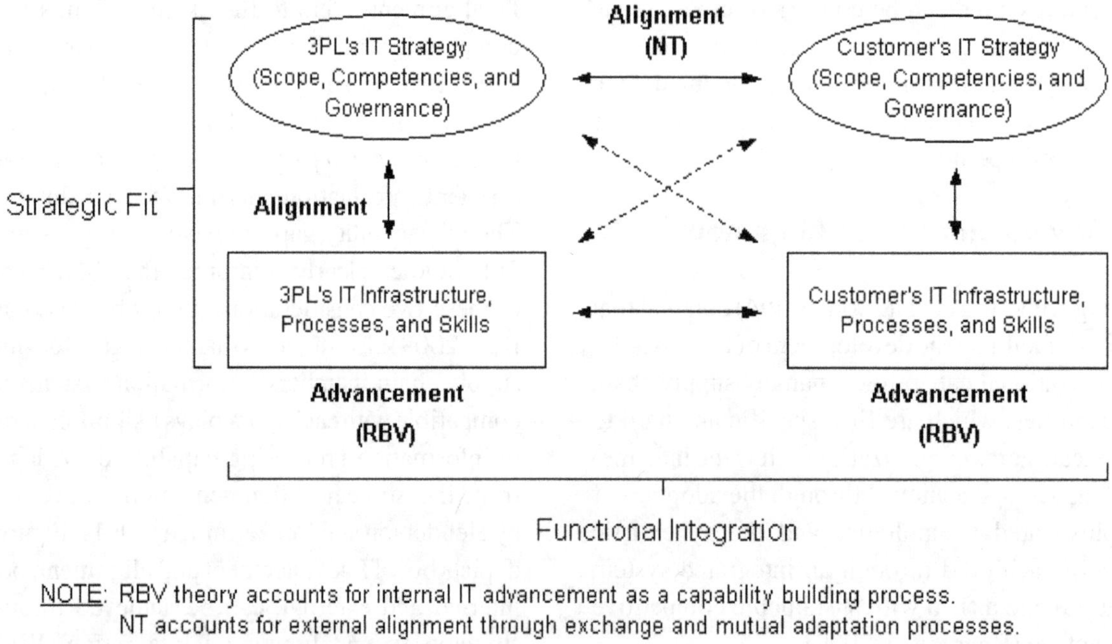

Source: Adapted from Henderson and Venkatraman (1991)

Figure 3. Total logistics cost in China, 2007 – 2011

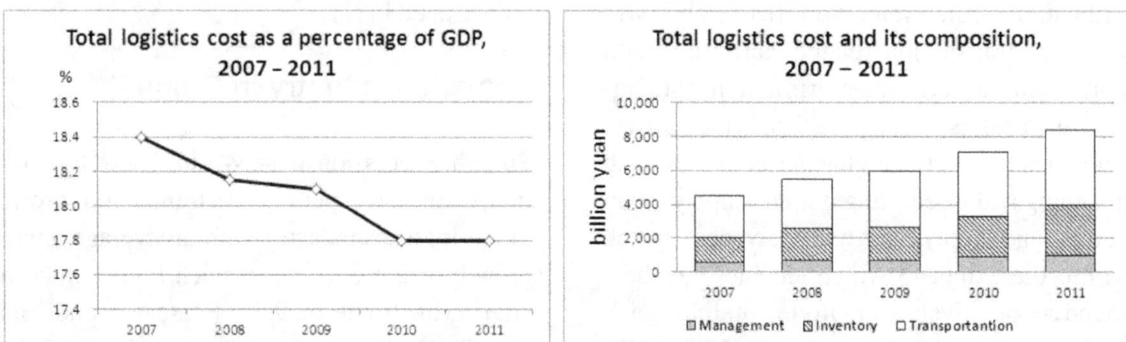

Source: China's Logistics Update 2012 (Li & Fung Research Centre, 2012, p. 7)

reflects the total demand for logistics services in a country, represents the size of the logistics market. In 2011, the total logistics value of China as reported by the China Federation of Logistics & Purchasing is 158,400 million yuan (or about 25,900 million US$) representing a growth of 12.3% in comparison with the value of 2010. Logistics demand coefficient, which is a ratio of logistics value to GDP, also indicates the state of the logistic industry. In 2011, the coefficient was 3.4, up from 2.9 in 2009 and 3.2 in 2010 denoting a growing logistics market (Li & Fung Research Centre, 2012). Although the booming economy of China has slowed down a bit in recent year, the logistics market is still developing and there is a strong incentive and scope for 3PL firms to invest in ICT to reduce logistics costs. Therefore, a status survey of the current ICT usage in the industry is useful to help reveal the position as well as point out the direction for development.

METHODOLOGY

To answer research questions 1 to 3, a self-administered questionnaire survey was conducted to collect data for analysis. A list of companies in the freight forwarding and the 3PL industries of China was first compiled using the information obtained from the Ministry of Communications (MOS) of China and the membership list of the China International Freight Association. There were more than 10,000 companies in the list that constituted the sampling frame, comprising large, medium-sized and small firms. Following the US standard (SBA, 2010), large firms refer to those that have 500 or more employees, while medium-sized ones are those with 101 to 499 employees. Companies with 100 or less employees are regarded as small firms. The survey focused on companies in the seven economically developed regions of the country – eastern, southern, south-western, central, northern, north-eastern, and north-western China. A sample of 800 companies was selected from the list through stratified random sampling based on the total numbers of companies in the different regions. Questionnaires were sent by e-mail to the selected companies and returned by e-mail upon completion. Follow-up telephone calls were made to solicit response and assist in the completion of the questionnaire where necessary. The survey was conducted and completed in October of 2009. A total of 92 valid responses were gathered with a response rate of 11.5%.

To facilitate response, close-ended questions were asked in the questionnaire. The questions were set on the basis of the literature review with reference to the 2007-2013 Annual 3PL Studies (Langley Jr., 2007, 2008, 2009, 2010, 2012, 2013) in design. To encourage response, a relatively short questionnaire was employed. There are a total of 12 questions covering the various aspects, such as factors contributing to the usage of ICT in the firm, current IT systems used, perceived significance of ICT, and benefits of the IT systems used, as identified in the literature. The first section (with 8 questions) collects background information of the company and the current level of ICT usage. The second section (with 4 questions) investigates the factors affecting the usage of ICT, perceived significance, and benefits of ICT utilization. A seven-point Likert scale ranging from 1 (Very Unimportant) to 7 (Very Important) is used to gauge the views. Since the data collected are in ordinal scale, non-parametric tests were used for data analysis.

FINDINGS AND DISCUSSIONS

Profiles of Companies Surveyed

For purpose of comparison, the sample was subdivided into three groups for analysis. Among the 92 freight forwarding and 3PL companies surveyed, 30 are small companies (with 100 or less employees), 30 are medium-sized (with 101 to 499 employees), and 32 are large organizations (with 500 or more employees) including some of the major 3PL service providers in China. Most of the surveyed companies have a long history of establishment. More than 60% of them have been in the business for more than 9 years. As expected, the majority of the firms (more than two-thirds) are serving the most economically developed regions, such as the eastern and the southern parts of China where their clients conglomerate. 65% of the companies surveyed are state-owned businesses.

The most common logistics services provided by all firms are air express (99%), land freight (99%), warehousing (100%), and packaging/repackaging (100%). Table 4 shows a summary of the profiles of the companies surveyed.

Table 4. Profiles of companies surveyed

Firm Size	Small (<=100)	Medium-Sized (101-499)	Large (>=500)	Total
Mailouts and Returns:				
Questionnaires issued	300	250	250	800
Valid returns	30	30	32	92
Response rate	10.0%	12.0%	12.8%	11.5%
Years Operating in China:				
Less than 3 years	2	0	0	2 (2%)
3 - 9 years	17	11	6	34 (37%)
More than 9 years	11	19	26	56 (61%)
Major Market Areas*:				
Eastern China	22	21	23	66 (72%)
Southern China	19	19	22	60 (65%)
South-western China	0	2	4	6 (7%)
Central China	1	2	6	9 (10%)
Northern China	5	2	10	17 (18%)
North-eastern China	10	9	15	34 (37%)
North-western China	3	8	1	12 (13%)
Ownership:				
State-owned	3	27	30	60 (65%)
Privately-owned	15	0	0	15 (16%)
Joint-venture	12	3	2	17 (18%)
Main Services Provided*:				
Air express	29	30	32	91 (99%)
Sea freight	18	29	32	79 (86%)
Land freight	29	30	32	91 (99%)
Warehousing	30	30	32	92 (100%)
Integrated logistics	23	28	32	83 (90%)
Consulting	12	7	32	51 (55%)
Packaging/ repackaging	30	30	32	92 (100%)

* Multiple answers permitted

Status of ICT Adoption

Table 5 shows the usage of ICT among the surveyed companies at the time of survey. It can be seen that large firms are basically using all commonly available ICT for business probably because they have more resources to invest in information systems and technologies. For small and medium-sized firms, adoption of more expensive technologies such as EDI, RFID, and ERP systems are relatively low in level in comparison with other ICT used. Traditional EDI is a relatively old technology for legacy systems operating on expensive private or value-added networks (Lu et al., 2001). Only large corporations could afford its use. Web-based EDI or WEDI building on XML protocol is now a more popular technology particularly among small and medium-sized firms because of its lower set-up cost (e.g., no need for a private network) and greater functionality (Nurmilaasko, 2008). RFID is a relatively new technology and is still in developing stage although it has already been applied in many industries such as manufacturing and

Table 5. Current status of ICT usage by companies surveyed

ICT Used	Small	Medium-Sized	Large	Total
Telephone	30	30	32	92 (100%)
Facsimile	30	30	32	92 (100%)
Internet access	30	30	32	92 (100%)
Electronic data interchange (EDI)	30	22	32	84 (91%)
Global positioning system (GPS)	30	30	32	92 (100%)
Barcode technology	30	30	32	92 (100%)
Customer relationship management system (CRM)	30	30	32	92 (100%)
Radio frequency identification technology (RFID)	0	11	32	43 (47%)
Local area network (LAN)	30	30	32	92 (100%)
Wireless LAN	30	30	32	92 (100%)
Enterprise resource planning system (ERP)	1	28	32	61 (66%)

retailing (Attaran, 2007; Jones et al., 2005). Owing to high tag price and considerable set-up cost as well as other challenges such as technical hurdles and multiple standards (Li et al., 2006; Wu et al., 2006), RFID is still not competitive with barcode and therefore not widely adopted by companies particularly the small and medium-sized firms. ERP is again a relatively expensive technology for integration and usually only affordable to large and medium-sized firms. The development of Web Services, a modular design to enable smaller firms to assemble their customized suites of e-business application software on the Internet, may however help change the situation (Vidgen et al., 2004). In addition, all the companies surveyed reported that they had their own websites, online transaction systems, and online tracking systems. They also exchanged data and information with their clients via the Internet on a regular basis.

Chi-square tests (Sharpe et al., 2010) were run to determine if adoption of EDI, RFID, and ERP is related to firm size. The test results suggest that there is a positive association between propensity to adopt the ICT in question and firm size in all the three cases with the calculated χ^2 greater than the critical value (Table 6). The correlation coefficient C for EDI is around 0.4 indicating that the association is a moderate one. For RFID and ERP, the coefficients are both greater than 0.6 suggesting that the association is relatively strong. Therefore, we can reject the null hypothesis of no association and conclude that large and medium-sized firms are more receptive to the RFID and ERP technologies and show a stronger tendency to adopt. The findings of the Marascuilo procedure (Marascuilo, 1966) also confirm that the differences in propensity of large firms to adopt RFID and ERP are significantly different from that of small and medium-sized firms.

Table 6. χ^2 test for association between propensity to adopt RFID and firm size

	Chi-Square Test		Marascuilo Procedure (Diff. Between)		
ICT Used	χ^2	C	S and M	S and L	M and L
EDI	18.1	0.41	√	×	√
RFID	64.0	0.64	×	√	√
ERP	79.3	0.68	√	√	×

$\alpha = 0.05$, d.f. = 2, critical $\chi^2 = 5.99$, √ - Significant, × - Not Significant

H_0: There is no association between propensity to adopt the ICT in question and firm size

H_A: Propensity to adopt the ICT in question and firm size are positively associated

Reasons for Using ICT

Respondents were asked to give their views on the significance of six common reasons for using ICT as identified in the literature. Responses on individual reason, measured in a 7-point Likert scale, were tested for significance using Wilcoxon signed rank test for single sample (Wilcoxon, 1945). Kruskal-Wallis tests (Kruskal & Wallis, 1952) were also used to determine if there are significant differences between the reasons for using ICT among the three groups of firms. The results are shown in Table 7. Overall speaking, companies consider all the six reasons relatively

Table 7. Reasons for using ICT

	Median Score of Group				Difference Between Groups		
Reason for ICT Usage	S	M	L	All	S and M	S and L	M and L
Internal integration	3*	5*	6*	5*	√	√	√
Customer satisfaction	5*	5*	5*	5*	×	×	×
Information sharing with clients	4	6*	6*	5*	√	√	×
Customer base	3*	5.5*	5*	5*	√	√	×
Organization competitiveness	5*	6*	5*	5*	√	×	√
Company image	3*	5*	6*	5*	√	√	√

Scale: 1 - Very Unimportant 2 - Unimportant 3 - Relatively Unimportant 4 - Neutral 5 - Relatively Important 6 - Important 7 - Very Important * Significant at $\alpha = 0.05$ √ - Significant × - Not Significant

important with customer satisfaction, information sharing with clients, and organization competitiveness slightly more important than the other reasons. This finding is expected as 3PLs are selling services in lieu of products. Therefore, customer satisfaction and frequent communication with clients to understand their needs are certainly important. Sub-sample analysis reveals that while small firms rely heavily on ICT to enhance customer satisfaction and increase competitiveness, medium-sized firms believe that ICT can also help them develop their customer base. Large firms, on the other hand, tend to use ICT more for internal integration, information sharing with clients, and promotion of company image. These findings may reflect an interesting characteristic of the 3PL industry in China. As small and medium-sized firms (most of them are privately owned or joint ventures) are struggling to take hold of their market share and customer base, large firms (most of them are state-owned) with well-established markets and clientele are more interested in securing their dominance through integration and company image enhancement using ICT. Comparisons between groups reveal that apart from customer satisfaction, which is of the same level of importance to all groups of firms, reasons for using ICT vary among groups to different degrees. Generally speaking, differences between medium-sized and large firms are moderate. However, small firms are basically different from their medium-sized and large counterparts in almost all the reasons for using ICT as revealed by the Dunn's multiple comparison procedure (Dunn, 1964).

Significance of Current ICT Usage

To find out how significant the various ICT is to the 3PL firms, all the surveyed companies were asked to give their views on the ICT they used. As shown in Table 8, most firms regarded low-tech ICT such as telephone, facsimile, and Internet access as important or very important tools. An interesting note is that company website was accorded the highest level of importance by small and medium-sized firms probably because it is the quickest and a less expensive way to market the company in a highly competitive business environment. All firms, regardless of firm size, also considered barcode and CRM important. The former is almost a must for modern-day logistics operations whereas the latter is paramount to understanding the needs of the customers hence ensuring their satisfaction. The other observation is that while medium-sized and large firms placed more or less equal importance on the different ICT, small firms tended to accord low importance to more expensive ICT such as EDI, RFID, and ERP probably because their scale of operation did not require the use of these technologies at all. Outcomes of the Kruskal-Wallis tests also confirm these observations. For the perceived significance of Internet access and company website, there is no significant difference among the three groups of firms. While there are significant differences between small and medium-sized, and between small and large firms in the perceived importance of many other ICT used, the views of

Table 8. Significance of current ICT usage

	Median Score of Group				Difference Between Groups		
ICT Used	S	M	L	All	S and M	S and L	M and L
Telephone and facsimile	7*	6*	6*	6*	√	√	×
Internet access	7*	6*	7*	6*	×	×	×
Company website	7*	7*	6*	7*	×	×	×
EDI	3*	5*	5*	5*	√	√	√
Barcode system	6*	6*	6*	6*	×	√	×
RFID	2.5*	5*	5*	4	√	√	×
ERP	2*	4*	5*	4	√	√	×
CRM	5*	5*	6*	5*	√	√	×

Scale: 1 - Very Unimportant 2 - Unimportant 3 - Relatively Unimportant 4 - Neutral 5 - Relatively Important 6 - Important 7 - Very Important * Significant at $\alpha = 0.05$ √ - Significant × - Not Significant

medium-sized and large firms in this regard are basically aligned except in the use of EDI which is an expensive legacy technology usually only affordable to large companies.

Significance of IT Systems Used

As revealed in the 2007-2009 Annual 3PL Studies, four IT systems, namely, transportation management system (TMS), warehouse management system (WMS), order management system (OMS), and inventory management system (IMS), are of particular importance and are widely used in supply chain management. They play a critical role in the logistics performance of the whole supply chain allowing better coordination and synchronization of activities, easier tracking and tracing of items, as well as effective management of orders and material flows. Used appropriately, they can help achieve greater efficiency as well as responsiveness across the entire supply chain. As such, large 3PL companies providing integrated logistics services will usually invest in these relatively expensive systems to better serve their customers as well as to use them as competitive weapons against competitors. In this survey, respondents were asked to give their views on the significance of these systems to their companies (as shown in Table 9). As expected, the findings reveal that large firms, which usually have more resources to invest in IT, tend to place more significance in these systems than their smaller counterparts. This is also confirmed by the Kruskal-Wallis test results. Among the four systems, TMS appears to be regarded as an important one by all companies probably because most of the 3PL firms, particularly the large ones, are providing freight forwarding or inbound/outbound distribution services, which will benefit significantly from a TMS for coordination and synchronization of activities. For small 3PL firms, IMS is also accorded some importance likely because management of inventory for clients is a common service they provided.

Table 9. Significance of IT systems used

	Median Score of Group				Difference Between Groups		
IT System Used	**S**	**M**	**L**	**All**	**S and M**	**S and L**	**M and L**
Transportation management system	4	5*	6*	5*	√	√	√
Warehouse management system	4*	5*	5*	5*	√	√	×
Order management system	4	6*	5*	5*	√	√	√
Inventory management system	4	6*	5*	5*	√	√	×

Scale: 1 - Very Unimportant 2 - Unimportant 3 - Relatively Unimportant 4 - Neutral 5 - Relatively Important 6 - Important 7 - Very Important * Significant at $\alpha = 0.05$ √ - Significant × - Not Significant

Benefits of Using ICT

All the 92 surveyed companies were asked about the benefits they perceived of using ICT and their views are summarized in Table 10. In general, they tend to agree that ICT can help them reduce

Table 10. Benefits for using ICT

	Median Score of Group				Difference Between Groups		
Benefit	**S**	**M**	**L**	**All**	**S and M**	**S and L**	**M and L**
Cost reduction	7*	6*	6*	6*	√	√	×
Business control	4*	5*	6*	5*	√	√	×
Customer integration	3*	6*	5*	5*	√	√	√
Improved competitiveness	5*	6*	6*	5*	√	√	×
Error reduction	6*	6*	6*	6*	×	×	×
Service differentiation	3*	5*	6*	5*	√	√	×

Scale: 1 - Very Unimportant 2 - Unimportant 3 - Relatively Unimportant 4 - Neutral 5 - Relatively Important 6 - Important 7 - Very Important * Significant at $\alpha = 0.05$ √ - Significant × - Not Significant

costs and errors in operation. The other relatively greater benefit is to improve competitiveness. For medium-sized firms, better business control and customer integration are also regarded as major benefits. For large firms, ICT permits and facilitates service differentiation that may further improve the competitiveness of the company. Again, these findings reveal the different focuses and strategies adopted by 3PL firms of different sizes. Owing to their shorter history of establishment, smaller market share, and nascent customer base, small firms tend to compete on price and therefore focus more on cost reduction and efficiency improvement through the use of ICT. Medium-sized firms, which have basically secured their foothold in the market with a relatively stable market share and customer base, tend to use ICT for both internal and external integration to improve their competitiveness. For large firms with accumulated knowledge and experiences due to longer history of establishment, they tend to make use of ICT to develop their capabilities to provide diversified and distinctive value-added services. If the small firms can be seen as using ICT and a Red Ocean strategy to compete in an existing market, the large firms may be evolving to employ ICT and a Blue Ocean strategy to develop new markets to enjoy greater first-mover benefits (Kim and Mauborgne, 2005). The literature shows that some 3PL companies in developed countries are actually moving in this direction (Kim et al., 2008). Findings of the Kruskal-Wallis tests again show that the perceived benefits of using ICT are significantly different between small firms and medium-sized/large firms. Apart from error reduction that is equally important to all groups, small firms tend to accord levels of importance to the various ICT benefits totally different from their larger counterparts. However, there is little difference between medium-sized and large firms in the accordance of importance suggesting that their perspectives in this regard are similar.

A summary of the survey findings on ICT adoption by the 3PL firms in China without showing the statistics are given in Table 11.

MANAGERIAL IMPLICATIONS

The findings of the survey clearly indicate that current ICT adoption in the 3PL industry of China varies with firm size (see Tables 5 and 6). Small 3PL firms with smaller customer base, limited scope of available services, and constraints in

Table 11. Summary of survey findings

Area of Investigation	Findings
Main services provided	• The most common services provided by the surveyed 3PL firms regardless of firm size are air express, land freight, warehousing and packaging/repackaging.
Current status of ICT adoption	• Large 3PL firms are using almost all ICT available whereas small and medium-sized 3PL firms are using more low-tech ICT such as telephone, fax, GPS, and barcode. • Expensive technologies such as traditional EDI, RFID, and ERP are adopted mostly by large 3PL firms only. • All firms have their own websites, online transaction, and online tracing systems.
Reasons for using ICT	• All firms consider customer satisfaction, information sharing with clients, and organization competitiveness as the major reasons for using ICT. • Small firms use ICT mainly to improve customer satisfaction and competitiveness. • Medium-sized firms use ICT more to develop customer base. • Large firms tend to use ICT for internal integration, information sharing with clients, and promotion of company image.
Significance of current ICT usage	• All firms regard telephone, Internet, and company website as the most important ICT. • Barcode and CRM are also considered as important tools by most firms. • EDI, RFID, and ERP are only considered important by large firms.
Significance of IT systems used	• All firms consider TMS important. • Small firms also regard IMS important whereas small and large firms accord more or less equal significance to TMS, WMS, OMS, and IMS.
Benefits of using ICT	• All firms consider cost saving and error reduction as the major benefits. • Small firms regard ICT as tools to improve their competitiveness. • Medium-sized firms benefit from ICT for internal and external integration. • Large firms capitalize on ICT for service differentiation to develop new market.

resources, tend to use ICT to cut costs and reduce errors so as to compete on price (see Tables 7 and 13). Using RBV (Barney, 1991; Conner, 1991; Grant, 1991; Wernerfelt, 1984), NT (Haakansson & Ford, 2002; Halldórssan et al., 2007; Johanson and Mattsson, 1987), and IT strategic alignment (Henderson & Venkatraman, 1991) as underlying theories to account for the observations, the managerial implications of the current approach taken by the small 3PL firms in ICT adoption can be foreseen. With emphasis on low-tech ICT and incapability to invest on more advanced IT technologies and systems to develop their unique capabilities, small firms may achieve short-term gains through ICT adoption but will not be able to attain long-term sustainable competitive advantage according to RBV. The failure to advance their IT systems through investment and to align their IT strategy, infrastructure and processes with that of their customers through exchange and mutual adaptation will also reduce their competitiveness in the long run as envisaged in NT. Therefore, for small 3PL firms, this is a critical managerial issue that needs to be addressed. One way to alleviate the problem will be to invest in less expensive open-source Web-based technologies such as WEDI (Adshead & Thomas, 2003; Nurmilaasko, 2008) and Web Services (Fensel & Bussler, 2002; Hagel, 2002; Vidgen et al., 2004; Wu, 2004). These technologies are designed to enable small and medium-sized enterprises to improve their Web-enabled operability so that they can link up with customers and larger business partners. Early investment in RFID technology to attain greater early adopter advantage is also important because it is almost certain that RFID will be fully adopted in supply chain operation and management in the near future (Attaran, 2007; Tajima, 2007). To minimize initial investment cost, small 3PL firms can apply RFID to a single process as pilot at the beginning to simplify data management and to accumulate operational experience. Free RFID middleware running on desktop computer can also be used for data storage and organization during pilot testing until the scale of operation requires an upgrade (Lau & Sirichoti, 2012).

For medium-sized and large 3PL firms, the findings of the survey reveal that they are adopting ICT to a much larger extent. Large firms are basically investing in all major IT technologies and systems to develop their unique capabilities in the business (see Table 5). The use of ICT by medium-sized firms for both internal and external integration reveal that, unlike their smaller counterparts, they are aiming more for long-term benefits than short-term gains (see Tables 7 and 13). The use of ICT for service differentiation by the large firms also ties in with the tenets of RBV and the Blue Ocean strategy (Kim and Mauborgne, 2005). In comparison with the small 3PL firms, the issue with medium-sized and large 3PL firms lies more in IT alignment with their customers. As the 2002-2012 Annual 3PL Studies reveal, there is a continuing expectancy-performance gap between shipper expectations of 3PL IT-based services and the actual experiences with these services (Langley Jr., 2013). It is clear that further endeavor in this regard is the top priority. Managerial implications of this misalignment can be profound. Study by Henningsson and Hedman (2010) shows that industry-wide information integration could render much greater benefits than enterprise-wide integration. However, the existing methods for enterprise-wide information integration such as business process re-engineering cannot be easily applied to industry-wide information integration. It means that IT alignment between 3PL firms and their customers cannot be achieved by the 3PL firms alone. Therefore, an inter-organizational managerial task force or technical team will need to be set up to understand customer expectation, explore scope of alignment and collaboration, and overcome any technical hurdles encountered in the alignment process. This will strengthen alliances among business partners as well as improve the competitiveness of the entire supply chain in the long run.

FOLLOW-UP RESEARCH

To determine if ICT adoption and utilization in the 3PL industry of China has improved since the 2009 questionnaire survey, a follow-up desktop research was conducted to comprehensively review the literature on ICT application in China. While the findings are diverging and show no significant increase in ICT application, the outcome suggests that awareness of the importance of IT capabilities in the logistics industry is growing and the positive impact of IT on the financial performance of 3PL firms is being increasingly recognized. The latest Annual 3PL Study also shows that the expectancy-performance gap between client expectations and the actual experiences with the services provided is decreasing indicating the 3PL industry worldwide is moving towards utilizing more advanced and sophisticated ICT to meet the needs of the customers.

Wang et al. (2008) investigated the impact of IT on the performance of 3PL firms in China and found a positive linkage. They hypothesized that IT importance (or strategic position of IT) in a 3PL firm can enhance its IT efforts (or investment in IT infrastructure and expertise), which improve its IT advantage (relative to its competitors' IT sophistication), and in turn, improves its financial performance. They also hypothesized IT involvement (or top executives' involvement in IT and business strategic planning) can have a positive effect on financial performance as well. Among the 760 firms in the Chinese logistics industry surveyed, 105 provided valid returns and the data confirmed the linkage. That is, the study found that greater IT advantage, together with greater IT involvement, can offer 3PL companies better financial performance. As the level of ICT adoption and utilization in China is still relatively low due to the country's traditional labor-intensive economy, the application of IT may therefore offer greater advantage to a 3PL firm over competitors in China than in developed countries where ICT has been widely adopted.

In contrast, Ghijsen et al. (2009) studied the contemporary 3PL market in China trying to relate trust, IT, 3PL usage to customer service performance. Based on the valid responses from 76 companies operating in the Yangzi and the Pearl River Delta regions where industry is more concentrated and the usage of 3PL service is more common, they found that there is no direct linear effect of IT usage on customer service performance from the customers' perspective. They attributed the outcome to the absence of supportive governmental regulations and the poor electronic infrastructure in China. The many information management platforms hence multiple standards and requirements, insufficient customer demand, and the lack of experience of the customers in managing the interface with 3PL providers are considered to be inhibitors to correctly understand the significant role of IT in modern-day supply chain management. Obviously, the lack of understanding of IT usage or the awareness of IT importance by the customers can be a major hindrance to ICT adoption in the 3PL industry.

In examining the reasons for the Chinese automotive industry to outsource its logistic function to 3PL providers, Zhang et al. (2012) identified, among others, the demand for advanced communication technology and effective information flow across the entire automotive supply chain as one of the major causes for outsourcing. For example, through outsourcing, small companies lacking sophisticated information systems can rely on 3PL providers to provide just-in-time delivery to supply parts to major automotive manufacturers. In fact, when analyzing the positive impacts of logistics outsourcing on the automotive industry, Zhang et al. (2012) recognized that procurement management, warehouse management, transportation management, order management and customer service management are the major areas where significant benefits can be harvested. This finding is in alignment with that of the annual 3PL surveys. To achieve maximum efficiency and responsiveness in these management functions and activities, full visibility across the entire supply chain via an Internet-based platform and dynamic optimization with real-time data and sophisticated applications are critical. This will require heavy investment in ICT infrastructure and software as well as IT capabilities from the 3PL providers. Therefore, a higher level and expedited rate of ICT adoption in the 3PL industry will certainly promote logistics outsourcing and meet the growing demand for integrated logistics service by the manufacturing industry.

In fact, findings of the 2013 Annual 3PL Study reveal that the expectancy-performance gap between client expectations and the actual experiences with 3PL IT capabilities (or the so-called "IT gap") is gradually closing (see Figure 1). Back in 2002, the gap was as wide as 62%. Throughout the years, there has been a gradual improvement in 3PL IT capabilities and for the last three years – 2010, 2011, and 2012 – the IT gap basically remains unchanged at about 40%. This finding indicates that awareness of IT importance in the global 3PL industry has certainly enhanced in the last decade. The narrowing IT gap also suggests that IT efforts of many 3PLs should have been increased to better meet the requirements and expectations of their customers. The trend is likely to continue and therefore proper IT advancement and alignment between 3PL firms and their customers as depicted in Figure 2 would become more and more critical.

CONCLUSION AND FURTHER RESEARCH

Through an analysis of the 2007-2013 Annual 3PL Study results (Langley Jr., 2007, 2008, 2009, 2010, 2012, 2013), this study has investigated the utilization of ICT in the 3PL industry from a global perspective. The observations are compared against those obtained through a questionnaire survey on the current status of ICT adoption and utilization by the 3PL firms in China. In general, the survey findings align with those obtained from the 2007-2013 Annual 3PL Studies. It is found that ICT is being extensively utilized in the 3PL industry for various reasons ranging from internal integration to improvement of organizational competitiveness. Probably due to limitation in resources, small 3PL companies are only able to use relatively low-tech ICT such as telephone, facsimile, Internet access, WEDI, GPS, and some less sophisticated logistics information systems for business. In contrast, medium-sized and large 3PL firms are using basically all available ICT including the more expensive RFID and ERP systems to better serve their customers. While small 3PL companies are using ICT at a tactical level mainly to cut costs and reduce errors in the day-to-day logistics operations, medium-sized and large firms have evolved to make use of ICT for strategic purposes such as business control, customer integration, and service differentiation. From a RBV perspective, small firms are still using ICT as resources to compete for short-term gains. In contrast, medium-sized and large firms tend to develop ICT as unique capabilities to help them achieve long-term competitive advantage (Barney, 1991; Conner, 1991; Grant, 1991; Wernerfelt, 1984). The use of all available ICT and expensive IT systems by the medium-sized and the large 3PL firms also suggest an awareness of and a propensity for IT advancement and alignment as advocated in NT. TMS is regarded to be the most important logistics information system by all surveyed firms while the levels of importance of other systems such as WMS, OMS, and IMS vary depending on the size of the company and the services provided. Again, for small 3PL companies whose services are usually related to transportation and management of inventory for clients, TMS and IMS stand out to be critical competing tools. For large companies, approximately equal significance is accorded to all systems suggesting an all-round development strategy and a long-term capability-building plan.

This observation can again be accounted for using the RBV theory. Follow-up research reveals that although there is no strong evidence to suggest any significant changes in ICT adoption in the 3PL industry of China, general awareness of IT importance and hence IT investments appear to be on the increase. These findings align with that of the latest Annual 3PL Study which shows that the expectancy-performance gap between client expectations and the actual experiences with 3PL IT capabilities is gradually closing.

This study is basically a desktop research augmented with a status survey using non-parametric statistical analysis to present a general picture of ICT utilization in the 3PL industry and the current ICT adoption and utilization by 3PL firms in China. Although it does have contributed to knowledge by providing a snapshot of the current situation, the relatively short questionnaire has prohibited the gathering of more in-depth information to explore the many factors affecting a firm's decision on ICT investment. Also, the questionnaire survey has focused only on the logistics service providers. The perspective from the users has yet to be investigated. Such investigation will enable an interesting comparison against the IT capability gap identified in the Annual 3PL Studies. Future research may extend the survey to cover a larger sample and to include customers as well as other business partners in the supply chain so as to investigate the benefits of ICT utilization across the whole supply chain. The level of IT advancement and alignment between 3PL firms and their customers as envisaged in NT can also be examined. Exploratory studies on the various considerations on ICT utilization and the critical success factors of ICT implementation can also be conducted using an in-depth case study approach. This will enable the revelation of the challenges faced by companies, particularly small firms with limited resources, experiences, and expertise, in using ICT for service coordination and collaboration with business partners. A framework or a set of guidelines may then be developed to assist the whole industry in achieving a supply-chain-wide standardized ICT implementation to maximize the overall efficiency and benefit.

ACKNOWLEDGMENT

The authors would like to thank the companies and their employees surveyed in this study for providing the information.

REFERENCES

Adshead, A., & Thomas, D. (2003, July 1). Web-based EDI forces suppliers to change tactics. *Computer Weekly*.

Angeles, R. (2000). Revisiting the role of internet-EDI in the current electronic commerce scene. *Logistics Information Management*, *13*(1), 45–57. doi:10.1108/09576050010306396

Attaran, M. (2007). RFID: An enabler of supply chain operations. *Supply Chain Management: An International Journal*, *12*(4), 249–257. doi:10.1108/13598540710759763

Barney, J. B. (1991). Firm resources and sustained competitive advantage. *Journal of Management*, *17*(1), 99–120. doi:10.1177/014920639101700108

Bharadwaj, A. S. (2000). A resource-based perspective on information technology capability and firm performance: An empirical investigation. *Management Information Systems Quarterly*, *24*(1), 169–196. doi:10.2307/3250983

Bhatnagar, R., Sohal, A. S., & Millen, R. (1999). Third-party logistics services: A Singapore perspective. *International Journal of Physical Distribution and Logistics Management*, *26*(9), 569–587.

Bowersox, D. J., Closs, D. J., & Stank, T. P. (1999). 21st century logistics: Making supply chain integration a reality. *Supply Chain Management Review*, *3*(3), 17–28.

Bowersox, D. J., & Daugherty, P. J. (1995). Logistics paradigms: the impact of information technology. *Journal of Business Logistics*, *16*(1), 65–80.

Bowersox, D. J., Daugherty, P. J., Dröge, C. L., Rogers, D. S., & Wardlow, D. L. (1989). *Leading edge logistics: Competitive positioning for the 1990s*. Oak Brook, IL: Council of Logistics Management.

Carr, N. (2003). IT doesn't matter. *Harvard Business Review*, *81*(5), 41–49. PMID:12747161

Chopra, S., & Meindl, P. (2012). *Supply chain management: Strategy, planning, & operations* (5th ed.). Upper Saddle River, NJ: Pearson.

Collis, D. (1994). Research note: How valuable are organizational capabilities. *Strategic Management Journal*, *15*(8), 143–152. doi:10.1002/smj.4250150910

Conner, K. R. (1991). A historical comparison of resource-based theory and five schools of thought within industrial organization economics: Do we have a new theory of the firm? *Journal of Management*, *17*(1), 121–154. doi:10.1177/014920639101700109

Davenport, T., & Brooks, J. (2004). Enterprise systems and the supply chain. *Journal of International Management*, *17*(1), 8–19.

Deloitte. (2003). *Chips with everything*. Retrieved March 2, 2010, from http://www.rfid4u.nl/downloads/deloittechipwitheverything.pdf

Dunn, O. J. (1964). Multiple comparisons using rank sums. *Technometrics*, *6*, 241–252. doi:10.1080/00401706.1964.10490181

Edwards, P., Peters, M., & Sharman, G. (2001). The effectiveness of information systems in supporting the extended supply chain. *Journal of Business Review*, *22*(1), 1–27.

Fensel, D., & Bussler, C. (2002). The web service modelling framework. *Electronic Commerce Research and Applications*, *1*(2), 113–137. doi:10.1016/S1567-4223(02)00015-7

Ghijsen, P., Semeijn, J., & Wang, A. (2009). Modern 3PL services in China: The role of trust. *Journal of International Business and Economy*, *10*(2), 103–135.

Grant, R. M. (1991). The resource-base theory of competitive advantage: Implications for strategy formulation. *California Management Review*, *33*(3), 114–135. doi:10.2307/41166664

Gunasekaran, A., & Ngai, E. W. T. (2004). Information systems in supply chain integration and management. *European Journal of Operational Research*, *159*(2), 269–295. doi:10.1016/j.ejor.2003.08.016

Haakansson, H., & Ford, D. (2002). How companies interact in business networks? *Journal of Business Research*, *55*(2), 133–139. doi:10.1016/S0148-2963(00)00148-X

Hagel, J. (2002). Edging into web services. *The McKinsey Quarterly*, *4*, 29–38.

Halldórsson, A., Kotzab, H., Mikkola, J. H., & Skjøtt-Larsen, T. (2007). Complementary theories to supply chain management. *Supply Chain Management: An International Journal*, *12*(4), 284–296. doi:10.1108/13598540710759808

Henderson, J. C., & Venkatraman, N. (1991). Understanding strategic alignment. *Business Quarterly*, *55*(3), 72–79.

Henningsson, S., & Hedman, J. (2010). Industry-wide supply chain information integration: The lack of management and disjoint economic responsibilities. *International Journal of Information Systems and Supply Chain Management*, *3*(1), 1–20.

Introna, L. D. (1991). The impact of information technology on logistics. *International Journal of Physical Distribution and Logistics Management*, *21*(5), 32–37. doi:10.1108/EUM0000000000387

Jarillo, P. (1988). On strategic networks. *Strategic Management Journal*, *9*(1), 31–41. doi:10.1002/smj.4250090104

Johanson, J., & Mattsson, L. G. (1987). Inter-organizational relations in industrial systems: A network approach compared with the transaction cost approach. *Inter-Organizational Studies of Management and Organization*, *17*(1), 34–48.

Jones, P., Clarke-Hill, C., Comfort, D., Hillier, D., & Shears, P. (2005). Radio frequency identification and food retailing in the UK. *British Food Journal*, *107*(6), 350–360. doi:10.1108/00070700510602156

Kathuria, R., Anandarajan, M., & Igbaria, M. (1999). Linking IT applications with manufacturing strategy: An intelligent decision support system approach. *Decision Sciences*, *30*(4), 959–992. doi:10.1111/j.1540-5915.1999.tb00915.x

Kim, C., Yang, K. H., & Kim, J. (2008). A strategy for third-party logistics systems: A case analysis using the blue ocean strategy. *Omega*, *36*(4), 522–534. doi:10.1016/j.omega.2006.11.011

Kim, W. C., & Mauborgne, R. (2005). Blue ocean strategy: From theory to practice. *California Management Review*, *47*(3), 105–121. doi:10.2307/41166308

KPMG. (2008). *Logistics in China: Property and infrastructure*. Retrieved March 2, 2010, from http://www.kpmg.com.hk/en/virtual_library/Property_Infrastructure/LogisticsChina.pdf

Kruskal, W. H., & Wallis, W. A. (1952). Use of ranks in one-criterion variance analysis. *Journal of the American Statistical Association*, *47*(260), 583–621. doi:10.1080/01621459.1952.10483441

Lai, F., Hutchison, J., & Zhang, G. (2005). Radio frequency identification (RFID) in China: Opportunities and challenges. *International Journal of Retail and Distribution Management*, *33*(12), 905–916. doi:10.1108/09590550510634639

Lai, F., Zhao, X., & Wang, Q. (2006). The impact of information technology on the competitive advantage of logistics firms in China. *Industrial Management & Data Systems*, *106*(9), 1249–1271. doi:10.1108/02635570610712564

Langley, C. J., Jr. (2007). *2007 third-party logistics: results and findings of the 12th annual study*. Retrieved October 7, 2013, from http://www.3plstudy.com

Langley, C. J., Jr. (2008). *2008 third-party logistics: results and findings of the 13th annual study*. Retrieved October 7, 2013, from http://www.3plstudy.com

Langley, C. J., Jr. (2009). *2009 third-party logistics: results and findings of the 14th annual study*. Retrieved October 7, 2013, from http://www.3plstudy.com

Langley, C. J., Jr. (2010). *2010 third-party logistics: Results and findings of the 15th annual study*. Retrieved October 7, 2013, from http://www.3plstudy.com

Langley, C. J., Jr. (2012). *2012 third-party logistics: Results and findings of the 16th annual study*. Retrieved October 7, 2013, from http://www.3plstudy.com

Langley, C. J., Jr. (2013). *2013 third-party logistics: Results and findings of the 17th annual study*. Retrieved October 7, 2013, from http://www.3plstudy.com

Lau, K. H., & Ma, W. L. (2008). A supplementary framework for evaluation of integrated logistics provider. *International Journal of Information Systems and Supply Chain Management*, *1*(3), 49–69. doi:10.4018/jisscm.2008070104

Lau, K. H., & Sirichoti, T. (2012). A study on RFID adoption in the grocery retailing industry of Thailand. *International Journal of Information Systems and Supply Chain Management*, *5*(3), 58–77. doi:10.4018/jisscm.2012070104

Li, S., Visich, J. K., Khumawala, B. M., & Zhang, C. (2006). Radio frequency identification technology: Applications, technical challenges and strategies. *Sensor Review*, *26*(3), 193–202. doi:10.1108/02602280610675474

Li & Fung Research Centre. (2012). China's logistics industry update 2012. *China Distribution & Trading*, *96*, 1–46.

Lin, C. Y., & Ho, Y. H. (2009). An empirical study on the adoption of RFID technology for logistics service providers in China. *International Business Research*, *2*(1), 23–36.

Lu, E. J. L., Tsai, R. H., & Chou, S. (2001). An empirical study of XML/EDI. *Journal of Systems and Software*, *58*(3), 271–279. doi:10.1016/S0164-1212(01)00044-9

Marascuilo, L. A. (1966). Large-sample multiple comparisons. *Psychological Bulletin*, *65*(4), 280–290. doi:10.1037/h0023189 PMID:5325892

Nurmilaasko, J. M. (2008). Adoption of e-business functions and migration from EDI-based to XML-based e-business frameworks in supply chain integration. *International Journal of Production Economics*, *113*(2), 721–733. doi:10.1016/j.ijpe.2007.11.001

Posey, C., & Bari, A. (2009). Information sharing and supply chain performance: Understanding complexity, compatibility, and processing. *International Journal of Information Systems and Supply Chain Management*, *2*(3), 67–76. doi:10.4018/jisscm.2009070105

Reed, R., & DeFillippi, R. J. (1990). Causal ambiguity, barriers to imitation and sustainable competitive advantage. *Academy of Management Review*, *15*(1), 88–102.

Research and Markets. (2009). *China logistics industry report, 2006-2010 brochure*. Retrieved March 2, 2010, from http://www.researchandmarkets.com/reports/357283/china_logistics_industry_report_2006_2010.pdf

Sharpe, R. S., De Veaux, R. D., & Velleman, P. F. (2010). *Business statistics*. Boston: Addison Wesley.

Sum, C. C., Teo, C. B., & Ng, K. K. (2001). Strategic logistics management in Singapore. *International Journal of Operations & Production Management*, *21*(9), 1239–1260. doi:10.1108/EUM0000000005926

Tajima, M. (2007). Strategic value of RFID in supply chain management. *Journal of Purchasing and Supply Management*, *13*(4), 261–273. doi:10.1016/j.pursup.2007.11.001

Topacan, U., Basoglu, A. N., & Daim, T. U. (2010). Exploring the adoption of technology driven services in the healthcare industry. *International Journal of Information Systems in the Service Sector*, *2*(1), 71–93. doi:10.4018/jisss.2010093005

U.S. Small Business Association (SBA). (2010). *Summary of size standards by industry*. Retrieved 7 October, 2013, from http://www.sba.gov/content/summary-size-standards-industry

Vaidyanathan, G. (2005). A framework for evaluating third-party logistics. *Communications of the ACM*, *48*(1), 89–94. doi:10.1145/1039539.1039544

Vidgen, R., France, D., Powell, P., & Woerndl, M. (2004). Web service business transformation: Collaborative commerce opportunities in SMEs. *Journal of Enterprise Information Management, 17*(5), 372–381. doi:10.1108/17410390410561008

Vijayaraman, B. S., & Osyk, B. A. (2006). An empirical study of RFID implementation in the warehousing industry. *International Journal of Logistics Management, 17*(1), 6–20. doi:10.1108/09574090610663400

Wang, Q., Lai, F., & Zhao, X. (2008). The impact of information technology on the financial performance of the third-party logistics firms in China. *Supply Chain Management: An International, 13*(2), 138–150. doi:10.1108/13598540810860976

Wernerfelt, B. (1984). A resource-based view of the firm. *Strategic Management Journal, 5*(2), 171–180. doi:10.1002/smj.4250050207

White, A., Daniel, E. M., & Mohdzain, M. (2005). The role of emergent information technologies and systems in enabling supply chain agility. *International Journal of Information Management, 25*(5), 396–410. doi:10.1016/j.ijinfomgt.2005.06.009

Wilcoxon, F. (1945). Individual comparisons by ranking methods. *Biometrics Bulletin, 1*(6), 80–83. doi:10.2307/3001968

Wu, C. (2004). A readiness model for adopting web services. *Journal of Enterprise Information Management, 17*(5), 361–371. doi:10.1108/17410390410560991

Wu, F., Yeniyurt, S., Kim, D., & Cavusgil, T. (2006). The impact of information technology on supply chain capabilities and firm performance: A resource-based view. *Industrial Marketing Management, 35*(4), 1317–1323. doi:10.1016/j.indmarman.2005.05.003

Zhang, L., Hou, Z., & Qiu, X. (2012). *Third party logistics in Chinese automobile industry: A case study of a Chinese automotive manufacturer*. (Unpublished Thesis). Department of Industrial Development, IT and Land Management, Faculty of Engineering and Sustainable Development, Högskolan, I Gävle.

Chapter 14
A Platform for Collaborative Outbound Logistics

Yves Wautelet
Katholieke Universiteit Leuven, Belgium

Isabelle Mirbel
University of Nice Sophia-Antipolis, France

ABSTRACT

Electronic collaboration in the field of Supply Chain Management notably allows enhanced visibility for each of the involved actors through information sharing and global optimization through coordination. The proper achievement of these advantages is nevertheless not trivial since it implies the development of rather complex IT systems. Each actor is indeed mostly dealing with his own data semantics and custom processes so that data centralization and sharing is seldom achieved. Moreover, even if there would be a willingness to share data, this inherent heterogeneity would constitute a serious burden for software developments. The present chapter presents the results of a research aimed to partially solve these issues. It indeed presents the analysis and design of a platform for e-collaboration among the main Outbound Logistic (OL) actors based on a common conceptualization (including unified data semantics) and a set of services (supporting business processes realization). The chapter briefly presents the generic processes but mostly focuses on the representation of these services at a strategic level through an analysis of their added value and risk for OL actors potentially adopting the software solution.

INTRODUCTION

In nowadays economic context, information systems development and management has become a determinant issue. New concepts and technologies continuously emerge allowing a better representation and more efficient data exchanges in complex software problems. The significance of these can notably be found in supply chain management (SCM). Indeed, modern supply chains involve series of actors incarnated by various collaborating or competing companies where several roles

DOI: 10.4018/978-1-4666-4506-6.ch014

played by lots of individuals are interacting to achieve common as well as individual goals. In such a context, one needs powerful tools to represent logistic flows, to deliver strategic information to supply chain partners and to ensure e-collaboration. Collaborative decision will tend to avoid local equilibriums (at actor level) and wastes in the global supply chain optimization, giving opportunities to achieve the greatest value that the chain can deliver at lowest cost (Pache and Spalanzani, 2007; Samii, 2004). This paper is part of the effort to combine advanced software engineering (SE) and project management concepts to structure complex collaborative software development and deliver added value information technology (IT) systems in SCM.

Outbound logistics (OL) is the part of the supply chain which focuses on the product delivery with consequently a strong highlight on stocking and transportation. The development of a collaborative platform for OL actors induces high complexity which implies the necessity of a software engineering (SE) framework presenting various levels of knowledge and views to:

- Furnish adequate documentation to each stakeholder;
- Allow strategic reasoning (opportunities/risk balance);
- Cover all of the traditional software development disciplines;
- Be integrated in a mature development life cycle;
- Ensure adaptability and flexibility;
- Envisage third party software data reuse.

To that extend, we propose a framework structured through three levels of knowledge:

- A *strategic* level where the applicative package is represented in terms of highly aggregated functionalities called *services* (with actors as service providers) providing a basis for opportunities/risk management. Besides being a view of the software problem every stakeholder can understand, this high level representation enables them to evaluate (with the services as scope elements) the opportunities and threats that would result from the use of the software application. This view is consequently useful for top-management;
- A *tactical* level where each of the services are detailed in a static manner using rich organizational concepts, i.e. *tasks*, *goals* and *resources* and their *depending actors*. This view is useful for middle managers in charge of the development of the objectives defined at strategic level;
- An *operational* level where each of the elements required for service fulfillment are operationalized through the behavior of software agents in charge of the successful realization of the sub-processes defined at tactical level through atomic actions.

The paper is structured as follows. Firstly, the context of this research is overviewed. The structure of the software development framework is then depicted using a pyramidal perspective. The concepts and technologies used into the software development framework are positioned. Afterwards, the application domain – outbound logistics – is outlined, followed by the description of the software applicative package development. Finally, a framework made of success factors for IT realizations in SCM is applied onto our contribution for validation purpose. Finally, the paper turns to the related work and conclusions are derived.

PAPER POSITION AND CONTRIBUTIONS

This section introduces the research context, focus and method to highlight its main contributions and limitations.

Research Context

In august 2005, the Belgian *Walloon Region* introduced the *Marshall Plan*, a vast economic investment to reinforce attractivity and competitiveness of Walloon's companies in order to raise employment rate. This plan notably integrates poles of competitiveness, and, among these poles, *Logistics in Wallonia*, created in July 2006, which is concerned with transportation and logistics. Into this latest pole, *TransLogisTIC*, proposed as a driving project, has officially started in March 2007.

Currently, the real-time visibility of information flows throughout the whole OL chain fails to ensure competitive integrated logistics. However, decisions at European level request for the development of multimodal transportation, notably through Eurocorridors as the "C" (Antwerp - Bale - Lyon) which would cross the Belgian Walloon area. For this reason, the TransLogisTIC project has been built up, with, as long-term main objective to develop combined, performing and complete transportations into the Walloon Region with transport by rail particularly promoted in accordance with the European policy (see Marco Polo program). The project has initially been planned on a 3 year basis with 14 million euro as overall budget and has involved several complementary actors including 10 private companies and 5 universities and research labs.

The analysis of the socio-economic context in which TransLogisTIC takes place has led to various concrete objectives, among which:

- The development of control systems tailored to the rail freight expectations;
- The development of real time innovative localizing solutions covering the whole OL chain based on a new generation of positioning systems as well as on the optimal and extended use of existing infrastructures;
- The development of an online collaborative logistic platform allowing shippers, carriers, infrastructure managers and final clients (i.e. the major OL actors) to share information for a better optimization of the logistic chain.

The current paper exclusively focuses on this last point especially through the work developed by the *Center for Supply Chain Management* (*CESCM*) at the *Louvain School of Management* (*LSM*) of the *Université catholique de Louvain* (*UCL*).

Research Focus

As overviewed in the previous section, the project is based on a series of complementary goals with lots of technical challenges for the participating partners. The focus of the study presented here is both on the development of the collaborative platform (*what must be developed, integrated, automated and how?*) coupled with a strategic analysis of the offered services (*are the innovations provided by the platform adoption adding value and/or threatening the business of each of the involved actors?*) represented through a custom knowledge representation framework.

Research Method

Interviews have been performed onto series of actors involved onto the OL chain. These interviews have mainly focused on two aspects: *business process modeling* on the one side and *evaluation of the interest of the technological innovations brought by the platform on the other side.*

At earliest stages of the project, the study of business processes allowed to distinguish a set of services and to describe them generically. In other words, OL as it is described in section 4 is the result of a generic composition of the specific processes modeled through interviews and technical documents analysis performed for various actors (see further). An iterative approach permitted us to validate the models and adapt them based on the actors' feedback. Thanks to the use of generic processes the collaborative platform could be adapted for all of the actors interested (this could lead to business process reengineering at actor level). Similarly, the level of importance of the technical innovations brought by the platform's adoption has been discussed and quantified with each of the actors during the interviews. The design of a Multi-Agent System (MAS) on the basis of the developed analysis models as well as the prototyping of the platform has been performed by the CESCM research team (5 full-time researchers). It was tested and validated in lab with standard scenarios. A technical report extensively describes business processes, MAS design and test protocols (Wautelet, 2009).

All interviews and site visits were conducted in Belgium. For confidentiality reasons, the participating companies cannot be cited, but note that the interviews involved:

- Shippers issued of the steel sector (4 companies), the glass sector (1 company), the chemical sector (2 companies) as well as on the food sector (1 company). Each of these companies are multinationals operating on multiple sites and delivering worldwide;
- Carriers using different modes of transport (4 companies). Focus has been put on railways and truck transport but an exclusive maritime transporter has also been involved. Each of these companies are multinationals operating on multiple sites and delivering internationally;
- The infrastructure manager in charge of the Belgian railways, the one managing the biggest Belgian port as well as a mixed private/public company developing the port activities at provincial level.

Contributions and Limitations

The paper introduces a multiple level framework for software development. The originality of the framework is that it maps the traditional managerial decision levels with models issued of SE. The models structured in the framework have been defined and formalized in other scientific publications but they are used here in an original way. The framework also allows strategic reasoning. In other words, on the basis of coarse-grained analysis elements located at strategic level, an opportunity and risk analysis can be performed guiding the adoption decision. The SE elements in question have the advantage of allowing an analysis on the basis of the services the platform offers without entering into too fine technical (i.e. behavioral) details. The framework thus combines SE knowledge representation with managerial decision taking.

Moreover, the OL processes have been described in a generic way based on the interviews, technical documentation and information received from the different participating partners. This generic description has served as a root model for process automation within the platform.

The knowledge representation framework is applied onto the development of a collaborative platform for OL actors. Besides being an illustration of the SE framework, this shows the concrete results of an applied research project – *TransLogisTIC* – which is of interest to every researcher in the field of collaborative SCM through information systems. Also, the results of the application of the strategic analysis for adoption decision can be used as a basis for better understanding the stakes of IT collaboration solutions.

Summarizing, the representation of the collaborative platform for OL within its multiple perspectives and the identification of the potentials usage it offers constitutes the main contribution of the paper.

The paper is however not a technical description of the OL processes as implemented into the platform (this contribution can hardly be published for industrial reasons), nor a description of semantics for OL network representation (which will be the subject of another communication), nor a description or manual of the use of the developed platform.

The illustration and assessment of the platform's performance using both software metrics, supply chain metrics as well as simulations (Terzi & Cavalieri, 2004; Cavaleri & Terzi, 2006; Gunasekaran et al., 2004) for comparison with non-collaborative software is under progress and not in the scope of this paper which mainly focuses on knowledge representation, innovation management and multi-agent system development for IT in SCM.

FROM MANAGERIAL DECISION LEVELS TO SOFTWARE DEVELOPMENT

Management sciences define several layers for decision making. Indeed, decisions do not have the same impact - from a marginal short term consequence to a major long term impact - so that their time horizon is variable. Software development methodologies are traditionally divided into a set of disciplines (often called *analysis*, *design*, *implementation*, *test*, etc).

We propose here to map the knowledge (represented through diagrams) required for software engineering (SE) to the organizational decision levels. The different decision levels are incarnated by individuals having different visions and expectations on the software product in development. Indeed, while the software application is mainly used by operators (at operational level), requirements are often supervised by middle management (tactical level) while top management only has to perceive the main functions as well as the long term implications of its IT decisions (strategic level).

We propose here a framework to deal with huge and complex software problems that cannot be captured at once but require multiple levels of organization to get an adequate view of the system to-be for each of the concerned stakeholders. That is why we propose to map the evoked management perspectives within software development models (a graphical representation is shown in Figure 1). More precisely:

- The *Strategic Level* is concerned with decisions including the general direction i.e. long term goals, philosophies and values. In a SE perspective, this covers the upper level problem *analysis* with only the main functionalities represented in the form of *services*. Those face *opportunities* and *threats* for the actors potentially adopting the software solution. This dimension is represented using the *Strategic Services Diagram* (see (Wautelet et al., 2008) for an exhaustive definition, justification and meta-model). Services are used here as high-level abstract entities for representing the software package (thus mad of service consumers, services and service providers). Each of them represents rather independent (loosely coupled) but complementary (thus communicating) software;
- The *Tactical Level* is concerned with more concrete decisions aimed at implementing the strategy defined at the upper level. In a SE perspective it concerns the problem *analysis* at a more detailed level. Services are decomposed following an *i* (i-star)*

Figure 1. Proposed software development framework

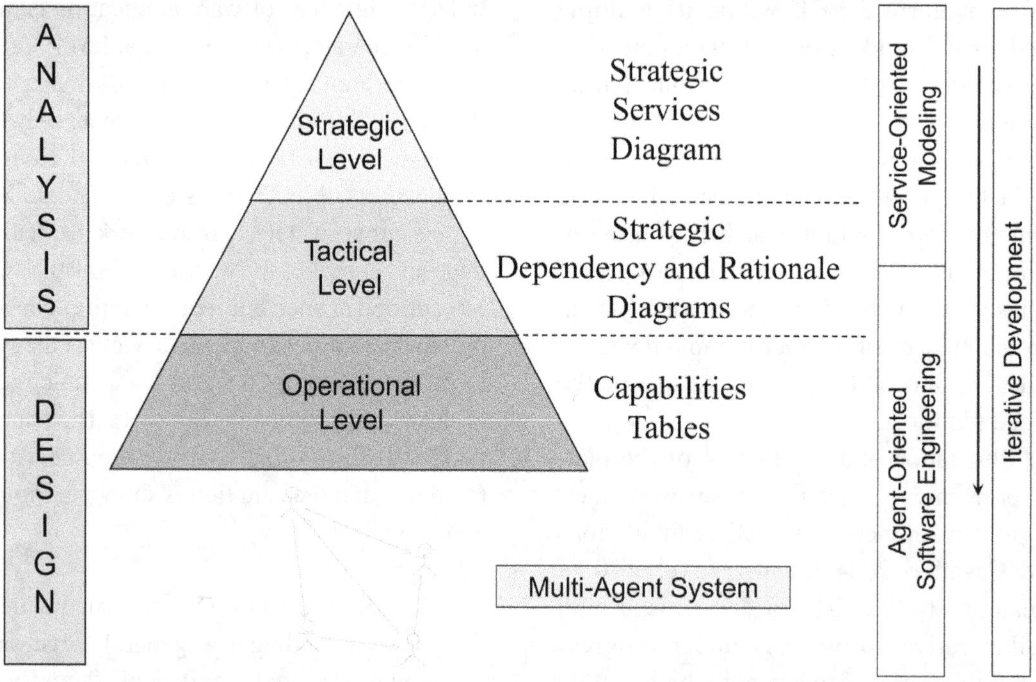

organizational approach (Yu et al., 2011; Yu, 1997; Yu 1995) in the form of *goals, tasks* and *resources* that specific actors are responsible for;

- The *Operational Level* is concerned with every day decisions, used to support the tactical ones. Their impact is immediate, short term, short range, and usually low cost. In a SE perspective it corresponds to introducing a solution to the described problem through a software *design.* Subprocesses modeled in i* models are turned into a set of capabilities constituting the skeleton of a *Multi-Agent System.*

Moreover, we advice to proceed iteratively in order to test the results of the software modeling and preliminary prototypes to correct the misunderstandings and get a software product in accordance with user requirements and business processes (see (Wautelet et al., 2013) for such an application into the development of a production management system in the steel industry).

Implementation issues using *object-oriented languages* as *Java*, *C++* or *agent-oriented ones* such as *JADE* (Bellifemine et al., 2007), *JaDEX* (Pokahr et al., 2005, Braubach et al., 2005) remain outside the scope of this paper and are not depicted here.

OUTBOUND LOGISTICS

The model presented hereafter is a process and interaction view of a supply chain with respect to outbound logistics (OL). It has been build-up generically on the basis of OL actors interviews performed in the context of TransLogisTIC. OL is the process related to the movement and storage of products from suppliers to end-users. The model incorporates processes realized at the actor's level and the interaction between the different actors.

Note that, in this section, an actor is considered to be an active entity that carries out actions to achieve services. The term actor is thus defined at highest aggregation level and can be decomposed on a lower level basis by roles played by individuals or groups of individuals active for the main actor. Similarly, a company can be "more than" a single actor; for example, a same company could sometimes be a shipper, when it fulfills orders from its clients for example, or a final client, when it has ordered shipments from its own suppliers.

Involved Actors

Four major actors can be identified in OL processes:

- **The shipper:** This actor is the one requiring transportation for a shipment (often the shipper actor is a company manufacturing a product). He has received an order from a client and has to deliver this order. The shipment has its own characteristic in terms of volume, transportation constraints, pick-up and delivery dates, transportation mode and the shipper has its own expectation on the price he should be charged. The shipper will turn towards one or several carriers to realize that shipment;
- **The final client:** This actor is the one who has ordered a shipment from a supplier and awaits its delivery. The client is generally not involved in the choice of the carrier, but is impacted by the planned delivery date and the delays that could result in the realization of the transportation;
- **The carrier:** This actor is the one owning transportation capacity; it receives transportation orders from shippers for the realization of shipments to final clients. The core business of this actor is transportation. As such, he has to realize these orders while maximizing his profit;
- **The (logistic) infrastructure manager:** This actor owns or acts on the behalf of the owner of the logistic infrastructure required by carriers such as hubs, rail tracks, airports, docks, etc. The utilization of this infrastructure is rented to the carriers while other services could be offered (transshipments, cross-docking, storage, etc.)

Collaboration in SCM

Collaboration consists of two mechanisms.

First, information sharing allows enhanced visibility between the actors in the supply chain. This information can generate value for each actor for planning their own operations and optimizing their own decisions. Information sharing can appear at every stage. At strategic level, announcing its own planned requirements and developments will permit the other actors to position themselves better in their own strategic decisions. Tactically, it raises the opportunity to better fit the needs of its clients and suppliers. Operationally, all actors are kept informed of the realization of one's operations and can adapt more rapidly to problems encountered by one actor that could affect the others.

A second and more subtle aspect of collaboration is coordination. If information sharing has been a current practice in a supply chain for a long period of time, and trust has emerged between the actors, they can consider moving further to coordinate decisions. In this case, decisions are not only taken based on internal and shared information in order to improve its own efficiency, but are discussed and taken in order to optimize some processes more globally. The goal is not only to take the decisions that are favorable for one actor but also decisions that favorable for its business partners. In this case, it is the whole efficiency of the chain itself that can be improved.

Process View

For each actor, the OL processes it is involved in can be segmented into different levels. These levels – *strategic*, *tactical* and *operational* – are the same as the ones used in the framework of Section 3. It is, however, from primary importance to note that the perspective is different. Indeed, they are used here for representing the levels of "decision making" of the actors involved in the OL processes (*what is the term at which the OL decisions are taken*) rather than knowledge representation for SE as it was previously (*what problem that the platform is intended to solve*). At strategic level, the actors take long term decisions determining their relations with other actors, fixing their long term objectives, aiming at matching their capacity to the market requirements and defining adequate

Figure 2. The outbound logistics workflow

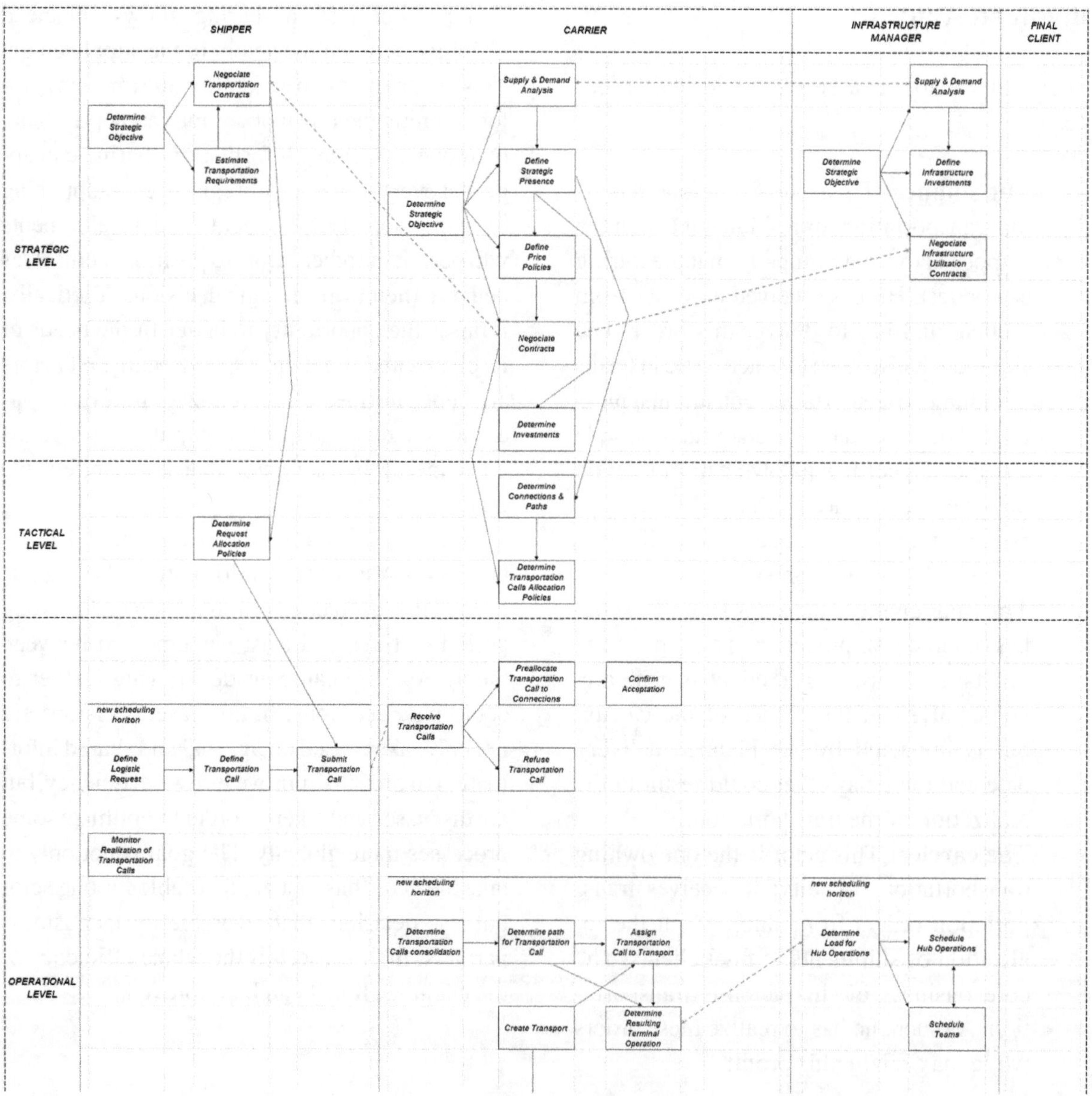

corporate policies. At tactical level, the actors take medium term decisions to achieve the strategic goals. Those decisions define the structure in which operations will take place. At operational level, day-to-day decisions are taken to direct the logistic flows. At each level, interactions occur between the actors through their (lower level) roles.

On the generic process and interaction diagram, the processes and interactions are defined in a generic way, which could be further extended by defining, for each process, the tasks involved. This "tasks definition" gives rise to an extended model which goes beyond the scope of this paper but is documented in (Wautelet, 2009). A global representation is given in Figure 2. The processes are categorized by levels – strategic, tactical and operational – and by actor – shipper, carrier, infrastructure manager and final client.

Decisions at Strategic Level

At strategic level, actors are involved in their own strategic analysis and define their strategic goals. They determine their offer, markets, objectives and future investments as well as their corporate strategies. Global supply and demand planning are realized and turned to long term business partnerships.

Shippers are involved with the evaluation of future transportation needs and the determination of the required type of transportation including characteristics in terms of *volume*, *specific constraints*, *reliability*, *quality*, *schedules*, *frequency* and *expected fares*. They are also scanning the market for appropriate carriers to take care of their transportation activities. Long term contracts can eventually be settled with them.

Carriers are involved in deciding on which market to be present and with which offer. In this model, the term «presence» is used to refer to a strategic offer made by a carrier. A presence is the expression of this offer which defines the destinations served, the capacity made available on specific destinations pairs or in specific regions, the involved transportation mode, the targeted type of clients and possibly the tariffs. *The presence allows the offer to be visible by the market.* Carriers can find shippers reacting to their presence definition with whom they may enter into long term contracts.

Infrastructure managers are involved in the evaluation of the capacity they plan to offer to the carriers as well as the policies for accepting them on their infrastructure. Similarly, they are busy with the determination of how the carriers use this capacity, under what conditions and how it is priced. Finally, decisions for infrastructure development are taken.

Final clients do generally have little impact on strategic OL decisions.

Decisions at Tactical Level

At tactical level, more concrete decisions are taken for implementing the strategy defined at corporate level. The business units adapt their strategy in terms of policies under which the operations will take place. For example, shippers may determine rules for allocating concrete requests for shipment to specific carriers, or to realize possible trade-off between reliability, speed and price of the transportation for each kind of shipment. Carriers are defining how to implement their announced presence, by defining *connections* and *paths*. Connections are links between two locations on which they plan to have transports. Defining connections structures the carriers' network allowing them to reserve the required capacity on those connections. Similarly paths are defined, which are decisions about how to route the shipments from the origin to the destinations by using a series of connections. The paths can be dependent on the type of shipments.

Decisions at Operational Level

The operational level deals with concrete shipments that have to be realized. The operational level itself can be structured into the *scheduling phase* and the *monitoring phase*.

In the scheduling phase, the players plan the short term realization of the OL. Shippers receiving orders from their clients will determine their day-to-day transportation needs. In our model, those needs are defined as *logistic requests*. Such requests identify the shipment, the origin and the destination, the requested pickup, delivery dates and other constraints specific to the shipment. Once the logistic requests have been identified for a scheduling horizon, shippers try to match those to a set of carriers, based on the tactical policies of the business unit. Shippers build, for each request, one or several *transportation calls*, which are calls that will be addressed to the carriers to realize a part of the journey of the shipment. Shippers then submit these calls to the relevant carriers.

Carriers receive the transportation calls, and either accept or refuse them. In the first case, they may want to reserve the required capacity on the path that will be used by the shipment. According to their tactical policies, they may already allocate the shipment to a series of connections. This allocation allows them to evaluate the delivery date which they can confirm to the shipper. As a new scheduling horizon begins, carriers' schedulers list the shipments allocated, or to be allocated, to each specific connection. They verify the feasibility of the allocation, and modify it if needed. They create « *transports* » to realize the shipments. In our model, a transport is a specific movement of a logistic unit (truck, train, barge, etc.) on a connection. It has a capacity, a schedule, a list of stops and the logistic unit is identified. The shipments are then allocated to specific transports. The path of each shipment is verified and required terminal operations, such as transshipments, are identified and communicated to the involved infrastructure managers.

Infrastructure managers receive an infrastructure load from the carriers, and schedules their operations and teams to meet such load. In an efficient supply chain, infeasibilities of one player are communicated to the others for schedule adaptations. When communication is absent, infeasibilities generally result in delays.

The monitoring phase consists of checking whether the timeline is respected or not, which is seldom the case. Shippers can be checking where their shipments are, if the delivery dates are still confirmed, if the shipments encounter problems on their route in order to determine in real-time any possible delay or loss of shipments. Final clients require the same visibility on awaiting shipments.

The logistic players, i.e. the carriers and the infrastructure managers, can react by taking correctives measures in case of problems in the realization of their schedules. However, this requires, first, an adequate visibility on the realization of the transports, second, enough flexibility in rescheduling, and, third, communication with the other logistic partners.

Third Party Information Providers

Finally, actors are already using a series of information systems (IS) for various purposes. For efficient integration of the platform, these need to be custom interfaced. Generically, we point to the following types of IS that the platform needs to interact with:

- The Fleet Management System (FMS) is computer software that enables people to accomplish a series of specific tasks in the management of a company's vehicle fleet. It can include vehicle telematics (tracking and diagnostics), driver management, fuel management, vehicle maintenance and so on;
- The Warehouse Management System (WMS) is a key part of the systems managing the supply chain. It helps controlling the movement and storage of materials within a warehouse and processing the associated transactions, such as shipping, receiving, putaway and picking;

- The Enterprise Resource Planning II (ERP) system is an enterprise-wide information system designed to support most of the business system. It maintains in a single database the data needed for a variety of business functions such as *Manufacturing*, *Supply Chain Management*, *Financials*, *Projects*, *Human Resources* and *Customer Relationship Management*;
- The Transportation Management Systems (TMS) is computer software designed to manage transportation operations. Its goal is the determination of the most efficient and most cost-effective way to execute the movement of product(s). The TMS will be further overviewed in the case study, it includes various functions such as:
    - Planning and optimizing of terrestrial transport rounds;
    - Transportation mode and carrier selection;
    - Management of air and maritime transport;
    - Real time vehicles tracking;
    - Service quality control;
    - Vehicle load and route optimization;
    - Transport costs and scheme simulation;
    - Shipment batching of orders;
    - Cost control, key performance indicators (KPI) reporting and statistics.

COLLABORATIVE PLATFORM DEVELOPMENT

This section introduces the application of the framework overviewed in Section 3 onto the development of an outbound logistics collaborative platform supporting the processes defined in Section 4. The development of such a package is however too extensive to be entirely described here. We will limit ourselves, at strategic level, with the use of the *Strategic Services Diagram* (*SSD*), which documents the application package in terms of organizational services and, at tactical level, focus on only one of them, *Manage Transports*. In other words, at tactical level, the i* *Strategic Dependency* and *Strategic Rationale* diagrams for the *Manage Transports* service only will be presented. Moreover, since complexity from one level to the lower evolves exponentially we concentrate on one task, *Select Most Adequate Transport*, when illustrating the MAS development at operational level.

Strategic Level

This section introduces the development of the OL collaborative platform at strategic level; firstly the model based on services is overviewed, secondly a reasoning perspective summarises the opportunities and threats of its use for each of the concerned actors.

The Collaborative Application's Services

The SSD of Figure 3 presents the application package through multiple *Services*. These highly aggregated elements are aimed at representing the problem to understand its main purpose, but also the potential problems (i.e. *Threats*) it faces; the potential improvements (i.e. *Opportunities*) and the social structure (i.e. the *Dependencies*) which govern *Actor* interactions.

The actors (hereunder in bold) and the data structures (hereunder in italic) have been defined in the process view of section 4. Each service represents a complex business process summarized here in a few sentences (a complete UML specification is available in (Wautelet, 2009)). Modeled services are:

- The *Treat Orders* service represents the process to plan the coming material flows by building up *Logistic Requests* on the basis of the client's orders. The *Shipper's*

Figure 3. The strategic services diagram

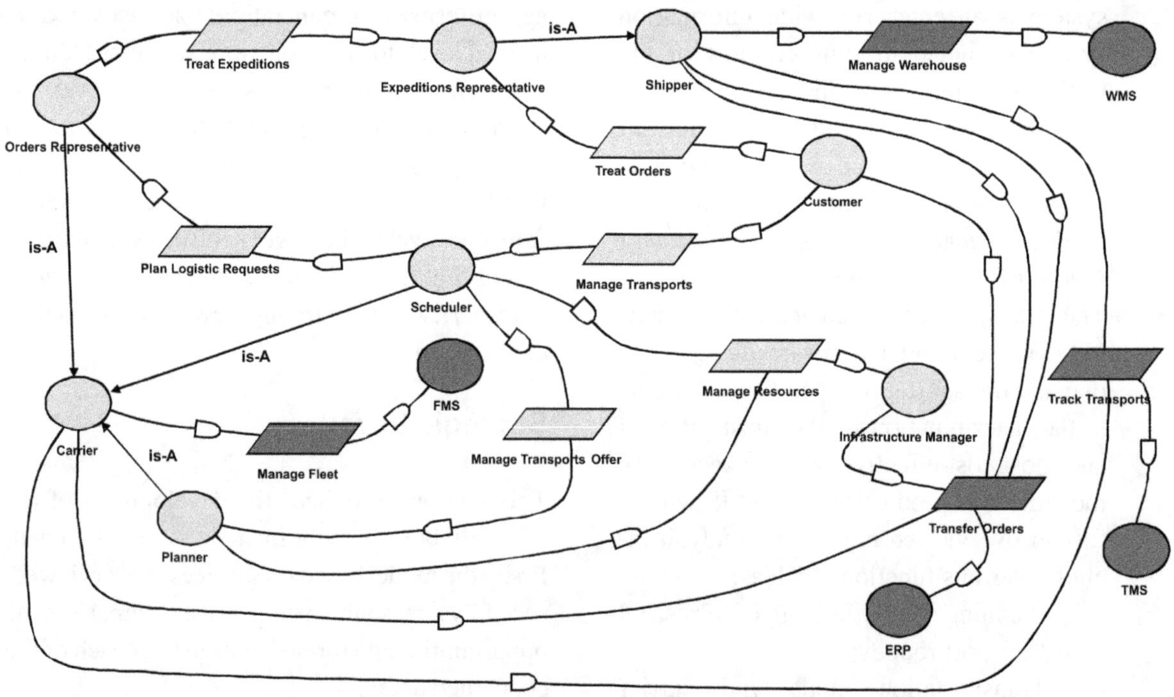

Expedition Representative is the responsible actor;

- The *Treat Expeditions* service represents the process to build *Transportation Calls* on the available *Logistic Requests*. The *Shipper's Expedition Representative* is the responsible actor;
- The *Plan Logistic Requests* service represents the process to evaluate the possibilities (eventually by relaxing constraints) of accepting the transmitted *Transportation Calls*. The *Carrier's Orders Representative* is the responsible actor;
- The *Manage Transports Offer* service represents the process to update the *Carriers' Transportation Offer* in function of the general environment (demand and capacity on particular origins and destinations, new transportation possibilities, etc). The *Carrier's Planner* is the responsible actor;
- The *Manage Transports* service represents the process to plan the *Carriers' (physical) Transports* linked to its Transportation Offer to answer to the accepted *Transportation Calls* demand. The *Carrier's Scheduler* is the responsible actor;
- The *Manage Resources* service represents the process to ensure an adequate planning of required resources (working teams, docks, cranes, etc.) for the planned *Transports*. The *Infrastructure Manager* is the responsible actor;
- The *Track Transports* service represents the process to real-time track the transported goods and perform a series of activities such as dynamic replenishments, alternative route calculation, etc. The *Transport Management System* is the responsible actor;
- The *Manage Fleet* service represents the process to accomplish a series of specific tasks in the management of a carrier vehicle fleet as for example telematics (tracking and diagnostics), driver management, fuel

management, vehicle maintenance and so on. The *Fleet Management System* is the responsible actor;

- The *Manage Warehouse* service represents the process to manage the movement and storage of materials within the warehouse and processing the associated transactions, such as shipping, receiving, putaway and picking. The *Warehouse Management System* is the responsible actor;
- The *Transfer Orders* service represents the process to manage all the input data flows that must be transferred from or to the actor ERP system. The *Enterprise Resource Planning* is responsible actor.

Strategic Analysis

Opportunities and threats of the collaborative software application package are discussed in this section. The analysis is generic in the sense that we adopt the point of view of the responsible actor (that can be visualized in the SSD) for each of the assessed opportunities and threats of a particular service.

A weight is first of all assigned to opportunities/threats on a discrete scale ranging from 1 to 5 (the highest, the most critical for the project). The added value/risk exposure for each particular service to each particular opportunity/threat is then evaluated on the basis of a low/medium/high/none scale. The overall opportunity and threat hierarchies are established through two matrices assigning to each service respectively a global added value to an opportunity and a global risk exposure to a threat. We basically define four values for opportunities added value/risk exposure probability:

- **Low:** "L" in a yellow filled cell, has a weight of 1;
- **Medium:** "M" in a orange filled cell, has a weight of 2;
- **High:** "H" in a red filled cell, has a weight of 4;
- **Non-existing:** an empty cell, this service is not concerned by this opportunity/threat, has a weight of 0.

To be adopted by the actors involved in the OL chain, the collaborative platform has to demonstrate its advantages. To that extend we distinguish the following weighted opportunities:

- **Real-Time Information Transmission:** Real-time information transmission represents a valuable tool for each of the collaborating actors since it notably allows dynamic replenishment. The applicative package thus represents an opportunity since it enables to considerably lower information transmission time. This opportunity has been given a weight of 5;
- **Business Process Optimization:** Reengineering business processes notably through standardized information transmission can lead to avoid waste of resources (notably human ones) so that time and money can be spared. This opportunity has been given a weight of 4;
- **Symmetric Information:** The use of the applicative package ensures that when an actor has a specific demand of services it is transmitted to whole of the concerned actors. For example the shippers' logistic requests are transmitted to whole of the carriers so that each of them can adapt their transport offer. Information symmetry tends to lower the eventuality that partners are unfairly (de)favored. This opportunity has been given a weight of 3;
- **Unified Data Structures:** The use of unified data structures based on common semantics representing one version of the truth eases the everyday communications between supply chain actors and constitutes a standardized formalization that can

serve as a legal basis. This opportunity has been given a weight of 2;

- **Integrated Information Portal:** The use of a single portal for managing all the aspects of an actors' business eases the everyday transactions and lowers maintenance costs so that resources can be spared. This opportunity has been given a weight of 1.

The opportunities/services matrix presented in Figure 4 determines a hierarchy among the services to rank them on the basis of added value. The *Manage Transports* and *Manage Resources* services are credited with an added value of 60 which underlies the solution could potentially be the most benefiting for carriers and infrastructure managers (which are the *depender* actors for these services). The matrix also shows the substantial added-value from the collaborative application package's services so that each of the collaborating actors can potentially have an interest by the adoption of the solution. The platform however makes sense only if it is adopted by a minimal number of concerned actors.

Similarly, to be adopted by the actors involved in the outbound logistic chain, the collaborative platform has to support risk management. To that extend we distinguish the following weighted threats:

- **Loss of Local Optimality:** Global optimization could lead actors of the supply chain to a worse situation that when they were only driven and optimal on the basis of their own business strategy. This threat has been given a weight of 4;
- **Loss of Local Autonomy:** The autonomy of an actor refers to its capability to make and influence decisions. The provided information of an actor to the other can lead its autonomy to be adjusted in one sense or the other so that it can be an opportunity or a threat (we nevertheless exclusively consider this as a threat here). In other words, increased transparency can lead to a loss of autonomy. This threat has been given a weight of 3;

Figure 4. The opportunities/services matrix

		Opportunity weight	SERVICES: Treat Orders	Treat Expeditions	Plan Logistic Requests	Manage Transports Offer	Manage Transports	Manage Ressources	Manage Warehouse	Track Transports	Manage Fleet	Transfer Orders
OPPORT.	Real-Time Information Transmition	5	H	M	H		H	H	M	H	M	M
	Business Process Optimization	4	H	H	H	L	H	H	H	L		
	Symetric Information	3	M	M	M	M	H	H				
	Unfiied Data Structures	2	L	H	H	M	H	H	M	M	M	H
	Integrated Information Portal	1	H	H	H	H	H	H	M	M	M	M
Overall Service Added Value			48	44	54	18	60	60	32	30	16	20

- **Transmission of Strategic Information:** By adding transparency onto the companies offer and demand, information that the company would prefer to have left hidden can be determined by third parties. This threat has been given a weight of 3;
- **System Intrusion:** An intrusion takes place when a user of an information system takes an action that he is not legally allowed to take. The intruder may come from outside, or it may be an insider exceeding his limited authority when taking action. The intruder's actions may be detrimental to the health of the system or to the services it offers. This threat has been given a weight of 2;
- **Data loss:** Facing huge data transfers between actors some required data can be lost or poorly transmitted. This threat has been given a weight of 2;
- **System Failure:** The information system may be down and unusable for a defined time period. This threat has been given a weight of 2.

The threats/services matrix presented in Figure 5 allows determining a hierarchy among the services to rank them on the basis of risk exposure. The *Manage Transports* service has an overall risk of 46 which underlies risks are potentially the highest for carriers (which is the *depender* actors for these services).

The strategic analysis of the identified services leads to the following conclusions:

- Adopting the collaborative software applicative package with the services defined in the SSD presents opportunities and risks for all of the identified actors;
- Carriers have the highest implication since they can take the highest added value of adopting the services but also face highest risks. Adopting the solution can thus be of primary interest for them but they have to set up operational solutions to avoid that threats become real which has an operational cost. A more sophisticated cost/benefit can then be run on an individual basis by each of the involved companies.

Figure 5. The threats/services matrix

		Threat weight	SERVICES: Treat Orders	Treat Expeditions	Plan Logistic Requests	Manage Transports Offer	Manage Transports	Manage Ressources	Manage Warehouse	Track Transports	Manage Fleet	Transfer Orders
THREATS	Loss of Local Optimality	4		L		H	H	H	H	H	M	L
	Loss of Local Autonomy	3		L		H	H	H	M	H	H	L
	Transmission of Strategic Information	3	M	H	H	H	M	L	M	M	L	L
	System Intrusion	2	M	M	M	M	H	M	M	M	L	M
	Data Loss	2		L			L	L			L	L
	System Failure	2	L	L	L		L	M	M	M	M	
Overall Service Risk Exposure			14	27	18	44	46	41	36	42	31	16

Tactical Level

The i* framework defines two models for analyzing the organizational setting, the actors composing it as well as their intentionality. Those models are:

- The *Strategic Dependency Diagram* (*SDD*) which describes the network of social relationships among actors;
- The *Strategic Rationale Diagram* (*SRD*) which describes and supports the reasoning through which each actor goes with respect to its relationships with other actors.

Those diagrams are, in our framework, used to further describe each of the services identified and modelled in the SSD. One SDD and one SRD are thus developed for each of the services of the SSD. Due to a lack of space, we however only present here the SDD and SRD of one service, *Manage Transports*, in Figure 6. The i* actors represent the OL ones as well as third party components providing information to the software system. Each OL actor depicted in section 3.1. (as well as the final client) are represented as actors; third party components we want to integrate in the global applicative package are also represented as actors (see for instance TMS at the bottom of the diagram). As illustrated on the SDD, the TMS is interfaced with our platform to achieve the goals *Vehicle Load and Route to be Optimized*, *Events to be Handled* and *Transports to be Followed Up*. Moreover, the SRD further documents the actors' rationale. Notably, we find the task *Select Most Adequate Transport* which is aimed to select the best fitting transport offer on the basis of some defined requirements and under defined constraints.

Operational Level

This section concentrates on operational aspects; the problem analyzed is thus based on an operational solution in the form of a computer system made of entities behaving to fulfill the (i*) tasks and goals identified at tactical level. To operationalize those, a MAS is used for flexible and active software development. Due to a lack of space we will however only illustrate resolution of the *Select Most Adequate Transport* task here. The MAS involves three main actors: the Shipper, Carrier and Infrastructure Manager (agents are the instance of these actors). The Shipper is responsible for the capabilities *createLogisticRequest* and *acceptResolutionSequence*; the Carrier is responsible for *handleTransportationCall*, *evaluateTransportationServiceOffer*, *relaxConstraints*, *computeRequiredOperations*, *handleProposalAcceptance* and *createTransport*; finally, the Infrastructure Manager is responsible for *evaluateInfrastructureAvailibility* and *bookRequiredResources*. Those are depicted and extensively described in Figure 7. The execution sequence is defined graphically in the UML sequence diagram of Figure 8. Capabilities are represented through arrows; the agent responsible for their handling is on the target of them. Following our model, to achieve the *Select Most Adequate Transport* task, the proper execution of this sequence of capabilities is required. Further implementation and technical MAS issues are outside the scope of this paper.

VALIDATION

The main contribution of the paper has been defined as the description of the collaborative application package for outbound logistics within its representation on both business process and software engineering perspectives. Gunasekaran & Ngai (2004) have developed a framework made of the critical success factors for evaluating IT applications and strategy in SCM. As a validation process, we suggest to instantiate the six success factors composing their framework onto our specific contribution. This section thus

Figure 6. The strategic dependency and strategic rationale diagrams of manage transports

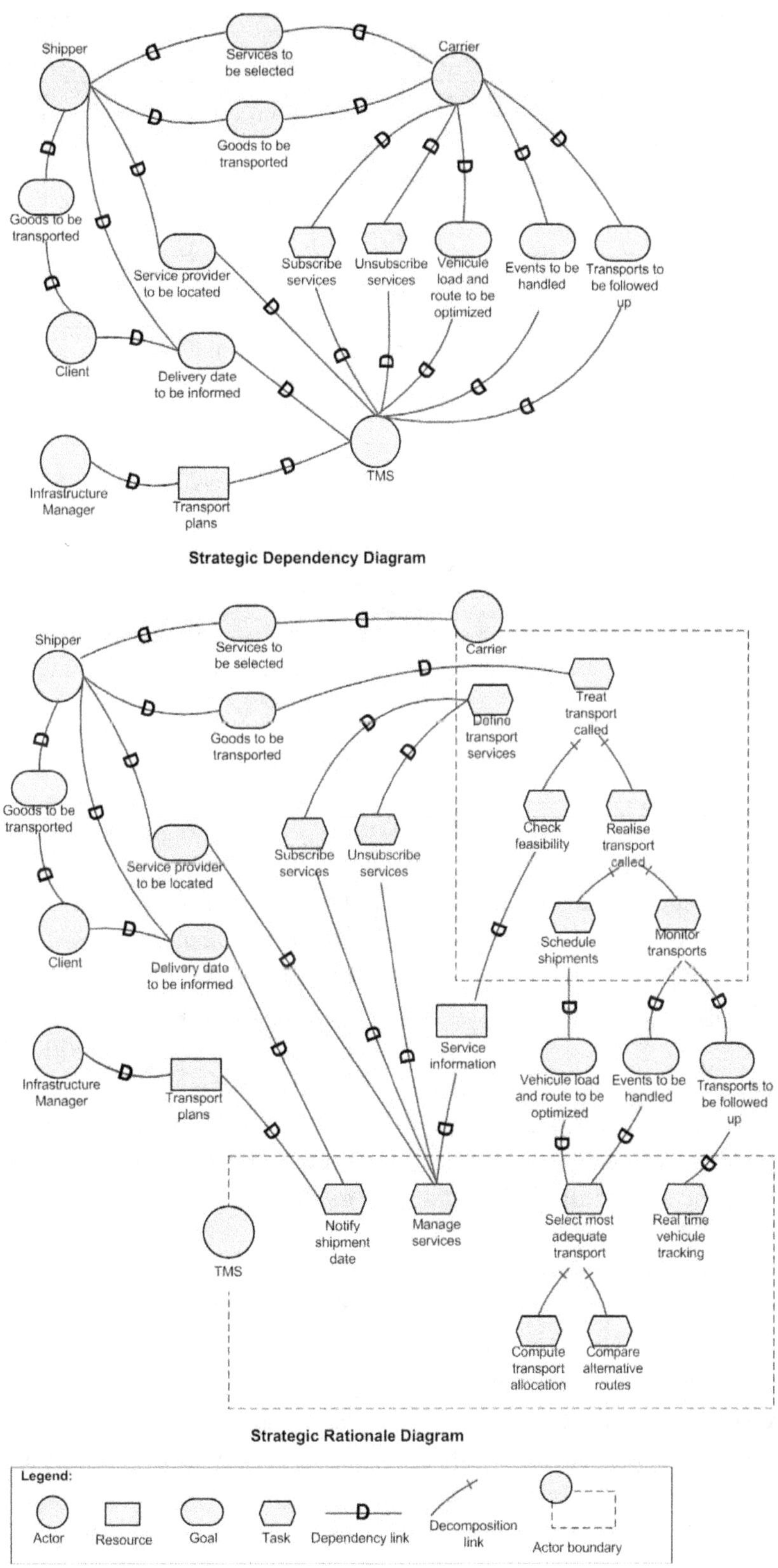

Figure 7. The Select Most Adequate Transport task's capabilities

No	Capabilities	Description	Responsible Agent	Ressource
1	createLogisticRequest()	When the Shipper requires to transport merchandise from a source to a destination (generally to fulfill an order or for internal reasons) it creates a logistic request (which remains an internal concept)	Shipper	logistic_request
2	handleTransportationCall(logistic_request)	The Shipper transmits a Logistic_Request in the form of a Transportation_Call to a Carrier which is is in charge of proposing a planning to realize them (note that the Logistic_Request is an abstract requirement, the Transportation_Call is a Logistic_Request with requested departures and arrival dates/times adressed to a Carrier)	Carrier	logistic_request
3	evaluateTransportationServiceOffer(transportation_call)	The Carrier disposes of a transportation offer in the form of Transportation Services. So it is in charge of evaluating the possible matches between the demand (the transportation call) and its own offer.	Carrier	service
4	relaxConstraints(transportation_call)	The Carrier eventually needs relaxing the constraints to mach transportation offer and demand.	Carrier	transportation_proposal
5	computeRequiredOperations(transportation_proposal)	The Carrier computes the required logistic operations needed to sucessfully achieve the possible transports	Carrier	transportation_proposal
6	evaluateInfrastructureAvailability(transportation_proposal)	The Infrastructure Manager manages a series of ressources to make transportations possible. Ressources are limited and should be booked; on request it answers on disposals.	Infrastructure_Manager	transportation_proposal
7	acceptResolutionSequence(transportation_call_resolution_sequence)	The Shipper is responsible for accepting the transportation call resolution sequence proposal submitted by the Carrier.	Shipper	transportation_call_resolution_sequence
8	handleProposalAcceptance(transportation_call_resolution_sequence)	When the Shipper has accepted the transportation call resolution sequence proposal, the Carrier is in charge of achieving the procedures required to adequately fulfill it.	Carrier	transportation_call_resolution_sequence
9	createTransport(transport_proposal)	The Carrier is in charge of creating the transport so that it will be part of the Carrier and other invovled agents future plannings.	Carrier	transport
10	bookRequiredRessources(operation_list)	Required ressources to adequately fulfill the transport are booked at the infrastructure manager	Infrastructure_Manager	operation_list

briefly introduces these success factors and then envisages their application onto our contribution.

Strategic Planning

The objective of strategic planning is to develop long-term plans and changes to the organization to improve the company's competitiveness. More and more effort is spent on that discipline and (Gunasekaran & Ngai, 2004) argues that *IT should support the long term objectives and goals of SCM both in terms of flexibility and responsiveness to changing market requirements*. He also emphasizes the need for organizational restructuring when adopting software systems and concludes of the essentiality to prioritize strategic dimensions that influence IT in SCM into a single organizational structure. To such an extent, the involvement of top management is essential:

The collaborative platform overviewed in this paper has been depicted in terms of highly aggregated elements and opportunities and threats have been overviewed in order to furnish a formal representation that can be used for decision taking and as a planning base for top management. The representation proposed in this paper therefore (at least partially) covers this particular success factor.

The Virtual Enterprise

The concept of virtual enterprise (VE) focuses on the development of a network of collaborative firms each of them owning core competencies allowing reaching the market on time with adequate products. (Gunasekaran & Ngai, 2004) point out that developing a VE becomes an important

Figure 8. The Select Most Adequate Transport capabilities resolution path

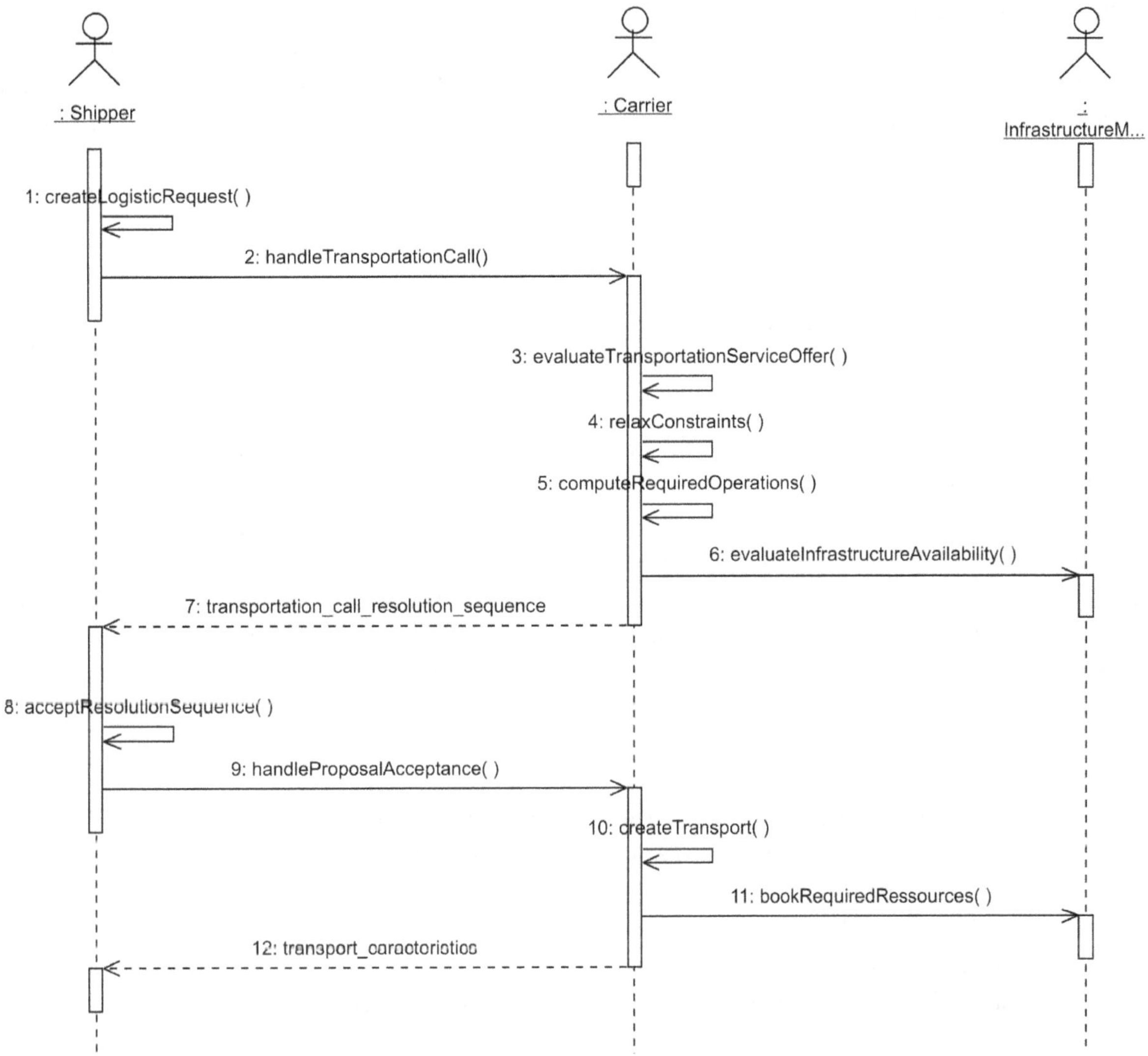

strategy for achieving agility in supply chain. They moreover highlight that this type of enterprise is made up of collaborating partners who are offering different products/services based on their core competencies:

As far as our collaborative platform is concerned, each of the adopting partners becomes, de facto, a member of a virtual enterprise composed of all the participating companies.

e-Commerce and SCM

(Kalakota and Whinston, 1996) define e-business as the establishment of a computer network to search and retrieve information in support of business decision making and inter-organizational cooperation. (Gunasekaran & Ngai, 2004) points out that it facilitates inter-organizational communication, reduces cycle times and develops collaborative work while (McIvor et al., 2003) argues that it is radically reshaping traditional supply chain structures in many industries and reduces costs of actors' integration:

e-commerce functionalities are provided by the software package. Indeed, data concerning demand, offer, orders, planning, localization, etc. is transmitted among the concerned users of the platform for efficient commercial relationships and e-collaboration.

Infrastructure for IT in SCM

(Gunasekaran & Ngai, 2004) emphasize that developing IT infrastructure requires investments in Internet services and other web development:

Technically speaking the collaborative platform deployment requires the installation of a "rich client" developed within the Eclipse platform with a centralized database. A theoretical study has concluded that even with an adoption of a huge amount of actors, data exchanges can easily be absorbed by a broadband internet connection. Collection of real time "ground" data has been the most important technical issue of the TransLogisTIC project because this involves GPS positioning and GPRS data transfers. Without entering into details, most of the technical problems have been addressed and real-time positioning information can be delivered as long as the tracked goods remain in an area covered by classical mobile phone networks.

Knowledge and Information Technology Management

Knowledge Management (KM) is concerned with recognizing and managing all of an organization's intellectual assets to meet its business objective:

The use of the collected data constitutes a "gold mine" for KM issues. However, this perspective has, so far, been poorly studied or discussed. Indeed, data availability issues are managed on an actor basis so that a single actor only has access to the information about its own business issues and the demand he can address i.e. the information that he legitimately has to dispose of. This information can already be used for KM issues but the use of whole of the information could give better results in the perspective of, for example, identifying key issues for global optimization. KM in the context of the collaborative platform remains thus an open issue.

Implementation Issues of IT in SCM

IT implementation issues appear to be critical:

The software platform development has been realized using Java object technology using the Eclipse platform but is based on a multi-agent engine. Each aspect of the software development has been documented using the i framework for social modeling but also classical UML models so that the documentation is complete, follows the industry standards and covers the whole development life cycle.*

RELATED WORK

Govindu and Chinnam (2007) propose a generic process-centered methodological framework called the *Multi-Agent Supply Chain Framework* (*MASCF*), to simplify MAS development for supply chains applications. Their idea is to map elements from the *Supply Chain Operations Reference* (*SCOR*) (Poluha, 2007, Huan et al., 2004) model to the Gaia development methodology (Wooldridge et al., 2000) for multi-agent supply chain systems development. Mapping the SCOR and the Gaia concepts is interesting for analysts but it supposes modeling supply chain processes using the semantics imposed by the SCOR. This leads to a custom development process in the particular context of developing MAS for SCM through the parallel adoption of the industry well known SCOR and the research-oriented Gaia methodology. So this would help integrating agent-oriented technologies in the development of SCM applications through already mature methodologies. The main positional difference with the framework presented in this paper is that we introduce a domain independent methodology that suits at best when developing large inter-organizational applicative packages. Our SE knowledge representation framework is not targeted to SCM since it does not directly incorporate a particular SCM semantic and can be applied to virtually any domain. Moreover, MASCF is essentially focused on the MAS design and (organizational) analysis is not covered. Consequently their approach for agent-oriented design of SCM applications covers too poorly the analysis of the software system.

Bechini et al. (2008) define SE patterns using UML and technologies for supply chain traceability when developing collaborative e-business solutions. Their conceptual framework allows to standardize the design process and the used communicational technologies when developing a collaborative platform in SCM as ours. Their contribution can be further used when bringing our developed prototypes to support an industrial load so that the platform finally meets ISO standards. Their framework is an interesting technological pattern for standardization purpose but does not constitute a finished or semi-finished software product.

(Premkumar, 2000) defines general guidelines for information systems (IS) development in SCM which do constitute a basic process for applications development. These guidelines however are located on various levels mixing strategic and operational issues. The approach is poorly structured but offers fragments of a process specially targeted to IS developments in SCM what the SE knowledge representation framework of this paper lacks to offer.

CONCLUSION

In the last 30 years, ERP systems ensured the integration of business processes among the functions of a single organization. A modern supply chain however involves series of actors issued of multiple companies sending and sharing the information required to properly manage and optimize the supply chain. To achieve the integration of these partnerships, a unique management IS going across a single organization's borders can be developed so that all of the participating actors are involved in a same virtual enterprise.

This paper has studied the development of a collaborative platform for OL through a framework for knowledge representation and strategic reasoning in SE. Lots of research work has been realized studying patterns for e-business and e-collaboration, defining data exchange standards or security norms but few work has illustrated the development of such a platform to indentify the concrete issues of induced technological innovations at various levels of abstraction. The work proposed in this paper is thus aimed at raising the level of integration of the OL actors to a higher extent by the adoption of a common technology. More precisely, we emphasize on:

- The *generic description of OL processes* (including strategic, tactical and operational aspects) that the platform is aimed at supporting;
- The platform's *supporting services to OL actors* description through the use of strategic, tactical and operational perspectives. This offers custom views to each of the involved stakeholders (top-managers, middle-managers, analysts, designers, developers, etc.). Among these views, the strategic one offers reasoning and planning elements for actors willing to adopt the platform to overview its potentials and threats.

Finally, a custom IT modeling framework has been used through the software development allowing reaching characteristics such as flexibility, openness and traceability. We point out here that:

- Service-oriented development promises to change the way business and technology vendors buy, sell, deploy, and manage application portfolios. For the first time, business users are be able to summon applications to support a business process rather than launch a business process constrained by the application. Indeed, the environment tends to be more and more competitive and rapidly evolving, flexibility in information systems functionalities is a natural trend of the market. That is why a supply chain actor requiring functionality to support any supply chain process directly accessible from a vendor's library in the form of a service could represent the business model for the coming years. Modeling the software problem in terms of high level services as performed here is from primary interest in such a context. By distinguishing such entities early on into the software project it is possible to reason about opportunities and threats of common and integrated IT support; this thus offers stakeholders a tool for evaluating the alignment of the supporting solution with the overall IT strategy;
- Agent-oriented modeling (notably through the i* framework) allows to represent the organization as well as stakeholders' expectations using concepts as tasks, goals and resources that actors are depender or dependee for. These concepts are closer to the organizational setting than purely computer science issued concepts. They consequently permits a better representation of the service that should be offered;
- Architecturally, the advantages of logically designing a MAS particularly in the supply chain field are better interaction handling through actors' autonomy and flexibility among service composition.

The prototype of the collaborative OL platform in the context of the TransLogisTIC project has reached its end. Further developments are required to bring this platform to support an industrial load, interface more third party information systems as for example each version of mySAP ERP. The platform prototype has nevertheless driven a lot of interest from various researchers in SE and its evaluation on the basis of SE metrics is still under progress. Similarly, a unified semantic proposal for OL will be proposed soon.

ACKNOWLEDGMENT

This work was carried out while the author was post doctoral researcher at the Louvain School of Management, Université catholique de Louvain, Belgium. The research has been financed by the Walloon region under the auspices of the TransLogisTIC project.

REFERENCES

Bechini, A., Cimino, M., Marcelloni, F., & Tomasi, A. (2008). Patterns and technologies for enabling supply chain traceability through collaborative information e-business. *Information and Software Technology*, *50*(4), 342–359. doi:10.1016/j.infsof.2007.02.017

Bellifemine, F., Caire, G., & Greenwood, D. (2007). *Developing multi-agent systems with JADE*. Hoboken, NJ: Wiley. doi:10.1002/9780470058411

Braubach, L., Pokahr, A., & Lamersdorf, W. (2005). Extending the capability concept for flexible BDI agent modularization. In *Proceedings of the 3rd International Workshop on Programming Multiagent Systems* (LNCS), (vol. 3862, pp. 139-155). Berlin: Springer.

Cavalieri, S., & Terzi, S. (2006). Proposal of a performance measurement system for the evaluation of scheduling solutions. *International Journal of Manufacturing Technology and Management*, *8*(1-3), 248–263. doi:10.1504/IJMTM.2006.008803

Govindu, R., & Chinnam, R. (2007). MASCF: A generic process-centered methodological framework for analysis and design of multi-agent supply chain systems. *Computers & Industrial Engineering*, *53*(4), 584–609. doi:10.1016/j.cie.2007.06.003

Gunasekaran, A., & Ngai, E. W. T. (2004). Information systems in supply chain integration and management. *European Journal of Operational Research*, *159*(2), 269–295. doi:10.1016/j.ejor.2003.08.016

Gunasekaran, A., Patel, C., Ronald, E., & McGaughey, R. (2004). A framework for supply chain performance measurement. *International Journal of Production Economics*, *87*(3), 333–348. doi:10.1016/j.ijpe.2003.08.003

Huan, S. H., Sheoran, S. K., & Wang, G. (2004). A review and analysis of supply chain operations reference (SCOR) model. *Supply Chain Management -. International Journal (Toronto, Ont.)*, *9*(1), 23–29.

Kalakota, R., & Whinston, A. B. (1996). *Frontiers of electronic commerce*. Reading, MA: Addison-Wesley.

McIvor, R., Humphreys, P., & McCurry, L. (2003). Electronic commerce: Supporting collaboration in the supply chain. *Journal of Materials Processing Technology*, *139*(1-3), 147. doi:10.1016/S0924-0136(03)00196-1

Pache, G., & Spalanzani, A. (2007). *La gestion des chanes logistiques multi-acteurs: Perspectives stratgiques*. Presses Universitaires de Grenoble.

Pokahr, A., Braubach, L., & Lamersdorf, W. (2005). Jadex: A BDI reasoning engine. In *Multi-agent programming: Languages, platforms and applications* (pp. 149–174). Berlin: Springer. doi:10.1007/0-387-26350-0_6

Poluha, R. G. (2007). *Application of the SCOR model in supply chain management*. Cambria Press.

Premkumar, G. P. (2000). Inter-organizational systems and supply chain management: An information processing perspective. *Information Systems Management*, *3*(17), 56–69.

Samii, A. K. (2004). *Stratgie logistique, supply chain management: Fondements - méthodes - applications*. Dunod.

Terzi, S., & Cavalieri, S. (2004). Simulation in the supply chain context: A survey. *Computers in Industry*, *53*(1), 3–16. doi:10.1016/S0166-3615(03)00104-0

Wautelet, Y. (2009). *Development of a multi-actor e-collaboration platform: business work unit 3, tasks 3.4.1 and 3.4.2*. Université catholique de Louvain.

Wautelet, Y., Achbany, Y., & Kolp, M. (2008). A service-oriented framework for mas modeling. In *Proceedings of the 10th International Conference on Enterprise Information Systems* (pp. 120-128). Barcelona, Spain: IEEE.

Wautelet, Y., Kolp, M., & Poelmans, S. (2013). Requirements-driven iterative project planning. In *Proceedings of Software and Data Technologies, 6th International Conference*. Berlin: Springer.

Wooldridge, M., Jennings, N., & Kinny, D. (2000). The gaia methodology for agent-oriented analysis and design. *Journal of Autonomous Agents and Multi-Agent Systems*, *3*(3), 285–312. doi:10.1023/A:1010071910869

Yu, E. (1995). *Modeling strategic relationships for process reengineering*. (Doctoral Dissertation). University of Toronto, Toronto, Canada.

Yu, E. (1997). Towards modeling and reasoning support for early-phase requirements engineering. In *Proceedings of the 3rd IEEE International Symposium on Requirements Engineering* (pp. 226-235). Annapolis, MD: IEEE.

Yu, E., Giorgini, P., Maiden, N., & Mylopoulos, J. (2011). *Social modeling for requirements engineering*. Cambridge, MA: MIT Press.

Chapter 15
Why Information Systems Replication Strategy Fails in Transnational Operation

Angela Lin
University of Sheffield, UK

Shin-Horng Chen
Jinwen University of Science and Technology, Taiwan

ABSTRACT

In search of cheaper resources and being able to serve nearby market more efficiently in order to maintain and improve their market position in a global economy, companies relocate, extend, or establish new production facilities overseas. Because of this, how to manage the Information Systems (IS) that support transitional activities within companies has become an important agenda not only in the business but also in the academic community. At a glance, replicating information systems and value chain that a company already has in the new facility overseas is seemingly a cheap and easy option. Nevertheless, in practice, it is not a trivial task, and at its worst, the IS can become obsolete. This chapter explains why replication strategy does not always work, and drawing on the practice lens approach, it posits that any changes in technology-in-use are the outcome of an ongoing structuring process where people constitute and reconstitute the structure of an IS in use. The chapter demonstrates that the structure of an IS is not defined a priori but emerges from daily use of the system, along with people's understandings of the system, and of its role in the context of organisational routines. On this view, any challenges arising from the attempt to replicate the IS can be regarded as largely inevitable, since the structures embedded and enacted in the use of system, along with the practices that in turn recursively structure the use of the system would also have changed.

DOI: 10.4018/978-1-4666-4506-6.ch015

INTRODUCTION

A distinctive characteristic of a global economy is the increasing number of transnational activities that are fuelled by fundamental economics of comparative advantage and by competition at both local and global level (King & Sethi, 2001). Companies relocate, extend, or establish new production facilities overseas in search of cheaper resources or being able to serve nearby markets more efficiently. Information systems (IS) no doubt play a significant role in supporting multinational corporations (MNCs) to perform transnational activities in particular coordinating value chain activities and information flows within and between companies. The strategic alignment literature emphasizes that the design of IS and the management of IS resources should reflect a company's transnational activities (Ives, Jarvenpaa, & Mason, 1993). For example, in the case where value chain configuration is high a company tend to replicate their existing value chain as well as the existing IS that have already been used to support the value chain activities in its overseas operation (Chow, 2004; Markus, Sia, & Soh, 2012; Niederman, Alhorr, Park, & Tolmie, 2012). Evidence however suggests that the replication of an IS from one location to another is seldom as straightforward as one might expect. The system interface may have to be changed for example, in order for it to be usable and accepted by local users (Burton-Jones & Gallivan, 2007; de Guinea & Markus, 2009); business practices supported by the system may have to be redesigned in order to accommodate local culture and work practices which in turn may affect system use (Hsiao, 2007); while specific rules and procedures built into the system may have to be modified, so that the system abides by local governmental directives (Zhu & Kraemer, 2005; Zhu, Dong, Xu, & Kraemer, 2006). Without addressing any required changes companies risk generating a quite different output from and ultimately non-acceptance of an information system in its new location.

This chapter aims to contribute to the existing knowledge and understanding of global IS management by demonstrating that managing and replicating value chain and an IS that supports it in MNCs is not a trivial task and at its worst the IS can become obsolete. To achieve this aim a qualitative case study was carried out to examine the process of replication of an e-procurement system within an MNC in Asia.

The chapter is organized as follows. The next section outlines the theoretical background to the study and introduces the concept of a practice lens. This is followed by a description of the research setting along with the research design, data collection and data analysis methods used. Section four presents the study's findings. The paper concludes with a discussion of the issues arising from the study and their implications for global information systems management.

RESEARCH PERSPECTIVES

Theoretical Background

Business-IS strategic alignment has long been argued as a critical success factor for successful IS adoption; and the studies of global IS management also stress the importance of such alignment (Ives, Jarvenpaa, & Mason, 1993; Katz & Townsend, 2000; King & Sethi, 2001; Mayasandra, Pan, & Leidner, 2011; Preston & Karahanna, 2009). Bartlett and Ghoshal's (1989) work on typology of MNCs is widely used in the research of global IS management to understand MNCs' overseas operation strategies and practices that are organized according to these strategies. Bartlett and Ghoshal (1989) categorise MNCs into four categories including: international, multinational global and, transnational corporations according to strategic role that overseas operations play in these companies. International corporations tend to view their overseas operations as distant outposts whose role is to support the parent company by contributing to sales to the parent company's manufacturing operations; and expanding parent's company's manufacturing operations in overseas. Technology and knowledge are transferred from the parent company to overseas sites; and decision making is usually done by the parent company in distance. Multinational corporation recognises and emphasizes the differences among national markets therefore they tend to develop more responsive strategies that reflect the local cultures and demands. The local operators are considerably more independent from the parent company and have freedom to make decisions that response to the local situations. Corporations with global strategy tend to view the world as a single market rather than multidomestic markets and assume that the similarities among national tastes are more than the differences. Global strategic approach requires more central coordination and control than others and research and development and manufacturing activities and decisions are managed from the centre. Transnational strategy is the hybrid of multinational and global strategies. It responses to the local needs yet the same time it captures the benefit of global efficiency, for example, each overseas operator can carry out its own tasks and make decision on a local-for-local basis but the headquarters may control and distribute the resources and manage specialised activities.

Building on Bartlett and Ghoshal's (1989) work, the typology of MNCs various IS strategic alignment model for MNCs have been put forwards to better understand the link between business and technology in light of the international business (Jarvenpaa & Ives, 1991, Katz & Townsend, 2000; King & Sethi, 2001; To, Liao, & Liu, 2004). Campbell et al. (2005) posit that for each type of MNCs there is a corresponding distinctive business strategy and IT strategy. Jarvenpaa and Ives (1991) identified a number of approaches for managing global IT which are aligned with the four business strategies proposed by Bartlett and Ghoshal (1989): independent global IT operations, headquarters driven global IT, intellectual cooperation in global IT, and integrated global IT. Jarvenpaa and Ives (1991) argue that independent global IT structure suits multinational strategy better because it allows local operators to choose the technology that reflects the influence of local software and hardware vendors and prevailing of the national communication standards and offerings. It focuses on local responsiveness and emphasizes the local requirements. Headquarters driven global IT would be an approach for global MNCs who strive for global efficiency through standardising procedures and centralising the management of the resources. Global MNCs expect to benefit efficiency from coordination and control through centralised IT. Intellectual cooperation in global IT seems to fit with international MNCs where the headquarters seems to influence rather than control the choice of IT in the overseas operator. The headquarters and overseas operator may jointly develop the system or the latter may ask the former to provide the system that is already in place. Finally, integrated global IT meets the needs of transnational MNCs by allowing some IT solutions to be tailored to meet the needs of local business and yet some solutions are developed, used, and shared across the all overseas subsidiaries. Using Bartlett and Ghoshal's (1989) and Jarvenpaa and Ives' (1991) work To et al. (2004) found that users are more likely to feel satisfied with the technology where there is business-IS fit and less so when there is misalignment between business and IS strategy and structure.

Business-IS strategic alignment in the multinational context helps identify the key issues for global IT management and draw attention to the differences in business strategy and organisational structure between a MNC and national company. Nevertheless some important aspects have been underplayed in the business-IS strategic alignment literature. First, the literature focuses more on achieving business and IS equilibrium within MNCs and less on the impact of contextual conditions (e.g. socio-economic, socio-political, and business environment) on IS adoption (Avgerou, 2008; Hsiao, 2007; Hsu, Kraemer, & Dunkle, 2006; Li, Lai, & Wang, 2010; Zhu & Kraemer, 2005). As a result the studies tend to assume that once business-IS alignment is achieved the equilibrium between the two will continue. However some further studies have advocated that business-IS strategic alignment is not a permanent but a provisional state (Furneaux & Wade, 2011; Majchrzak, Rice, Malhotra, King, & Ba, 2000; Strong & Volkoff, 2010). In practice, it is being argued that a decision to adopt any information system is situated within a specific business environment and reflects organizational as well as user expectations of the system at a particular point in time. In order to respond to a dynamic business environment MNCs need to readjust their strategy and practice constantly and these adjustments for example change the requirements for the information system (Furneaux & Wade, 2011; Karahanna, Straub, & Chervany, 1999; Kim & Malhotra, 2005; Zammuto, Griffith, Majchrzak, Dougherty, & Faraj, 2007). This explains why it is, in general, difficult to implement business-IS alignment and even more difficult to do so in a multinational context. And it also explains why an IS can become obsolete before the end of its life cycle because it is no longer being able to meet a company's strategic needs or support a company's transnational activities.

Second, the importance of structural properties of a system in global IS management has not been addressed in the global business-IS strategic alignment literature. The structural properties of a system include a set of specific social practices, and rules, and once they are built into the system they exist independently of the users of the system and cannot be changed by the users. The structural properties of a system can be in conflict with those of the adopting organizations in a different socio-political or socio-economic setting (DeSanctis & Poole, 1994; Soh & Sia, 2004; Volkoff, Strong, & Elmes, 2007). For example problems may arise when there are conflicts between a system designed in one cultural setting and the use of the system in another cultural setting (Kumar, van Dissel, & Bielli, 1998). Hence it would be naïve to believe that a system designed in one socio-economic context and for a specific set of business practices can simply be emulated and implemented in another context; with the hope that the system will achieve the same results (Avgerou, 2001).

Third, although it is recognised that transnational operation is complicated the studies seem to take the status of a company as given and assume that it would remain the same. In reality companies may grow in size and inevitably change the way in which they engage in transnational activities (Sambharya, Kumarawamy, & Banerjee, 2005). For example when a company takes a decision to move their production facilities overseas, its requirements for both value chain management and IS that supports the value chain may have already changed; therefore the decision to replicate its IS in overseas site although may be a quick solution it may not be appropriate and could cause the gaps between the new practices and practices supported by the system.

From the above it is apparent that global IS management should not focus simply on business-IS strategic alignment but beyond. To better manage global IS one should consider context in which IS is adopted, practice which IS support, and structural properties embedded in IS which impose constraints on practice. These three elements should be examine as a whole rather than as parts so that a fuller picture of how an IS that works well previously can become obsolete can be understood beyond business-IS misalignment. This study chose a practice lens approach to IS use as a theoretical foundation for investigating the process of IS adoption in the multinational context and how the changes in practices can lead to IS obsolescence.

Practice Lens: Technology-in-Practice

The practice lens analysis to IS use was first developed to examine how when interacting with a technology, people enact structures that shape their actions towards and situated use of that technology (Feldman & Orlikowski, 2011; Orlikowski, 2000). Informed by structuration theory (Giddens, 1984) the practice lens approach stresses the duality of social structure and the recursive nature of social practice. On this view the structure of an IS is understood as a set of rules and resources instantiated in recurrent social practice (Orlikowski, 2000); and is a structure that is not external to social action. Hence the practice lens analysis posits that people's interaction with and making sense of system draws not only on the technical properties of the system itself but also on the interpretive schemes that people bring to the interaction e.g. their knowledge, assumptions, and expectations of the system. As a consequence, people's interaction with the current system and the meaning that the system will have for them often depend on their previous experiences with other similar systems (Orlikowski & Gash, 1994). People also draw on their experiences and knowledge of the institutional context in which they work, and the norms associated with acting in that context. Any situation of system use is therefore structured not only by the technical properties of the system, but also by the meanings that the system has for people, and their experience and knowledge of organizational norms. Each time people interact with the system this structuring process enacts a specific set of rules and resources, with each interaction structuring future use.

Structure is enacted within practice and is not only 'constituted by' but is also 'constitutive of' social organization. The structure of an information system is "constituted recursively as humans regularly interact with certain properties of a technology and thus shape the set of rules and resources that serve to shape their interaction" (Orlikowski, 2000, p. 407). Accordingly the structure of an IS is not a static property but as a part of social practice which is continuously produced and reproduced over time. On this view the practice lens analysis views that the use of IS is malleable, with the nature of this use potentially altered by the day-to-day actions of those involved in interacting with the system. On the other hand the stability of a system can be established and its continued use maintained by repeated interaction with the system and by the use of recurrent social practices. Technology-in-practice can therefore emerge and act as a behavioral and interpretive template for the situated use of an information system. Over time the use of the system can become taken for granted and invisible yet structuring people's interaction with the system. Any stable state should be regarded however as provisional with any stable template for use subject to changes resulting from people's, experience and knowledge of the system, and their interpretations of the system in use, and of the organizational context within which they work. In response to any of these changes, people may either alter or request alterations in the system; they may draw on different interpretive schemes in the enactment of the system, or less likely adapt organizational routines. In each case the structure of the information system i.e. the nature of the technology-in-practice, changes in accordance with any changes in technical system, interpretive schemes or organizational routines.

The practice lens analysis may appear to emphasize the local context of system use. One should not mistakenly think however that it is only the local context that is considered to have any influence on the technology in use. Other contexts that also play a role in the enactment of system are overlooked. Monteiro and Rolland (2012) argue for example that the idea of context in the practice lens approach is open-ended and includes not only interpretive and organizational contexts but also historical circumstances, and environmental influences (e.g. technological, political, cultural, competitive). These different contexts can trigger changes in the situated use of an information system. As mentioned earlier, the practice lens approach also assigns considerable importance to a temporal dimension. On the practice lens view time is defined neither objectively (e.g. clock time) nor subjectively (e.g. personally constructed), but is a property that emerges from varied ongoing social practices and which is used to make sense of, regulate, coordinate, and account for people's activities (Orlikowski & Yates, 2002). The importance of the temporal dimension is further emphasized by Chu and Robey (2008) who argue that it is only with reference to the temporal dimension that we can explain what and when which aspect of the context plays a role in the use of the system.

This practice lens analysis was used in this research to study the process of replicating a value chain and IS that support it for two main sets of reasons. First, the approach directs attention to the process of structuring, where people constitute and reconstitute the structure of the information system i.e. the system in use. In doing so the stability of the technology-in-practice is provisional and can be changed when people experience changes in local circumstances or external environments. This notion also provides an analytical tool with which to identify the specific changes that trigger shifts in the structure of the information system and which can lead to its subsequent obsolescence. Although these changes can be identified individually, their influence on the technology in use should be taken as a whole and considered holistically; due to the fact that "the enactment of a technology-in-practice is situated within a number of nested and overlapping social systems, people's interaction with technology will always enact other social structures along with the technology-in-practice" (Orlikowski, 2000, p. 411).

Second, the argument that any current template of technology-in-use is structured by previous practices of technology-in-use is significant as it enables the former to be viewed within a 'historical' context, and within which a trajectory of technology-in-use can be placed. Rather than making a simple inference of obsolescence from observed circumstances of non-use, the obsolescence of an IS can instead be analyzed as and being seen to be the endpoint of a discontinuous series of changes within and external to the practice of technology in use. The temporal and contextual dimensions of the practice lens approach enhance such analysis. The temporal dimension of the practice lens approach helps us explain not only what and when which aspect of the context plays a role in the use of the system, but also the significance and impact on the shared practices and use of the system. The temporal dimension also enables us to display the structuring process and to identify a trajectory of technology-in-use. The contextual dimension takes not only local organizational contexts into account, but also external contexts such as sectorial, national, and international contexts. Both these contexts are considered to contributing to the norms that people draw on in their interactions with the system; and which gives form to other social structures enacted by the users of the system. In sum the structure of an IS, the technology in use, is not 'independent' of its environment, but is and structured and enacted in relation to other social practices.

From the above it is apparent that the practice lens provides analytical capacity to assist us to capture the emergence of structure of system use which is formed and reformed through people's constant interaction with the system within a number of nested and overlapping social systems. The practice lens analysis addresses the issues that have been overlooked in business-IS strategic alignment approach to global IS management and captures and offers the details of how the same technology that has been used in the company cannot be simply replicated in a different location.

RESEARCH METHODS

This study uses the case study method. This is for a number of reasons. First, the case method is a preferred method for answering 'how' and 'why' questions (Yin, 2009); and as such is appropriate for answering how and why an IS, previously proved to be useful, is unable to achieve similar results, or even become obsolete, after replication. Second, answering the research questions guiding this study requires an in-depth description of the structuring process of the system in use, and therefore a method that has ability to deal with a variety of evidence is preferred. Third, the case method enables the researcher to study and understand a social phenomenon in its own context and from the participants' point of view (Klein & Myers, 1999; Walsham, 1995; Walsham, 2006). A single case study was conducted to investigate how a well-established and used information system later fell into disuse after being replicated and reintroduced to other sites within a company. The remainder of this section describes the research setting and discusses the procedures used to collect and analyze the data.

Research Setting

This case study is concerned with an e-procurement project initiated, implemented, adopted, and replicated within NoteCom (all names presented here are pseudonyms). NoteCom is a notebook manufacturer based in Taiwan which provides original design manufacture (ODM) and original equipment manufacture (OEM) services to global computer companies such as Acer, Dell, HP, IBM and Toshiba. NoteCom's decision to implement an e-procurement system that would computerize its supply chain management and connect its suppliers was influenced by an increasingly intensive competitive environment emerging from other countries. Instead of developing a system from scratch and customizing it to support its procurement practices NoteCom chose an internationally renowned standard software package called SupplyLink. The initial introduction and implementation of SupplyLink within NoteCom and among its suppliers was successful. Since NoteCom hoped to continue to use the system to coordinate procurement activities between itself and its suppliers, the system was also introduced to its new operational sites in China. Although the system was successfully adopted in the headquarters in Taiwan the replication of the system in China did not work as efficiently as expected and was only used by most suppliers to coordinate administrative matters with NoteCom on an ad hoc basis. NoteCom was also chosen as the case for a number of reasons. First, NoteCom is a leading computer manufacturer and has abundant experience and knowledge of its sector. Second, it has been using a supply chain management system for sometime. The experience of using the system for a lengthy period of time is important since it allows the study to examine the organization's use of the system over time. Third, after adopting SupplyLink NoteCom became internationalized by expanding its operations across border to China. The strategic role of its overseas operation is simple: to support the headquarters' production. Fourth, NoteCom is a good example of an IS that, after being extensively used for the purposes for which it was designed, has since fallen into disuse, upon attempted replication across the company.

Data Collection and Analysis

The study reported here is part of a broader ongoing research project investigating the relationship between business practices and IS use within NoteCom. Data collection began in March 2004 and continued until 2009. Semi-structured interviews were the primary data collection method. Interviews were conducted either face-to-face or via teleconference. Interviewees consisted of employees working in the Research and Development (R&D), Procurement (both in Taiwan in China), and Management Information Systems

(MIS) departments of NoteCom. Interviews with the procurement staff focused on procurement practices and related issues before and after cross-border operations; interviews with the R&D staff focused on supplier selection and the process of product development in NoteCom. The MIS department provided information about SupplyLink's adoption and implementation at its production sites and in its suppliers in China as well as information about significant changes in procurement practices. A Vice President of the Global Logistics Centre in NoteCom, who managed procurement operations in China, also provided additional insights into a number of issues relating to procurement practices both within the company and among its suppliers in China. Key suppliers were also interviewed to understand their own interactions with NoteCom. Inventory management specialists were also interviewed as a data source for exploring the Vendor Management Inventory model (VMI), a business model currently employed by those logistics providers who offer services to NoteCom's Chinese operations. In addition to the formal interviews that were conducted, participatory observations were also undertaken through attendance and participation in NoteCom's procurement meetings. Participation in these meetings enabled the researchers to learn and understand about how the procurement system incorporated a Bill of Materials (BOM), how procurement conflicts were resolved, and how inter-organizational collaboration occurred *in situ*. Organizational documents were also collected in order to understand the project better.

Data was analyzed iteratively in three steps. The first step consisted of identifying and understanding events critical to the implementation, adoption, and attempted replication of SupplyLink. The second step of the data analysis focused on examining actual procurement activities. Guided by a practice lens approach, the study identified three core activities in the procurement practices: supplier selection, material purchasing, and delivery. The third step of the data analysis consisted in examining how changes in procurement practices impacted on the use of the SupplyLink system. This entailed the researchers identifying the pattern of practices within each phase, and demonstrating incompatibilities which occurred between procurement practices and technology-in-use.

DATA ANALYSIS AND FINDINGS

This section is organized into three parts. The first two parts identify the contexts within which NoteCom operated and the significant changes in those contexts that NoteCom encountered. The third part identifies the changes in technology-in-use within NoteCom.

Before Cross-Border Operations

Prior to 2003 NoteCom's business practices and operations were structured by the global consumer computer market and by sectorial conditions. At the time the consumer computer market was largely dominated by personal computers (PCs) and most innovation and design of hardware components and peripherals were developed for standard PCs. At the time NoteCom was faced with at least two technology-related challenges which underpinned their interactions with, and which led to close collaboration with, their suppliers. The first challenge was to overcome a gap in technical knowledge. Generally speaking, notebook technologies were still in their infancy and NoteCom received limited technical know-how for product design and development through knowledge transfer from its customers. As a result, NoteCom was bound to collaborate with its suppliers in order to develop the necessary technical know-how and product design needed to bridge the company's technical knowledge gap. The second challenge came from the lack of hardware components that were specifically designed and developed for notebooks. Since most hardware components used for notebook production were for PCs, engineers from within the R&D department and from the suppliers spent a good deal of time working together to repeatedly test each notebook component until it could be used. In addition to this they also needed to collaborate on developing hardware interface protocols that enabled communication to take place between the components.

Given the immature stage of development in the notebook sector the procurement procedures at NoteCom were structured by product research and development activities; with NoteCom choosing their suppliers on the basis of their technical competence rather than any other criterion e.g. cost. Suppliers were expected to provide not only hardware components but also to collaborate with NoteCom working on technical solutions and developing prototypes. As one R&D engineer explained:

When developing a product prototype, we do not need 'cheap' (i.e. offering low costs) but 'competent' suppliers. This is important because it is suppliers' technical competence and the quality of their products that will ultimately determine the quality of our products. Besides, they need to be able to work with us, understand our designs, and respond quickly to any engineering changes as problems emerge. (R&D Engineer A.)

The R&D department therefore played the role of assessing and selecting suppliers in the procurement process; with engineers within the R&D department charged to visit suppliers on a regular basis in order to assess their production capacity, quality control, operations management, materials distribution, technical services etc. Three suppliers were then selected for each hardware component. The Procurement Department's responsibility was then to rate the suppliers chosen by the R&D department and to decide the quantities that would be purchased from them. By not relying on any one supplier NoteCom could thus ensure continuity of supply.

NoteCom's procurement model 'purchasing via forecast and delivering just in time (JIT)' was built on the general built-to-order business practices current in the sector. This model required suppliers to work closely with NoteCom, forecasting the required inventory for example in preparation for subsequent production capacity planning and scheduling. The lead-time was one month for standard components and three months for special components. With this procurement model NoteCom aimed to fulfill an order as soon as possible, and in any case before its customers had an opportunity to revise their orders (e.g. to reduce quantities). NoteCom's procurement model also enabled the company to keep a minimum level of inventory on site and to shift responsibility for inventory management on to its suppliers. Suppliers were asked to assure the quality of their supplies in order to avoid any delays in production caused by poor quality hardware components. The forecast management function was therefore built into SupplyLink and played an important role in ensuring that appropriate levels of inventory were maintained, and suppliers could provide hardware components in time for production. This process was supported by the delivery management function in SupplyLink which assisted suppliers in managing the logistics of an order e.g. shipping, delivery notice, and certificate of receipt. In order to ensure that supplies could be delivered to NoteCom in time for production, suppliers were also encouraged to base their production facilities close to the company.

Figure 2 illustrates how SupplyLink was used to support NoteCom's JIT procurement practice; and how the structure of the information system, the system in use, was structured and reinforced in practice.

After Cross-Border Operations

Post 2003 the notebook sector experienced a technology breakthrough and growing consumer demand. This growing demand for notebooks encouraged major microchip fabricators such as Intel, AMD, VIA, and SIS to invest in the research and development of hardware components specifically for notebooks. The breakthrough in technologies included new central processing units (CPU) for notebooks, standardized hardware interface protocols (and therefore fewer problems incurred in integrating chipset drivers), along with the introduction of modular chipset designs. The microchip fabricators also issued official guidelines for the design of notebooks and their hardware components. One consequence of the issuing of these guidelines was the standardization of hardware components, including peripheral devices, across the sector. These general breakthroughs in notebook technology helped to improve a situation that had previously been lacking in identifying the appropriate components for notebook production and address the knowledge gap in the sector. These changes in the business environment subsequently triggered a change in NoteCom's customers' requests, and led to a restructuring of NoteCom's R&D activities, as well as alterations in NoteCom's relations with their suppliers, and the roles within the procurement process of the R&D and Procurement Departments.

As mentioned, the industry standardization in notebooks and the surge in market demand triggered a change in NoteCom's customer requests. Prior to this, its customers' main concern had been product reliability; after standardization their main concern was NoteCom's production capacity and whether NoteCom was able to meet an increasing market demand for its products. Besides this the price of NoteCom's offerings became a key consideration for its customers, as the quality and design of notebooks in the sector gradually became standardized. In an effort to scale up its production, to meet market demand and to keep its prices competitive, NoteCom took the decision to expand their production facilities overseas, and thereby access cheaper supplies and labor.

The standardization of hardware components and the interface protocols reduced technical complexities thus instead of focusing on troubleshooting the R&D department could concentrate on product design and innovation. Perhaps the most significant change brought by standardization and the new generation of CPUs was the change in NoteCom's relations with its suppliers. As a result collaboration with its suppliers on problem solving and product development became less important, because breakthroughs in the technology and the design guidance issued by the microchip fabricators solved most of the technical problems that had previously been encountered. Further to this, NoteCom also received additional technical support from the microchip fabricators addressing problems of chipset design. NoteCom's partnership with its suppliers therefore shifted from being an innovation and collaboration-based relationship focused on product development and knowledge creation to a transaction-based relationship focusing on product quality and price. This subsequently led to a change in NoteCom's criteria for assessing and selecting its suppliers: from a single criterion emphasizing technical competence to multiple criteria and an overall score of TQRDC (T: technical; Q: quality; R: responsiveness; D: delivery; C: cost). This change in the method of selecting suppliers led to further changes within NoteCom. First, responsibility for assessing and choosing suppliers was re-allocated to the Procurement Department. Second, some activities in the company were as a result redefined. For example due to the complications arising from the use of a different component the replacement of a hardware component prior to 2003 would have entailed a series of intensive interactions between R&D engineers and suppliers. Post 2003, the replacement of a hardware component simply entailed the finding of an alternative supplier who could offer the same product at a lower price.

NoteCom introduced and continued to use its procurement model, along with SupplyLink, at its new overseas production facilities. Existing procurement practices had been organized centrally where the Procurement Department consolidated, placed, and arranged delivery with its suppliers. It was soon evident however that such centralized procurement practices could not efficiently and effectively supply the series of distributed production sites in China. As a result a decentralized procurement practices emerged in which responsibility for procurement, e.g. arranging deliveries with suppliers and monitoring their performance, was delegated to particular production sites, These decentralized procurement practices gave greater flexibility and scope to each of the production sites, enabling them to make purchasing decisions and to arrange deliveries with suppliers in response to local needs.

NoteCom also found that in response to the changing business environment it also had to modify its own procurement practices so as to be able to support the operation of its JIT procurement model in the new environment. Assumptions about forecasting and delivery times for example which were embedded in SupplyLink had been based on previous conditions where the distance between NoteCom and its suppliers was short. These built-in assumptions became unrealistic and started to impede the procurement process. Consequently NoteCom began to use a mix of 'near-delivery' and 'far-delivery' modes as a way of managing their inventory. In near-delivery mode NoteCom worked with those of its suppliers who were based near to its production sites to control 20 per cent of its inventory and to ensure a one-day inventory stock level on each site. In far-delivery mode NoteCom outsourced its inventory management to logistics providers who in turn manage the remaining 80 per cent of NoteCom's inventory. NoteCom worked with two logistics providers and used their Vendor Management Inventory (VMI) system to monitor inventory levels. The VMI system further changed the dynamics between NoteCom and its suppliers, with any interactions between NoteCom and its Suppliers regarding inventory replenishment now mediated through the VMI system and its logistics providers.

Figure 3 illustrates the changes brought about in the technology used to support procurement in NoteCom; from SupplyLink to support JIT practice to the VMI system used to support the VMI business model. A further significant change is NoteCom's use of third party warehouses to support their activities.

Technology-in-Practice

Technological breakthroughs within the notebook sector and demand the in global consumer computer market have since 2003 triggered changes in NoteCom's business practices in particular its procurement practice and which in turn changed its collaborative relationship with suppliers. Table 1 summarizes the environmental changes that led to changes in NoteCom's own business practices and the latter further led to changes in its relationship with suppliers.

Changes in Collaborative Relationship

Prior to 2003 the collaboration between NoteCom and its suppliers can be described as obligational relationship. The selection for suppliers empha-

Table 1. Environmental changes leading to changes in NoteCom's business practices

Changes	Led to the Expansion of Production Capacities Overseas	Led to Changes in Business Practice After Cross-Border Operation
Environment	• Surge in consumer demand for notebooks; • Standardization of products; • Market competition	• Distributed production sites
Procurement practice	• Shift from technological to cost driven; • Need for increase in production capacities	• Shift from centralized to decentralized organization; • Use of intermediary: third party logistics providers; • Use of VMI inventory management; • Use of near-delivery and far-delivery modes
Collaborative relationship	• Obligational relationship; • Tight-coupling	• Arms-length relationship; • Loose-coupling

sized suppliers' technical know-how, and capacity and ability to help NoteCom with product development. Suppliers on the other hand recognize their role as a collaborator helping NoteCom with product development. Due to the technology breakthrough in notebook computers the collaborative relationship between NoteCom and the suppliers changed from obligational to arm-length relationship (Klein & Rai, 2009; Schultze & Orlikowski, 2004). The new selection process and criteria emphasized the objective conditions of suppliers instead. A procurement staff stated:

The relationship between suppliers and NoteCom is different now. In the past, procurement was linked with R&D. Suppliers were selected by R&D staff and. Suppliers were decided if their products were used by NoteCom for R&D, and then followed by price negotiation. But now things have changed. Since there are so many suppliers offer the same products and they would all be very keen to do business with us, we TQRDC ranking to select the best suppliers and assess their performance every six months or even every quarter.

It is interesting to note that arm-length relationship focuses on objective criteria and emphasize on short-term and contractual relationship. Table 2 summarizes the changes in the relationship.

Coordination between NoteCom and its suppliers was also redefined because of the new business practice. Prior to 2003 NoteCom employed zero inventory policy, JIT, to manage its production materials to reduce the cost. Then coordination between NoteCom and suppliers in terms of supplying materials for production can be described as tight-coupling which means that if there is any changes in the production plan the procurement staff had to inform the suppliers immediately in order for them to respond to the change accordingly (Danneels, 2003; Rank, Robins, & Pattison, 2010). Suppliers were also expected to take the initiative to learn about NoteCom's production status and make quick response

Table 2. Structure of supplier selection

Period	Structure Nature	Notes on Procurement Activity
Prior to 2003	Obligational relationship	• Suppliers were chosen by R&D staff. The procurement relationship is based on suppliers' technical know-how and capacities not just their products; • NoteCom depended on suppliers' collaboration on product development and relied on their assistance for problem solutions; • NoteCom established the long-term collaborative relationship with suppliers to assist them with product development and problem solving.
Post 2003	Arms-length relationship	• NoteCom chose suppliers on the basis of objective criteria; • NoteCom could choose from a large pool of suppliers rather than relied on few; • NoteCom established short-term relationship with suppliers to supply them with hardware components. Suppliers might be replaced if they failed to make to the top of NoteCom's TQRDC ranking.

to the situation and dispatch the needed materials. One of the procurement staff in NoteCom mentioned:

When we had a rush order from our customer we would immediately inform our suppliers to prepare the materials on standby and ship them to us when we needed.

After cross-border operations NoteCom employed VMI to manage its raw material inventories. Suppliers' were alerted by the VMI system when the inventory level is below the threshold and they would deliver the materials to the VMI warehouse; and NoteCom would get the materials needed whenever it needed for the production. The direct and tight-coupling relationship between NoteCom and suppliers hence became loose-coupling which coordinated by a third-party logistics provider (see Table 3).

Technology-in-Use

Due to the fact that SupplyLink was designed in the pre-2003 organizational and business context, it naturally imposed constraints on any

Table 3. Structure of the coordination between manufacturing and logistics

Period	Structure Nature	Notes on Procurement Activity
Before cross border operation	Tight-coupling	• NoteCom employed JIT model to manage its raw material inventory; • Timely coordination between procurement staff and suppliers was important; • Suppliers were required to take the initiative to monitor NoteCom's inventory level and respond to it accordingly.
After cross border operation	Loose-coupling	• VMI model was adopted to manage raw material inventory; • Suppliers were alerted and responded by the VMI system to replenish the materials.

new post-2003 business practices. In the new procurement practices selection of suppliers was based for example on a TQRDC score and was cost driven. The system was also expected to allow the Procurement Department at any time to use alternative suppliers who could provide quality products at a lower cost. In practice the information architecture, flow, and contents remained much the same as they were prior to 2003 when system was initially designed and adopted. For example the selection for suppliers were still technology focused and emphasized obligational relationship between NoteCom and the suppliers in the system; and it only allowed the placing of orders with those who were listed on the bill of materials (BOM). Adding new suppliers to the BOM was not a trivial task since each time the name of a new supplier was added to the BOM, the MIS Department would need to switch off SupplyLink and make a series of changes in its data structure and metadata. Mistakes were sometimes made during these changes and as a result of frequent in accurate changes SupplyLink gradually lost its ability to generate accurate information. Users began to complain that due to these inaccuracies SupplyLink increased their workload and impeded their productivity. Procurement staff would need to contact their suppliers directly via telephone and fax machine for example, in order to confirm their orders.

SupplyLink, which was designed to support centralized procurement, practices and a single production site was unable to support the new decentralized procurement practices. Assumptions built into the system such as centralized placement of orders and the delivery of all supplies to the same production site came to act as constraints on the new procurement procedure. As one purchasing engineer described it:

SupplyLink was not designed to provide support to individual sites to allow each site to handle its own procurement. (Purchasing Engineer)

Previously considered and proven to be useful SupplyLink was now considered to be 'a waste of time'. Why was this? First, this was due to the fact that use of the VMI service implied that the replenishment warning was no longer issued by SupplyLink at NoteCom but by VMI system at logistics providers. In addition the advance shipping notice (ASN) functionality, which computed the quantity of materials consumed during production and which issued the replenishment notice, was only used for near-delivery mode while the far-delivery mode used the VMI model. Second, it was due to the fact that the procurement practice embedded in the SupplyLink system was different from that embedded in VMI system. It is apparent that the SupplyLink failed to recognize that the coordination between NoteCom and suppliers was mediated through VMI and had shifted from tight-coupling to loose-coupling. As a staff member from one of NoteCom's suppliers remarked:

SupplyLink requires us to obtain an ASN number for each batch before it enters the warehouse. But the ASN number is only available for three days, and if we fail to deliver the goods within three days we will have to obtain another ASN number from the system again! Now, how can we guarantee delivery within three days? The customs clearance alone would take at least three days or more. Even if the goods were delivered to the VMI warehouse within three days, the ASN number is not required by the logistics providers' system. Really, we don't see the point of obtaining an ASN number from SupplyLink anymore!

DISCUSSION AND CONCLUSION

This study investigates why an IS, previously proven to be useful and fulfilling the purpose for which it was designed, falls into disuse and obsolescence after an attempt is made to replicate the system in a new location. Drawing on the practice lens approach this study posits that any changes in technology-in-use are the outcome of an ongoing structuring process where people constitute and reconstitute the structure of an IS in use. The structure of an IS not defined *a priori* but emerges from daily use of the system, along with people's understandings of the system, and of its role in the context of organisational routines (Boulus & Bjor, 2001; Jasperson, Carter, & Zmud, 2005; Orlikowski & Yates, 2002; Orlikowski, 2000). On this view any challenges arising from the attempt to replicate the IS can be regarded as largely inevitable, since the structures embedded and enacted in the use of system, along with the practices that in turn recursively structure the use of the system would also have changed.

There are a number of implications of the study for IS management. First, any structural changes both inside and outside an organization have an impact on technology-in-use. The role of environmental contexts in IS adoption has been identified and discussed in the literature; however they are often regarded as peripheral and external to an organization and its use of IS (Avgerou, 1996, 2001, 2008; Hsiao, 2007). This study has shown that the environmental contexts that lie outside of organizational boundaries do indeed give form to the structures that exist within an organization, as well as structuring their work practices (Calderia & Ward, 2003; Hsiao, 2007). Distinguishing between contexts, separating them into different layers, and examining the influence of different layers on organizational practice can give the impression that the influence of 'outer layers' is less direct than that of 'inner layers' (Rosemann, Recker, & Flender, 2008). In practice the influence that internal and external contexts have on system use is sometimes intertwined and difficult to be separate. NoteCom is a good illustration of how an organization's business practices, especially in this instance its procurement process and relations with suppliers were structured and subsequently influenced by global competition, broader sectorial trends and demand changes in the consumer computer market. This study therefore suggests that instead of distinguishing between and analyzing internal (organizational) and external (environmental) contexts as separate entities, all contexts should be regarded and analyzed as 'other structures' enacted in technology-in-use (Figure 1). In doing so all contexts and their impact can be given equal consideration with regard to the role that they play in the structuring and restructuring the system in use. This is particularly important to the study of IS use in organizations like NoteCom whose operations and strategies are sensitive to changes in their surrounding environments. To do otherwise would be to simplify the analysis and lose sight of what is important.

While the contextual dimension can identify what gives form to the structuring of system use, the temporal dimension gives meaning to these contexts and can help to explain why changes occur in technology-in-use (Orlikowski & Yates, 2002). The temporal dimension can help to explain for example why a particular decision to adopt a system is taken and in what context; and how over time use of the system changes in accordance with changes in the structural properties that shape and are shaped by people's interactions with the system. Hence it can be argued that any decisions about the design and adoption of SupplyLink were taken based on a shared understanding of a business situation obtaining in a specific time and place. This partly explains why a system may no

Figure 1. Practice lens: Technologies-in-practice (Orlikowski, 2000)

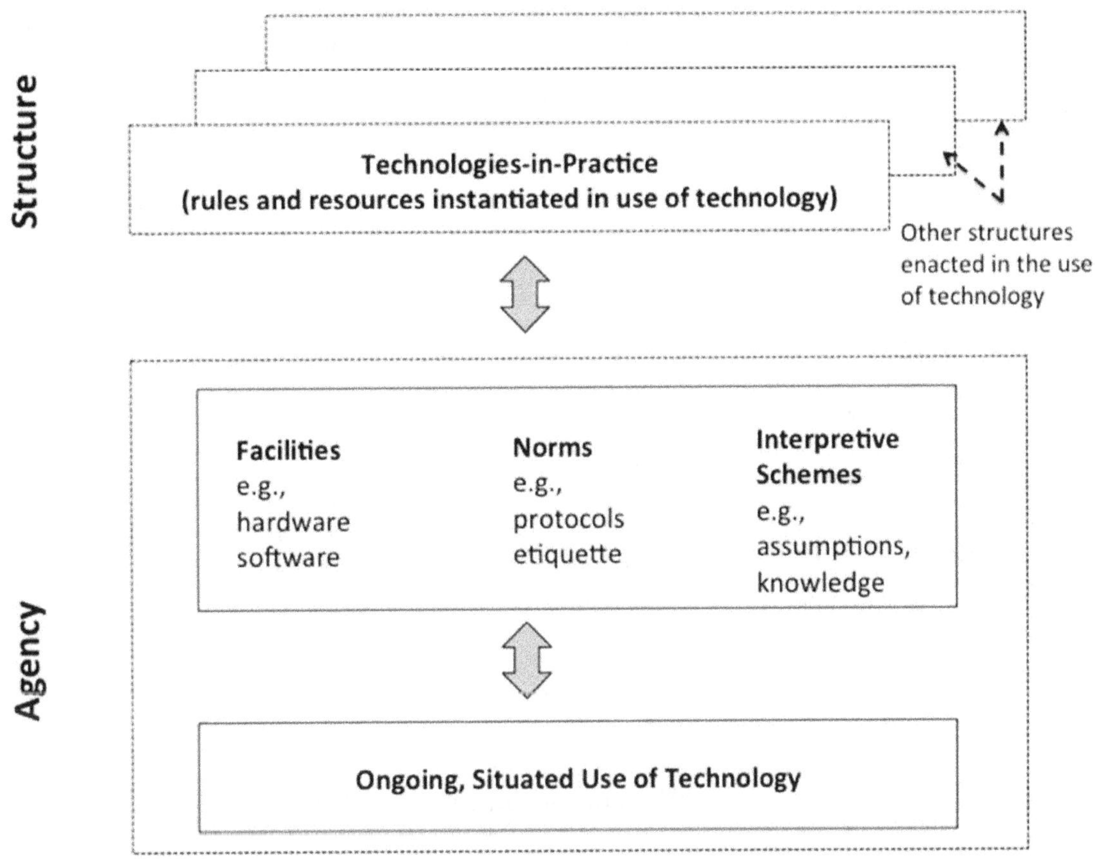

longer be perceived as useful when both time and circumstances have 'moved on'. The awareness that time and circumstances always 'move on' sheds light on the fact that the information system and the use of an information system in an organization can only be fully understood when they are examined in an historical context, a context in which the trajectory of the use of the system can be identified and illuminated. A further implication of a temporal dimension for studying IS use in a changing environment is that it helps to identify what is significant about the IS for an organization and its structures through examining the changes in shared practices among people and different parties. With this notion changes in practice occur in relation to changes in other practices that taken together provide conditions for the structuring of the system in use.

Third, the case study reveals that the use of an IS changes in relation to changes in organizational and business conditions; and in people's experience of practice, technology, knowledge, and other structures. Therefore if a state of equilibrium is reached (e.g. organizational-, contextual-, structural-fit) this state is to be considered a provisional one and always subject to change. It has been assumed that equilibrium can be restored if people and organizations adapt to change by subsequently appropriating the system to support the current structures and practices. In some

Figure 2. Just-in-time technology-in-practice enacted by NoteCom before cross-border operations

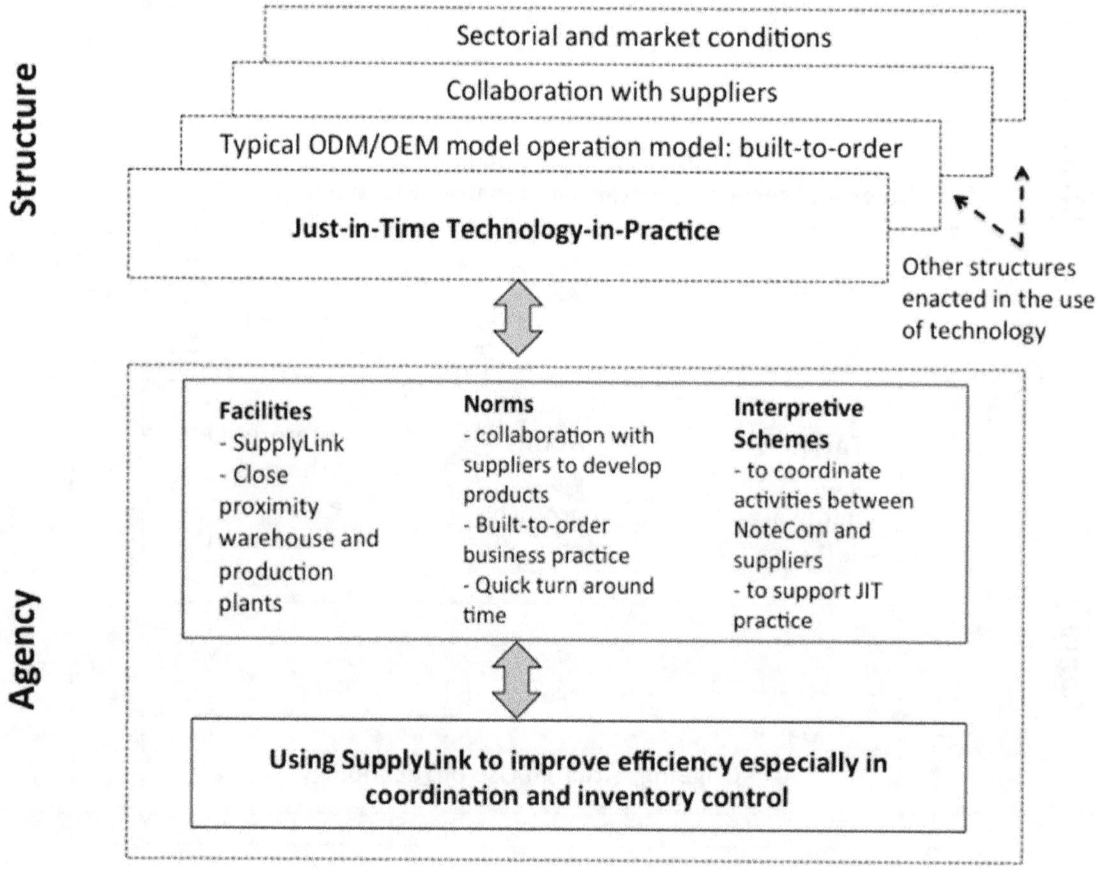

cases a new equilibrium can be achieved after the system is appropriated and continues to be used. In other cases however the system may fall into disuse, or be used for a purpose other than that for which it was designed; either because the initial system is 'out of date' or because the system is for some reason unable to support the new practices even it is appropriated (Furneaux & Wade, 2011; Kelly, Gibson, Holland, & Light, 1999; Schmidt, Lyytinen, Keil, & Cule, 2001). It would therefore be misleading to give the impression that after replication and once the system is localized use of the IS will achieve an equilibrium state and necessarily generate similar results. Instead energies should be focused on the new context in which the system is replicated, on how the system is perceived, and on what the system will be used for. One should regard and manage the system as a new rather an old one because the structures that the newly replicated system enacts will be different and people's interactions with the system will also be different. At the same time one should also bear in mind that use of the newly replicated system will also be structured to some extent by other people's prior use of and interactions with, and if pertinent previous replications of, the system. Therefore people's knowledge and experience of the system in any prior contexts of use, and how this is communicated to new users will need to be managed too.

Figure 3. VMI technology-in-practice enacted by NoteCom after cross-border operations

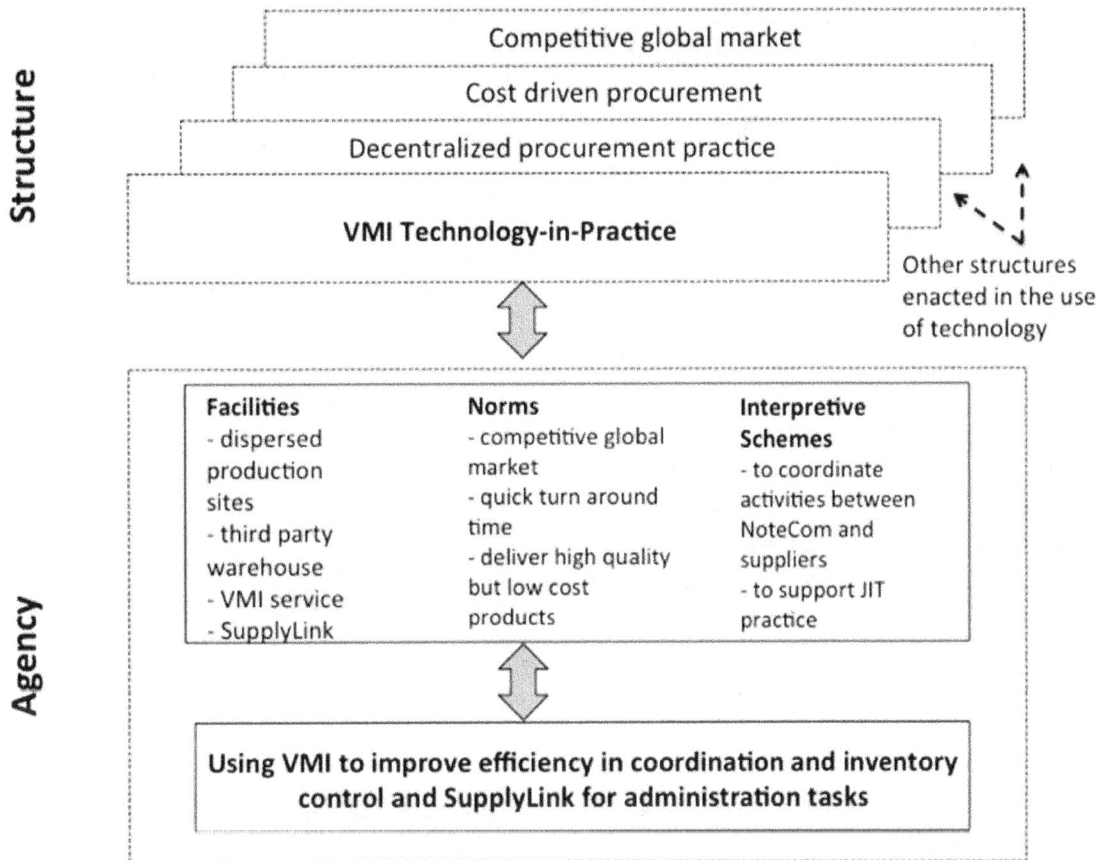

Furthermore, this study draws attention to the importance of facilities other than IS. Although other facilities are acknowledged and identified they are seldom discussed in any extensive way in the existing literature (Orlikowski, 1992, 2000). This study has demonstrated that any changes in other facilities, especially those supporting and being supported by IS can either make existing practices obsolete or create conditions for new practices. In order to sustain its existing JIT procurement practices NoteCom had for example to outsource its inventory management and use third party warehouse services to overcome problems caused by the close spatial proximity of its suppliers and NoteCom's production sites. Coupled with this new facility (e.g. third party warehouses) was a new inventory management model (e.g. VMI) as well as a new inventory management system. The use of the VMI system later created a situation where use of the SupplyLink system was perceived by its suppliers to be 'a waste of time'. It is therefore apparent that in order to manage IS adoption managers and practitioners have to give equal consideration to all of the facilities, and not just IS, that may be involved in influencing a specific operation. This is especially important in situations where IS are involved in integrating a company's global supply chain, since even a minor change in existing arrangements can have a major impact on the overall supply chain process and its related business practices.

This study also has a number of theoretical implications. First, the study has demonstrated the usefulness of the practice lens in studying the replication of IS. The analytical capacity of the framework captures rich details of how system is used in complicated nested social systems and offers explanations for why replication may fail in a new location. Second, the practice lens has been used to study continuity of IS use in organizations (Chu & Robey, 2008; Monteiro & Rolland, 2012; Orlikowski, 2000) but this study has demonstrated that the same framework can be used to examine discontinuity of system use. The contextual and temporal dimensions of the framework enable one to identify the changes in system use in relation to the changes in contexts and through such examination one can explain how an IS become disused. Third, the study shows that though system itself may have interpretive flexibility technical properties can still impose structure on system use and make system adaptation to particular practice difficult. Therefore it is important to bear in mind and examine what and how technical limitations embedded in the system have impact on the success of system replication.

REFERENCES

Avgerou, C. (1996). Transferability of information technology and organisational practices. In M. Odedra-Straub (Ed.), *Global information technology and socio-economic development* (pp. 106–115). Nashua, NH: Ivy League.

Avgerou, C. (2001). The significance of context in information systems and organisational change. *Information Systems Journal, 11*(1), 43–63. doi:10.1046/j.1365-2575.2001.00095.x

Avgerou, C. (2008). Information systems in developing countries: A critical research review. *Journal of Information Technology, 23*(3), 133–146. doi:10.1057/palgrave.jit.2000136

Bartlett, C. A., & Ghoshal, S. (1989). *Managing across borders: The transnational solution.* Cambridge, MA: Harvard Business School Press.

Boulus, N., & Bjorn, P. (2010). A cross-case analysis of technology-in-use practices: ERP-adaptation in Canada and Norway. *International Journal of Medical Informatics, 79*(6), 97–108. doi:10.1016/j.ijmedinf.2008.06.008 PMID:20053582

Burton-Jones, A., & Gallivan, M. J. (2007). Toward a deeper understanding of system usage in organizations: A multilevel perspective. *Management Information Systems Quarterly, 31*(4), 657–679.

Calderia, M., & Ward, J. (2003). Using resource-based theory to interpret the successful adoption and use of information systems and technology in manufacturing small and medium-sized enterprises. *European Journal of Information Systems, 12*(2), 127–141. doi:10.1057/palgrave.ejis.3000454

Campbell, B., Kay, R., & Avison, D. (2005). Strategic alignment: A practitioner's perspective. *Journal of Enterprise Information Management, 18*(6), 653–664. doi:10.1108/17410390510628364

Chow, W. S. (2004). An exploratory study of the success factors for extranet adoption in e-supply chain. *Journal of Global Information Management, 12*(1), 60–67. doi:10.4018/jgim.2004010104

Chu, T. H., & Robey, D. (2008). Explaining changes in learning and work practice following the adoption of online learning: A human agency perspective. *European Journal of Information Systems, 17*(1), 79–98. doi:10.1057/palgrave.ejis.3000731

Danneels, E. (2003). Tight-Loose coupling with customers: The enactment of customer orientation. *Strategic Management Journal, 24*(6), 559–576. doi:10.1002/smj.319

De Guinea, A. O., & Markus, M. L. (2009). Why break the habit of a lifetime? Rethinking the roles of intention, habit, and emotion in continuing information technology use. *Management Information Systems Quarterly*, *33*(3), 433–444.

DeSanctis, G., & Poole, M. S. (1994). Capturing the complexity in advanced technology use: Adaptive structuration theory. *Organization Science*, *5*(2), 121–147. doi:10.1287/orsc.5.2.121

Feldman, M. S., & Orlikowski, W. J. (2011). Theorizing practice and practicing theory. *Organization Science*, *22*, 1240–1253. doi:10.1287/orsc.1100.0612

Furneaux, B., & Wade, M. (2011). An exploration of organizational level information systems discontinuance intentions. *Management Information Systems Quarterly*, *35*(3), 573–598.

Giddens, A. (1984). *The constitution of society: Outline of the theory of structure*. Berkeley, CA: University of California Press.

Hsiao, R. (2007). Misaligned market: The importance of industry context in technology-mediated exchanges. *Journal of Global Information Management*, *15*(3), 69–87. doi:10.4018/jgim.2007070104

Hsu, P., Kraemer, K. L., & Dunkle, D. (2006). Determinants of e-business use in U.S. firms. *International Journal of Electronic Commerce*, *10*(4), 9–45. doi:10.2753/JEC1086-4415100401

Ives, B., Jarvenpaa, S. L., & Mason, R. O. (1993). Global business drivers: Aligning information technology to global business strategy. *IBM Systems Journal*, *32*(1), 143–161. doi:10.1147/sj.321.0143

Jarvenpaa, S. L., & Ives, B. (1991). Applications of global information technology: Key issues for management. *Management Information Systems Quarterly*, *15*(1), 33–50. doi:10.2307/249433

Jasperson, J. S., Carter, P. E., & Zmud, R. W. (2005). A comprehensive conceptualization of post-adoptive behaviors associated with information technology enabled work systems. *Management Information Systems Quarterly*, *29*(3), 525–558.

Karahanna, E., Straub, D. W., & Chervany, N. L. (1999). Information technology adoption across time: A cross-sectional comparison of pre-adoption and post-adoption beliefs. *Management Information Systems Quarterly*, *23*(2), 183–204. doi:10.2307/249751

Katz, J. P., & Townsend, J. B. (2000). The role of information technology in the fit between culture, business strategy and organizational structure of global firms. *Journal of Global Information Management*, *8*(2), 24–35. doi:10.4018/jgim.2000040102

Kelly, S., Gibson, N., Holland, C. P., & Light, B. (1999). A business perspective of legacy information systems. *Communications of the Association for Information Systems*, *2*(7), 1–27.

Kim, S. S., & Malhotra, N. K. (2005). A longitudinal model of continued IS use: An integrative view of four mechanisms underlying postadoption phenomena. *Management Science*, *51*(5), 741–755. doi:10.1287/mnsc.1040.0326

King, W. R., & Sethi, V. (2001). Patterns in the organization of transnational information systems. *Information & Management*, *38*(4), 201–215. doi:10.1016/S0378-7206(00)00050-1

Klein, H. K., & Myers, M. D. (1999). A set of principles for conducting and evaluating interpretative field studies in information systems. *Management Information Systems Quarterly*, *23*(1), 67–94. doi:10.2307/249410

Klein, R., & Rai, A. (2009). Interfirm strategic information flows in logistics supply chain relationships. *Management Information Systems Quarterly*, *33*(4), 735–762.

Kumar, K., van Dissel, H. G., & Bielli, P. (1998). The merchant of Prato-revisited: Toward a third rationality of information systems. *Management Information Systems Quarterly*, *22*(2), 199–226. doi:10.2307/249395

Li, D., Lai, F., & Wang, J. (2010). E-business assimilation in China's international trade firms: The technology-organization environment framework. *Journal of Global Information Management*, *18*(1), 39–65. doi:10.4018/jgim.2010091103

Majchrzak, A., Rice, R. E., Malhotra, A., King, N., & Ba, S. (2000). Technology adaptation: The case of a computer-supported inter-organizational virtual team. *Management Information Systems Quarterly*, *24*(4), 569–601. doi:10.2307/3250948

Markus, M., Sia, S. K., & Soh, C. (2012). MNEs and information management: Structuring and governing IT resources in global enterprise. *Journal of Global Information Management*, *20*(1), 1–17. doi:10.4018/jgim.2012010101

Mayasandra, R., Pan, S. L., & Leidner, D. (2011). Examining the strategic alignment and implementation success of a KMS: A subculture-based multilevel analysis. *Information Systems Research*, *22*(1), 39–59. doi:10.1287/isre.1080.0214

Monteiro, E., & Rolland, K. (2012). Trans-situated use of integrated information systems. *European Journal of Information Systems*, *21*(1), 1–13.

Niederman, F., Alhorr, H., Park, Y. H., & Tolmie, C. R. (2012). Global information management research: What have we learned in the past decade? *Journal of Global Information Management*, *20*(1), 18–56. doi:10.4018/jgim.2012010102

Orlikowski, W. J. (1992). The duality of technology: Rethinking the concept of technology in organizations. *Organization Science*, *3*(3), 398–427. doi:10.1287/orsc.3.3.398

Orlikowski, W. J. (2000). Using technology and constituting structures: A practice lens for studying technology in organizations. *Organization Science*, *11*(4), 404–428. doi:10.1287/orsc.11.4.404.14600

Orlikowski, W. J., & Gash, D. C. (1994). Technological frames: Making sense of information technology in organizations. *ACM Transactions on Information Systems*, *12*(2), 174–207. doi:10.1145/196734.196745

Orlikowski, W. J., & Yates, J. (2002). It's about time: Temporal structuring in organizations. *Organization Science*, *13*(6), 684–701. doi:10.1287/orsc.13.6.684.501

Preston, D. S., & Karahanna, E. (2009). Antecedents of IS strategic alignment: A nomological network. *Information Systems Research*, *20*(2), 159–179. doi:10.1287/isre.1070.0159

Rank, O. N., Robins, G. L., & Pattison, P. E. (2010). Structural logic of intraorganizational networks. *Organization Science*, *21*(3), 745–764. doi:10.1287/orsc.1090.0450

Rosemann, M., Recker, J., & Flender, C. (2008). Contextualisation of business processes. *International Journal of Business Process Integration and Management*, *3*(1), 47–60. doi:10.1504/IJBPIM.2008.019347

Sambharya, R. B., Kumaraswamy, A., & Banerjee, S. (2005). Information technologies and the future of the multinational enterprise. *Journal of International Management*, *11*(2), 143–161. doi:10.1016/j.intman.2005.03.005

Schmidt, R. C., Lyytinen, K., Keil, M., & Cule, P. (2001). Identifying software project risks: An international Delphi study. *Journal of Management Information Systems*, *17*(4), 5–36.

Schultze, U., & Orlikowski, W. J. (2004). A practice perspective on technology-mediated network relations: The use of internet-based self-serve technologies. *Information Systems Research*, *15*(1), 87–106. doi:10.1287/isre.1030.0016

Soh, C., & Sia, S. (2004). An institutional perspective on sources of ERP package-organization misalignments. *The Journal of Strategic Information Systems*, *13*(4), 375–397. doi:10.1016/j.jsis.2004.11.001

Strong, D., & Volkoff, O. (2010). Understanding organization-enterprise system fit: A path to theorizing the information technology artifact. *Management Information Systems Quarterly*, *34*(4), 731–756.

To, P. L., Liao, C. C., & Liu, Y. C. (2004). The impact of the fit between global organizational configurations and information technology configurations on user information satisfaction. *Journal of International Management*, *11*(2), 109–138.

Volkoff, O., Strong, D. M., & Elmes, M. B. (2007). Technological embeddedness and organizational change. *Organization Science*, *18*(5), 832–848. doi:10.1287/orsc.1070.0288

Walsham, G. (1995). Interpretive cast studies in IS research: Nature and method. *European Journal of Information Systems*, *4*(2), 74–81. doi:10.1057/ejis.1995.9

Walsham, G. (2006). Doing interpretive research. *European Journal of Information Systems*, *15*(3), 320–330. doi:10.1057/palgrave.ejis.3000589

Yin, R. K. (2009). *Case study research: Design and methods* (4th ed.). Thousand Oaks, CA: Sage Publications.

Zammuto, R. F., Griffith, T. L., Majchrzak, A., Dougherty, D. J., & Faraj, S. (2007). Information technology and the changing fabric of organization. *Organization Science*, *18*(5), 749–762. doi:10.1287/orsc.1070.0307

Zhu, K., Dong, S., Xu, S. X., & Kraemer, K. L. (2006). Innovation diffusion in global contexts: Determinants of post-adoption digital transformation of European companies. *European Journal of Information Systems*, *15*(6), 601–616. doi:10.1057/palgrave.ejis.3000650

Zhu, K., & Kraemer, K. L. (2005). Post-adoption variations in usage and value of e-business by organizations: Cross-country evidence from the retail industry. *Information Systems Research*, *16*(1), 61–84. doi:10.1287/isre.1050.0045

APPENDIX

The Interview Protocol

The interviews conducted in this study contained four parts.

In Part 1, we interviewed MIS personnel on overall system implementation and basic system setting, which comprised (1) implementation of the procurement system; (2) the operating range of the system support in procurement activities; (3) the logic and ideals embedded by the system in handling procurement activities; and (4) adjustments or corrections of the system functions when it was relocated.

In Part 2, we interviewed the R&D department, the procurement department, the suppliers, and the logistics companies regarding specific activity content (questions regarding how) for two phases in overall procurement activities (e.g., product development, material procurement, production, and logistics). For example:

1. The questions asked of the R&D personnel included "During the first phase, how are appropriate suppliers identified? For new product development, what interaction do you have with suppliers?"
2. The questions asked of the procurement personnel included "During the second phase, what operations must be performed when you procure materials from suppliers? How do you evaluate supplier performance?"
3. The questions asked of the suppliers included "During the second phase, how are materials delivered to this company? What operations require coordinating with this company?"

In Part 3, we interviewed the relevant company departments on the reasons, background, and primary considerations (questions regarding why) of the procurement activities conducted for two phases. For example:

1. The questions asked of the R&D personnel included "During the first phase, why do you interact so closely with the suppliers? How important are the suppliers to your product development?"
2. The questions asked of the procurement personnel included "During the second phase, why do you communicate with the suppliers regarding Operation A? What is the significance of Operation A? Is this practice the same as it was previously? Why do you use this indicator to evaluate supplier performance?"
3. The questions asked of the suppliers included "During the second phase, why is this practice employed to supply materials to this company? Did this company make any requests? What were these requests? Why did this company make such requests?"

In Part 4, we interviewed the users of the procurement system regarding the system use to understand whether inconsistencies occurred between the procurement system and the procurement activities.

For example, we asked the system users "What is the procedure for purchasing materials using the procurement system? What are the steps of system operation? How is the system used to process material shipments? What difficulties are encountered when using the system to process material shipments?"

Chapter 16
Framework Based on Benefits Management and Enterprise Architecture:
The Private Cloud in the Business Strategy

António Rodrigues
Instituto Universitário de Lisboa (ISCTE-IUL), Portugal

Henrique O'Neill
Instituto Universitário de Lisboa (ISCTE-IUL), Portugal

ABSTRACT

In 2010, a framework aiming to address strategic investment decisions on IT infrastructure was developed. It was based in Benefits Management principles and Enterprise Architecture concepts, being inspired by the emerging public cloud technological trend. Meanwhile, the public cloud concept did not materialise at the expected pace and other alternatives have emerged in the market, in particular the private cloud-based solutions. This fact required the framework to be updated to cope with the business and technological requirements of the private cloud concept. A new version of the framework has been developed and was used to help managers to address IT investment decisions on private cloud in an international bank.

INTRODUCTION

In the "distant" year of 2010 a research project was set-up to address two pressing issues in the IT business strategy domain (Rodrigues & O'Neill, 2012):

DOI: 10.4018/978-1-4666-4506-6.ch016

- **Benefits management in IT investments:** At that time the economic climate already stressed the need for the firms to thoroughly reflect on the planned investments. At such it became imperative to provide a distinctive reason for each new IT project, often only seen as a cost centre. The recommendations then presented stressed the need to provide an effective liaison between the project and the added value the investment brings to the business;

- **The cloud based enterprise architecture:** The paradigm adopted by companies to view the management of their IT was clearly in a moment of change. The traditional data center as it was known until then, was starting facing the competition of alternative solutions, from managed services, outsourcing and possibly the most disruptive cloud computing. The offerings in this field were multiplying and coming up on the most varied forms. Interestingly, most of the new solutions were being introduced by the traditional suppliers of the past on-premise solutions. This fact alone helped to justify the opportunity of the research subject.

The research enabled to establish the framework presented in Figure 1.

The 6 phases that comprise the model may be described as follows:

- **Alignment Business-IT:** To assure that the investment made in IT is aligned with the business objectives and that the project is able to achieve the expected benefits. This phase is materialized through a Dependencies and Benefits Network that enables to define the relationships between IT→Stakeholders→ Business changes →Business benefits;

Figure 1. First version of the framework, presented in 2010 (Rodrigues & O'Neill, 2012)

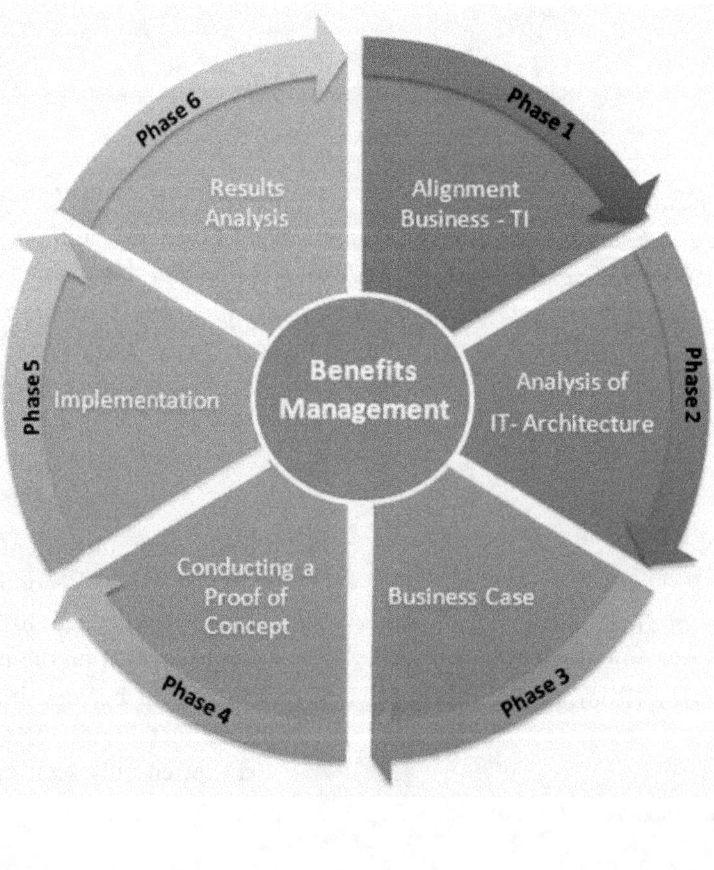

- **Analysis of IT-Architecture:** To identify what is the most suited solution for the IT architecture. The model considers on premise, cloud and hybrid solutions. This phase results are of vital importance as the generated architecture will decisively determine the project subsequent phases;
- **Business Case:** The data collected will be integrated in a decision tree model that enables to assess the project, not just from a financial perspective but also considering other non financial benefits. Sometimes projects and business initiatives are justified using just a cost-benefits financial analysis. This approach may be misleading. Although being an undoubtedly important factor, the financial dimension should not be the only one to take into account as a decision making criteria;
- **Proof of Concept:** This phase is very important, especially when the option to outsource an application is being considered. The proof of concept enables to anticipate service levels and quality of response, among other key performance indicators;
- **Implementation:** The implementation phase must be carefully managed (plan, execute, control, assess) in order to minimize the effect of future constraints;
- **Results Analysis:** This phase enables to analyse results to validate all the premises and assumptions previously made. This phase becomes also very important to confirm the benefits that usually are announced in the early conception stages of the project.

This model neither aimed to become a new project management methodology nor to present itself as an alternative to widely used enterprise architecture frameworks (Zackman, TOGAF, etc.). The framework just aimed to be a guideline to enable companies to choose the best option among alternative IT investment projects. It is common knowledge that there are no readymade solutions that may be indistinctly applied to all situations. Each project has to find its own way. If the framework would help to choose, frame and implement the best solutions, then the objective would have been achieved.

Three years later, with the economical crisis worsening, especially in the European southern countries, we are curious to verify what are the firms doing to lower costs in IT infrastructures. In particular, we want to assess the level of adoption of the promising cloud computing architectures that were then emerging.

The Adoption of the Cloud Concept has not Occurred at the Expected Pace

In the business domain, the transition from the on-premises model to the cloud model has not occurred at the expected pace. According to Frost & Sullivan's 2012 survey of IT decision-makers: "*Only 15% of U.S. businesses currently use cloud infrastructure services, and even then, they only use them intermittently*" (Frost & Sullivan 2012).

The issues that have threatened to overshadow the "clouds" in the recent past were not based on technical constraints but rather result from resistance to change factors. One of these factors is clearly the security (Gartner, 11-2012). As shown in Figure 2, it continues to be a discouraging factor when managers consider adopting the cloud concept.

Being an expected reaction, it justifies some of the conclusions of the previous study where an hybrid model was proposed, combining characteristics of the on-premise and the cloud architectures.

However the costs incurred with the systems management and administration continue being very significant, as represented in Figure 3 (Gartner, 11-2012). Since 2006 the number of

Figure 2. What to consider when moving a business application to a public cloud

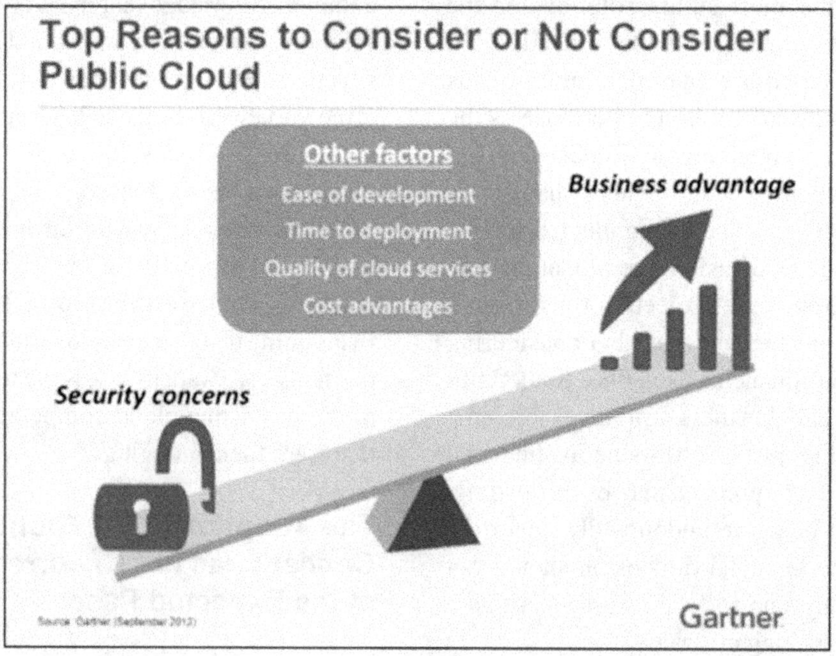

the virtualization solutions has increased while the number of physical servers has declined. The strategy adopted by most firms brought all the dynamics of the virtual world. However it was unable to reduce costs, in particular those related with the systems management and administration processes, which have steadily increased. This is a deep business concern specially for those departments who have to pay for these services.

This graphic must be carefully analysed. If the adoption of virtualization solutions has not been able to reduce costs, then the IT managers should might have considered other viable alternatives like the cloud computing based solutions.

However, it was not quite what happened. Although there is a growing level of acceptance, especially in the areas of email and collaborative tools, the general consensus around the cloud platforms has not materialized in a wide usage.

If the public cloud solutions did not represent a widely disseminated alternative, so the question that arises is:

How do companies think reducing costs related to the management and administration of IT infrastructure?

PRIVATE CLOUD

The answer to the above question may be given through the emerging concept of Private Cloud. It may be of great importance for companies and organizations as Private Cloud may serve as an intermediate stage between the concepts of on-premise and public cloud computing. However it is important to contextualize the concepts. In this regard, it is understood that the definition advocated by IDC (2012a) reflects quite well the reality of the distinct cloud models. According to this consultancy firm there are three types of cloud, as shown in Figure 4, which can be characterized as follows:

Figure 3. Expenditure on data centers goes far beyond the server costs themselves (Gartner, 11-2012)

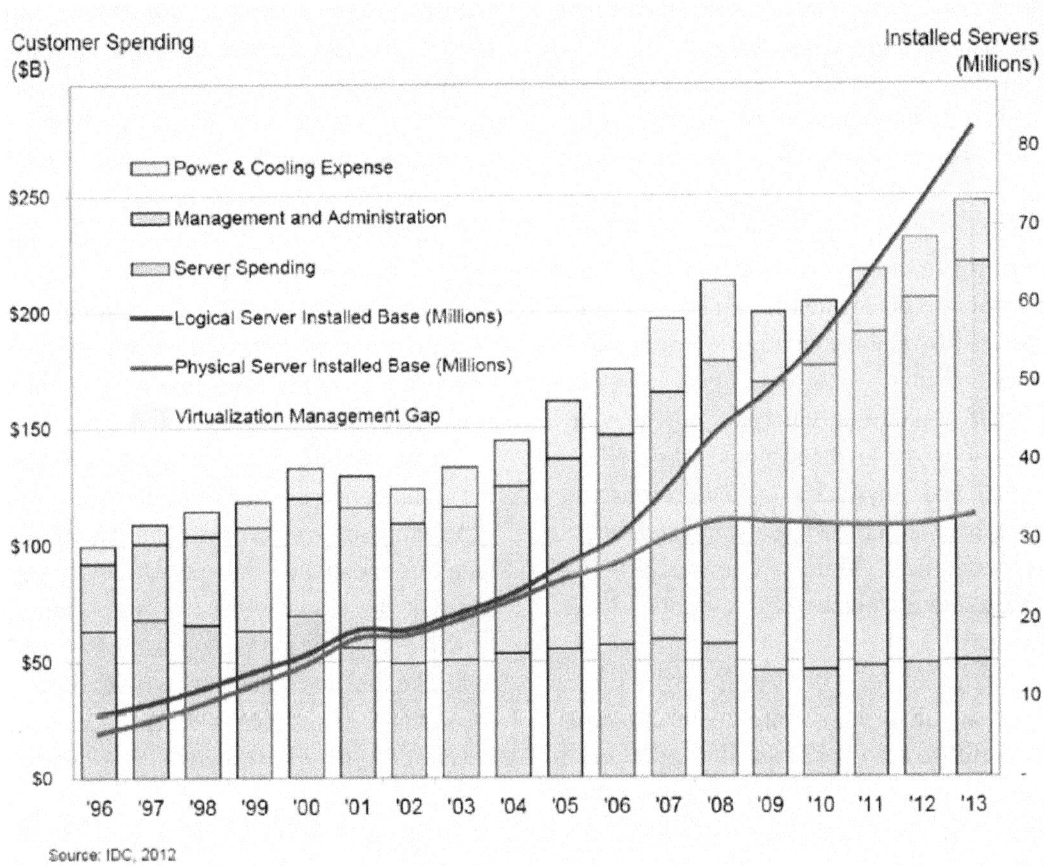

Figure 4. IDC: Cloud Models definition (IDC, 2012a)

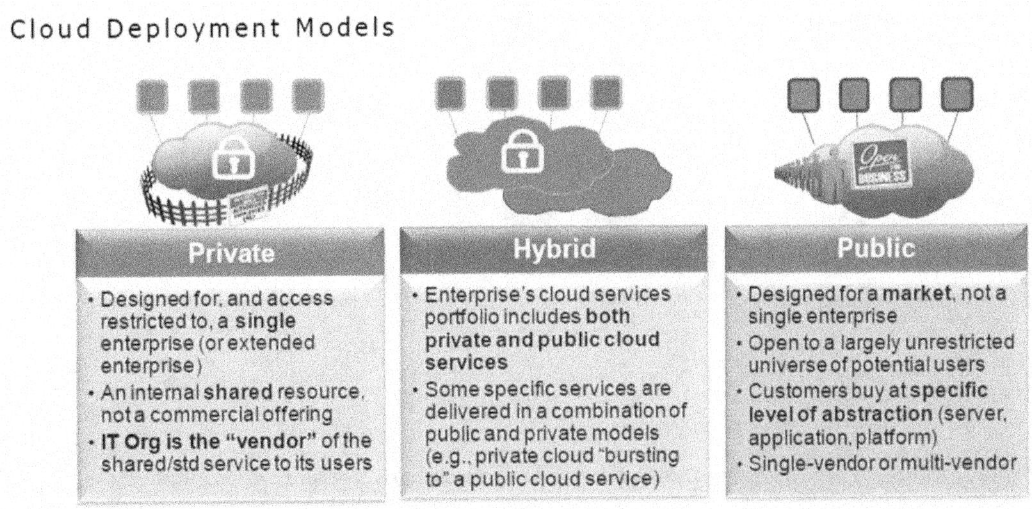

Source: IDC, 2012

- **Public:** Open to a largely unrestricted universe of potential users and designed for a market, not a single enterprise;
- **Private:** Designed for, and with access restricted to, an individual company. It is an internal shared resource, not a commercial offering, where the IT organization is the "vendor" of the shared service to its users. Private clouds can be deployed inside a corporate datacenter or can be built and operated by an outsourcer or a hosting services provider;
- **Hybrid:** Where an enterprise has private and public deployment models, as well as virtual and physical noncloud resources, and where some services may be delivered in a combined public/private model (e.g., private cloud "bursting" to a public cloud service).

Having established the concepts of cloud, it may be useful to step back, leaving the clouds and returning down to the earth, remembering that most companies and organizations still hold their IT infrastructure based on the on-premises model. However, there is evidence based on different published studies, that companies that currently still have their systems based in the traditional model are looking forward to develop cloud-based solutions. As shown in the Figure 5 the cloud market is desirable from the point of view of turnover and there is increasing, but its figures are still far from the traditional market.

Until recently the world seemed to be divided between the on-premises and off-premises concepts, as mirrored in Figure 6.

As previously noted, the pace of adoption of the cloud concepts has not yet achieved its initial expectations. However, we should neither get the idea that the cloud concepts are no longer valid nor that the cloud will be no more than a fad. Much less we should infer that the Private Cloud will take the space that was not yet occupied by the Public Cloud. These concepts do not replace, but rather complement each other. A private cloud

Figure 5. Cloud versus Enterprise software spending (Gartner, 2012a)

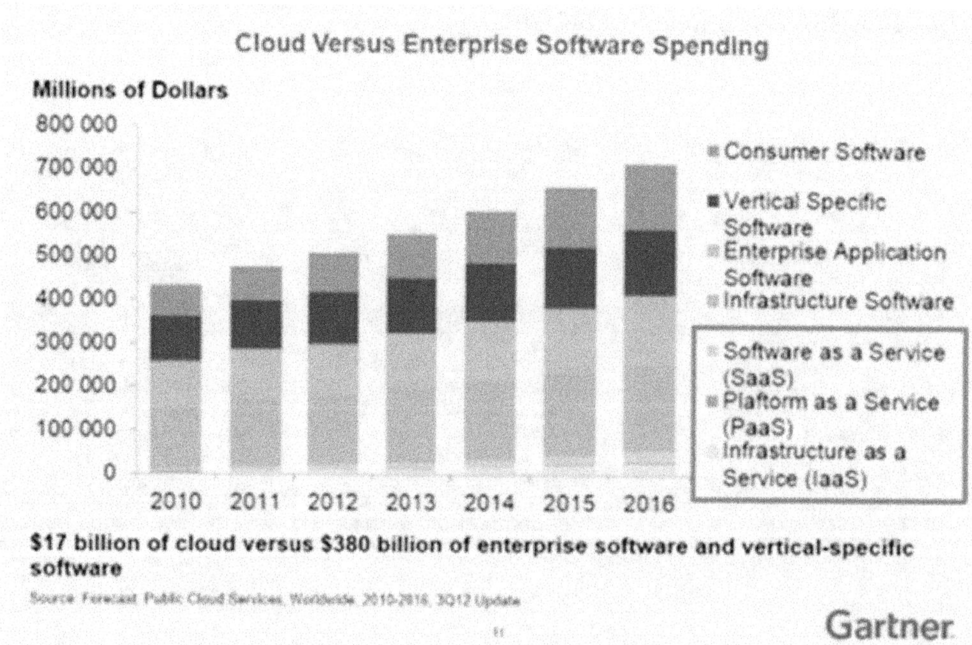

Figure 6. The world appeared divided into on-premises and off-premises

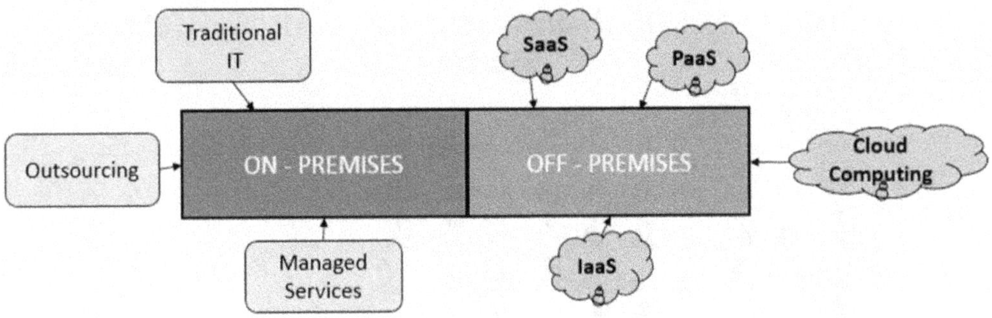

solution may be seen as a facilitator who can help spreading and consolidate the cloud concept.

Nevertheless, we are apparently in the presence of a nonsense: on the one hand the traditional model is being abandoned because it is expensive (Figure 3), on the other hand firms are facing the possibility of adopting a model in which all the burden becomes the responsibility of the organization, including the costs. So, it is legitimate to ask: "What is gained by this change, both from a technical standpoint (architecture) and from a financial and operational standpoint (benefits)?"

There is no single answer to this question, since every company or organization has its own peculiarities which must be analyzed to assess what may be the best solution to be applied in each case. One purpose of this document is to help clarifying these relevant questions.

PRIVATE CLOUD ARCHITECTURES

The cloud solutions are being considered in a very positive as shown in Figure 7, which reflects the views from heads of IT (Gartner Webinars, 2012).

Whenever there is a new concept, there are always some unclear ideas that must be clarified to overcome resistance to change. The Gartner article titled: "*5 things Private Clouds are not!*" (Gartner, 2012b) helps to explain some very common inaccuracies:

- **Private Cloud is not virtualization:** Virtualization has certainly an important role in private cloud architectures, however a set of virtual servers do not become a private cloud;
- **Private Cloud is not just about cost reduction:** A private cloud does not just represents a cost saving in HW. The adoption of the concept should certainly achieve other benefits that will translate into cost savings, resulting from increased efficiency in resources allocating, standardization, among others;
- **Private Cloud is not necessarily on-premises:** Although it is a fairly common practice, not always the private cloud platforms will have to stay at the company that created it. Alternatively they can stay on the hosting system a skilled operator;
- **Private Cloud is not only Infrastructure as a Service (IaaS):** There may be the idea that the role of a private cloud is just to "supply" an infrastructure layer (IaaS), however the current architectures allow to offer a broader set of end-to-end services, ranging from the storage to the application;

Figure 7. Organization versus Cloud models (Gartner Webinars, 2012)

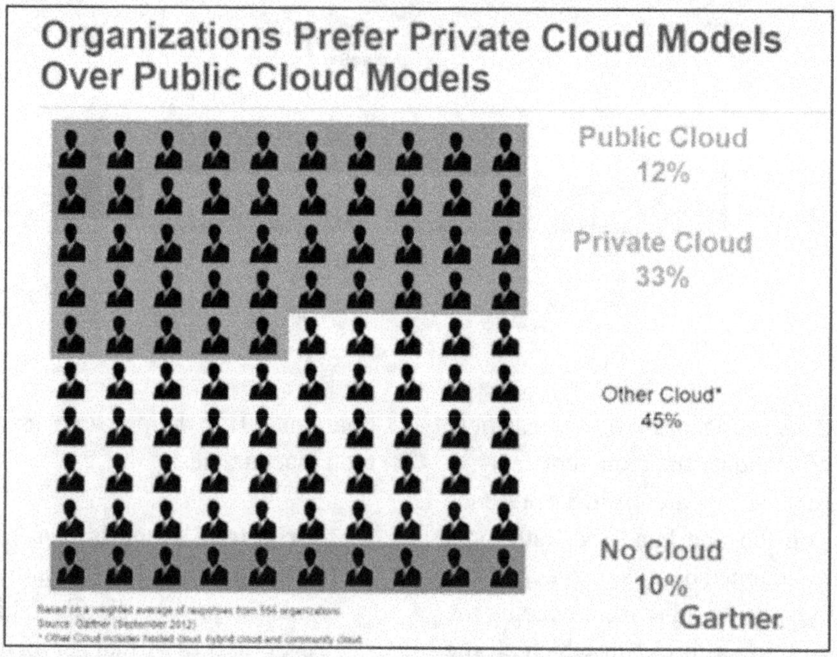

- **Private Cloud is not always going to be private:** A private cloud will not be the solution for the rest of the company's life. It may be seen as an intermediate stage where concepts, standards and acceptance may be tunned, preparing the move towards a public cloud.

IDC (2012a) defines some of the characteristics that a private cloud presents:

- An assembly of standardized shared services that enables highly automated turnkey service provisioning and operation across shared infrastructure resources;
- Self-service provisioning by end users via browser-based portals using standardized service catalogs and menus;
- Dynamic and elastic scaling of resource consumption based on changing demand and business requirements (frequently enabled by virtualization, automated provisioning and workload balancing migration technologies);
- Consumption-based metering to support pay-as-you-go pricing models for public cloud services and consumption-based chargeback/showback for private cloud.

To be competitive when compared with the traditional model, this type of architecture must be based on the following basic principles:

- Reducing the cost of IT-Hardware (server, storage, network);
- Reduced operating costs;
- Enable to change from capex to opex;
- Faster service delivery;
- Reduce infrastructure management.

The adoption of a Private Cloud solution just becomes possible by the development of all-in-one type architectures, as illustrated in Figure 8:

- **Network:** Contrary to the traditional models, where each server has one or more network connections, this concept relies on the ability to consolidate all the network requirements, using a minimum number of high speed links (10Gb / s);
- **Storage:** Here the processes are much more streamlined, allowing all servers to access to the same storage;
- **Compute Servers:** A set of servers provide the architecture with a considerable number of CPUs and Memory, which are critical resources for the remaining layers. Typically blade servers are used, due to their greater ratio "capacity versus space";
- **Hypervisor:** The hypervisor enables the creation and management of virtual servers that take advantage of the resources provided by the compute servers layer;
- **Virtual Servers:** Finally, conditions are created to setup the servers. This type of solution, which is based on the use of templates that guide the deployment of the new servers, can be difficult to set-up. Customers of this type of solution should avoid creating multiple templates as they can certainly satisfy more than 90% of their software applications just by responding to the requirements of the presentation layer, of the business layer and of the database layer;
- **Software Applications:** Materialization of the applications on the servers. These servers are in principle organized by layers, depending on the needs of the distinct applications;
- **Unified Management & Administration:** This layer is perhaps the icing on the top of the cake. It aims to simplify the whole orchestration required by the various layers. It is supposed to provide easy to use administration tools that do not require deep knowledge of all technical topics in question.

Figure 8. Typical architecture of private cloud

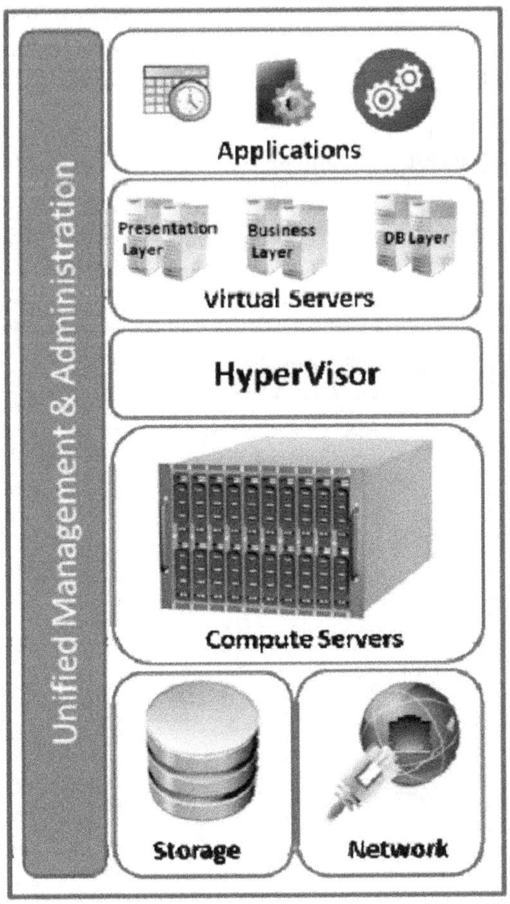

COMMERCIAL SOLUTIONS

The architecture described above is commercially embodied in solutions named Integrated Systems, which Gartner defines as: "an emerging class of data center systems that deliver a combination of server, storage or network devices in a pre-integrated stack" (Gartner Webinars, 2012).

This definition enables to consider three segments of solutions:

- **Integrated Application Systems:** Server, storage and network integrated with application software to provide appliance or appliance-like functionality;
- **Integrated Platform Systems:** Server, storage, and network integrated with platform software to provide a cloud-like management environment;
- **Integrated Infrastructure Systems:** Server, storage and network integrated to provide a shared computing infrastructure.

The Figure 9 illustrates some of the solutions currently available on the commercial market (Gartner Webinars, 2012). The wide range of offerings that have emerged enable us to deduce that the key providers are foreseeing a business opportunity and no one wants to miss.

The purpose of this article is not to analyze the strengths and weaknesses of each one of the solutions. It only wants to show evidence that the IT market is highly aware of the Private Cloud phenomenon. Moreover, the proliferation of the cloud concept is fostered by the existence of these solutions. They have a low installation cost when compared with the cost of traditional IT. The solutions for public and private cloud allow small businesses to think in a different way. As David Chappel stated that:

Cloud platforms let even small organizations to dream big dreams. (Chappell, 2011)

Investiment versus Benefits in the Integrated Systems Solutions

The initial investment is practically nonexistent in the Public Cloud solutions and the costs tend to increase or decrease depending on the use. The Private Cloud financing model is different and it resembles more to the on-premise models. As can be seen from Figure 10, there is an initial investment required to set-up the infrastructure, which has to be later recovered. However the initial investment is significant and may not be easy to all companies.

Figure 9. Integrated systems (Gartner Webinars, 2012)

System Type	Vendor	Model
Integrated Application	Oracle	Exadata
	HP	App System
	IBM	PureApplication
Integrated Platform	Oracle	Exalogic
	HP	CloudSystem
Integrated Infrastructure	HP	VirtualSystem
	Dell	vStart
	VCE	Vblock
	IBM	PureFlex

Figure 10. Integrated systems: Investments and benefits over time (IDC, 2012b)

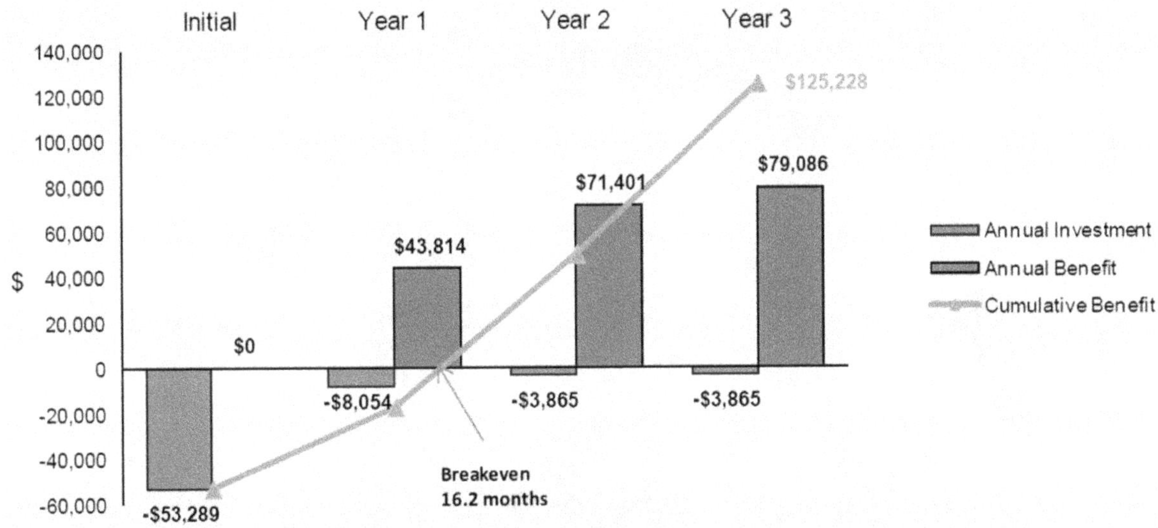

Source: IDC, 2012

THE PROPOSED FRAMEWORK

Assuming the development and acceptance of the Private Cloud concept it becomes important to update the framework initially proposed according to the requirements of this new reality. Like the previous version of the proposed framework, the current is based also on two core disciplines:

- **Benefits Management:** The proposals made in the framework to manage benefits, are inspired by the work of the Information Systems Research Center (ISRC) of the Cranfield School Management (Ward et al, 2007). Obviously the work that we have developed in recent years have introduced elements that will provide the uniqueness of the required framework (Rodrigues, 2009);(Rodrigues & O'Neill. 2010), (Rodrigues & O'Neill. 2012);
- **Enterprise Architecture:** This issue gains a special importance. When today we look at most of our businesses we find that the provision of IT solutions is performed by quite different and heterogeneous systems, and are often provided by multiple vendors. The proposed framework considers the enterprise architecture as the common thread to all the solutions, being able to provide the required alignment between business and IT (Ross et al, 2006).

When compared with the previous one, the new version of the framework presented in Figure 11, has the advantage of bearing all cloud solutions, whether private or public. Obviously some changes were made to "accommodate" the new requirements, which will be detailed phase by phase. To keep both frameworks consistent there was an attempted to materialize the new model in a format similar to the previous one. In order to facilitate the explanation of each step, we highlight the following points:

Figure 11. New framework

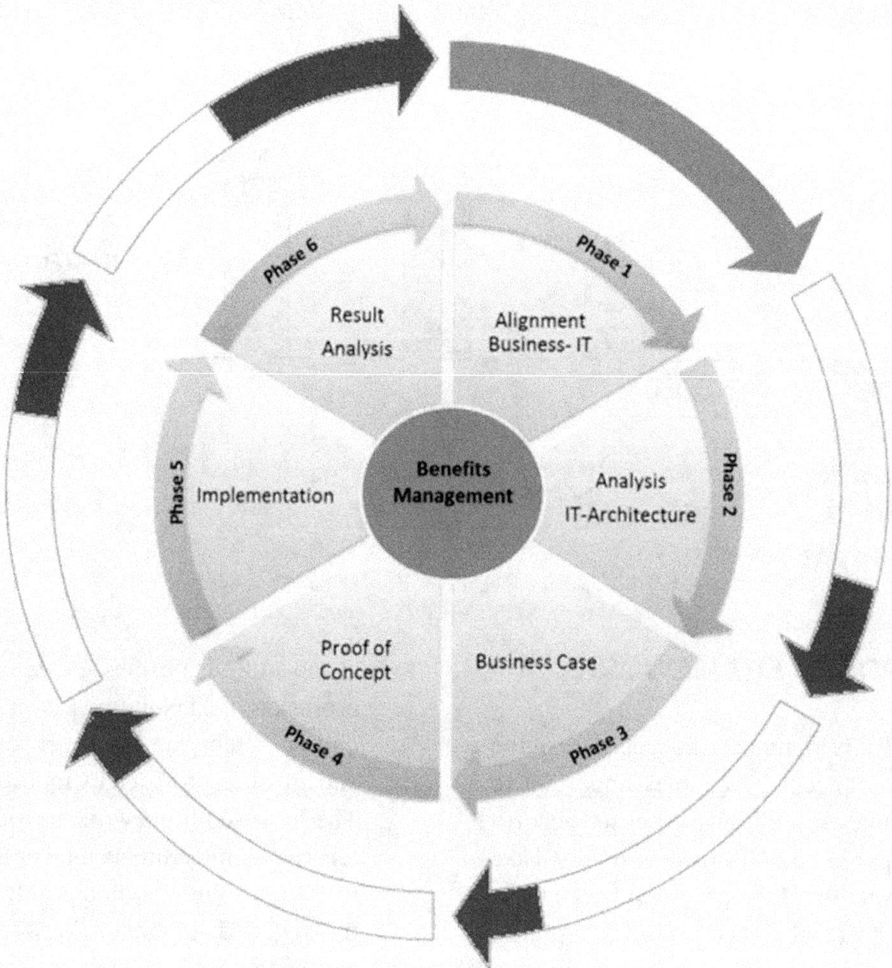

- **Centre:** The center of the figure is occupied by the key element of the model, the benefits management approach;
- **Phases:** The phases are arranged as sectors of a circle, from the first until to the sixth. The inner arrows in the phases intend to represent that the cycle may have as many iterations as the number of applications or projects that we intend to implement;
- **Duration of the phases:** The outer arrows relate to the implementation time of each phase if the Private Cloud solution is choosen. With the exception of the first phase, where an identical time frame is considered, it is expected that all the other phases will have a lower completion period.

The six stages that make up the framework of the new model will be detailed in the following sections. Wherever appropriate, the differences between Private and Public Clouds will be highlighted in each one of the sections.

Alignment Business - IT

The objective of this first phase is to ensure that there is a proper alignment between the business, which usually triggers the idea, and IT, which typically is responsible for the implementation. To achieve this alignment it is suggested to find an answer to three simple questions:

- **Why:** Why is the company making this investment?

To answer to this question we will have to define what the business drivers are by searching to the root causes of the project. These may have their origin in sources similar to those listed in Table 1:

- **What:** What benefits does the organization expect to get?

This means an investment in a project that aims to implement an application, an improvement or a new infrastructure. An investment implies a return which should be identified at this stage and later materialized in phase # 3. The return may not be monetary, it may be of a qualitative nature (better service, reduced response time, improved customer satisfaction, etc.):

- **How:** To achieve the identified benefits how will be applied the changes to the IT resources?

Table 1. Types of business drivers (Ward et al, 2007)

Infrastructure	Related developments in the field of IT as, for instance, a project to carry out an infrastructure that enables greater mobility for employees.
Context	Projects aiming to enable the company to work according to a directive set by a regulatory body, such as the Nationals Bank for organizations in the financial sector.
Result-Oriented	Projects aiming to achieve a specific purpose, such as cost reduction or the integration of a new set of features to make it possible to provide an advanced service.

The answer to this question is given by the benefits and dependence net shown in Figure 12, which combines the key elements of the process in a graphical form: IT→Stakeholders→Business Changes→Business Benefits→Investment Objectives→Business Drivers.

At this stage it is considered that the technical component should be agnostic, so there will be no significant implications if at a later stage the option will be for a public or private cloud.

Analysis IT – Architecture

At this stage the technical requirements of each project must be fitted according to the definitions established by the enterprise architecture implemented in the company, which should be as clear and comprehensive as possible. Following this way it will be avoided the need to rethink the whole strategy for each new project.

It is important to assure that there will be an element called *integration architecture*, as shown in Figure 13. This component should be unique and dynamic enough to be able to provide answers, independently of the cloud solution that will be adopted. The basic principles of such integrator element should be:

- **Identity and Access Management:** Implementation of mechanisms for authentication and authorization, to regulate who accesses the data and how it is done;
- **Data Management:** It is concerned with data protection and ensures they are accessible in the best possible way. Data access should be done through Web Services thus becoming transparent to the applications which data repository is being used or even where it is located;
- **Regulation:** Ensuring that all legal requirements are met, measuring the service level agreements (SLA) and providing audit, preferably carried out by specialized

Figure 12. Benefits and dependence net (Ward et al, 2007)

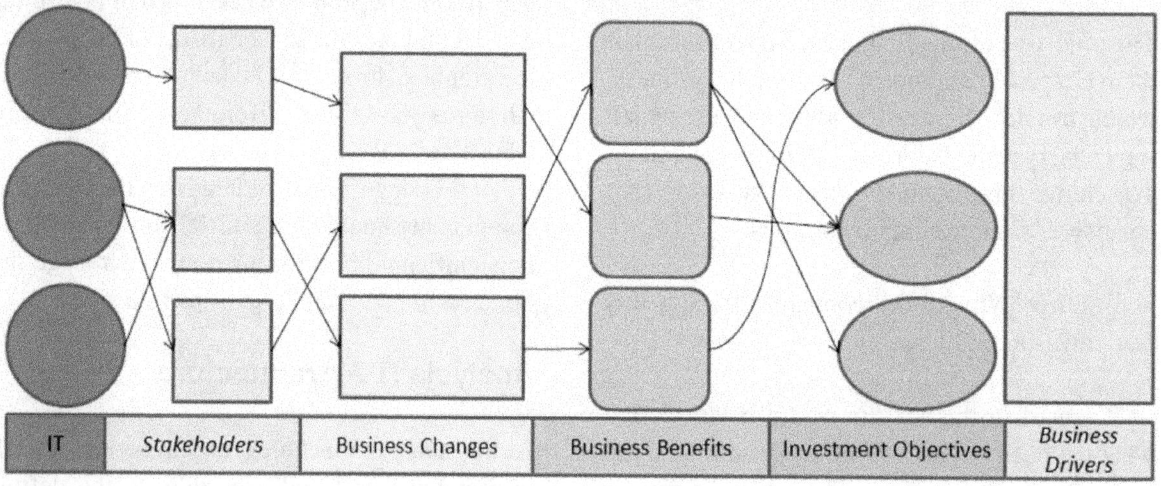

Figure 13. Integration architecture

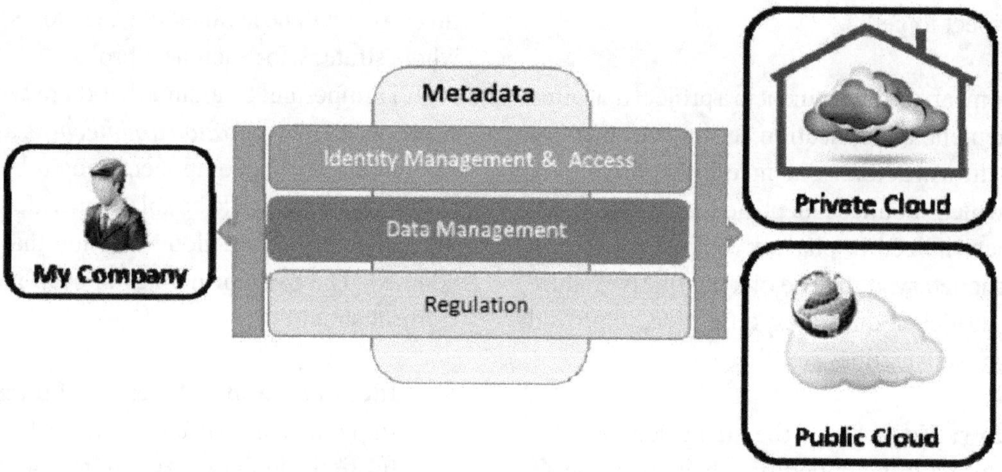

independent companies to ensure the consistency of the system as a whole;

- **Meta-Data:** Dictates the transformation and adaptation rules necessary for two different systems to communicate.

It may seem peculiar the proposal to adopt the same principles of architecture, whether it may be a public or a private cloud. This justified because the architecture principles must be common. In this way it is possible to consider the private cloud functions as an intermediate step for reaching a public cloud.

It is expected that in the case of a private cloud, the implementation of regulatory requirements or the need to improve service levels tend to be much quicker, because there will be no need to negotiate with external entities.

Business Case

It is necessary to carefully outline the projects and justify the gains that they aim to bring to the organization. The framework suggests the categories of metrics defined in Table 2 as the basis for designing a business case. The dimensions presented in the table have obvious advantages when compared with simpler financial indicators. Obviously the framework recognizes the usefulness of financial indicators, as they are also included in this new version.

By placing the different perspectives on the same table the framework broadens its scope of application and enables to compare the alternative solutions:

- **Traditional IT:** Analysis of all perspectives (not just financial) if the project is implemented in the traditional way;
- **Private Cloud and Public Cloud:** The same type of analysis but in this case when applied to the prospective cloud, public or private.

Proof of Concept

The objective of this phase is to implement the architecture defined in Phase # 2. The Figure 14 suggests a possible approach to be followed during this phase:

- **Integration:** Identifying a set of application tests as well as integration tests with other existing subsystems. Responsiveness must also be measured;
- **Service Level Agreement:** Comparison of the performance measures assessed in the previous point, compared with the agreed goals. Even in the case of the adoption of a private cloud, there must be defined performance references;
- **Auditing:** Validating the access logging mechanisms, for the application or for administration purposes;
- **Incident Management:** To assess the way the incident management processes address the respective incidents is an important requirement; in particular, when it involves external entities, it is important to understand the responsiveness of the supplier;
- **System Administration:** Testing of the management functions, with specific requests that are not usually available to the ordinary user. When testing a Private Cloud solution it must be assured that the members of the testing team are distinct from those in charge of the cloud administration.

At this stage, it was also decided not to to create different conditions, depending on the cloud solution. There is no reason to distinguish the cautions and recommendations to have before the implementation phase. In the case of Private Cloud, where teams are internal, will surely accelerate the process steps. Even so, it is not recommended to "cut corners".

Table 2. The "New" business case

	Traditional IT	Private Cloud	Public Cloud
Financial	By applying a cost-price analysis or a different financial formula, it becomes possible to reveal a financial benefit		
Quantifiable	There are enough evidences that enable to quantify in advance the expected benefit of the changes		
Measurable	A way of measuring performance may be applied but it is not possible to anticipate how much the improvements will be at the end of the implementation of changes		
Observable	By having a large amount of subjectivity this is the most problematic approach. Agreed criteria are used by individuals or specific groups which, based on their experience, will decide to what extent the benefit was achieved		

Implementation

In the implementation phase, there are two points to consider. The first relates to the care that should

Figure 14. Steps to be performed during the proof of concept

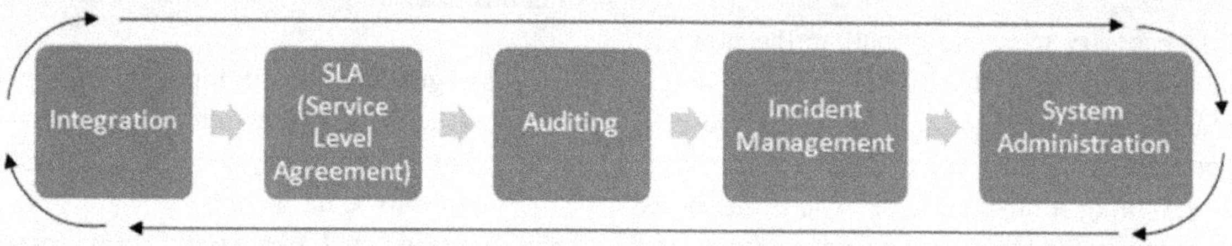

be taken whenever a new application is available, suggesting the initial adoption of a pilot project and progressively enlarging the universe of use. This approach has the advantage of reducing the exposure to risk that is dependent of the architecture complexity.

The second point relates to the need for an appropriate applications portfolio. These may be integrated in the respective quadrant, as shown in Figure 15.

This table follows the principle advocated by George Moore (Moore, G, 2002), which classifies the applications in the following categories:

- **Core Applications:** Which are considered vital to the business. According to the same author, such applications should always be managed by the company itself;
- **Peripheral Applications:** Are those identified as less important and without which

Figure 15. Classification of applications according to their potential to progress to the cloud computing concept

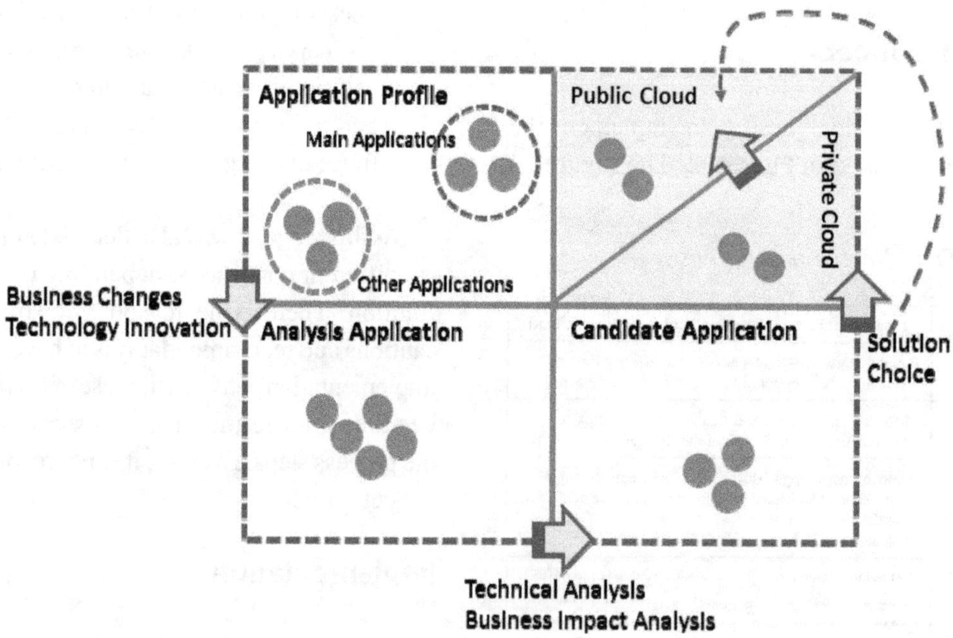

the business can survive. George Moore considers these applications as eligible to be included in a cloud solution.

The above table also suggests the possibility of an application to be moved to a new quadrant, depending on the following principles:

- **Business Changes or Technology Innovation:** An application can be moved if its utility to the business changes, or if it starts to be used by a different universe of users that make the company rethink the application context. Also any technological innovation may cause a change in the whole application scenario;
- **Technical Analysis or Business Impact Analysis:** When a change of quadrant occurs it is necessary to ensure that there is a technical analysis and that there are no identified constraints. It is also necessary to identify the impact to the business;
- **Solution Choice:** Obviously it depends on the number of cloud solutions that the company adopted during the architecture definition. Should the company have both solutions (private and public) now it is possible to make the correct choice, based on the technical analysis suggested in the previous paragraph;
- **Private to Public Cloud:** As in the other quadrants, also in the quadrant dedicated to the cloud is possible a transition to a public cloud-based solution, as has been argued throughout this document. Change may be triggered by factors such as: responsiveness of the platform requirements increase, peak usage, among others.

Results Analysis

The goal of this last phase of the framework is to analyze the results. Each stage produces outputs that must be interpreted according to Figure 16.

Figure 16. Combining the output of the different phases of the model in order to analyze the results

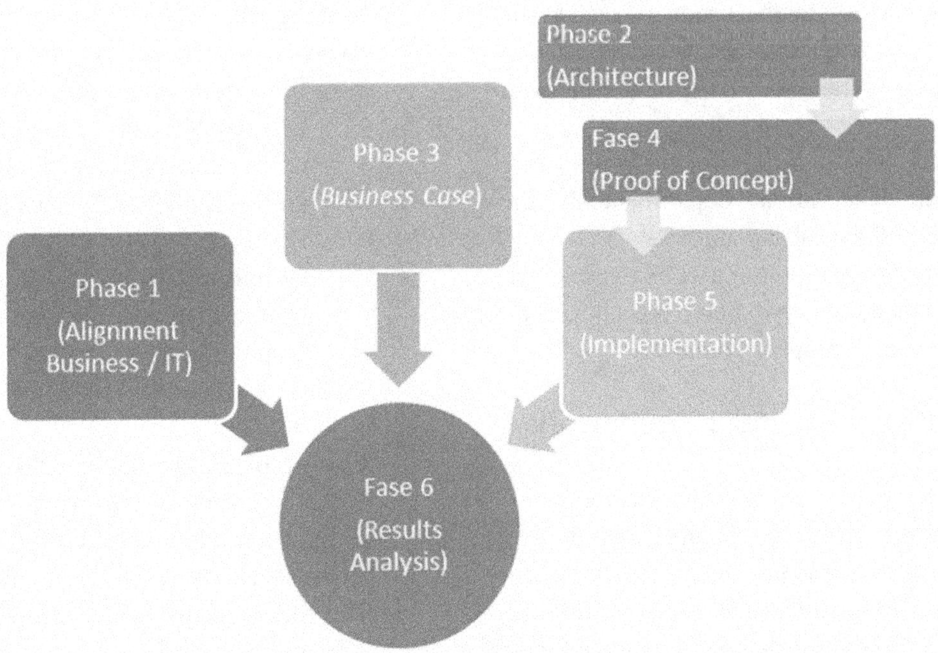

Each phase has its own control mechanisms. This phase combines the results, to confirm the goals that have been promised in the initial phases. This exercise of "self-reflexion" is very important to enable managers to rethink the way they commit themselves with the results, especially in the first and third phases.

THE FRAMEWORK IN REAL LIFE

The authors have been given the opportunity to put into practice the ideas presented in the paper. This experience has been very beneficial as it gives us a clear sense of the reality and enabled to test the framework. Our "case study" scenario has been developed in a private bank from the Portuguese financial system, which has the following characteristics:

- It is a retailing bank, offering a wide range of financial products;
- In terms of financial ranking it is among the top 5 largest Portuguese Banks;
- It has about 900 branches spread throughout the country;
- It provides alternative channels, such as the presence on the Internet, where clients are allowed to carry out most of the banking services;
- It has approximately 6,000 employees, spread over various departments of the Bank in Portugal and abroad;
- It has a presence in countries like Venezuela, Mozambique, Libya, South Africa, as well as in some European countries.

The First Version of the Framework

Some of the ideas present in the first version of the framework have been applied and showed the power to bring improvements in the life cycle of projects. There was the recognition that the projects could better managed to improved their alignment with the business. It was also recognized that, in the final stage of each implementation, the necessary conclusions should be taken in order to evaluate what went well, less well and, more important, what lessons to take for the future. The initiatives undertaken to foster the required improvements were:

- **Business Alignment:** Two distinct groups were created with complementary objectives. The first group starts by assessing the idea behind the project as suggested by the framework's phase # 1.If the project gets the necessary approval, then a minimum budget is made available to enable the different teams to estimate the costs, as well as to predict the time required to develop the respective task. Upon completion of the previous phase is possible to have an estimate in terms of finance and time. These values will be evaluated by the second group, which decides whether or not to approve the project;
- **Applications implemented in the Public Cloud:** Over the last three years the use of cloud computing has been a reality, but banking is a very conservative business environment. The use of cloud was mostly done in the following applications:
    - **Electronic mail:** It has been an optimal alternative for employees who are geographically located outside Portugal;
    - **Collaborative tools:** Allows sharing of information regardless of the geographic location or the group company where the employee carries out its activity;
- **Review of Data:** Another change to the normal life cycle of the project was the creation of a group whose mission is to audit all projects whose budget is over € 25K. The following points are subject to strict scrutiny:

- **Budget Deviation:** Comparison between the original budget and the actual budget. The project manager is interested in minimizing deviations, because the quality of budgets is a factor in the individual performance assessment;
- **Equipment and Software Invoices:** All expenses are audited so there is assurance that the budget will be followed.

The Second Version of the Framework

Going back to 2008 the Bank decided to make a partnership with an outsourcing services provider to manage all the infrastructure component.

As mentioned earlier, at the beginning of 2012 the bank faced some challenges which required management decisions to be made:

- **Infrastructure costs reduction:** It was found that the amounts charged by outsourcing in some infrastructure components, mainly in development environments, did not correspond to the gains introduced, so it is mandatory to reduce these costs;
- **Greater dynamics in the deployment of solutions:** All outsourcing companies have their own certified operational procedures which are required to manage the services they provide to the customers. However, these principles are not always compatible with the speed that sometimes is needed to develop or adapt the applications, so it was necessary to accelerate the deployment;
- **Alternative to the outsourcing provider:** Competition is always healthy, so it was identified the need to create alternative to offerings that were previously only within the reach of the outsourcing services provider.

The answer to these challenges was associated with the implementation of a Private Cloud. This project gave an opportunity to validate the second version of our framework. After about a year the project situation is reflected in Figure 17. As can

Figure 17. Environment distribution by solution

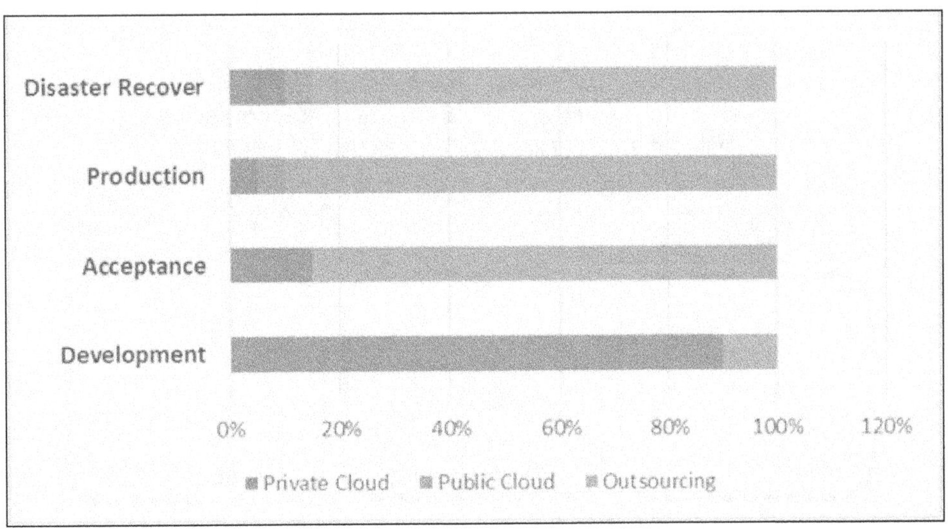

be seen by the distribution of environments, most of the component development is already supported by the Private Cloud. The Private Cloud is also being considered as an alternative solution to replace the other environments, in particular those that are still under the current outsourcing contract.

It has been advocated over the article that a Private Cloud component should be treated in a manner similar to the Public Cloud.

In the case study that is being described there is a single platform, operated by the same teams, to address some of the ITIL processes: Help Desk, Incident and Problem Management, Configuration and Change Management, Service Level Management. That is represented in Figure 18.

Two advantages are evident in this option. Firstly, there is just an operational procedure whatever the solution that supports the application. Thus, it is possible to change the application platform and all processes remain unchanged. Secondly it allows to streamline the processes with the private cloud solution, before eventually moving to the public cloud on a larger scale than exists today. As it has been mentioned, it is believed that private clouds are a stepping stone to public clouds.

CONCLUSION

We ended the previous version of the article by promising that we would continue to work to improve the framework then presented (Rodrigues & O'Neill, 2012). With this new version we aim to have fulfilled the promise.

From the standpoint of management benefits we consider that the framework resisted well to the time that has passed. The ideas that were then defended remain actual. The crisis situation we are experiencing all over the world allowed to emphasize that each investment must be planned, supervised and it must ensure that provides an adequate return.

However this new version was "born" earlier than we had foreseen. From the point of view of architecture, we must confess that we were surprised by the rapid rise of the Private Cloud

Figure 18. Single model to address the ITIL processes

phenomenon. Nevertheless we consider that it does not question the public cloud concept previously advocated. Rather than, both concepts complement one another. We would venture to say that the consolidation of the Private Cloud is sustaining the foreseen growth of the emerging Public Cloud. Making an analogy with our day-to-day, if we imagine that we want to cross a creek, just someone with great legs may get through it, however, if there is a stone in the middle, then everyone may get across the creek.

At this moment we confirm our commitment to continue watchful and wherever necessary the appropriate changes will be introduced in the framework.

REFERENCES

Chappell, D. (2011). *The benefits and risks of cloud platforms: A guide for business leaders*. David Chappell & Associates.

Frost & Sullivan. (2012). *U.S. cloud user survey: Evolving usage of public, private, and hybrid clouds*. Author.

Gartner. (2012a). *High-tech tuesday webinar: The AIM market: Beyond SOA and toward cloud*. Author.

Gartner. (2012b). *5 things private clouds are not!* Retrieved April 2, 2013, from http://www.businesscloud9.com/content/best-practice-5-things-private-clouds-are-not-according-gartner/11855

Gartner Webinars. (2012). *Integrated systems: High-tech Tuesday webinar: Cloud or fog: Clearing the air around cloud services*. Author.

IDC. (2012a). *Cloud transition series: Cloud definitions and market opportunities*. Author.

IDC. (2012b). *Converging the datacenter infrastructure: Why, how, so what?* Author.

Moore, G. (2012). *Living on the fault line*. New York, NY: HarperCollins Publishers.

Rodrigues, A. (2009). *Methodology for implementing software as a service - The analysis of benefits management*. (Master's Thesis). Lisbon University Institute (ISCTE-IUL), Lisbon, Portugal.

Rodrigues, A., & O'Neill, H. (2010). *Framework based on benefits management and enterprise architecture*. Paper presented at CENTERIS 2010 – Conference on Enterprise Information Systems. Viana do Castelo, Portugal.

Rodrigues, A., & O'Neill, H. (2012). Framework based on benefits management and enterprise architecture. *Information Resources Management Journal, 25*(2), 34–51. doi:10.4018/irmj.2012040103

Ross, W. J., Weill, P., & Robertson, C. D. (2006). *Enterprise architecture as a strategy*. Cambridge, MA: Harvard Business School Press.

Ward, J., & Daniel, E. (2007). *Benefits management delivering value from IS & IT investments*. London: Wiley.

Chapter 17

Performance of Firms: A Study of Offshore Outsource Service Providers

Soni Agrawal
Indian Institute of Technology, Kharagpyr, India

Kishor Goswami
Indian Institute of Technology, Kharagpyr, India

Bani Chatterjee
Indian Institute of Technology, Kharagpyr, India

ABSTRACT

With the evolution of information technology, firms offshore outsource services to developing and low service cost countries to have cost as well competitive advantages. This is a growing practice, but there has been limited empirical attention in understanding the outsourcing phenomenon, particularly from the perspective of service provider firms that execute important business processes for their overseas clients. This shows the need to study the factors that play a significant role in the growing trend to outsource and why only a few service provider firms report success. In this chapter, the authors try to find factors that influence performance of service provider firms. Multiple regressions using four indicators of firm performance are carried out to see the influence of certain factors on Information Technology Enabled Service (ITES) firms' performance.

INTRODUCTION

This paper investigates factors influencing performance of ITES provider firms. Traditionally, value chain activities related to a particular business are done in-house. However, with the development of information technology (IT) and the availability of services at competitive price, offshore outsourcing of different value chain activities is taking place. Firms go offshore to have competitive advantages, access to unique resources, early entry into market, etc. A firm needs to take care

DOI: 10.4018/978-1-4666-4506-6.ch017

of certain factors that are important in its survival and growth. Each firm possesses fundamentally different predisposition to evaluate their value, assimilate, and apply them into final goods. In order to attract clients, a service provider firm has to exhibit such values to their client firms. In offshore outsourcing of services, clients want certain quality or cost related advantages from a third party service provider firm. Moreover, relationship between the service provider and the service receiver plays an important role to be successful, as the locations of the service provider and the receiver are different in this case. For improving the quality of relationship, service provider firms often go beyond rules, agreements, and exceptions. Several Asian countries, notably India, China, and Philippines had made rapid strides in projecting themselves reliable and low-cost destinations that are capable of serving the world's IT-related sourcing needs. India with favourable government policies i.e., ten-year tax holiday, rebate on custom duties, etc., provide a more conducive regulatory environment to global corporations. This helps India in becoming leader with a considerable number of offshore service provider firms. The contribution of IT-BPO sector revenue in national GDP was 1.2 per cent in 1988 and 6.4 per cent in 2010. It is estimated that Indian IT-BPO sector generated US$ 76 billion revenue in the financial year 2010. It is also estimated that the sector generated employment of nearly 2.5 million during this period (NASSCOM, 2011). Considering the immense growth potential of ITES and the active role of the Indian service provider firms, it becomes essential to analyze the factors that influence performance of ITES firms in India to remain sustainable. It is also challenging because source of firm-level data in the context of Indian BPO sector are mostly middle-of-the-road private firms and are cautious in sharing data. The other challenge is to gather required information from the top level executives of these firms, who are usually having a very busy schedule.

Factors that influence performance of ITES firms are many as these firms have to remain sustainable in the global competitive market. These ITES firms provide services to overseas clients, who are having different language, culture, and even their time-zones are not same in most of the cases. Although, demand for these services is increasing as overseas firms in developed countries are getting services at a relatively lower price from developing countries. On the other side, service provider firms in developing countries are responding to the needs of the overseas client firms by enhancing their products and services, and they are also trying to utilize location based advantages in executing various business activities and processes. These ITES firms are trying hard to improve their performance. This needs a careful attention to look into the issue to understand the factors that are important for these ITES provider firms for their performance.

FACTORS INFLUENCING PERFORMANCE OF FIRMS

Review of studies from the perspective of service receiver firms show that performance of offshore outsourcing receiver firms depend on a variety of factors. These factors are their readiness, superior strategy, selection of suitable processes for outsourcing, sound contracts, and good relationship management (Feeny & Willcocks, 1998; Willcocks & Lacity, 2006). Client firms want the provider firm to take them into account with their special technological and business characteristics (Lacity & Hirchheim, 1993). There are other factors also that are identified important in earlier studies. Moreover, a detailed discussion of the identified factors of firms' performance is discussed below.

Lee et al. (2007) explained that organizational learning and knowledge management capabilities are important antecedents of e-business value. Findings of Hussein et al. (2007) indicated that

IT related factors of firms are significantly correlated with IS outsourcing success. It is found that offshore outsourcing service provider firms must possess qualities such as technical competence and experience (Goles, 2001; Gonzalez et al., 2005; Mouhalis, 2007; Martı´nez-Noya et al., 2012). Some researchers emphasized on quality standards and services such as IT capabilities of firms (Levina & Ross, 2003; Han et al., 2008; Khan et al., 2011). It is also found that the involvement of top management in negotiating the contract results in better payoffs (Bandyopadhyay & Pathak, 2007). Davern & Kauffman (2000) emphasized that the engagement of superior executives make it sure that adequate resources are allocated properly and timely. Roberts & Dowlings (2002) found the positive relationship between corporate reputation (brand image) and financial performance.

Service provider firms should have the ability to identify, acquire, develop, and deploy human resources to achieve its organizational objectives (Desarbo, et al., 2005; Lahiri & Kedia, 2009; Khan et al., 2011). Researchers suggested that the quality of human capital plays an important role to achieve above-average performance, which ultimately results in generation of competitive advantages (Hatch & Dyer, 2004). Contractor & Mudambi (2008) examined human capital and other factors for offshore outsourcing performance of 25 nations in service and manufacturing industry over a period of 14 years and found that human capital significantly influenced both services and goods. The results showed that offshore outsourcing of product development is partially explained by the shortage of highly skilled technical talent in US (Lewin et al., 2009). Jain et al. (2005) postulated that highest proportion of skilled labour is located in developing countries. Outsourcing service provider firm's distinctive resources and innovative knowledge assets contribute to firm performance (He & Wang, 2009). Outsourcing service provider firms continue to offer a wide variety of value-added services to contribute effectively in rapidly growing markets. They use expertise arising from and embedded in their different internal assets and employees. In this process, they generate revenues, learning, and increased competitiveness (Budhwar et al., 2006). Bresnahan et al. (2002) found that IT and human skills with educated and experienced employees together influence performance of the firm. Grover et al. (1996) also found a direct relationship of performance of service providers' employees (having capabilities such ability to work in different culture, commitment, etc.) to the relationship between the service provider and the receiver firms.

Training of employees is also found important in earlier studies because of continuously changing needs of service receiver firms (Grant, 1991; Valle & Castillo, 2009). Earlier, only tangible factors were considered as important influencing factors for firm performance. Gradually, importance of intangible resources such as training of employees of firms was also found important (Grant, 1991; Valle & Castillo, 2009). It was found that very few earlier studies had measured the direct influence of training on performance due to technical difficulties associated with its measurement (Valle & Castillo, 2009). The universalistic approach says that there is a direct relationship between training and performance (Valle & Castillo, 2009). The logic behind this idea is that return on training will be more than investment. The justification lies on the fact that trust is built between the employer and the employee, subsequently resulting in long term retention of employees. Employees perform their duties with greater effort and efficient manner. It may be because such performance creates a moral debt in employees (Tzafrir, 2005). Most of the researches measure a set of human resource practices combined into a single factor but not training separately as an independent influencing factor. Hatch & Dyer (2004) found that existing experienced employees with up gradation of skills cannot be compared with the gaining of human capital by means of recruitment. The influence of training is also found controversial. This was found positive as well as negative in a very few studies. In the study of Bartel (1994), it was found that implementation of formal training programs increased labour productivity by 17%. It was also found that there is a positive influence of training on satisfaction with highly internationalized firms (Downes et al., 2001).On the other hand, negative effect of training was found on performance, in case, employee is not able to exhibit reciprocal commitment. Moreover, it can be risky for a company, if an employee leaves the company after getting trained. That will incur loss, not only to the investment made on the training of the human capital (Tzafrir, 2005) but also the loss because of transferability of trained human resource to other organizations.

Several researchers discussed a wide range of risks in offshore outsourcing from client's perspective, resulting in undesirable outcomes such as unexpected escalated costs, disputes and litigation, loss of organizational competencies (Auburt et al., 1999; Bahli & Rivard, 2003), and risks related to the security and confidentiality of data (Evaristo et al., 2005). Some other risks are difficulties in contract enforcement, uncertainty due to imperfect information regarding market demand as well as technological changes and breakthroughs (Gopal et al., 2003; Swinarski et al., 2006).

Service provider firms generally face problems in quoting price in the competitive environment as they have to give quality services at a very competitive price. According to Sohn et al. (2009), effective pricing models can surely result in the financial success of a firm (Claver et al., 2002; Khan et al., 2011). There are several methods for pricing such as cost based pricing method, in which a firm decides a price for a product based on costs either fixed or variable, or a combination of both. This rule is popular in industrial firms as it is simple and safe as well as it covers all the costs (Morris & Joyce, 1988). Pricing strategy becomes highly complicated in export related products and services, where competition is high. A cost plus strategy makes the exporter unresponsive to demand and competitive actions (Piercy, 1981). Another concept is strategic pricing, or managerial pricing orientation (Smith, 1995). The managerial pricing orientation is defined as 'the pattern of policies, activities, or behaviours that business units typically engage in with regard to information gathering and processing; objectives, decision rules and beliefs; organizational decision processes; and organizational responsiveness relating to setting or changing price' (Smith, 1995). Teng & Thompson (1996) provided a general quality–price decision model where pricing is mainly based on quality. In summary, firms that are more competent in pricing, deal better with the complexity of a pricing process. This suggests that there is no generally 'best' or 'worst' practice but a contingency approach seems appropriate.

It is likely that firms need to focus on sharing of knowledge or information and to achieve better performance to improve organizational profitability (Han et al., 2008; Winkler et al., 2008; Khan et al., 2011), and to gain access to other firms' capabilities (Youndt et al., 1966). This process becomes more important when business environment is dependent on in-depth knowledge of technologies and processes. Moreover, some articles discussed issues related to client-supplier relationship (Kern & Blois, 2002; Lee et al., 2004). The relationship intensity includes measures of trust, dependence, commitment, and coordination (Morgan & Hunt, 1994). Trust between both the parties (Clarck et al., 1995), understanding of others' environment rather than merely sharing information, etc., were found important in earlier studies. It was also found that joint training on interdependent tasks, joint planning sessions, and formation of cross-functional teams are important in firms, where shared work is done (Rai et al., 2009). The level of communication in which firms try to understand client's style, standard, and culture (Hurst & Hanessian, 1995), and ability to give good value for client's money were also found important (Ang & Straub, 1998).

Gonzalez et al. (2005) analyzed the factors responsible for the success of large Spanish firms and concluded that service provider firms should try to understand clients' objectives. Frequent client-vendor contacts are important in building superior working relationships, confidence, comfort, and trust (Clarck et al., 1995). Good outsourcing contract ensures that the expectations of clients towards customized services, information security, flexibility in delivery of services, etc., will be realized (Willcock et al., 1995; Lee, 2001).

Enlow & Irtel (2006) found that majority of buyers in both public and private sectors and the majority of service providers (at least 30% of the annual contract value of their deals) value the way the relationship has been managed. The study of Lee et al. (2004) suggested that partners would be interdependent through socialization and will exhibit reciprocal behaviour because they will be morally obligated to serve each other's purpose (Kingshott, 2006). Partnership quality in earlier studies was based on trust, cultural proximity, knowledge sharing, etc. (Lahiri et al., 2011), and influences positively the performance of firms (Goles, 2001; Lee, 2001; Han et al., 2008). The study of Lahiri & Kedia (2009) tested the role of partnership quality in three different models. Partnership quality was measured as independent variable, mediating variable as well as moderating variable in three different models. The other factors considered in the study were human capital, organizational capital, and management capital as independent variables. Firm performance was taken as the dependent variable. Thus, the discussion shows that relationship between partnership quality and firm performances is dealt in different ways in different studies.

Institutional environment was conceptualized as an environment comprised of normative, regulative, and cognitive structures that constrain and legitimize firm's behaviour (Scott, 1995). The element of environment was defined as accepted rules of conduct that are reflected in laws, governance mechanisms (North, 1990), and in economic behaviour (North, 1990; Estrin et al., 2009). The elements of legal system governs both organizational and industry actions by establishing the base for production, exchange, and distribution with the intent of establishing order and legitimacy (Yiu & Makino, 2002). Environment is considered relevant, when physical and social factors outside the organization influence the decision-making process positively (Duncan, 1972). Researchers found the relationship between infrastructure such as real estate, availability of telecommunication facilities, etc., on the performance of firms (Khan et al., 2011).

Demand of offshore outsourcing of services is increasing globally because of aging of population in developed countries, time separation, and organizational differences that would subsequently influence decision making. Kearns & Lederer (1999) investigated environmental uncertainty and found that this is influencing firms' performance. It was suggested that firms can get benefitted in the global environment even in uncertainty with the coordination of subsidiaries, which are geographically dispersed (Kogut & Kulatilaka, 1994). Even in economic downturn, firms (ITES) can get the opportunity to restructure or acquire new lines of business while valuations remain attractive (NeoIT, 2008). This indicates that firms should act to leverage the benefits of globalization and optimize their global supply chain, if they are getting support from their home countries (Espinosa et al., 2006).

From the above discussion it is found that, environment outside the firm may be both favourable as well as unfavourable in the ongoing global scenario. Although this is an important factor that influences the performance of a firm, it is difficult to assess the influence of this factor on ITES firm's performance.

MEASUREMENT OF PERFORMANCE

Performance of firm is the indicator of an organization's success (Tan & Litschert, 1994). There are many studies that advocate financial based economic measures to capture performance. They prefer perception based data for independent variables and financial based data for dependent variable (Venkatraman & Ramanujam, 1986; Ichniowski & Shaw, 1999). They advocate that financial measures are more suitable or perhaps the best indicators to measure organizational success and sustainability. On the other hand, numbers of studies are also available, which used perception based data in their studies to capture performance both for independent variables and dependent variables (Youndt et al., 1996; Katou & Budhwar, 2006). However, perception based data was found more suitable to measure performance in case of emerging economies (Bae & Lawler, 2000). It was justified that there is a strong relationship between perceptual measures and objective measures of organizational performance. Performance of firm is measured by 'market share', 'profitability', 'growth', and 'customer satisfaction' in the study of Khandwalla's (1977) using perception based measures.

In any industry, both clients and service suppliers play important roles. The magnitude of change introduced by IT offshore outsourcing, presents a significant positive economic impact on nations, firms, and employees of both the service provider and the service receiver countries (Jain et al., 2005). It is observed that in international outsourcing, there is a need of dedicated studies from the point of view of service provider firms also (Parkhe, 2007). Limited empirical attention in the understanding, from the perspective of the service provider firms that execute important business processes for their overseas clients, is an identifiable gap here.

There are many factors that were identified by service receiver firms as important and they considered those factors before finalizing their service providers. Those factors might be important for service providers firms that can lead ultimately to their (service provider firm) performance. This shows the need to study the influence of factors on the performance of ITES provider firms.

THEORETICAL UNDERPINNING

Success in offshore outsourcing industry is based on the expertise of the service provider firms in providing valuable services, and to satisfy their service receiver firms (Sen & Shiel, 2006; Lahiri & Kedia, 2009). As the locations of the service provider and the service receiver firms are different in case of offshore outsourcing, it is required that service provider firm should understand the needs of the service receiver firms and should try to fulfil their requirements in the desired manner. Although, expectations might be different in both the cases, the source of success for both the partners hinges on service providers' resource-base and how the expertise arising from such resource-base is suitably utilized. In this context, firm's RBV (Barney, 1991; Armstrong & Shimizu, 2007) seems appropriate to explain and justify both the model as formulated for this study.

RESEARCH METHODOLOGY

Data and Sample

Data for this study were collected mainly on the basis of perceptions of the executives of ITES firms, commonly used for behavioural and managerial studies (Ramanujan et al., 1986; Deshpande et al., 1993; Mennon et al., 1999). Reason behind using perception based data was that, factors identified from the review of literature were tangible as well as intangible in nature. It was difficult to quantify all the desired data on the identified factors. Unavailability of quantitative data from the ITES firms was another reason. It is found that financial reporting was not available in most of these ITES firms. It becomes more difficult in case of emerging economies (Hoskisson et al., 2000). However, collection of perception based data from individual ITES firms was possible, and such were used by earlier researchers also for measuring firms' performance.

For collection of data, a detailed questionnaire was developed to obtain perceptions of the executives. The questionnaire was developed for this study due to non-availability of such instruments in Indian context. To design the questionnaire, initially important factors were identified from the review of literature. Thereafter, based on definitions, questionnaire was designed. As there are limited studies available, the study had taken only a few questions from earlier studies that were fit to the definition of a particular factor with or without modifications or minor variations in language and response categories to suit Indian condition. Management experts' opinions were also taken on the draft of the questionnaire to assure content validity.

For explanatory factors, response categories against each item were on a five point Likert scale ranging from '1' influencing very less, to '5' influencing very much. In formulating the questionnaire 5 point Likert scale is used and respondents were asked to mark their views. This justifies that the answers of the categories are mutually exclusive as the respondents had to select not more than one choice against an item. Attempt was also made to minimize the bias in the language of the questions.

Finally, data from 142 firms were collected. Out of this, 35 organizations are from Kolkata, 62 organizations are from Delhi (NCR), and rest 45 organizations are from Bengaluru. The criterion for retaining responses was to ensure that all the questions had been responded so that there were no missing data. Official designation of the respondents included president, vice president, director, chief administrative officer, chief operating officer, and senior manager. Reliability and validity tests were carried out and it was found that both reliability and validity were satisfactory for all the factors.

IDENTIFICATION OF THE FACTORS

In this study, factors influencing performance are derived from the review of literature based on studies related to service receiver as well as service provider firms. The factors considered by a service receiver firm while selecting a service provider firm, give a clear idea of service receivers' requirements as well as the business opportunity to the service provider firm. The factors such as training facilities provided by a service provider firms to their employees and lower rate of attrition were indicated important by service receiver firms. These factors were considered in earlier studies also from service providers' perspective, and were considered as a single factor under human capital or employees (Lahiri & Kedia, 2009). In the present study, 'human capital' and 'training' are considered two independent factors as found important after a discussion with ITES provider firms. Similarly, there are different dimensions related to 'pricing strategies' and 'risk' that influence performance of the firms. In the present study, conceptualization of the factors is based on the understanding and feedbacks collected from the executives of the ITES firms.

Another factor, 'partnership quality' was tested in earlier studies of offshore outsourcing from service provider's perspective. In the present study, 'partnership quality' is expressed through two independent factors, viz., 'service level agreement' and 'sharing of information'. Although partnership quality was used widely in offshore outsourcing literature, it seems that knowledge sharing and understanding, joint working and partnership contracts are mixed together. Certain dimensions of partnership quality such as contracts, business requirements, and issues related to services delivery talk about contract related concepts. Whereas, knowledge sharing is based on interpersonal trust and working together attitude. Therefore, in the present study, partnership quality is expressed into two independent factors. The assumption here is that an organization is likely to be successful in the long run, if it constantly offers its partners bigger and better benefits than its competitors. In return, they gain trust and get long term relationship. The rationale here is that, the factor is influenced more as process oriented nature rather as resource base. Therefore, this factor along with basic resources and capabilities of an organization and components of corporate strategy or synergistic combinations thereof contribute to the performance of the company (Kavcˇar & Tavcˇar, 2008).

Both the service provider and service receiver firms work to improve the processes to remain competitive ultimately. In this process, sharing of information between partners is essential. This study takes an assumption that stronger the agreements between the service provider and service receiver are, better the performance would be achieved by the service provider firms. Better understanding, based on trust and commitment in sharing of information between service provider and service receiver firms, would be propelled by better utilization of resources and capabilities of both the firms that ultimately will influence the firm performance.

It is difficult to develop and define measures for unobservable factors particularly if the industry is new (Armstrong & Shimizu, 2007). In the present study, these measures were modified, wherever needed to make them appropriate for this research context. After setting up scales, the whole questionnaire was discussed with academicians and their comments were analyzed and incorporated accordingly. Nine independent and four dependent factors were identified on the basis of literature review as presented in Figure 1. Perception based measures are used in this study to measure firm's

Figure 1. Factors influencing performance of ITES firms

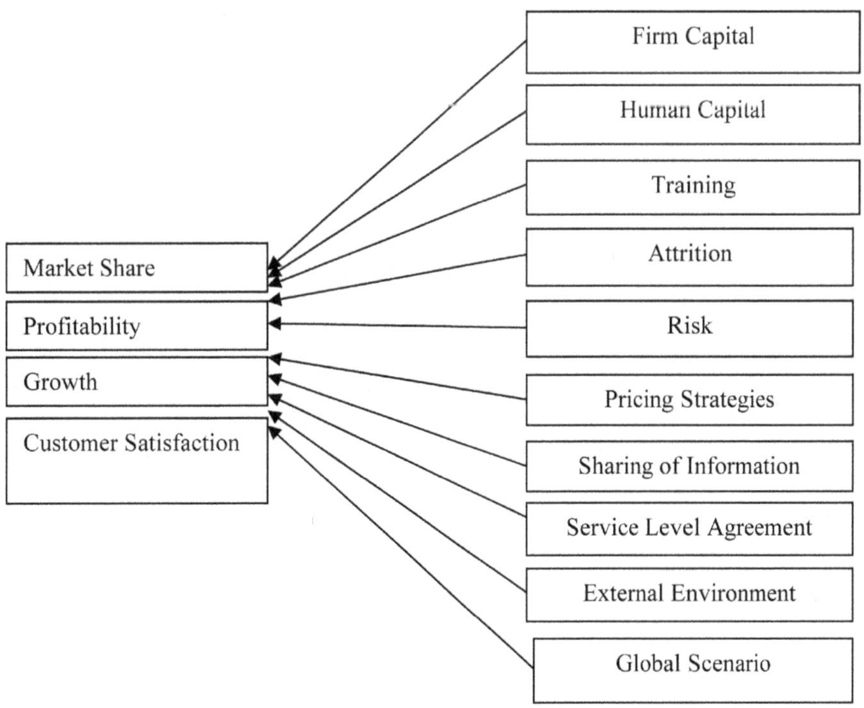

performance. The reason behind using perception based data is that, the industry is new and quantification of data on the identified factors is not possible either because of non-availability of data or intangible nature of the identified factor.

Multiple regression analysis is used here, as the aim of the present objective is to see the direct influence of all the independent factors on the performance of ITES firms. Some other similar statistical models such as step-wise regression has been used in earlier studies when the requirement was to see the influence of independent factors as one step to another or to see the forms of sequence of different independent factors on dependent factors. Hierarchical regression model compares the levels of interest factors across samples one after another. For this research objective, multiple regression technique seems more suitable, as the motive of this objective is to analyze the influence of independent factors on the performance of ITES firms.

Statistical Model

$$Y_i = \alpha_i + \sum_{j=1}^{10} \beta_j X_j$$

$+ u_i$ for all $i = 1, 2, 3,4$

where:

α_i = Constant
β_j = Regression coefficient
u_i = Disturbance term
The dependent factors in the model are:
Y_1 = Market share
Y_2 = Profitability
Y_3 = Growth
Y_4 = Customer satisfaction

The independent factors in the model are:

X_1 = Firm capital
X_2 = Human capital
X_3 = Training
X_4 = Risk
X_5 = Pricing strategies
X_6 = Sharing of information
X_7 = Service level agreement
X_8 = External environment, and
X_9 = Global scenario

Dependent or criterion factors are identified as 'market share', 'profitability', 'growth', and 'customer satisfaction'. Reason behind using four different factors for measuring performance is because the understanding and strategy of all the firms in measuring performance might not be the same. Some firms may emphasize more on customer satisfaction, while some firms might be more concerned for profitability. At the same time, new firms might be interested in the growth of the firm in the initial years of the establishment. This justifies that performance of a firm is measured by different indicators, which may derive some useful information for a group of factors influencing performance. Here both independent as well as dependent factors are measured with the help of perception based data, collected by individual firms. Multiple regression analysis (MRA) is carried out to derive the factors that influence the performance of ITES firms. Initially, it was conceptualized that financial data on the dependent factors will also be included in the study, but finally they are excluded as most of the ITES firms were not comfortable in disclosing their actual financial data with the researcher.

Reliability tests in the study are carried out using Cronbach's alpha (α). For testing internal consistency of all the factors considered in this study, correlation among the factors is analyzed.

Inter-item correlation is measured to test the internal consistency of each factor. Validity tests are performed with the help of convergent validity, construct validity, and internal validity. Convergent validity is tested using item-total correlation (correlation of each item to the sum of the remaining items). This method is used to check uni-dimensionality and internal consistency of measures (Gerbing & Anderson, 1988). Exploratory factor analysis (EFA) with varimax rotation technique is used to see the construct validity of the factors. Apart from that, internal validity is tested with the help of item-total regression technique (Malhotra, 2007).

Before analysis, an attempt is made to determine whether the key statistical assumptions such as linearity of relationship and normal distribution of data are met. Linearity is tested using scatter plot method, which revealed apparent linearity. Normal distribution is verified by derivation of a normal probability plot through kurtosis and skewness. As several ITES firms did not respond to the questionnaire survey, t-test is carried out to derive the non-response bias. However, t-test did not reveal any significant difference ($p< 0.05$) between the respondents and non-respondents. Comparisons of early and late respondents also did not reveal any significant difference ($p< 0.05$), indicating that the differences between respondents are not related to non-response bias.

RESULTS

Reliability Test and Descriptive Statistics

Cronbach's α is used to determine the reliability of the scales. It is based on average correlation or covariance between questions in a scale and was extensively used by Nunally (1978). The analysis used a value of 0.50 or higher to ensure reliability of the measures. A few questions in the questionnaire scale were deleted on the basis of low alpha value of 0.40, a cut off point for deletion of questions, suggested by Gaski & Etzel (1986). Nunally (1978) suggested that reliability over 0.5 is adequate for basic research but under apparent reconsideration. The descriptive statistics and Cronbach's α value of the factors influencing performance of ITES firms are presented in Table 1.

Mean values of all the questions range from 3.35 to 4.54 with standard deviation (SD) from 0.35 to 0.93 as presented in Table 1. Correlation coefficient values are calculated among all the factors to see internal consistency of the factors, and results are presented in Table 2. Correlation coefficient values are not showing any significant correlation among the factors considered in this study. This shows that all the factors considered are independent, though related with each other. The results indicate positive relationship among the factors in most of the cases, except a few negative relations.

INTER-ITEM CORRELATION OF THE FACTORS

Pearson correlation is used to test 'Inter-item correlation' to see the internal consistency of the individual factors. Low to moderate correlation is found among different dimensions of each factor and values of correlation coefficients (r) range from 0.23 to 0.94 ($p<0.01$). Values of correlation coefficient amongst the four dimensions of the factor 'firm capital' range from 0.64 ($p<0.01$) to 0.94 ($p<0.01$), the four dimensions of 'human capital' range from 0.67 ($p<0.01$) to 0.81 ($p< 0.01$), the two dimensions of the factor 'training' are 0.71 ($p<0.01$), and for factor 'risk' it is 0.36 ($p<0.01$). Correlation coefficient value between the two dimensions of the factor 'pricing strategies' is 0.76 ($p<0.01$), for the factor 'sharing of information' is 0.49 ($p<0.01$), and for the factor 'external environment' is 0.53 ($p<0.01$).

Table 1. Cronbach's α with original and retained items and descriptive statistics of factors influencing performance of ITES firms

Independent Factors	Original Items	Items Retained	Cronbach's Alpha	Mean	Standard Deviation (SD)
Firm capital	5	4	0.93	4.13	0.76
Human capital	4	4	0.92	4.13	0.59
Training	2	2	0.83	4.14	0.35
Risk	4	2	0.50	4.43	0.67
Pricing strategies	5	2	0.83	4.24	0.76
Sharing of information	5	2	0.65	4.34	0.53
Service level agreement	6	4	0.62	4.54	0.50
External environment	5	2	0.66	3.94	0.87
Global scenario	4	3	0.62	3.53	0.93

Table 2. Pearson correlations coefficients of factors influencing performance of ITES firms

	Firm Capital	Human Capital	Training	Risk	Pricing Strategies	Sharing of Information	Service Level Agreement	External Environment	Global Scenario
Firm capital	1	-0.02	0.03	0.12	0.14	0.07	0.01	0.06	-0.09
Human capital		1	0.04	-0.09	0.05	0.17	0.18	0.13	0.19
Training			1	0.29	-0.09	0.09	0.17	0.12	0.63
Risk				1	0.54	0.13	0.02	0.23	0.15
Pricing strategies					1	0.46	0.02	0.17	0.13
Sharing of information						1	0.24	0.12	-0.01
Service level agreement							1	0.32	0.15
External environment								1	0.39
Global scenario									1

Correlation coefficient values amongst the three dimensions of the factor 'service level agreement' are from 0.23 to 0.44 ($p<0.01$), and for the factor 'global scenario' range from 0.35 to 0.37 ($p<0.01$).

VALIDITY TESTS OF THE FACTORS

Item-Total Correlation of the Factors

'Item-total correlation' is used to test the validity of the factors. Results are consistent with the results of earlier studies, opinions of academic researchers, and pilot testing (Han et al., 2008; Lahiri & Kedia, 2009). Thus, the convergent validity is also established. Use of item-total correlation (correlation of each item to the sum of the remaining items) has long been advocated as a useful method to check for uni-dimensionality and internal consistency of measures (Gerbing & Anderson, 1988). Researchers such as Lee (2001) and Lahiri & Kedia (2009) have also used this test as a way to assess convergent validity of the measures. The item-total correlation values obtained as part of the validity analysis. Correlation values obtained for the factors ranged from 0.87 to 0.93 (firm capital), 0.88 to 0.91 (human capital), 0.18 to 0.25 (training), 0.19 to 0.28 (risk), 0.15 to 0.18 (pricing strategies), 0.37 to 0.54 (sharing of information), 0.12 to 0.36 (service level agreement), 0.06 to 0.07 (external environment), and 0.02 to 0.09 (global scenario). These values indicate that the individual items of a factor are consistent with the overall measures for a particular factor. Moreover, construct validity of the factors is also tested further.

Construct Validity Test

An EFA is carried out using principle components analysis with varimax rotation to assess the construct validity of the scales used in the study. Rotated factor loadings ranged from 0.71 to 0.93 and there exists no cross loadings (Table 3). Results of EFA show that in each rotation, all the items are loaded in the single factor only. This suggests that the measurement model fits data reasonably well and proves the construct validity. Besides convergent and construct validity, internal validity is also measured (Malhotra, 2007).

Internal Validity Test

For testing internal validity, scores of all the questions of each factor are summed up for each respondent to get the scores of all the nine in-

Table 3. Coefficients of exploratory factor analysis of factors influencing performance of ITES firms

Independent Factors	No. of Items	Factor Loading	% of Variance Explained
Firm capital	4		
-Firm size		0.89	
-Firm brand		0.92	
-Firm technological capability		0.87	
-Firm management capability		0.93	
Human capital	4		
-Employee degree		0.92	
-Employee enthusiasm		0.88	
-Employee experience		0.89	
-Employee culture		0.89	
Training	2		
-Training to make fit for pursuing goals		0.92	
-Training continuous		0.92	
Risk	2		
-Risk related to projects		0.83	
-Risk on payments		0.83	
Pricing strategies	2		
-Fixed or variable price		0.94	
-Flexible pricing		0.94	
Sharing of information	2		
-Sharing information		0.86	
-Sharing experience		0.86	
Service level agreement (SLA)	4		
-SLA related to projects		0.71	
-SLA flexibility in delivery		0.48	
-SLA related to IPR		0.78	
-SLA for timely flexibility		0.73	
External environment	2		
-Government policies		0.87	
-Infrastructure		0.87	
Global scenario	3		
-Aging population		0.76	
-Global recession		0.77	
-Global policy		0.75	

dependent factors. Then, a multiple regression analysis is carried out to identify the importance of each factor scores for development of the total questionnaire scale. In this analysis, total factor scores are used as the dependent factor and the scores of the ten independent factors are treated as independent factors. The results are presented below:

Total Score = 0.48 (firm capital)
+ 0.44 (human capital) + 0.16 (training)
+ 0.23 (risk) + 0.12 (pricing strategies)
+ 0.17 (sharing of information)
+ 0.24 (service level agreement)
+0.21 (external environment)
+ 0.15 (global scenario)

$R^2 = 0.89$, F = (118.94) ($p<0.001$)

The above R^2 and all the independent factors considered in this study are statistically significant at 1% level of significance. This result indicates that the above scales are internally valid as all the sub-scales significantly contributed to the total scale.

All the reliability and validity tests show that questionnaire data is reliable as well as valid. Therefore, further analysis is done to derive the influence of independent factors on dependent factors.

INFLUENCE OF THE FACTORS ON PERFORMANCE OF ITES FIRMS

This section explains the influence of firm's internal resources and capabilities on the performance of ITES firms. This study contributes to the growing body of literature and by identifying factors to measure performance. Earlier studies do not confirm any performance indicator that is taken unanimously as a measure of ITES firms' performance. In the absence of that, four measures, which were used earlier by other researchers to measure firms' performance (Murphy et al., 1996; Georgius & Dibbern, 2005), are considered in this study. Results of this study reveal the appropriateness of the indicator(s) to measure ITES firms' performance. The influence of the independent factors on dependent factors is presented in Table 4 to Table 6.

For testing the direct effect of independent factors on ITES firms' performance, OLS method was used. In order to enhance the clarity of the results certain restrictions were imposed, which are mentioned below:

1. The identified combination of factors should explain the maximum amount of variance in the criterion as measured by the adjusted coefficient of determination (Adjusted R^2). The standardized regression coefficient (β weights) of each factor in the equation should be statistically significant;

Table 4. Regression results of factors influencing profitability of ITES firms

Independent Factors	Beta	t-Values	Significance
(Constant)	-----	1.83	0.07
Firm capital	0.17	2.09	0.38
Human capital	0.03	0.31	0.76
Training	0.19	2.12	0.04
Risk	-0.01	-0.13	0.90
Pricing strategies	-0.03	-0.22	0.83
Sharing of information	0.18	1.79	0.07
Service level agreement	0.11	1.26	0.21
External environment	0.01	0.03	0.98
Global scenario	0.04	0.49	0.63
R^2	0.13		
Adjusted R^2	0.07		
D-W	1.72		
F	2.19		
Significance	0.03		

Table 5. Regression results of factors influencing growth of ITES firms

Independent Factors	Beta	t-Values	Significance
Constant	-----	5.62	0.01
Firm capital	0.10	1.62	0.22
Human capital	0.05	0.59	0.56
Training	0.06	0.71	0.48
Risk	-0.26	-2.46	0.02
Pricing strategies	0.04	0.36	0.72
Sharing of information	-0.11	-1.08	0.28
Service level agreement	0.16	1.76	0.08
External environment	-0.02	-0.27	0.79
Global scenario	-0.06	-0.70	0.48
R^2	0.10	-----	------
Adjusted R^2	0.04	-----	------
D-W	1.80	-----	------
F	1.74	-----	------
Significance	0.08	-----	------

Table 6. Regression results of factors influencing customer satisfaction of ITES firms

Independent Factors	Beta	t-Values	Significance
Constant	-----	1.28	0.20
Firm capital	0.21	2.75	0.01
Human capital	0.16	1.93	0.05
Training	0.24	2.92	0.01
Risk	-0.01	-0.02	0.99
Pricing strategies	-0.04	-0.36	0.72
Sharing of information	0.01	0.05	0.96
Service level agreement	0.18	2.17	0.32
External environment	-0.11	-1.25	0.21
Global scenario	0.19	2.26	0.03
R^2	0.22	-----	------
Adjusted R^2	0.17	-----	------
D-W	1.75	-----	------
F	4.16	-----	------
Significance	0.01	-----	------

2. In this study, MRA is used to predict the influence of independent factors on dependent factors taking the overall sample into consideration.

Factors Influencing Market Share of ITES Firms

An attempt is made here to analyze the influence of the independent factors on the dependent factor 'market share' (Model 1). 'Market share' is frequently used to measure firm performance (Tsenga et al., 2009). However, the model is not significant, which indicates 'market share' cannot be taken as an efficient tool to measure performance of ITES firms with these influencing factors.

The results with insignificant dependent factor 'market share' as performance indicator show that, 'market share' as a performance indicator is not conceived well by ITES firms. Probable reason might be that, demand of offshore outsourcing is growing rapidly. ITES firms might be trying to get more and more projects from offshore clients, but they are not able to estimate their market share. Moreover, global slowdown influenced the demand for offshore outsourcing services temporarily. There was an adverse effect of backlash issues in developed countries with respect to offshore outsourcing to developing countries. As Friedman (2005) has written in his book that offshore outsourcing of services from developing countries is giving an adverse effect to American citizens as lots of jobs are going offshore.

Factors Influencing Profitability of ITES Firms

'Profitability' is used here as the dependent factor (Model 2). Results in Table 4 show that the model is statistically significant and 7% variation in 'profitability' is explained by the independent factors ($F = 2.19$, $P<0.03$). The positive standardized beta coefficient found between 'human capital' and 'profitability' is 0.19 ($p< 0.04$).

The influence of the factor 'training' is positive and statistically significant and beta coefficient of factor 'training' is 0.19 ($p<0.04$). This reveals that training of employees is very important for better performance of ITES firms. ITES firms need to have competitive staff. Continuous training strengthens employers' resources into firms, which is an important factor for firm performance. ITES firms should try to improve skills of their existing employees. This helps service provider firms to provide customized solutions to their clients. Thus, training and up gradation of skills becomes crucial for firms. Training might help companies to become competent for changing requirements, and might help in reducing attrition, which is one of the major problems faced by these industries as employees easily switch over to a new firm or new job for a better salary or some other incentives. Therefore, how to utilize and manage the existing resource portfolio, and what resource configurations would provide sustained above-average returns, should be the two key concerns of top managers.

The positive and significant value of 'sharing of information' indicates that trust and desire to share experiences can satisfy customer, as it helps in fulfilling the real motive of offshore outsourcing. As found in the earlier studies, trust between partners is suggested as an important factor for ITES firms' performance (Currall & Inkpen, 2002; Inkpen & Currall, 2004).

The importance of training facilities and sharing of information of service supplier firms indicate how well a firm is able to exploit resources in improving their performance. Other factors, viz., 'risk' and 'pricing strategies' are found negative and insignificant. The factors, which were found important from the review of literature, are not found significant to the performance of ITES firms measured through profitability. The influence of factor 'risk' is towards the desired direction, but not significant.

The influence of the factor 'service level agreement' is positive and statistically significant and beta coefficient of factor 'service level agreement' is 0.16. This indicates that favourable service level agreements influence the growth of the firms positively. Moreover, negative and significant beta values of the factor 'risk' are in the desired direction that indicates that lower the risk, the higher the growth of the firms.

Factors Influencing Growth of ITES Firms

'Growth' is taken here as the dependent factor (Model 3). The results in Table 5 show that the model is statistically significant. The R^2 value is 0.10. This indicates that the independent factors together explain 4% variation in firms' growth ($F = 1.74$, $P< 0.08$).

Factors Influencing Customer Satisfaction of ITES Firms

In this analysis, 'customer satisfaction' is considered as dependent factor (Model 4). Results in Table … show that the value of R^2 is 0.22, thus the model explains 17% of variation on the dependent factor 'customer satisfaction'. The F-value indicates that the model is significant ($F=4.16$, $P<0.01$). The standardized beta coefficient of the factor 'firm capital' is 0.21 ($p<0.01$). It is 0.16 ($p<0.05$) for 'human capital', 0.24 ($p<0.01$) for 'training', 0.18 ($p<0.03$) for 'service level agreement', and 0.19 (0.03) for 'global scenario'.

The standardized beta coefficient of factor 'firm capital' is 0.21 ($p< 0.01$). The results indicate that the factor is significantly influencing the customer satisfaction of ITES firms. Probable reason might be that, firms with desired qualities and capabilities might be doing better in the long run in getting projects continuously. This shows that the attributes of a firm such as experience, brand image, size, firm technology, and managerial IT capability play crucial role on firms' performance measured by customer satisfaction. The significant effect of the factor 'firm capital' suggests that codified knowledge-base embedded within firms along with the existing systems and processes, and experiences of firm are crucial determinants to firm performance. Brand image significantly influences to get new projects and in developing trust with the clients. This is also supported by earlier study of Raman et al. (2006), which argued that organizational capabilities play important role in success of firms. Saini & Johnson (2005) also examined firms' capabilities (such as information technology capability, strategic flexibility, and trust-building capability) and explained that these capabilities are critical for superior firm performance in electronic commerce (e-commerce). They further stated that clients, who become partner with Indian service providers, would not only appreciate the organizational resources including various capabilities and the qualities possessed by these service provider firms, but also the quality of manpower.

The positive and significant beta coefficient of factor 'human capital' is 0.16 ($p<0.05$). This proves the importance of employees with the customer satisfaction. The positive standardized beta coefficient value indicates that the factor 'human capital', which considers employees' education, experience, enthusiastic attitude towards work, and familiarity with handling different cultures, influences the customer satisfaction of ITES firms. This highlights the importance of knowledge level of employees, which also includes familiarity of firms' employees with technology. If employees understand internet technologies, they are more likely to interact with their customers or business partners, and carry on business over the internet (Zhu & Kraemer, 2002). Absorptive capacity or familiarity of employees with technology, which is also a dimension of quality of skilled employees, is a critical component in understanding IT management practices. The availability of this capacity is very useful in implementing new technology (Boynton et al., 1994; Harrington & Guimaraes, 2005). A service provider firm with qualified staff and required technology becomes viable and more likely to get success in business (Baldwing et al., 2001). The influence of factor 'human capital' suggests that it is the quality of employees that matters most in achieving desired performance.

Apart from that, the positive beta coefficient of the factor 'training' is 0.24 ($p<0.01$), which confirms the importance of training of employees enable an organization in providing customized, quality services in time and ultimately influences customer satisfaction. Positive and significant value of service level agreement towards customer satisfaction indicates that service receiver firms value contracts that are flexible and conducive. That positively builds the image of the suppliers and these results in repeat customers as proved by significant positive beta value towards customer satisfaction, which is a measure of performance.

Another positive and significant factor 'global scenario' ($\beta = 0.19$, $p<0.03$) indicates that favorable global conditions are important for customer satisfaction of the firms too. Policies of government and global conditions influence the decision of client firms for offshore outsourcing of services. The results support the argument that favorable global environment for offshore outsourcing in developed countries gives an opportunity to their firms to offshore outsource services from developing countries. This ultimately influences the demand, and subsequently growth of the service provider firms.

The positive and significant values for 'firm capital', 'human capital', 'training', 'service level agreement', and 'global scenario' prove that firm capabilities, talented employees, continuous training to the employees, favorable global scenario, and service level agreement with the receiver firms satisfies the customers. Other independent factors included in the analysis are found insignificant. Factors, viz., pricing strategies', 'sharing of information', 'external environment', and 'risk' indicate that, they are not the major drivers for customer satisfaction. The results show that these factors are not directly influencing performance of firms measured by the factors in the study.

The results show that there are some factors that are not found significant to the performance of ITES firms measured by four indicators. Those factors might be having some different connotation that could not be captured in the present study. Therefore, it is suggested that in future scholars may attempt to study the reasons to derive the variation in the influence of the factors.

The present study contributes to the broad realm of international business by drawing attention to global sourcing in general and focusing on BPO or KPO service providers firms in particular. How best to utilize and manage the existing resource portfolio and what resource configurations would provide sustained and above-average returns are the two key concerns of the top managers of service provider firms. Ideally, service provider firms are moving from arms' length tactical relationship to transformational relationship. This could happen because there is a mutual interest of value creation and competitive advantage between the service provider and the service receiver firms. The study sheds additional light on the factors influencing performance of ITES firms with the help of empirical analysis and discussions.

The study suggests that the top level executives of BPO or KPO service provider firms need to focus on their pool of firm capital and human capital as it exerts positive influence on the performance of firms. Service provider firms should focus not only on quality of manpower or organizational resources they possess, but also they should consider the total resource-base including various capabilities and quality of cooperative behavior. The results of this study suggest the importance of intangible assets of the firms as important determinants to ITES firms' performance. This is confirmed in this study as 'training' is found a significant independent factor influencing profitability and customer satisfaction. For another factor 'sharing of information', findings suggest that it depends on the strategy of a firm. Earlier research on inter-organizational knowledge management showed that knowledge sharing is particularly difficult between organizations (Ghosh, 2011). There are many issues involved such as privacy assurance, credibility and information asymmetry, and reliability on the other party (Vaidyanathan, 2008). However, the positive value of this factor with respect to profitability of firms implies that clients are interested in sharing work related information with their service provider firms. Thus, this ultimately influences the performance of firms as service receiver firms gain confidence with their suppliers. This gives them credibility and business in the long run.

FUTURE SCOPE OF STUDY

No study can be complete as there is always a scope of future research that remains untapped and can be taken in future for a better understanding of a phenomenon in the changing environment. Offshore outsourcing industry is on evolutionary stage and different environmental and economic changes in the global world are making the future of offshore outsourcing unpredictable. There is a future scope to study and analyze the influence of the factors on the performance of ITES firms using both perceptions based as well as financial based metrics, provided financial data is available. In future, researchers can analyze the impact of the strategic measures taken by ITES firms such as strategies related to reducing attrition, the impact of the strategies on the retention of employees in their firms, etc.

REFERENCES

Ang, S., & Straub, D. (1998). Production and transaction economies and is outsourcing: A study of the US bank industry. *Management Information Systems Quarterly*, *22*(4), 535–352. doi:10.2307/249554

Armstrong, C. E., & Shimizu, K. (2007). A review of approaches to empirical research on the resource-based view of the firm. *Journal of Management*, *33*(6), 959–986. doi:10.1177/0149206307307645

Aubert, B., Dussault, S., Patry, M., & Rivard, S. (1999). *Managing the risk of it sourcing*. Paper presented at the 32nd Hawaii International Conference of System Sciences (HICSS). Hawaii, HI.

Bae, J., & Lawler, J. J. (2000). Organizational performance and HRM strategies in Korea: Impact on firm performance in an emerging economy. *Academy of Management Journal*, *43*(3), 502–517. doi:10.2307/1556407

Bahli, B., & Rivard, S. (2003). The information technology outsourcing risk: A transaction cost and agency theory-based perspective. *Journal of Information Technology*, *18*(3), 211–221. doi:10.1080/0268396032000130214

Baldwing, L. P., Irani, Z., & Love, P. (2001). Outsourcing information systems: Drawing lessons from a banking case study. *European Journal of Information Systems*, *10*(1), 15–24. doi:10.1057/palgrave.ejis.3000372

Bandyopadhyay, S., & Pathak, P. (2007). Knowledge sharing and cooperation in outsourcing projects - A game theoretic analysis. *Decision Support Systems*, *43*(2), 349–358. doi:10.1016/j.dss.2006.10.006

Barney, J. (1991). Firm resources and sustained competitive advantage. *Journal of Management*, *17*, 99–120. doi:10.1177/014920639101700108

Bartel, A. P. (1994). Productivity gains from the implementation of employee training programs. *Industrial Relations*, *33*(4), 411–425. doi:10.1111/j.1468-232X.1994.tb00349.x

Boynton, A. C., Zmud, R. W., & Jacobs, G. C. (1994). The influence of IT management practice on IT use in large organizations. *Management Information Systems Quarterly*, *18*(3), 299–318. doi:10.2307/249620

Bresnahan, T., Brynjolfsson, E., & Hitt, L. (2002). Information technology workplace organization, and the demand for skilled labor, firm level evidence. *The Quarterly Journal of Economics*, *117*(1), 339–376. doi:10.1162/003355302753399526

Budhwar, P. S., Luthar, H. K., & Bhatnagar, J. (2006). The dynamics of HRM systems in Indian BPO firms. *Journal of Labor Research*, *27*(3), 339–360. doi:10.1007/s12122-006-1027-7

Clarck, T. D., Zmud, R. W., & McCray, G. E. (1995). The outsourcing of information services: Transforming the nature of business in the information industry. *Journal of Information Technology*, *10*(4), 221–237. doi:10.1057/jit.1995.26

Claver, E., Gonzales, R., Gasco, J., & Llopis, J. (2002). Information systems outsourcing: Reasons, reservations and success factors. *Logistics Information Management*, *15*(4), 294–308. doi:10.1108/09576050210436138

Contractor, F. J., & Mudambi, S. M. (2008). The influence of human capital investment on the exports of services and goods: An analysis of the top 25 services outsourcing countries. *Management International Review*, *48*(4), 433–445. doi:10.1007/s11575-008-0025-9

Currall, S. C., & Inkpen, A. C. (2002). A multilevel approach to trust in joint ventures. *Journal of International Business Studies*, *33*, 479–495. doi:10.1057/palgrave.jibs.8491027

Davern, M. J., & Kauffman, R. J. (2000). Discovering potential and realizing value from information technology. *Journal of Management Information Systems*, *16*(4), 121–143.

Desarbo, W. S., Di Benedetto, C., Song, M., & Sinha, I. (2005). Revisiting the miles and snow strategic framework: Uncovering interrelationships between strategic types, capabilities, environmental uncertainty, and firm performance. *Strategic Management Journal*, *26*(1), 47–74. doi:10.1002/smj.431

Deshpande, R., John, U. F., & Webster, F. E. (1993). Corporate culture, customer orientation, and innovativeness in Japanese firms: A quadratic analysis. *Journal of Marketing*, *57*, 23–27. doi:10.2307/1252055

Downes, M., Thomas, A. S., & Singley, R. B. (2001). Predicting expatriate job satisfaction: The role of firm internationalization. *Career Development International*, *7*(1), 24–36. doi:10.1108/13620430210414847

Duncan, R. B. (1972). Characteristics of organizational environments and perceived environmental uncertainty. *Administrative Science Quarterly*, *17*(3), 313–327. doi:10.2307/2392145

Enlow, S., & Ertel, D. (2006). Achieving outsourcing success: Effective relationship management. *Benefits Review*, *38*(3), 50. doi:10.1177/0886368706288216

Espinosa, J., DeLone, W., & Lee, G. (2006). Global boundaries, task processes and IS project success: A field study. *Information Technology & People*, *19*(4), 345–370. doi:10.1108/09593840610718036

Estrin, S., Baghdasaryan, D., & Meyer, K. M. (2009). The impact of institutional and human resource distance on international entry strategies. *Journal of Management Studies*, *46*(7), 1171–1196. doi:10.1111/j.1467-6486.2009.00838.x

Evaristo, R., Audy, J. L. N., Prikladnicki, R., & Avritchir, J. (2005). Wholly owned offshore subsidiaries for IT development: A program of research. In *Proceedings of the 38th Hawaii International Conference on System Sciences*. IEEE.

Feeny, D. F., & Willcocks, L. P. (1998). Core IS capabilities for exploiting information technology. *Sloan Management Review*, *39*(3), 9–21.

Friedman, T. L. (2005). *The world is flat: A brief history of the twenty first century*. New York, NY: Farrar, Straus & Giroux.

Gaski, R. M., & Etzel, M. J. (1986). The index of consumer sentiment toward marketing. *Journal of Marketing*, *50*, 71–81. doi:10.2307/1251586

Georgius, & Dibbern. (2005). Vendor strategies in the German market for information technology and business process outsourcing. In R. Hirschheim, A. Heinzl, & J. Dibbern (Eds.), *Information system outsourcing* (2nd Ed.), (pp. 381-386). New York: Springer-Verlag.

Gerbing, D. W., & Anderson, J. C. (1988). An updated paradigm for scale development incorporating unidimensionality and its assessment. *JMR, Journal of Marketing Research*, *25*(2), 186–192. doi:10.2307/3172650

Ghosh, B. (2011). Cross-cultural knowledge management practices to support offshore outsourcing. In M. Al-Shammari (Ed.), *Knowledge management in emerging economies: Social, organizational and cultural implementation* (pp. 249–260). Denver, CO: Metropolitan State College of Denver. doi:10.4018/978-1-60960-783-8.ch623

Goles, T. (2001). *The impact of the client/vendor relationship on outsourcing success*. (Ph D dissertation). Department of Decision and Information Sciences, University of Houston, Houston, TX.

Gonzalez, R., Gasco, J., & Llopis, J. (2005). Information systems outsourcing: Reasons in the largest Spanish firms. *International Journal of Information Management*, *25*(2), 117–136. doi:10.1016/j.ijinfomgt.2004.10.002

Gopal, A., Sivaramakrishnan, K., Krishnan, M. S., & Mukhopadhyay, T. (2003). Contracts in offshore software development: An empirical analysis. *Information Technology & People*, *49*(12), 1671–1683.

Grant, R. B. (1991). A resource-based theory of competitive advantage: Implications for strategy formulation. *California Management Review*, *33*(3), 114–135. doi:10.2307/41166664

Grover, V., Cheon, M. J., & Teng, J. T. C. (1996). The effect of service quality and partnership on the outsourcing of information systems functions. *Journal of Management Information Systems*, *12*(4), 89–116.

Han, S. H., Lee, J. N., & Seo, Y. W. (2008). Analyzing the impact of a firm's capability on outsourcing success: A process perspective. *Information & Management*, *45*, 31–42. doi:10.1016/j.im.2007.09.004

Harrington, S. J., & Guimaraes, T. (2005). Corporate culture, absorptive capacity and IT success. *Information and Organization*, *15*(1), 39–63. doi:10.1016/j.infoandorg.2004.10.002

Hatch, N. W., & Dyer, J. H. (2004). Human capital and learning as a source of sustainable competitive advantage. *Strategic Management Journal*, *25*(12), 1155–1178. doi:10.1002/smj.421

He, J., & Wang, H. (2009). Innovative knowledge assetsand economic performance: The asymmetric roles of incentive and monitoring. *Academy of Management Journal*, *15*, 919–938. doi:10.5465/AMJ.2009.44633414

Hoskisson, R. E., Eden, L., Lau, C. M., & Wright, M. (2000). Strategy in emerging economies. *Academy of Management Journal*, *43*(3), 249–267. doi:10.2307/1556394

Hurst, I., & Hanessian, B. G. (1995). Navigating IT channels: Integrate or outsource? *The McKinsey Quarterly*, *3*, 103–110.

Hussain, Z., Wallace, J., & Cornelius, N. E. (2007). The use and impact of human resource information systems on human resource management professionals. *Information & Management*, *44*(1), 74–89. doi:10.1016/j.im.2006.10.006

Ichniowski, C., & Shaw, K. (1999). The effects of human resource management systems on economic performance: An international comparison of US, and Japanese plants. *Management Science*, *45*(5), 704–721. doi:10.1287/mnsc.45.5.704

Inkpen, A. C., & Currall, S. C. (2004). The co-evolution of trust, control, and learning in joint ventures. *Organization Science*, *15*(5), 586–599. doi:10.1287/orsc.1040.0079

Jain, K., Manson, N., & Sankhe, S. (2005, February). The right passage to India. *The McKinsey Quarterly*. Retrieved March 2, 2006, from http://www.mckinseyquarterly.com/article_page.aspx?ar=1581&L2=16&L3=17

Katou, A., & Budhwar, P. (2006). The effect of human resource management systems on organizational performance: Test of a mediating model. *International Journal of Human Resource Management*, *17*(7), 1223–1253. doi:10.1080/09585190600756525

Kavcˇicˇ, K., & Tavcˇar, M. I. (2008). Planning successful partnership in the process of outsourcing. *Kybernetes*, *37*(2), 241–249. doi:10.1108/03684920810851140

Kearns, G. S., & Lederer, A. L. (2000). The effect of strategic alignment on the use of IS-based resources for competitive advantage. *The Journal of Strategic Information Systems*, *9*(4), 265–293. doi:10.1016/S0963-8687(00)00049-4

Kern, T., & Blois, K. (2002). Norm development in outsourcing relationships. *Journal of Information Technology*, *17*(1), 33–42. doi:10.1080/02683960210137174

Khan, S. U., Niazi, M., & Rashid, A. (2011a). Factors influencing clients in the selection of offshore software outsourcing vendors: An exploratory study using a systematic literature review. *Journal of Systems and Software*, *84*(4), 686–699. doi:10.1016/j.jss.2010.12.010

Khandwalla, P. (1977). *The design of organizations*. New York: Harcourt Brace Jovanovich.

Kingshott, R. P. (2006). The impact of psychological contracts upon trust and commitment within supplier–buyer relationships: A social exchange view. *Industrial Marketing Management*, *35*(6), 724–739. doi:10.1016/j.indmarman.2005.06.006

Kogut, B., & Kulatilaka, N. (1994). Operating flexibility, global manufacturing, and the option value of a multinational network. *Management Science*, *40*(1), 123–139. doi:10.1287/mnsc.40.1.123

Lacity, M., & Hirschheim, R. (1993). The information systems outsourcing bandwagon. *Sloan Management Review*, *35*(1), 73–86.

Lahiri, S., & Kedia, B. L. (2009). The effects of internal resources and partnership quality on firm performance: An examination of India BPO providers. *Journal of International Management*, *15*(20), 209–224. doi:10.1016/j.intman.2008.09.002

Lahiri, S., & Kedia, B. L., & Mukherjee. (2011). The impact of management capability on the resource–performance linkage: Examining Indian outsourcing providers. *Journal of World Business*, *47*(1), 145–155. doi:10.1016/j.jwb.2011.02.001

Lee, J. (2001). The impact of knowledge sharing, organizational capability and partnership quality on IS outsourcing success. *Information & Management*, *38*(5), 323–335. doi:10.1016/S0378-7206(00)00074-4

Levina, N., & Ross, J. W. (2003). From the vendor's perspective: Exploring the value proposition in information technology outsourcing. *Management Information Systems Quarterly*, *27*(3), 331–364.

Lewin, A. Y., Massini, S., & Peeters, C. (2009). Why are companies offshoring innovation? The emerging global race for talent. *Journal of International Business Studies*, *40*(6), 901–925. doi:10.1057/jibs.2008.92

Martín-de-Castro, G., Navas-López, J., López-Sáez, P., & Alama-Salazar, E. (2006). Organizational capital as competitive advantage of the firm. *Journal of Intellectual Capital*, *7*(3), 324–337. doi:10.1108/14691930610681438

Menon, A., Sundar, G., Bharadwaj, P. T. A., & Stenen, W. E. (1999). Antecedent & consequences of marketing strategy making a model & a test. *Journal of Marketing*, *63*, 18–40. doi:10.2307/1251943

Morris, M. H., & Joyce, M. L. (1988). How marketers evaluate price sensitivity. *Industrial Marketing Management*, *17*(2), 169–176. doi:10.1016/0019-8501(88)90019-3

Mouhalis, J. (2007). *The details behind a successful offshore BPO programme*. Oxford, UK: Zackin Publications.

Murphy, G. B., Trailer, J. W., & Hill, R. C. (1996). Measuring performance in entrepreneurship research. *Journal of Business Research*, *36*(1), 15–23. doi:10.1016/0148-2963(95)00159-X

National Association of Software and Services Company. (2011). *Indian IT-BPO industry*. Retrieved December 12, 2001, from http://www.nasscom.org/indian-itbpo-industry

Neo, I. T. (2008). *Navigating economic uncertainty: Leveraging the services globalization advantage*. Retrieved November, 12, 2008, from http://www.neoadvisory.com/whitepapers.php#WPFeatured

North, D. C. (1990). *Institutions, institutional change and economic performance*. Cambridge, MA: Cambridge University Press. doi:10.1017/CBO9780511808678

Nunnally, J. C. (1978). *Psychometric theory*. New York: McGraw-Hills Book Company, Inc.

Parkhe, A. (2007). International outsourcing of services: Introduction to the special issue. *Journal of International Management*, *13*(1), 3–6. doi:10.1016/j.intman.2006.12.004

Piercy, N. F. (1981). British export market selection and pricing. *Industrial Marketing Management*, *10*, 287–297. doi:10.1016/0019-8501(81)90038-9

Rai, A., Maruping, L. M., & Venkatesh, V. (2009). Offshore information systems project success: The role of social embeddedness and cultural characteristics. *Management Information Systems Quarterly*, *33*(3), 617–641.

Raman, P., Wittmann, C. M., & Rauseo, N. A. (2006). The role of organizational capabilities in successful CRM implementation. *Journal of Personal Selling & Sales Management*, *26*(1), 39–54. doi:10.2753/PSS0885-3134260104

Ramanujan, V., Venkatraman, N., & Cammilus, J. C. (1986). Multi-objective assessment of effectiveness of strategic planning: A discriminant analysis approach. *Academy of Management Journal*, *29*(2), 347–372. doi:10.2307/256192

Roberts, P. W., & Dowling, G. R. (2002). Corporate reputation and sustained superior financial performance. *Strategic Management Journal*, *23*(12), 1077–1093. doi:10.1002/smj.274

Saini, A., & Johnson, J. L. (2005). Organizational capabilities in e-commerce: An empirical investigation of e-brokerage service providers. *Journal of the Academy of Marketing Science*, *33*(3), 360–375. doi:10.1177/0092070305276150

Scott, W. R. (1995). *Institutions and organizations*. Thousand Oaks, CA: Sage.

Sen, F., & Shiel, M. (2006). From business process outsourcing (BPO) to knowledge process outsourcing (KPO), some issues. *Human Systems Management*, *25*, 145–155.

Smith, G. E. (1995). Managerial pricing orientation: The process of making pricing decisions. *Pricing Strategy and Practice*, *3*(3), 28–39.

Sohn, S. Y., Moon, T. H., & Seok, K. J. (2009). Optimal pricing for mobile manufacturers in competitive market using genetic algorithm. *Expert Systems with Applications*, *36*, 3448–3453. doi:10.1016/j.eswa.2008.02.051

Swinarski, M. E., Kishore, R., & Rao, H. R. (2006). *Impact of IT service provider process capabilities on service provider performance: An empirical study*. Paper presented at the 39th Hawaii International Conference on System Sciences (HICSS-39). Kauai, HI.

Tan, J. J., & Litschert, R. J. (1994). Environment–strategy relationship and its performance implications: An empirical study of the Chinese electronics industry. *Strategic Management Journal*, *15*(1), 1–20. doi:10.1002/smj.4250150102

Teng, J. T., & Thompson, G. L. (1996). Optimal strategies for general price–quality decision models of new products with learning production costs. *European Journal of Operational Research, 93*(3), 476–489. doi:10.1016/0377-2217(96)00028-8

Tsenga, F.-M., Chiub, Y.-J., & Chenc, J.-S. (2009). Measuring business performance in the high-tech manufacturing industry: A case study of Taiwan's large-sized TFT-LCD panel companies. *Omega, 37*, 686–697. doi:10.1016/j.omega.2007.07.004

Tzafrir, S. S. (2005). The relationship between trust, HRM practices and firm performance. *International Journal of Human Resource Management, 16*(9), 1600–1622. doi:10.1080/09585190500239135

Vaidyanathan, J. (2008). Business process outsourcing: An approach to gain access to world-class capabilities. *Business Process Management Journal, 14*(1), 23–38. doi:10.1108/14637150810849382

Valle, I. D. D., Castillo, M. A. S., & Rodriquez-Duarte, A. (2009). The effects of training on performance in service companies: A data panel study. *International Journal of Manpower, 30*(4), 393–407. doi:10.1108/01437720910973070

Venkatraman, N., & Ramanujam, V. (1986). Measurement of business performance in strategy research: A comparison of approaches. *Academy of Management Review, 11*(4), 801–814.

Willcocks, L., & Lacity, M. (2006). *Global sourcing of business and IT services*. New York: Palgrave Macmillan. doi:10.1057/9780230288034

Willcocks, L., Lacity, M., & Fitzgerald, G. (1995). Information technology outsourcing in Europe and the USA: Assessment issues. *International Journal of Information Management, 15*(5), 333–351. doi:10.1016/0268-4012(95)00035-6

Winkler, J. K., Dibbern, J., & Heinzl, A. (2008). The impact of cultural differences in offshore outsourcing—Case study results from German–Indian application development projects. *Information Systems Frontiers, 10*(2), 243–258. doi:10.1007/s10796-008-9068-5

Yiu, D., & Makino, S. (2002). The choice between joint venture and wholly owned subsidiary: An institutional perspective. *Organization Science, 13*(6), 667–683. doi:10.1287/orsc.13.6.667.494

Youndt, M. A., Snell, S. A., Dean, J. W., & Lepak, D. P. (1996). Human resource management, manufacturing strategy and firm performance. *Academy of Management Journal, 39*(4), 836–866. doi:10.2307/256714

Zhu, K., & Kraemer, K. L. (2002). E-commerce metrics for net-enhanced organizations: Assessing the value of e-commerce to firm performance in the manufacturing sector. *Information Systems Research, 13*(3), 275–295. doi:10.1287/isre.13.3.275.82

Chapter 18
Evolution of Supply Chain Collaboration:
Implications for the Role of Knowledge

Michael J. Gravier
Bryant University, USA

M. Theodore Farris II
University of North Texas, USA

ABSTRACT

Increasingly, research across many disciplines has recognized the shortcomings of the traditional "integration prescription" for inter-organizational knowledge management. This research conducts several simulation experiments to study the effects of different rates of product change, different demand environments, and different economies of scale on the level of integration between firms at different levels in the supply chain. The underlying paradigm shifts from a static, steady state view to a dynamic, complex adaptive systems and knowledge-based view of supply chain networks. Several research propositions are presented that use the role of knowledge in the supply chain to provide predictive power for how supply chain collaborations or integration should evolve. Suggestions and implications are suggested for managerial and research purposes.

INTRODUCTION

Recent research indicates that the era of inter-organizational collaboration and knowledge-sharing has arrived under the guise of supply chain management. Investments in supply chain management provide a competitive advantage to business. For example, AMR's "Supply Chain Top 25" grew revenue an average of 29% over the prior year (Hofman and Aronow, 2012). Perhaps more telling for the scholar of knowledge-based organizations are the conclusions by the World

DOI: 10.4018/978-1-4666-4506-6.ch018

Economic Forum (2012) that countries could grow their GDP six times more by using policies that address the management of supply chain processes instead of relying on tariffs. Supply chain processes tend to be knowledge-intensive and depend upon international collaboration, coordination with government entities such as customs, reliable physical infrastructure for transportation and communications, and standardizing inter-organizational procedures. In a very real sense, supply chain management really represents knowledge exchange management between firms, with most pundits espousing "integration" of knowledge and processes.

Integration of inter-organizational processes has long comprised the prescription for success in supply chain management and related literature (Frohlich & Westbrook, 2001; Gustin, Daugherty, & Stank, 1995; Stock, Greis, & Kasarda, 1999). Information sharing and various levels of coordination and collaboration have traditionally found strong empirical support (c.f., Daugherty, Ellinger, & Gustin, 1996; Lee, Padmanabhan, & Whang, 1997; Lummus & Vokurka, 1999; Narasimhan & Jayaram, 1998; Stank, Keller, & Closs, 2001; Tan, Kannan, & Handfield, 1998). Notwithstanding the oft-cited works that support the "integration prescription," one systematic review of the literature revealed the link between integration and supply chain performance as shaky (Fabbe-Costes & Jahre, 2008). One simulation study found that information sharing may have no value at all or may even increase costs, depending on demand patterns (Jonsson and Mattsson, 2013). The mixed evidence suggests an incomplete theoretical understanding of integration and inter-firm collaboration.

One issue may be the implicit assumption that all collaboration is the same. Empirical studies that distinguish among the different manners of collaborating and the different outcomes to the various supply chain members remain relatively few, with Frohlich and Westbrook's (2001) landmark article representing a sort of genesis of this body of literature. More recent work has found that supply chain strategies vary but seem to become more sophisticated the closer the firm is to the market (Bourlakis, et al., 2012), despite the additional economies of scale and other benefits that firms higher up the supply chain often have. This may reflect the presence of hypercompetitive environments characterized by the rapid rise and fall of firms (McNamara, Vaaler, & Devers, 2003; Wiggins & Ruefli, 2005); retail firms that are not responsive to customer needs do not last long. Hypercompetitive environments act to "dis-integration" supply chains as firms seek higher quality, lower cost or other product characteristics desired by the market.

Another important issue has been the realization that the traditional conceptualizations of supply chain management implicitly assume a "steady-state" condition. Increasingly turbulent economic and global systems mean that supply chains must be adaptable and resilient to manage their risks—yet methods for assessing and managing inter-organizational network change remains in a nascent status (Pettit, et al., 2013). Supply chain agility comprises a well-established conceptualization of responsive and adaptable inter-organizational networks of firms; however, supply chain agility primarily focuses on descriptive or normative theories rather than predictive capability (c.f., Gligor, et al., 2012).

In the past few years, supply chain literature has begun to treat supply chains as knowledge-based constellations of organizations. Researchers have increasingly focused on the question of when, how and why supply chain integration works (or doesn't). Recent work has evaluated with whom companies integrate (Huo, 2012), the relationship of information flows to material flows (Prajogo and Olhager, 2012), short-term knowledge sharing vs. deeper knowledge generation (Jayaram and Pathak, 2013), product and process strategies as antecedents of supply chain integration strategies (Droge, et al., 2012), and the technological platforms for sharing information and knowledge (Bendoly, et al., 2012). Implicit (and often explicit) to this work are supplier and customer relationship management, and the benefits of long-term vs. short-term relationships.

Interestingly, little research has attempted to map supply chains, and then relate supply chain macro-level characteristics to knowledge needs. Methodological challenges mean that most past work has been static in nature or relied on assumptions that environmental conditions remain constant. Few studies have assessed longitudinal evolution of supply chains despite the important potential public policy and strategic contributions of such research. In a sense this work represents an extension of the intra-firm organizational influence on knowledge management proposed by Nickerson and Zenger (2004).

This research assesses the patterns of collaboration and firm mortality in supply chains by use of simulation. The pace of technological advance, sensitivity to economies of scale, and various market scenarios provide the environmental control variables; evolutionary outcomes and degree of interfirm collaboration comprise the observations at each level of a four-tier supply chain. The resulting patterns of supply chain evolution provide the basis for developing implications for different degrees of inter-firm knowledge management.

LITERATURE REVIEW

Studying and modeling supply chain evolution from a knowledge-exchange perspective requires a multi-disciplinary approach. Classical supply chain governance theory, adaptation, and evolution literatures provide the conceptual model for the subsequent simulation of inter-organizational evolution. Particular emphasis is paid to the integration between firms using a newly developed measure called the collaboration index.

Supply Chain Theory

Ronald Coase's 1937 work on transaction costs offers the theoretical roots of supply chain management, prompting one scholar to describe the reduction of transaction costs as "the heart of the interest in supply chain management (Hobbs, 1996, p.26)." Williamson, the researcher most associated with transaction costs, stated, "...whereas TCE examines individual transactions, SCM introduced a broader systems perspective in which related transactions are grouped and managed as chains (Williamson, 2008, p. 5)." Transaction cost economics (TCE), sometimes referred to as transaction cost analysis (TCA), offers perhaps the most extensive empirical support of any of the extant exchange governance theories, with between 250 and 500 citations appearing annually in scholarly works since the early 1990s (David and Han, 2004).

Many scholars trace the origins of supply chain management to Forrester's 1958 *Harvard Business Review* article where he stated (p. 37): "Management is on the verge of a major breakthrough in understanding how industrial company success depends on the interaction between the flows of information, materials, money, manpower, and capital equipment." Williamson's conceptualization of TCE emphasizes the adaptability of the interactions between firms as "the central problem of economic organization" (1991, p. p. 278)—and it is this adaptive response to environmental pressures that motivates this study of supply chain evolution. Williamson (1975, 1986) founded his theoretical framework on two primary assumptions of human behavior (bounded rationality and opportunism) and two dimensions of transactions (frequency and asset specificity).

Bounded Rationality and Opportunism

Bounded rationality and opportunism are the two primary dimensions of uncertainty. Uncertainty embodies any unanticipated change to the circumstances surrounding an exchange (Rindfleisch & Heide, 1997). Bounded rationality describes the behavioral uncertainty resulting from the cognitive limits of managers who try to anticipate every contingency in a market exchange (Leiblein, 2003). Opportunism refers to the behavioral uncertainty resulting from self-interest seeking behavior. According to TCA, high environmental uncertainty increases transaction costs of due to the need to adapt contractual agreements to compensate for unanticipated variations in volumes, technology/product design, sources of supply, doubtful customer loyalties, and other unforeseen circumstances.

The dimensions of uncertainty are manifestations of information asymmetries, unaligned goals, or lack of commitment. Industry standardization may reduce the risks from lack of commitment; the low cost of tracking electronic trails in technology intensive environments diminish the occurrence of information asymmetries (Garicano and Kaplan, 2001). Networks and other hybrid forms of governance are characterized by standards or norms and a high degree of shared information. This leads to reduced uncertainty without incurring the costs associated with vertical integration while avoiding most opportunity costs in turbulent environments that require switching partners.

In terms of predictive outcomes, TCA claims that firms employ vertical integration as a means of easing the burden of performance evaluation. This follows from TCA's assertion that evaluation problems give rise to measurement costs. However, Ouchi (1979) provides an alternate view that measurement costs are incurred in order to distribute rewards across parties in an equitable fashion. If equitable distribution does not occur, an individual firm may eventually reduce its individual efforts, incurring opportunity costs resulting from the productivity losses. Ouchi's insight harkens to the classical team production problem (Alchian & Demsetz, 1972) wherein labor that requires cooperative production. In this perspective, vertical integration will occur when equitable distribution becomes difficult—such as when margins are low for a highly commoditized item (resulting in larger players using bargaining power to increase their profits at the expense of suppliers or customer) or when the contribution of individual players becomes expensive or difficult to measure relative to the value of productivity (again leading to paying below market value). In either case, performance will eventually suffer as either the firm acquiring the difficult-to-measure inputs or the buyer/coordinator of the inputs will reduce their level of effort to match their reward. Additionally, uncertainty may also lead to more market-based exchanges due to the increased flexibility in partner-selection (Rindfleisch & Heide, 1997).

Empirical evidence demonstrates mixed results regarding TCA's predictions of the effects of uncertainty, probably largely due to the difficulty of measuring the construct (David & Han, 2004; Rindfleisch & Heide, 1997). Because uncertainty is a multidimensional construct, it is best studied via longitudinal or dynamic methods such as the current study that take into account factors such as environmental dynamism, environmental heterogeneity, and innovation. Any insights gathered with regard to uncertainty would be an important contribution.

TCA's Exchange Factors

Two kinds of exchange factors appear in the model, derived from TCA's exchange factors that predict the optimal interfirm governance form. Exchange factors include 1) frequency and 2) asset specificity.

Frequency

Generally speaking, as two firms conduct more exchanges with each other, they find more efficient ways of conducting the exchanges, driving down the transaction costs on a per unit basis.

Frequency of transactions has different effects on the governance form depending on asset specificity and behavioral uncertainty. The costs of monitoring behavior in close partnerships mean that as long as there is no behavioral uncertainty, open market transactions are very efficient. When the costs of motivating partners to align their interests outweigh the costs of more integrated forms of governance, then firms move away from open market transactions. This insight suggests a U-shaped relationship between economies of scale and frequency of transactions—relatively few players at the early and late stages of an industry's life cycle mean fewer options for switching and more "lock in" leading to more frequent transactions between the remaining firms. During the highly competitive rapid growth phase, the appearance of many firms provides many opportunities to switch to a better partner, and less incentive to stick with the same partner, thus lowering frequency. Additionally, a rapid rate of technological advance forces firms to keep up with successful innovators or perish.

Asset Specificity

Asset specificity provides a means for limiting the effects of bounded rationality and technological uncertainty. By investing in resources customized for the specific conditions of exchange between two firms, more information becomes available for managerial decision making. Increased information reduces the coordination costs between firms, as described under previously under bounded rationality (Garicano and Kaplan, 2001). In addition to reducing coordination costs between firms, asset specificity becomes desirable as a means of "locking in" sources of supply or buyers in the presence of environmental turbulence for two reasons. Firstly, successful innovation flourishes under conditions of frequent interaction and interdependence between firms (Varadarajan & Cunningham, 1995); market transactions do not allow for interfirm learning and exchange of ideas, but asset specificity ensures closer or even exclusive relationships. Secondly, greater leaps in innovation increase the costs of failure while also increasing the incentives for achieving success (Chesbrough & Teece, 1996); increasing asset specificity reduces the risk to each firm while increasing the chances of successful innovation.

Shortcomings of TCA

As Williamson (2008) himself pointed out, TCE limits itself to individual transactions, rather than grouping transactions and considering them as part of a system. TCE represents essentially a contractual approach to explaining inter-firm transactions, which leads to several theoretical shortcomings (for a more detailed expose, see Deitrich, 2012). Overlooked in most current TCA scholarship is the importance of production factors. Coase's (1937) original vision involved balancing the marginal contribution of owning production versus managing via exchanges on the open market. The balancing act occurs in "an outside network of relative prices and costs" (Coase, 1937: p. 389). It is a relatively recent development that firms now have the tools to actively manage this balancing act across multiple firms. The active collaboration across multiple firms in order to lower transaction costs or take advantage of lower external production costs is the essence of supply chain management.

The means of managing collaboration across multiple firms may vary from purely open market, one-time exchanges to exclusive, long-lasting relationships. These two classifications form the anchors for what has been described as a conceptually useful continuum of interfirm exchange governance from purely "transactional" to purely "relational" exchanges (Heide, 1994). In essence, transactional and relational exchanges offer different strategies for cost tradeoffs between transaction and production costs (see Figure 1). Transactional exchanges make the explicit assumption that as a firm's capacity grows it must deal with more

Figure 1. Conceptualization of total cost curves for transactional and relational exchanges

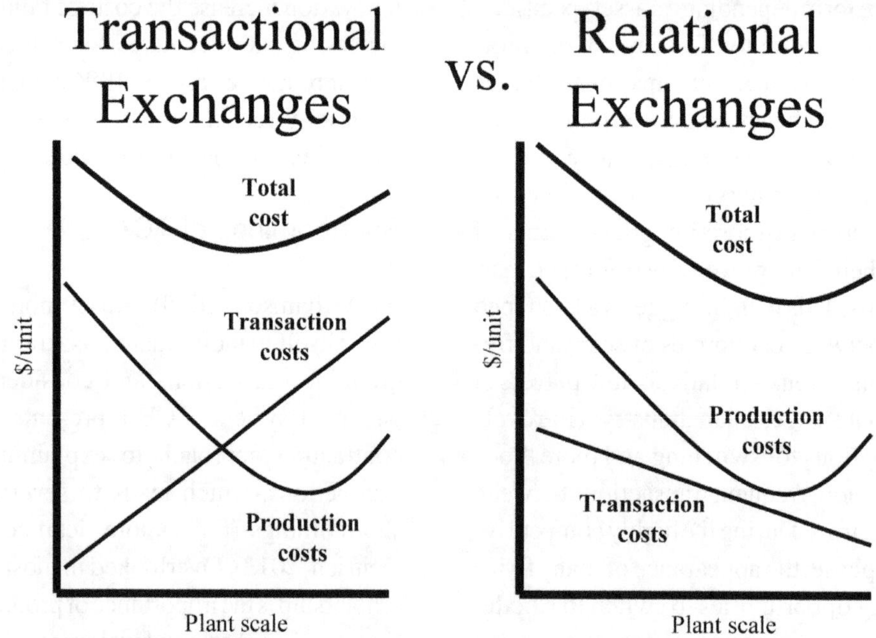

open market transactions, and transaction costs will eventually increase per unit as the result of having more trading partners whose needs must be tracked, and more information to be gathered and compared before making purchasing decisions, and generally increased costs of greater managerial scope. Relational exchanges limit interactions to one or very few trading partners, thus lowering transaction costs on a per unit basis as the result of greater efficiencies—analogous to economies of scale or learning curve effect, but applied to interfirm collaboration. At some point, production will reach diseconomies of scale, raising unit production costs and ultimately raising total costs.

Despite a strong empirical record, TCA theory does suffer some drawbacks (Joshi and Campbell, 2003). It lacks explanatory power for how firms organize themselves in a network then adapt their exchange behavior based upon changes to the production and demand environments. TCA has also been criticized for focusing on exogenous market factors and failing to explain firm-level decisions (Hunt and Morgan, 1995). As a theory of interfirm collaboration TCA suffers from being firm-centered (or at best dyadic) and static in nature (Rindfleisch and Heide, 1997)—a far cry from the supply chain reality of networks with highly dynamic firm interactions. Recent theoretical work into complex adaptive systems offers a new paradigm for studying systems characterized by adaptive agents who repeatedly interact and adapt to their environment (Surana, et al., 2005).

Adaptation in Supply Chains

Recently several new theoretical perspectives of supply chain management have been proposed that incorporate the complex adaptive systems (CAS) perspective (c.f., Mena, et al., 2013; Pilbeam, et al., 2012; Schoenherr, et al., 2011). CAS theory offers a way of studying supply chain networks of firms that adapt their behavior based upon experience and the outcomes of interactions (Choi, et al., 2001; Surana, et al., 2005; Vargo and Lusch, 2004). Complex adaptive systems developed from complexity science and arose out of the study of open systems. In the organizational context, a system consists of interconnected components that interact; such systems are "open" because they exchange resources with the environment. When the members of a system have many interactions that result in a whole that is interdependent with the environment, they comprise a complex system (Anderson, 1999). The idea of complex systems underlies Forrester's (1958, 1961) dynamic models now commonly used as the bases for studying the beer game or the bullwhip effect in supply chain analysis (Lee, Padmanabhan, & Whang, 1997a; Lee, et al., 1997b).

Complex adaptive systems build on the idea that "adaptation builds complexity" (John H. Holland, 1995). In a CAS, members of a system are called entities (Surana, Kumara, Greaves, & Raghavan, 2005). Each entity communicates with other entities and the environment, accumulating experience (learning), continuously interacting, and changing its behavior, and its own structure as well as the system's structure. The CAS perspective has been applied successfully for many years to the study of socio-economical processes, to include economics (J. H. Holland & Miller, 1991; Limburg, O'Neill, Costanza, & Farber, 2002; Markose, 2005), organizational learning (Chiva-Gomez, 2003; McElroy, 2000; Morel & Ramanujam, 1999), psychology (Dooley, 1997; Goldstone & Sakamoto, 2003), linguistics (Kirby, 2000), anthropology (Abel, 1998), military strategy (Ilachinski, 2000), innovation (Eisenhardt & Tabrizi, 1995), and strategy (Bettis & Prahalad, 1995).

There are six characteristics that make the CAS paradigm particularly appropriate for the study of supply chains: 1) interactions, 2) interdependencies, 3) high non-linearity, 4) self-organization, 5) evolution, and 6) dynamism. Interestingly, CAS possess the most salient qualities that both Coase (1937) and Williamson (1991) ascribed to market exchanges: movement from one equilibrium point to another, continuously dynamic, firm dependence on an outside network of other firms and prices, and made up of autonomous but not entirely independent actors. As such, the CAS paradigm offers a strong foundation for the study of supply chains as both markets and systems.

Complex adaptive systems theory describes increasingly complex networks as more expensive and fragile to maintain, but their diversity of structures provides greater robustness to environmental fluctuations. While the increased diversity and complexity of structures increase resilience in the face of random attacks or environmental shifts, they also increase vulnerability to targeted attacks or in the event that certain key system nodes succumb to environmental shifts (c.f., Holme, Kim, Yoon, & Han, 2002). This happens as the result of the high number of interactions that occur with certain key members of the system.

Supply Chain Evolution

"Evolution" as a word has become commonplace in contemporary society, so a precise definition is required for supply chain networks. Van de Ven and Poole (1995) provided perhaps the most cited definition in organizational studies for evolution: "cumulative changes in structural forms of populations of organizational entities across communities" (p. 517-518). They further elaborate that, "evolution explains change as a recurrent, cumulative, and probabilistic progression of variation, selection, and retention of organizational entities" (Van De Ven & Poole, 1995, p. p. 518).

This definition suggests three primary processes shape evolution: variation, selection, and retention. *Variation* creates novel forms of organizations. *Selection* results from the allocation or appropriation of scarce resources amongst competitors. *Retention* describes how certain organizational forms perpetuate. Figure 2 summarizes the conceptual relationships of evolutionary forces when a supply chain network is modeled as a CAS.

Figure 2. Modeling framework

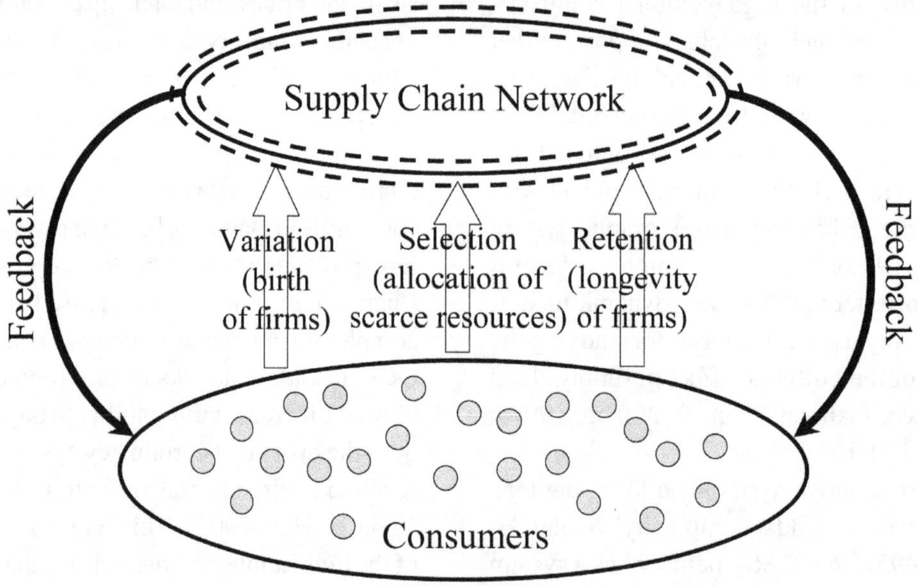

Selection

From the perspective of supply chain evolution as interpreted via the TCA lens, the allocation or appropriation of scarce resources occurs via the choice of both the right trading partners and the right mode of conducting inter-firm exchanges. A dominant form of inter-firm governance can be referred to as a market information regime. Market information regimes arise as a way of making socially agreed upon information routinized and widely available in order to reduce market uncertainty (Anand & Peterson, 2000). Market information regimes provide a means of control for information asymmetries resulting from uncertainties in actions and intentions by competitors or trade partners (Heimer, 1985). Market information regimes make others' actions more predictable, but in markets with powerful externalities such as a rapid rate of technological advance, they can become an obstacle as they routinize information that rapidly obsolesces or is otherwise overcome by events. Under these circumstances, larger companies often reduce their uncertainty by obligating another supply chain member to bear it (Heimer, 1985). Managers at the individual firm level cannot guarantee their exchange partners will continue to be the best trading partner into the future, and striking the right balance of commitment and flexibility in a relationship is a critical strategic decision. This translates into a problem of bounded rationality for the managers.

Evolution of interfirm collaboration implies change in firm relationships. This research effort refers to the selective force on supply chain relationships as transience. In order to measure transience in the simulation, a collaboration index was developed which kept track of the degree of permanence or loyalty of relationships in the supply chain network. Transience is defined as the inability to predict who exchange partners will be over time. High transience makes predicting who will be the future exchange partners for a given firm more difficult. As described subsequently, transience was assessed by maintaining a collaboration index which kept track of the degree of permanence or loyalty of relationships in the supply chain network.

Retention

Retention is defined as the continued existence of the same firms in a network and serves as the basis of stability in a supply chain network. Destabilizing effects that reduced longevity include a rapid pace of technology, volatility resulting from competition, mortality of partner firms, and switching of exchange partners. Retention is defined as the perpetuation of existing firms in the supply chain network and is measured as longevity of the individual firms and assessed using survival analysis.

Variation

Variation is the appearance of new combinations of firms that produce and distribute a final product co-exist in the supply chain network. Increased variation implies increased complexity of the supply chain network as both new firms and different forms of inter-firm governance rise and fall until more effective and efficient forms appear and are replicated. The extreme example of a supply chain network with low complexity is a single vertically integrated supply chain serving the entire end market. Increasingly complex supply chain networks exhibit more varieties of firms and wide variation in governance mechanisms between firms.

Complex adaptive systems theory describes increasingly complex networks as more expensive and fragile to maintain, but their diversity of structures provides greater robustness to environmental fluctuations. The simulation investigated the resilience of more complex networks in heterogeneous environments characterized by a rapid rate of technological advance.

FRAMEWORK AND RESEARCH METHOD

This section describes the methodology, experimental design, model validation and verification, and the statistical analysis. The supply chain modeled consists of a manufacturer who originates most of the product innovations, an assembler, and a retailer who serves the end market (see Figure 3). The methodological roadmap provided by Davis, et al. (2007), while analogous to roadmaps provided in other simulation texts (Banks, II, & Nelson, 1999; Law & Kelton, 2000; Maisel & Gnugnoli, 1972), is tailored to the application of simulation to theory building and extension (Table 1). The introduction presented step 1 which identified the research question as, "How does interfirm collaboration evolve in a supply chain

Table 1. Theory building roadmap

Step
1. Begin with a research question
2. Identify simple theory
3. Choose a simulation approach
4. Create computational representation
5. Verify computational representation
6. Experiment to build novel theory
7. Validate with empirical data

Figure 3. Basic supply chain

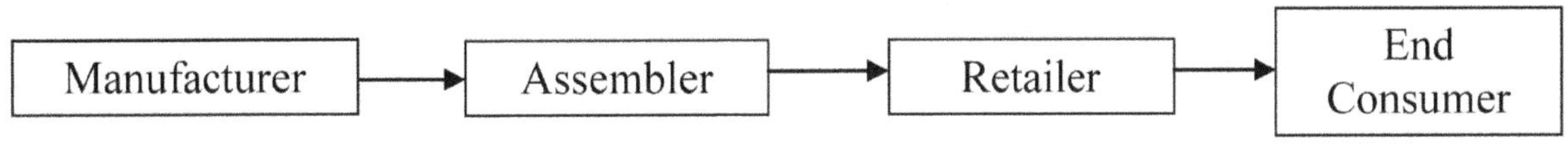

network under different product-market conditions?" Step 2, identification of "simple theory," appears as a literature review of transaction cost analysis, complex adaptive systems, and organizational evolution theories. Steps 3 through 5 are described in the methods section. The outcomes of Step 6 appear in the results section. Step 7 consists of empirical validation and is beyond the scope of this paper; empirical validation appears as an area of future research.

Figure 4 presents the simulation framework (for a more complete description of the simulation model, see Appendix A). It synthesizes the interrelationships of the simulation variables in order to extend TCA theory into a dynamic network environment by using the CAS paradigm. A brief description of the predicted relationships follows.

Simulation Approach

Simulation models have studied supply chains as complex adaptive systems using the systems dynamics approach (Akkermans, 2001; Kim, 2003; Lin, Huang, & Lin, 2002; Parunak, 1998; Pathak, 2005). Past researchers have used an agent-based approach to study how markets consisting of semi-autonomous entities self-organize in a way that characterizes complex adaptive systems. The systems dynamics approach to simulation was used. It is useful when research focuses on the influence of causal relationships among constructs on the behavior of the system (Davis, et al., 2007) and allows the researcher to specify several simple processes with circular causality—such as lower price leading to higher sales,

Figure 4. Simulation framework

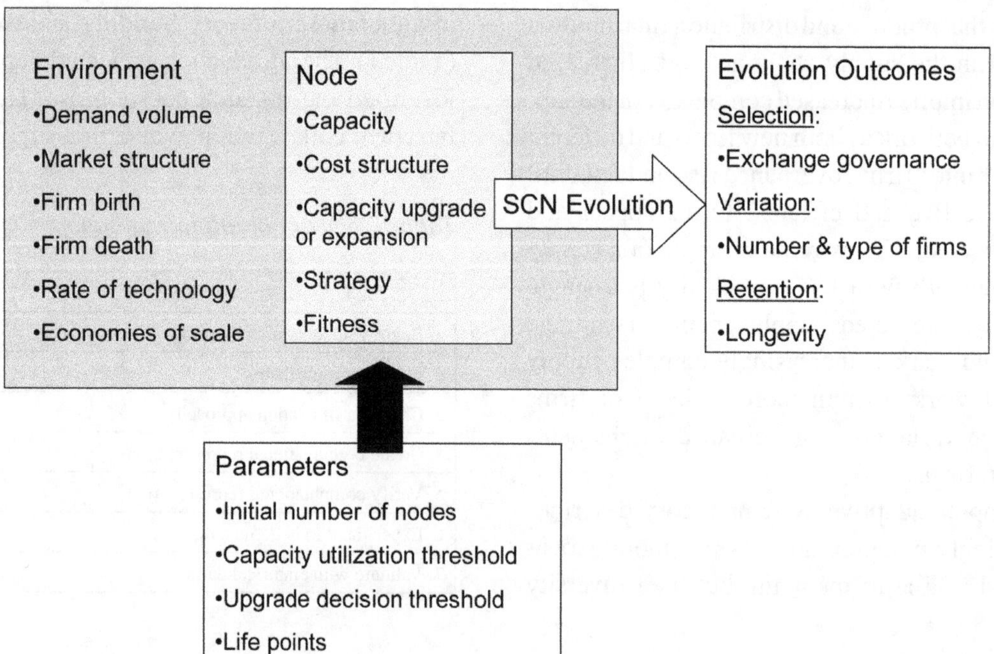

which provides positive feedback to lower price again. These processes of circular causality also intersect with other constructs. For example, lower prices may result from increased economies of scale. The simple theories underlying the modeling framework dictate the sources of stochasticity in the modeled system.

Experimental Design

A full factorial experimental design was implemented with a 2 x 3 x 2 experimental design consisting of two levels of capacity cost (high vs. low), three levels of consumer end markets (heterogeneous or homogeneous with high and low price sensitivity), and two levels of rate of technological change (rapid vs. slow), with a total of 12 experiments. In order to achieve an adequate sample size for each experiment, an *n* of 30 simulations runs was selected, making for a total of 360 samples.

Exogenous Factors

The exogenous factors in the simulation were the per unit cost, demand curve, capacity utilization threshold, capacity upgrade decision threshold, starting life points (for firm "health"), and the rate at which firm health degrades or improves in the face of poor or strong performance.

Per unit cost was the outcome of economies of scale and transaction costs. Inefficient use of capacity caused firms to die off. Increasing capacity utilization increased firm health and opportunity to increase capacity. Health was indicated by life points assigned to each firm. Exceptional capacity utilization increased life points. Low capacity utilization reduced life points and also reduced the likelihood of increasing capacity. When life points reached zero, the company was marked inactive to indicate that it had "died." Companies also gained production expertise over time as reflected by lowering cost per unit as a function of experience tied to company age.

Unfulfilled demand spawned the process of firm birth. A new firm formed when there was enough unfulfilled demand to support it. As the result of a demand curve with rapid early growth, many births appeared early in the simulation. When each company was created it started at the average capacity of other firms of the same type and would elect either a high quality or a low quality strategy. High quality strategy meant harvesting higher prices but lower production and lower optimum economy of scale; low quality strategy meant the converse. When confronted with a small price differential for high and low quality products, consumers preferred the high quality product. Prices for both high and low quality products decreased linearly over time, but price for low quality products decreased more rapidly. As the quality differential between the two classes of products diminished, more customers will prefer savings over a high quality product.

Demand recreated the product life cycle with volume following a normal curve, with rapid demand growth followed by decline. A stochastic element was implemented to emulate random demand fluctuation. The simulation started with a dozen firms at each level of the supply chain. At the beginning these firms practiced exclusive relationships with one buyer and one supplier. Growing demand prompted the appearance of new firms; when demand diminished, firms were forced out of business.

Dependent Variables

Retention was measured as the longevity, in simulation time steps, of individual firms, and analyzed via survival analysis, a specialized form of The dependent variable was the collaboration index, as previously described in detail.

The collaboration index was a scaled variable (from 0 to 1) that described the exchange behavior for a given company in terms of how exclusive the relationships are that the company maintains with its buyers or suppliers. If E(x,y) represented the total number of units sold by company x to company y, E was approximately the product of the number of exchanges that occurred and the number of units sold per exchange (x times y). If I_i represented the collaboration index, then the collaboration index was calculated as:

$$I_i = \frac{\sum_{j=1}^{n} E(x_i, y_j)^2}{\left(\sum_{j=1}^{n} E(x_i, y_j)\right)^2}$$

where n is the number of companies j with whom company *i* does business.

The collaboration index provides a single measure of both the magnitude and duration of the relationship between company x and buyers or suppliers y. This means that a manufacturer would only have one collaboration index for all of its relationships with its downstream buyers, but assemblers had separate collaboration indices for their suppliers and their buyers.

This formulation for a collaboration index had the interesting property that for *n* equal companies that company *x* did business with, the collaboration index would be $1/n$. This may be demonstrated by assuming that for n equal companies that all had the units exchanged per combination of (i, j),

$$I_i = \frac{\sum_{j=1}^{n} b^2}{\left(\sum_{j=1}^{n} b\right)^2} = \frac{nb^2}{(nb)^2} = \frac{nb^2}{n^2b^2} = \frac{1}{n}$$

The reciprocal of the collaboration index had the interesting property of equaling the approximate number of companies that were company x's primary suppliers or buyers. In other words, a simple calculation that estimates a "virtual number of companies" that company i does business with may be calculated by simply taking $1/I_i$.

Time-series analysis was used to assess the processes of evolution of exchange governance (Pathak, 2005; Surana, et al., 2005). In this section, analyses of the experimental results are presented in several stages with regard to their evolutionary implications. Thirty runs of 1,000 time steps were made of each of the 12 experimental scenarios. The resulting 3,600,000 data points (300,000 per experiment) provided a rich dataset for analysis.

Proportional time series analysis assessed transience as reflected by the evolution of interfirm relationships using the collaboration index. First, the simulation was divided into 5 time steps of 200 demand cycles each based upon the stages of the product life cycle (Figure 5). Operationalization of selection was based on changes to the collaboration index. Using PROC GLM in SAS, simple univariate analysis created 95% confidence intervals of the collaboration index at each time step. Binary encoding based upon statistically significant changes to the collaboration index between successive time periods was the basis for classifying as unstable, with a 1 indicating that a statistically significant change had occurred from one time step to the next and a 0 indicating that no statistically significant change to the collaboration index had occurred. Then a technique frequently applied to this type of analysis (Fokianos & Kedem, 2003)—Agresti's (1990) multinomial logit model—was used to analyze the series of encoded 1's and 0's.

RESULTS

In this section, analyses of the experimental results are presented in several stages with regard to their evolutionary implications. Thirty runs of 1,000 time steps were made of each of the 12 experimental scenarios. The resulting 3,600,000 data points (300,000 per experiment) provided a rich dataset for analysis.

Figure 5. Demand volume curve

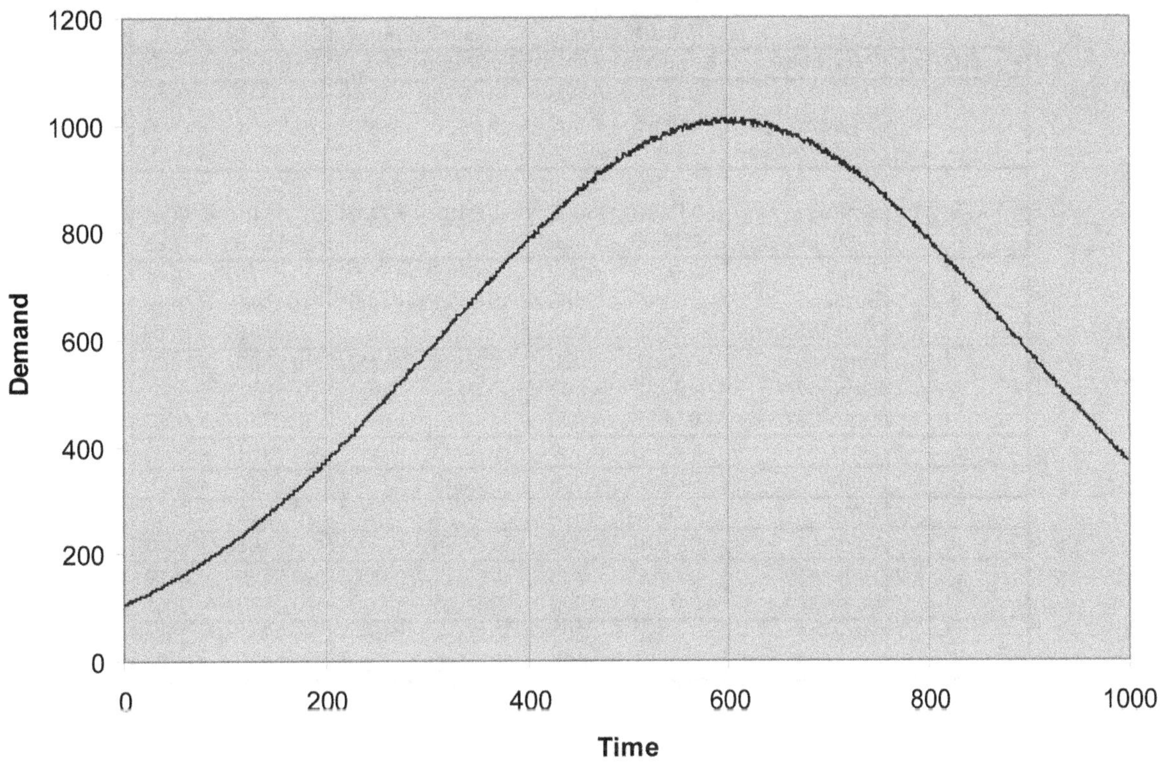

First, time series analysis provide insight into processes of variation, describing types of firms and strategies that propagated in the different supply chain network scenarios. Second, general linear models and survival analysis of firm longevity were used to assess retention of firms under different experimental conditions. Third, proportional time series analysis assessed selection as reflected by the evolution of interfirm relationships using the collaboration index.

Variation

Analysis of variation was conducted by use of differencing and generalized linear modeling (GLM). Differencing was a special case of time series analysis based on the difference in the number of firms between the current and the previous time steps (i.e., a positive difference indicates an increase in the number of firms). Differencing commonly appears as a way of focusing on change in time series analysis of non-stationary processes (Granger & Newbold, 1977; Nelson, 1973). The GLM procedure in SAS version 9.1 was used to model the differencing scores as the dependent variable with time and strategy as the independent variables individually for each experiment except the homogeneous high price scenarios in which no low price firms were spawned. All the GLM models returned significant overall F-scores. Table 2 displays the parameter estimates and their significance for manufacturers, assemblers, and retailers across all 12 experiments. Time cycles were divided by 1000 to put it closer to the same relative scale as the average differences.

A significant parameter estimate for time indicates that differences increased over time—either the number of firms increased at first than

Table 2. Variation analysis results

<table>
<tr><th></th><th></th><th colspan="2">Manufacturers</th><th colspan="2">Assemblers</th><th colspan="2">Retailers</th></tr>
<tr><th>Experiment</th><th>Parameter</th><th>Estimate</th><th>Pr > |t|</th><th>Estimate</th><th>Pr > |t|</th><th>Estimate</th><th>Pr > |t|</th></tr>
<tr><td rowspan="3">1</td><td>Time</td><td>-0.0112</td><td>0.0066</td><td>-0.0275</td><td><.0001</td><td>-0.0315</td><td><.0001</td></tr>
<tr><td>Strategy</td><td>0.0246</td><td><.0001</td><td>-0.0119</td><td><.0001</td><td>0.0176</td><td><.0001</td></tr>
<tr><td>Time*Strategy</td><td>-0.0681</td><td><.0001</td><td></td><td></td><td>-0.0602</td><td><.0001</td></tr>
<tr><td rowspan="3">2</td><td>Time</td><td>-0.0096</td><td>0.0185</td><td>-0.0272</td><td><.0001</td><td>-0.0290</td><td><.0001</td></tr>
<tr><td>Strategy</td><td>0.0232</td><td><.0001</td><td>-0.0136</td><td><.0001</td><td>0.0146</td><td>0.0015</td></tr>
<tr><td>Time*Strategy</td><td>-0.0699</td><td><.0001</td><td></td><td></td><td>-0.0673</td><td><.0001</td></tr>
<tr><td rowspan="3">3</td><td>Time</td><td>-0.0259</td><td><.0001</td><td>-0.0412</td><td><.0001</td><td>-0.0379</td><td><.0001</td></tr>
<tr><td>Strategy</td><td>0.0213</td><td><.0001</td><td>-0.0121</td><td><.0001</td><td>0.0194</td><td><.0001</td></tr>
<tr><td>Time*Strategy</td><td>-0.0618</td><td><.0001</td><td></td><td></td><td>-0.0638</td><td><.0001</td></tr>
<tr><td rowspan="3">4</td><td>Time</td><td>-0.0215</td><td><.0001</td><td>-0.0388</td><td><.0001</td><td>-0.0353</td><td><.0001</td></tr>
<tr><td>Strategy</td><td>0.0222</td><td><.0001</td><td>-0.0141</td><td><.0001</td><td>0.0163</td><td>0.0013</td></tr>
<tr><td>Time*Strategy</td><td>-0.0696</td><td><.0001</td><td></td><td></td><td>-0.0710</td><td><.0001</td></tr>
<tr><td>5</td><td>Time</td><td>-0.0978</td><td><.0001</td><td>-0.0453</td><td><.0001</td><td>-0.1151</td><td><.0001</td></tr>
<tr><td>6</td><td>Time</td><td>-0.0986</td><td><.0001</td><td>-0.0386</td><td><.0001</td><td>-0.1240</td><td><.0001</td></tr>
<tr><td>7</td><td>Time</td><td>-0.1134</td><td><.0001</td><td>-0.0615</td><td><.0001</td><td>-0.1259</td><td><.0001</td></tr>
<tr><td>8</td><td>Time</td><td>-0.1094</td><td><.0001</td><td>-0.0508</td><td><.0001</td><td>-0.1348</td><td><.0001</td></tr>
<tr><td rowspan="2">9</td><td>Time</td><td>-0.0400</td><td><.0001</td><td>-0.0319</td><td><.0001</td><td>-0.0636</td><td><.0001</td></tr>
<tr><td>Strategy</td><td>-0.0115</td><td><.0001</td><td>-0.0126</td><td><.0001</td><td>-0.0163</td><td><.0001</td></tr>
<tr><td rowspan="2">10</td><td>Time</td><td>-0.0356</td><td><.0001</td><td>-0.0304</td><td><.0001</td><td>-0.0627</td><td><.0001</td></tr>
<tr><td>Strategy</td><td>-0.0135</td><td><.0001</td><td>-0.0148</td><td><.0001</td><td>-0.0222</td><td><.0001</td></tr>
<tr><td rowspan="2">11</td><td>Time</td><td>-0.0534</td><td><.0001</td><td>-0.0471</td><td><.0001</td><td>-0.0703</td><td><.0001</td></tr>
<tr><td>Strategy</td><td>-0.0120</td><td><.0001</td><td>-0.0130</td><td><.0001</td><td>-0.0157</td><td><.0001</td></tr>
<tr><td rowspan="2">12</td><td>Time</td><td>-0.0496</td><td><.0001</td><td>-0.0439</td><td><.0001</td><td>-0.0710</td><td><.0001</td></tr>
<tr><td>Strategy</td><td>-0.0136</td><td><.0001</td><td>-0.0152</td><td><.0001</td><td>-0.0216</td><td><.0001</td></tr>
</table>

decreased over time (negative parameter estimate), or it decreased at first then increased (positive parameter estimate). A zero (statistically non-significant) time estimate indicated that there was no trend in the differencing over time—in other words, neither proliferation nor diminishment occurs systematically. In this last situation, a trend of variation was not occurring, compared to the situation of a significant parameter estimate indicating that a trend of variation was occurring. Overall, variation appears to depend strongly on end market conditions. Presentation of experimental results are grouped by end market scenarios.

Heterogeneous Demand (Experiments 1 through 4)

Analysis of the parameter estimation results of the variation analysis for manufacturers and retailers revealed statistically significant parameter estimates for time, strategy and the interaction term. Additionally, for both manufacturers and retailers, the parameter estimates for all four experiments were negative for the effect of time while they were positive for the effect of strategy. A negative slope on the time estimate indicated that the number of firms grow then decrease until it becomes negative, at which point firms begin dying out. The positive strategy parameter estimates indicated that the differencing for the high price firms cross zero later than for low price firms—initial population growth was faster for high price firms, but so was later population decline.

The interactive effect was significant and negative for manufacturers and retailers for all four experiments. High price firms experience relatively greater negative population growth over the course of the simulation compared to low price firms. Overall, this paints a picture for heterogeneous end markets prompting growth early in the simulation for both low and high price firms. After demand peaks, the number of both high and low price firms declined, but much more so for high price firms. The magnitude of the parameter estimates indicated by the slopes of the differences from time step to time step, and thus were commensurate with the amount of growth then decline. Since manufacturers and retailers each had the same starting populations, the greater parameter estimates suggested that retailers propagated to a higher population than manufacturers before rapidly dying out.

Assemblers had statistically significant negative parameter estimates for time and strategy, but once the interaction term was added strategy lost its significance. These results describe a life cycle for assemblers of initial growth that eventually diminish in keeping with the pattern of demand growth and shrinkage. High price assemblers gradually declined in population over the course of the simulation, the population of low price manufacturers gradually increased to serve the burgeoning low price market. The lack of significance for the interaction term indicated that assembler population does not have a significant growth-then-decline like manufacturers and retailers—instead, the total population of assemblers remained relatively steady as the population of high price assemblers declined while low price assemblers proliferated.

Homogeneous High Demand (Experiments 5 through 8)

All four experiments with a homogeneous high price demonstrated statistically significant negative parameter estimates for time for all three firm types. This indicated that all firms profligate rapidly at first with a steady decline in the rate of population growth followed by an increasing loss in population from time step to time step. The parameter estimates indicated the greatest variation (magnitude) for retailers and the least for assemblers. Assemblers apparently enjoy some benefit from being in the middle of the supply chain—retailers face the brunt of variability in demand, and manufacturers were impacted by variation in the success of their products, but assemblers feel these effects only secondarily. Since the population of manufacturers and retailers buffers the volatility of end market demand, assemblers only have to respond to their buyer and supplier markets.

Homogeneous Low Demand (Experiments 9 through 12)

For all firm types in all four experiments with a homogeneous low price demand setting, significant negative parameter estimates resulted for time and strategy. Introducing the interactive term reduced the overall R-square and resulted in a non-significant strategy coefficient. Based on this evidence, the simpler model was retained.

The parameter estimates for retailers exhibited the greatest magnitude for both time and strategy. This revealed that retailers exhibit the greatest increase at the beginning followed by greater declines in numbers of retailers. The negative value for strategy indicated that as high strategy firms begin to decline in numbers, low strategy firms increase in numbers. This relationship was relatively stronger for retailers than it was for manufacturers and retailers. The lack of a significant interaction term indicated that the increasing rate of population growth for low price firms does not change during the course of the simulation.

Retention

Since the supply chain network under study evolved over time, retention was assessed using parametric survival analysis since it specifically measured the time-dependence (Harrell, 2001). SAS 9.1 was used to conduct all retention analyses. Survival analysis addresses the positively skewed distribution of time to occurrence of an event and supplies a probability of surviving past a given time, which was often more useful than an expected lifespan. Analytical results of firm longevity are summarized in Table 3.

Rate of Technological Advance

Under conditions of rapid technological advancement manufacturers and assemblers displayed 11% and 12% reductions in longevity, respectively. Markets characterized by rapid technological advancement also reduced retailer longevity by a statistically significant but small (<3%) amount; otherwise rate of technological advance demonstrated no significant effect on retailers. These results followed intuitive expectations since manufacturer production capacity served as the starting point for technological advances in the supply chain, and assemblers relied on a small number of manufacturers. Assemblers choosing poorly found their suppliers going out of business. The

Table 3. Longevity parameter estimates by experiment

				Manufacturer		Assembler		Retailer	
Market	RTA	EOS	Experiment	*Strategy*	*Max Capacity*	*Strategy*	*Max Capacity*	*Strategy*	*Max Capacity*
Heterogeneous	*Fast*	*Fast*	1	0.7024	0.0035	0.1291	0.0029	0.1715	0.0159
				<.0001	<.0001	0.0384	<.0001	0.0021	<.0001
		Slow	2	0.7517	0.0048	0.0026	0.0255	-0.3849	-0.0718
				<.0001	<.0001	0.9705	<.0001	<.0001	<.0001
	Slow	*Fast*	3	0.7238	0.0108	0.1579	0.0028	0.254	0.0272
				<.0001	<.0001	0.0073	<.0001	<.0001	<.0001
		Slow	4	0.7599	0.014	0.1307	0.0246	-0.3452	-0.0522
				<.0001	<.0001	0.0862	<.0001	<.0001	<.0001
Homogeneous High	*Fast*	*Fast*	5	.	0.0226	.	0.0073	.	-0.0106
				.	<.0001	.	<.0001	.	<.0001
		Slow	6	.	0.0206	.	0.0264	.	-0.0958
				.	<.0001	.	<.0001	.	<.0001
	Slow	*Fast*	7	.	0.0235	.	0.0039	.	0.0116
				.	<.0001	.	<.0001	.	0.0006
		Slow	8	.	0.0182	.	0.0067	.	-0.0923
				.	<.0001	.	0.0002	.	<.0001
Homogeneous Low	*Fast*	*Fast*	9	0.7938	0.0046	0.328	0.0046	0.032	0.0205
				<.0001	<.0001	<.0001	<.0001	0.336	<.0001
		Slow	10	0.6796	0.004	0.4256	0.0217	-0.4116	-0.0328
				<.0001	<.0001	<.0001	<.0001	<.0001	<.0001
	Slow	*Fast*	11	0.6604	0.013	0.2464	0.0042	0.1081	0.0279
				<.0001	<.0001	<.0001	<.0001	<.0001	<.0001
		Slow	12	0.839	0.0172	0.0622	0.0044	-0.3391	-0.021
				<.0001	<.0001	0.0901	<.0001	<.0001	0.0013

strength of relationship with manufacturers and retailers also contributed to assembler longevity. As discussed subsequently, retailer longevity was more influenced by economies of scale.

For manufacturers price strategy (high vs. low quality product) exhibited a strong influence on longevity regardless of the rate of technological advance, with manufacturers following the high price strategy displaying significantly greater longevity. Assemblers following a high price strategy under conditions of fast rate of technological advance and homogeneous, low price end markets exhibited 39% and 53% greater longevity, respectively, than low price assemblers in response to fast and slow increases to economies of scale. This counterintuitive result was the outcome of intense inter-firm competition among low-price strategy firms prompted by rapid growth in that market; stagnant growth of the high end market resulted in stable conditions that allowed high end assemblers to persist. Under conditions of slow rate of technological advance and growth of economies of scale, the slow pace of change made for a relatively small difference in longevity for high vs. low price strategy assemblers. Assemblers lived longer in environments with homogeneous pools of suppliers and buyers; if existing manufacturers and retailers switched assemblers or went out of business, assemblers benefited from having more substitutes available.

Economies of Scale

Rapidly growing economies of scale prompted statistically significant increases to longevity for all three firm types with 9%, 7.5%, and 18.7% greater longevity, respectively, for manufacturers, assemblers, and retailers. Under conditions of rapid demand growth and rapidly increasing economies of scale, larger companies achieved lower production costs and the additional stability resulting from serving larger market shares. Retailers that invested in greater capacity also benefited from additional demand stability of having larger market share.

Analysis of the individual experimental results, economies of scale had a similar positive effect on longevity for manufacturers across all experimental conditions. For assemblers, the high price strategy related positively to increased longevity for the heterogeneous end market when the rate of increase of economies of scale was high (longevity for these assemblers was 13-14% greater); this effect was even greater for the homogeneous low price end market with a fast rate of technological advance (where high price assemblers lived 53% longer). Parametric regression results were statistically non-significant for assemblers in both a homogeneous low price end and heterogeneous markets with a slow rate of technological advance.

Economies of scale exhibited a strong effect on retailers. For heterogeneous and homogeneous, low price markets, conditions of fast-paced increase of economies of scale increased longevity of retailers following a high price strategy (from 11-29%). The exception was under conditions of a homogeneous low price end market with a fast rate of technological advance, results of which did not achieve statistical significance. Slowly increasing economies of scale factor provoked the opposite effect, reducing longevity of high price retailers by 29-34% compared to low-price retailers serving either heterogeneous or homogeneous low price end markets. Compared to low price retailers, the high price retailers failed to establish a stable pool of end market demand to survive fluctuations and competition. These results reflect the real-life consequence of economies of scale where under the right conditions large retailers quickly drive small retailers out of business.

Selection

When examining the results of the experiments, it was important to keep in mind that all experiments started with 10 firms at each level of the supply chain, each with one-on-one, exclusive relationships with suppliers and customers, and all firms dedicated to serving the high end market. Results are discussed for each end market condition.

Analysis of the evolutionary process of selection was implemented with binary encoding. The thousand time cycle duration of the simulation was divided into 5 time steps of 200 demand cycles each. Simple univariate analysis (via PROC GLM in SAS) was used to create 95% confidence intervals of the collaboration index for each time step. Successive time steps were assessed for significant changes to the collaboration index, with a 1 encoding indicating that a statistically significant change had occurred from one time step to the next; a 0 indicated no significant time change had occurred, and therefore evolution of the exchange relationships had remained stable. The series of 1's and 0's were then analyzed using a multinomial logit model defined by Agresti (1990) and frequently used for this type of analysis (Fokianos & Kedem, 2003). Table 4 presents a

Table 4. Significant trends in inter-firm collaboration

		PLC Transition to:			
	Pricing Strategy	**Growth**	**Maturity**	**Early Decline**	**Late Decline**
Manufacturer-to-Assembler					
Heterogeneous End Market	Low	0	-	0	0
	High	0	0	0	0
Homogeneous Low Price End Market	Low	+	-	-	-
	High	0	0	0	0
Assembler-to-Manufacturer					
Heterogeneous End Market	Low	+	+	0	-
	High	0	0	0	0
Rate of Technology Advance with a Heterogeneous End Market	Low	+	+	-	-
	High	0	0	0	0
Homogeneous Low Price End Market	Low	+	-	0	-
	High	-	-	-	0
Assembler-to-Retailer					
Heterogeneous End Market	Low	-	-	-	+
	High	-	-	-	0
Homogeneous Low Price End Market	Low	-	-	-	+
	High	-	-	-	0
Retailer-to-Assembler					
Heterogeneous End Market	Low	+	+	-	-
	High	0	0	0	0
Rate of Technology Advance with a Heterogeneous End Market	Low	-	+	0	0
	High	0	0	0	0
Homogeneous Low Price End Market	Low	+	-	0	-
	High	-	-	-	0

simplified depiction of the trends in inter-firm collaboration for each of the exchange relationship perspectives between manufacturers, assemblers, and retailers. Inter-firm collaborations evolved differently depending on the level in the supply chain, the product price strategy, and the rate of technological advance. Only variables that demonstrated statistically significant effects appear in Table 4; complete results appear in Appendix B.

Heterogeneous End Market

High price strategy perpetuated strong inter-firm collaborations across most interfirm exchanges throughout the demand life cycle. A shrinking high end market meant a shrinking pool of high end businesses with whom to conduct exchanges; surviving firms were those with a history of strong collaboration. The exception was the assembler-to-retailer relationship, which exhibited trend of diminishing collaboration throughout the demand life cycle until leveling off at late decline. This contrasts with unchanging retailer-to-assembler collaboration, indicating that even as assemblers expanded market opportunities, retailers stayed with established suppliers.

The rapid growth of a low end market prompted much more evolution of inter-firm collaboration for companies following the low price strategy with a slow pace of technological advance. Low price manufacturers exhibited a decline in collaboration during the transition to rapid growth, but thereafter remained unchanged. Low price assemblers exhibited little change in the collaboration index throughout the life of the simulation for their relationships with manufacturers. On the sell-side, low price assemblers appear to expand their markets as indicated by a declining collaboration index throughout most of simulation until late decline, when collaboration intensifies with the remaining retailers. Retailers on the other hand exhibited increasing collaboration as demand progressed to the growth and maturity stages, and then declining collaboration during the stages of declining demand. Retailers were forced to prolong supplier relationships when demand outstripped supply, but were more likely to switch suppliers as increasingly price sensitive customers declined in number relative to available supply.

High rate of technological advance exhibited a significant effect only for the buying side of interfirm relationships for low price companies. Assembler-to-manufacturer relationships demonstrated increased collaboration into both the growth and maturity stages, with declining collaboration thereafter. During the first half of the simulation, manufacturing capacity trailed end market demand, leading assemblers to leverage collaborative relationships to ensure a source of supply. Retailer-to-assembler collaboration initially declined as demand entered the growth phase, then increased during maturity, remaining level thereafter. In conjunction with the analysis on variation and retention, this indicates that under conditions of rapid technological advance that low price retailers at first struggled to find enough sources of supply to satisfy demand, with surviving retailers strengthening the relationships discovered during the growth phase which offered the benefit of reduced prices to the end consumer.

Homogeneous Low Price Demand

Homogeneous low price end market demand led to the same evolutionary path for interfirm collaboration for three out of four of supply chain relationships which exhibited declining collaboration throughout the simulation until remaining steady during the transition from early to late decline. High price manufacturers were the exception to this pattern as they exhibited unchanged collaboration indices throughout the demand life cycle.

Low price supply chains exhibited different evolutionary trajectories depending generally on whether firms are selling or buying. Manufacturers initially increased collaboration with assemblers, and then decreased collaboration for the rest of the demand life cycle. Assemblers decreased collaboration with retailers until the late decline stage when the collaboration index increased. On the sell-side, low price assemblers exhibited decreasing collaboration with retailers through growth, maturity, and early decline followed by an increase to collaboration during late decline. Low price companies exhibited a common pattern of evolution for buying side collaboration. Assembler-to-manufacturer and retailer-to-assembler collaboration increased during the growth phase, then decreased at maturity and remained steady until decreasing again at late decline.

Homogeneous High Price Demand

The experimental conditions of homogeneous, high end market demand prompted the appearance of additional numbers of firms to satisfy growing demand, but failed to prompt evolution of interfirm collaboration, as would be expected under conditions of low environmental pressure. Firms appear to maintain their exclusive ties, and the greater expense of expanding or upgrading capacity for high end product meant a slower pace of growth at the firm level; it also meant that aggregate production capacity continuously lagged end market demand. Firms maintained exclusive relationships under this setting largely due to the lack of available trading partners with unclaimed output, and lack of competitive pressures meant low mortality rates.

IMPLICATIONS FOR THE ROLE OF KNOWLEDGE IN SUPPLY CHAIN NETWORKS

Business literature may be re-orienting toward a dominant logic founded on collaboration and knowledge as the sources of competitive advantage (Vargo and Lusch, 2004a; Inkpen and Tsang, 2005; Grant and Baden-Fuller, 2004; McEvily and Chakravarthy, 2002; Kogut, 2000). For example, in marketing Service-Dominant Logic (Vargo and Lusch, 2004b) posits that knowledge, and the knowledge processes surrounding products and product management, are key sources of competitive advantage (Lusch and Vargo, 2006; Vargo and Lusch, 2004a). In essence, rather than focus on its physical properties and features, the product embodies the knowledge and collaborative abilities of all those who contributed to its creation. Furthermore, "...knowledge as the basis for competitive advantage can be extended to the entire supply chain" (Vargo and Lusch, 2004b, p. 9). In this view, knowledge exchange underlies successful organizational and collaborative forms.

But not all knowledge is the same, and different forms of collaboration will interact to affect the value and outcomes of inter-organizational knowledge management. What follows are propositions regarding how the evolution of inter-organizational collaboration will affect or be affected by the different roles of knowledge in the supply chain: knowledge generation, knowledge sharing, and knowledge implementation (Gravier, et., 2008). Some empirical research supports these propositions, but in the absence of a predictive theory for the role of knowledge in inter-organizational supply chains, it is hoped that these propositions can provide some insights or guide future research.

Much of the literature included derives from the work in strategic alliances, which is rooted in the premise that firms collaborate in order to access the knowledge and competences of other firms. But not all accessions of knowledge are created equal. Hamel distinguished "internalizing" as opposed to merely "accessing" knowledge endemic to another firm (Hamel, 1991). Real-life challenges to inter-organizational collaborations lead to a "collaborative membrane" that acts as a filter between organizations in alliances. The collaborative membrane can profoundly influence the magnitude, content, and direction of inter-organizational knowledge flows. In effect, the collaborative membrane defines the collaborative relationship that exists between alliance (and supply chain) members. Based upon these insights, inter-organizational interaction falls into one of two approaches to inter-organizational interaction: collaboration ("internalizing" knowledge) or modularization ("accessing" knowledge). The underlying premise rests on the assumption that the role that inter-organizational knowledge plays determines the degree of collaboration or modularity in the supply chain (see Figure 6).

Collaboration

Kahn's (1996) research into integration equated "collaboration" with continuous interaction, often informal in nature, most often without clearly defined structure. Collaboration attains collective goals via resource sharing and a common vision. Collaboration would represent the outcomes of long-term relationships characterized by trust and many interactions (a high collaboration index in the previously described simulation) Collaboration buffers volatility for firms in highly unstable environments, which enhances the opportunity to learn-by-doing (Sorenson, 2003). Collaboration characterizes well-functioning alliances but also many high-performing supply chains whose firms depend upon each other and interact frequently and effectively. Supply chain integration approaches such as vested outsourcing (Vitasek, 2011) and performance-based logistics (Randall, et al., 2010) increase collaboration.

Collaboration benefits supply chains with operations characterized by more complex interactions. Products involving optimization of highly interdependent or complex interdependencies of design and manufacture, or diverse design choices (such as a microprocessor), tend to benefit more from collaboration.

Modularity

Modularity represents a strong focus on the individuality of each firm and less inter-operational dependence (represented by lower levels of the collaboration index in this study's simulation). Traditional contract-based outsourcing represent one possible outcome of modular supply chains, although modularity also includes discrete organizational nodes or clusters held together by standards of member performance and conformance to design rules (Langlois, 2002). Modular supply chain architectures lower transaction costs while preserving the independent identities of the firms (Kahn, 1996). For many supply chains, modularity offers benefits over collaborations as

Figure 6. Role of knowledge in determining degree of interfirm collaboration

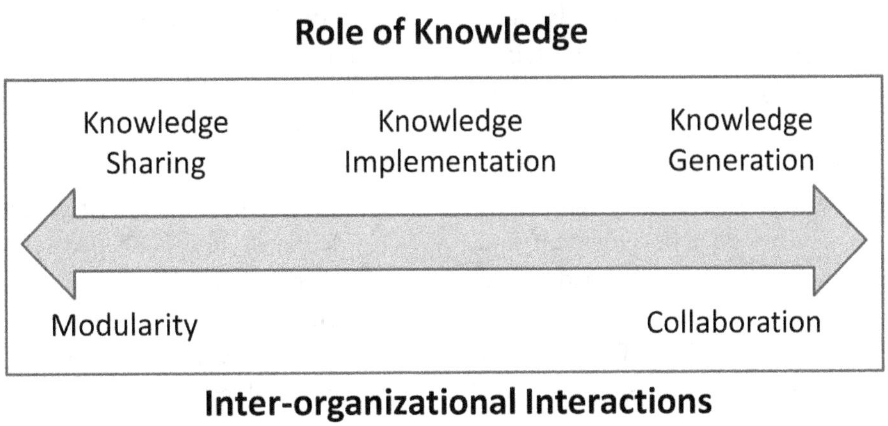

it allows redundant parallel operations by several network members, thereby improving speed and efficiency (Ethiraj and Levinthal, 2004). Compared to a collaborative supply chain with sole sourcing or a reduced supplier base, modularity has the potential to reduce idle time, prompting more completed work in the same amount of time compared to situations where the same operations are all conducted by one or a small number of firms.

Modularity most benefits supply chains that provide modular products with limited diversity that have independent markets for each of the modules; such markets also benefit from the recombinant possibilities endemic to these supply chains.

Computer memory represents one supply chain that benefits from modularity. Computer memory has many applications in a variety of electronic products, both for consumers and as components for more complex assemblies such as automobiles or aircraft. On the other hand, supply chains that require frequent or complex interactions, such as testing and integrating new product innovations, bring to light the difficulties of optimizing highly interdependent processes in a modular network (Ethiraj and Levinthal, 2004).

Knowledge Generation

In keeping with the knowledge problem-solving perspective of the knowledge-based theory of the firm (Nickerson and Zenger, 2004), and also with the information acquisition views of Rindfleisch and Moorman (2001), knowledge generation focuses on the acquisition of information that pertains directly to the development of new products or services. Knowledge generation in its various guises such as new product development and process innovation has a well-established record for boosting firm performance (c.f., Sethi et al., 2001; Kohli and Jaworski, 1993). Research has also found evidence of knowledge generation's effectiveness at developing new products in the inter-organizational context (Rindfleisch and Moorman, 2001; Sivadas and Dwyer, 2000; Im and Workman, 2004).

Inter-organizational knowledge generation could benefit from either modular or collaborative knowledge exchanges. For example, alliances have been called "...the most important source of new ideas and information that result in performance-enhancing technology and innovations" (Dyer and Singh, 1998, p. 665). Evidence indicates that close inter-organizational collaborations in alliances bolsters innovation generation (Roy et al., 2004; Nielsen, 2005). However, intra-firm technological and strategic flexibility capabilities may limit the benefits of inter-organizational collaborations (Zhou and Wu, 2010). Specifically, at the intra-firm level, technological capability can create "lock-in" or path dependencies. An inverted-U relationship appears to define technological capability's influence on innovation (Zhou and Wu, 2010), with too little technological capability resulting in a firm incapable of taking advantage of opportunities, and too much technological capability impeding adaptability. This evidence jibes with this study's simulation findings. The collaboration index increased most in situations of intense innovation and diminishing profit margins—two powerful real-life motivators to collaborate innovatively. Among the requirements for successful intra-organizational innovativeness are flexibility and departures from planned objectives; both of these factors destabilize close relationships (Sivadas and Dwyer, 2000). These observations lead to the following proposition (see Figure 7):

P1: The degree of knowledge generation will demonstrate a U-shaped relationship with inter-organizational knowledge exchanges, with the highest need for knowledge generation leading to optimal rather than maximal levels of collaboration.

Figure 7. Graphical representation of the proposed relationships between role of knowledge and supply chain collaboration

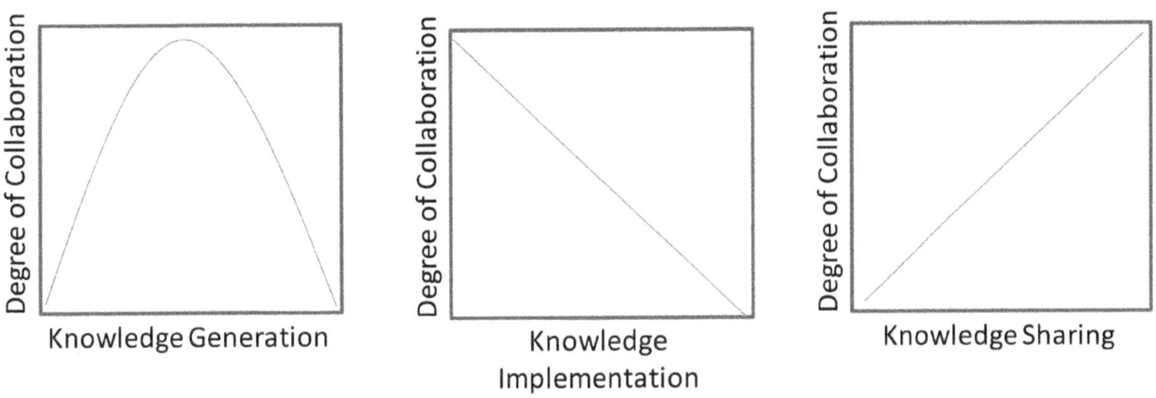

This proposition implies that the most innovative knowledge-exchanges will not be the closest collaborations, nor the arm's-length transactions, but rather the hybrid forms that blend the benefits of each of the extremes, especially the flexibility to change partners if necessary. In keeping with the simulation's findings, recent empirical work tends to support that close collaborations disintegrate less often from their failures and more often from the desirability of finding other partners (Greve, et al., 2012). On the surface, this proposition appears to fly in the face of common wisdom, but the high failure rate for strategic alliances implies that at worst close collaborations are useful only so long as they serve a purpose, and at best the benchmark organizations strive to enter alternative inter-organizational relationships that provide the benefits of alliances while minimizing impacts on flexibility.

Knowledge Sharing

Knowledge sharing leverages information systems and other to support sharing resources, competencies, personnel and other knowledge resources already possessed by at least one alliance member but not by at least one other (Baker and Sinkula, 1999; Kohli and Jaworski, 1993). Also known as knowledge transfer or interfirm learning, knowledge sharing refers to the extent that organizations are able to access each other's established know-how and critical information (Appleyard, 1996).

Distinct from knowledge generation, knowledge sharing merely accesses knowledge rather than internalizing it across the inter-organizational collaborative membrane (Hamel, 1991). Knowledge sharing of course generally occurs in many instances of organizational learning and other forms of knowledge generation (Roper and Crone, 2003; Appleyard, 1996). However, as a process, knowledge sharing represents the antithesis of knowledge generation, focusing on information sharing rather than mutual learning and creation of new knowledge (Rindfleisch and Moorman, 2001; Dyer and Singh, 1998).

Knowledge sharing often consists of horizontal alliances seeking to reduce environmental uncertainty (Bucklin and Sengupta, 1993). One example is SEMATECH, a group of semiconductor manufacturing firms that shares information for the purposes of setting standards and tracking industry trends. The semiconductor firms also notoriously guard their secrets with regard to innovations to both process and product.

Research has linked the sharing of knowledge at the organization level with higher short-term financial benefit (Moorman and Miner, 1997). Sharing complementary knowledge resources on supplier or customer markets can prove beneficial and may lead to competitive advantage; indeed, that knowledge sharing may constitute the primary advantage that firms accrue through alliances and close collaborations (Grant and Baden-Fuller, 2004). Toyota's production supply chain is often cited as an example of a knowledge sharing network (Dyer and Nobeoka, 2000).

Importantly, one study has found that knowledge sharing can have different effects on innovation depending on the breadth vs. depth of knowledge. Firms with a broad knowledge base tend not to benefit as much from external knowledge sharing, whereas firms with a deep knowledge base tend to benefit from sharing information externally (Zhou and Li, 2012). This follows the insights from this study's simulation that revealed that collaborations tend to be asymmetric, with firms such as retailers and assemblers with little need to innovate tending to collaborate more. On the other hand, firms such as manufacturers that need to innovate a lot tend not to collaborate outside of their boundaries. This and the previous observations lead to the following proposition:

P2: The degree of knowledge sharing will demonstrate an increasing relationship with inter-organizational knowledge exchanges, with more knowledge sharing leading to more collaboration.

Knowledge Implementation

The implicit and explicit costs associated with Generating and sharing knowledge generates certain costs and require varying levels of commitment and action on the part of the participating organizations. In order to avoid these costs or commitments, organizations may exercise the option to delegate certain activities to another organization (Kogut and Zander, 1992). Firms that elect to divest themselves of a non-core competence are putting knowledge implementation to use (Prahalad and Hamel, 1990). In supply chains, logistics knowledge frequently falls into the role of knowledge implementation, as evidenced by the rise of third party (3PL) and fourth party (4PL) logistics providers. As an example of knowledge exchanges occurring purely to allow another organization implement what they know how to do best, the latest evidence indicates that 3PL providers often coordinate not just logistics but all supply chain functions, allowing firms to focus on their core competences (Zacharia, et al., 2011). Firms avoid knowledge sharing with a logistics partner because of the volume of information that must be shared on package contents, origins, destinations, truck license number, air cargo flight numbers, shipping costs, hazardous material routing and the like, which generally requires that expensive information systems be developed for activities that occur outside of the firm. 3PLs and other logistics service providers function as "turnkey" service providers.

Knowledge implementation benefits both parties most when their knowledge requirements are relatively self-sufficient and independent. Firms add value when they implement or execute specialized knowledge, processes, and capabilities. The complexities of modern products and services often rely on recombinatorial capabilities with many components in a way that often becomes exceedingly complex; increasingly, firms rely on outside specialists to make subcomponents or provide specialty services as housing all production under one roof has become not just intractably complex but economically untenable. Just imagine trying to manage all the manufacturing for a common and relatively simple consumer product such as a cell phone, which involves more than 18 subassemblies (not including software) and a minimum of a dozen manufacturers across at least seven countries—and this list does not include many manufacturers nor any ancillary service providers (such as 3PLs) (*Economist*, 2011). More complex supply chains such as for automobile, pharmaceutical, medical equipment, and aircraft quickly become much more complicated.

Knowledge implementation benefit inter-organizational networks by enhancing execution or consolidating common knowledge and expertise in order to pool risk and reduce investment in additional production capacity or processes (Roper and Crone, 2003). Knowledge implementation focuses on efficiency rather than creating or accessing knowledge. The outcome is compartmentalized knowledge in the supply chain, but done in such a way that certain processes and services are readily provided, as required.

Knowledge implementation embodies the strategic blending of the unique capabilities of each organization in the network (Kogut and Zander, 1992). Creating new inter-organizational outcomes depends on mixing and matching firms, and the reliance on architectural capabilities at the interfirm level (Henderson and Cockburn, 1994). For example, Microsoft produces its eponymous office software for both Apple and the PC market. Despite the importance of the software to the usability of their final products, neither Apple nor Dell need to interact heavily with Microsoft; rather they simply purchase Microsoft's finished product and related services "ready to go." By facilitating economies of scale and compartmentalized competence, knowledge implementation enables leveraging the broader capabilities throughout the supply chain.

In the context of the earlier presented simulation, under experimental treatments of rapidly expanding markets collaboration declined rapidly, especially under scenarios of low price elasticity. Low price elasticity meant little incentive to innovate or collaborate, and with a rapidly growing market firms shopped from any source in order to meet demand. These insights lead to the following proposition:

P3: The degree of knowledge implementation will demonstrate a negative relationship with inter-organizational knowledge exchanges, with more reliance on knowledge implementation leading to less collaboration.

CONCLUSION

Assessing the role that knowledge plays in the supply chain holds promise for deepening scholarly understanding and practitioner management of supply chain evolution. Viewing supply chains as inter-organizational networks characterized by dynamism, adaptability, and knowledge-based exchanges allows some ability to predict patterns of collaboration, and could also explain why the "integration prescription" provides such mixed outcomes.

Past research has highlighted the importance of balancing transaction and production costs in accordance with Coase's original conceptualization of TCE (Gravier and Farris, 2012). The imbalance or balance of production vs. transaction costs may explain which supply chain strategies prove most effective or evolve out of a given set of circumstances (see Figure 8). If knowledge-based exchanges do indeed form the basis of supply chain evolution, then measuring the amount of knowledge inherent to the supply chain, as well as the ability of the supply chain to communicate knowledge efficiently, become important predictors of supply chain performance. Information theory's entropy provides precisely such a measure. Entropy—usually embodied by Shannon's entropy—measures the amount of information content; additionally, the measurement also shows the absolute limit to the amount of information that can be carried in a given channel (Shannon, 1948). As such, entropy holds the promise to measure the maximum complexity that a given supply chain can reasonably manage. Current ef-

Figure 8. Using production and transaction costs to predict supply chain designs

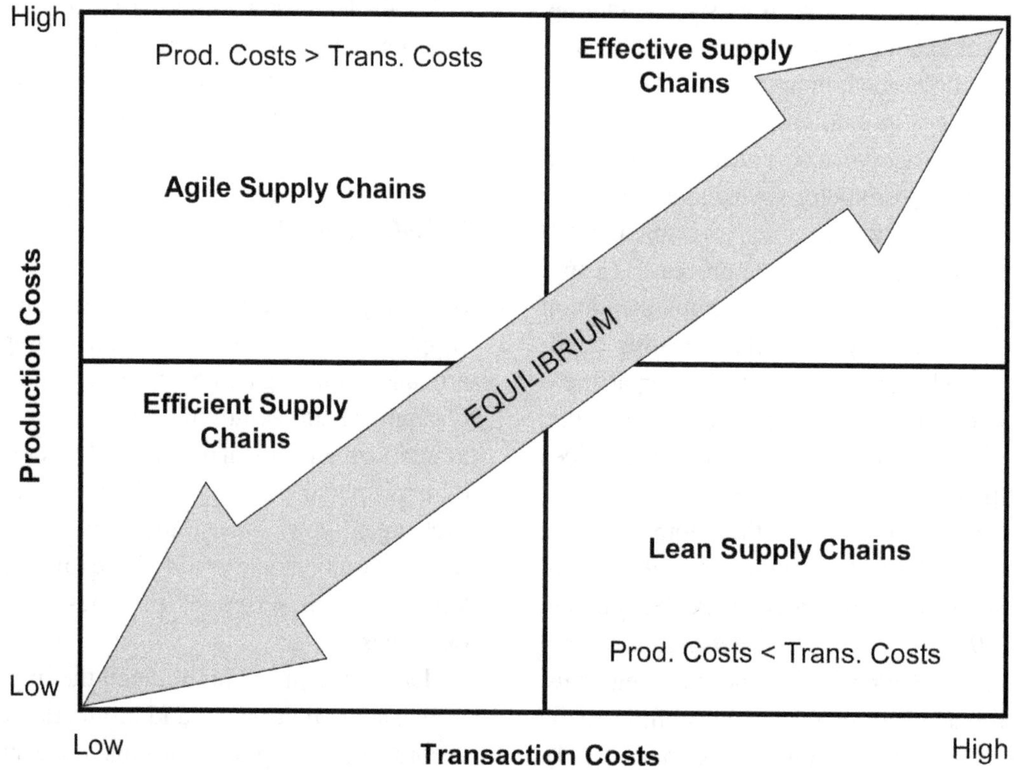

forts to measure entropy and apply it as a means of managing inter-organizational processes and operations demonstrate promise (c.f., Gravier and Kelly, 2012; Liu and Zhang, 2011). Assessing entropy would enable managers to assess whether collaboration and integration are on an upward or downward trend, and to assess trends against the propositions presented in this study in order to predict outcomes or explain shortcomings and success. Researchers could also apply the entropy measure to test this study's propositions for empirical validity.

With regard to changing paradigms in supply chain and inter-organizational knowledge management research, the dominant paradigm resting on an essentially static or steady state suffers shortcomings in a world characterized by increasing change. In order to escape from a reactive approach, organizations need tools to anticipate dynamic trends. Rather than "agile" supply chains, organizations should evolve into adaptive networks whose members maintain enough autonomy to respond to their environments. Rather than providing prescriptive mandates for managing entire networks of firms—a process that can be slow and limits firm agility—more and more researchers are suggesting that the right inter-organizational architecture to allowed to "emerge" so that supply chains effectively manage themselves. Empirical evidence hints that inter-organizational information technology and information sharing lead to improved supply chain performance outcomes, especially in situations where the firms work together over time and can evolve solutions (Prajogo and Olhager, 2012). Of course, it does appear that individual firms must first master integration within their own boundaries before they can effectively integrate with other firms (Huo, 2012), so this insight has as many implications for internal as it does for external decision-making and strategizing.

As a last thought, this research effort suggests that certain roles of knowledge can actually reduce integration and collaboration, while others may increase them. Managers and researchers should consider investing more resources to investigating methods that combine the benefits of both highly collaborative and modular supply chain networks. Research suggests that a dual network structure may provide the benefits of both close collaboration and flexibility. Capaldo's (2007) propositions suggest that a strong network near the hub of value creation with a weaker network to carry out distribution and other functions that do not add directly to the core value proposition may provide the best structure for certain supply chains. This suggests a boundary analogous to the customer decoupling point but instead of materials flows or customer information determining the "push-pull" boundary, the amount and type of collaboration determines a collaborative intensity boundary. From a theoretical standpoint, a collaborative intensity boundary offers a more pro-active approach to supply chain design since determining deficiencies and abundances in collaborative intensity ideally should precede material flows or movement of customer information. Adopting a collaborative intensity perspective of supply chain networks finds support in the interfirm problem-solving research that also found that complex problems (slower rate of technological advance) favored integrated (highly collaborative) interfirm boundaries, whereas simpler problems that were well structured favored lower levels of integration (Macher, 2006).

REFERENCES

Abel, T. (1998). Complex adaptive systems, evolutionism, and ecology within anthropology: Interdisciplinary research for understanding cultural and ecological dynamics. *Georgia Journal of Ecological Anthropology*, *2*, 6–29. doi:10.5038/2162-4593.2.1.1

Agresti, A. (1990). *Categorical data analysis*. New York: John Wiley and Sons.

Akkermans, H. (2001). *Emergent supply networks: System dynamics simulation of adaptive supply agents*. Paper presented at the 34th Hawaii International Conference on System Sciences. Hawaii, HI.

Alchian, A. A., & Demsetz, H. (1972). Production, information costs, and economic organization. *The American Economic Review*, *62*(5), 777–795.

Anand, N., & Peterson, R. A. (2000). When market information constitutes fields: Sensemaking of markets in the commercial music industry. *Organization Science*, *11*(3), 270–284. doi:10.1287/orsc.11.3.270.12502

Anderson, P. (1999). Complexity theory and organization science. *Organization Science*, *10*(3), 216–232. doi:10.1287/orsc.10.3.216

Appleyard, M. M. (1996). How does knowledge flow? Interfirm patterns in the semiconductor industry. *Strategic Management Journal*, *17*, 137–154.

Baker, W. E., & Sinkula, J. M. (1999). The synergistic effect of market orientation and learning orientation on organizational performance. *Journal of the Academy of Marketing Science*, *27*(4), 411–427. doi:10.1177/0092070399274002

Banks, J. S. C. II, & Nelson, B. L. (1999). *Discrete-event system simulation* (2nd ed.). Upper Saddle River, NJ: Prentice Hall.

Bendoly, E., Bharadwaj, J., & Bharadwaj, S. (2012). Complementary drivers of new product development performance: Cross-functional coordination, information system capability, and intelligence quality. *Production and Operations Management, 21*(4), 653–667. doi:10.1111/j.1937-5956.2011.01299.x

Bettis, R. A., & Prahalad, C. K. (1995). The dominant logic: Retrospective extension. *Strategic Management Journal, 16*(1), 5–14. doi:10.1002/smj.4250160104

Bourlakis, M., Maglaris, G., & Fotopoulos, C. (2012). Creating a best value supply chain? Empirical evidence from the Greek food chain. *International Journal of Logistics Management, 23*(3), 360–382. doi:10.1108/09574091211289228

Bucklin, L. P., & Sengupta, S. (1993). Organizing successful co-marketing alliances. *Journal of Marketing, 57*(2), 32–46. doi:10.2307/1252025

Bureau, U. S. C. (2005). *Survey of plant capacity*. Author.

Capaldo, A. (2007). Network structure and innovation: the leveraging of a dual network as a distinctive relational capability. *Strategic Management Journal, 28*(6), 585. doi:10.1002/smj.621

Chesbrough, H. W., & Teece, D. J. (1996). When is virtual virtuous? Organizing for innovation. *Harvard Business Review, 74*(1), 65–73.

Chiva-Gomez, R. (2003). The facilitating factors for organizational learning: Bringing ideas from complex adaptive systems. *Knowledge and Process Management, 10*(2), 99–114. doi:10.1002/kpm.168

Choi, T. Y., Dooley, K. J., & Rungtusanatham, M. (2001). Supply networks and complex adaptive systems: Control versus emergence. *Journal of Operations Management, 19*, 351–366. doi:10.1016/S0272-6963(00)00068-1

Coase, R. H. (1937). The nature of the firm. *Economica, 4*(16), 386–405. doi:10.1111/j.1468-0335.1937.tb00002.x

Daugherty, P. J., Ellinger, A. E., & Gustin, C. M. (1996). Integrated logistics: Achieving logistics performance improvements. *Supply Chain Management, 1*(3), 25–33. doi:10.1108/13598549610155297

David, R. J., & Han, S.-K. (2004). A systematic assessment of the empirical support for transaction cost economics. *Strategic Management Journal, 25*, 39–58. doi:10.1002/smj.359

Davis, J. P., Bingham, C. B., & Eisenhardt, K. M. (2007). Developing theory through simulation methods. *Academy of Management Review, 32*(2), 480–499. doi:10.5465/AMR.2007.24351453

Deitrich, M. (2012). *Transaction cost economics and beyond: Toward a new economics of the firm*. London: Routledge.

Dooley, K. J. (1997). A complex adaptive systems model of organization change. *Nonlinear Dynamics Psychology and Life Sciences, 1*(1), 69–97. doi:10.1023/A:1022375910940

Droge, C., Vickery, S. K., & Jacobs, M. A. (2012). Does supply chain integration mediate the relationships between product/process strategy and service performance? An empirical study. *International Journal of Production Economics, 137*(2), 250–262. doi:10.1016/j.ijpe.2012.02.005

Dyer, J. H., & Nobeoka, K. (2002). Creating and managing a high-performance knowledge-sharing network: The Toyota case. *Strategic Management Journal*, *21*(3), 345–367. doi:10.1002/(SICI)1097-0266(200003)21:3<345::AID-SMJ96>3.0.CO;2-N

Dyer, J. H., & Singh, H. (1998). The relational view: Cooperative strategy and sources of interorganizational competitive advantage. *Academy of Management Review*, *23*(4), 660–679.

Economist (2011). *Slicing an apple*. Retrieved on June 24, 2013, from http://www.economist.com/blogs/dailychart/2011/08/apple-and-samsungs-symbiotic-relationship

Eisenhardt, K., & Tabrizi, B. N. (1995). Accelerating adaptive processes: Product innovation in the global computer industry. *Administrative Science Quarterly*, *40*(1). doi:10.2307/2393701

Ethiraj, S. K., & Levinthal, D. (2004). Modularlity and innovation in complex systems. *Management Science*, *50*(2), 159–173. doi:10.1287/mnsc.1030.0145

Fabbe-Costes, N., & Jahre, M. (2008). Supply chain integration and performance: A review of the evidence. *International Journal of Logistics Management*, *19*(2), 130–154. doi:10.1108/09574090810895933

Fokianos, K., & Kedem, B. (2003). Regression theory for categorical time series. *Statistical Science*, *18*(3), 357–376. doi:10.1214/ss/1076102425

Forrester, J. W. (1958). Industrial dynamics: A major breakthrough for decision makers. *Harvard Business Review*, 37–66.

Forrester, J. W. (1961). *Industrial dynamics*. Cambridge, MA: MIT Press.

Frohlich, M. T., & Westbrook, R. (2001). Arcs of integration: An international study of supply chain strategies. *Journal of Operations Management*, *19*(2), 185–200. doi:10.1016/S0272-6963(00)00055-3

Garicano, L., & Kaplan, S. N. (2001). The effects of business-to-business e-commerce on transaction costs. *The Journal of Industrial Economics*, *49*(4), 463. doi:10.1111/1467-6451.00158

Gligor, D. M., & Holcomb, M. C. (2012). Understanding the role of logistics capabilities in achieving supply chain agility: A systematic literature review. *Supply Chain Management: An International Journal*, *17*(4), 438–453. doi:10.1108/13598541211246594

Goldstone, R. L., & Sakamoto, Y. (2003). The transfer of abstract principles governing complex adaptive systems. *Cognitive Psychology*, *46*(4), 414–466. doi:10.1016/S0010-0285(02)00519-4 PMID:12809681

Granger, C. W. J., & Newbold, P. (1977). *Forecasting economic time series*. New York: Academic Press.

Grant, R. M., & Baden-Fuller, C. (2004). A knowledge accessing theory of strategic alliances. *Journal of Management Studies*, *41*(1), 61–84. doi:10.1111/j.1467-6486.2004.00421.x

Gravier, M. J., & Farris, M. T. (2012). Integrative vs. disintegrative forces in the evolution of supply chain networks. *International Journal of Knowledge-Based Organizations*, *2*(3), 1–32. doi:10.4018/ijkbo.2012070101

Gravier, M. J., & Kelly, B. (2011). Measuring the cost of complexity in supply chains: Comparison of weighted entropy and the bullwhip effect index. In *Modelling value*. Berlin: Springer. doi:10.1007/978-3-7908-2747-7_13

Gravier, M. J., Randall, W. S., & Strutton, D. (2008). Investigating the role of knowledge in alliance performance. *Journal of Knowledge Management*, *12*(4), 117–130. doi:10.1108/13673270810884291

Greve, H. R., Mitsuhashi, H., & Baum, J. A. C. (2012). Greener pastures: Outside options and strategic alliance withdrawal. *Organization Science*, *24*(1), 79–98. doi:10.1287/orsc.1110.0733

Gustin, C. M., Daugherty, P. J., & Stank, T. P. (1995). The effects of information availability on logistics integration. *Journal of Business Logistics*, *16*(1), 1–21.

Hamel, G. (1991). Competition for competence and interpartner learning within international strategic alliances. *Strategic Management Journal*, *12*(S1), 83–103. doi:10.1002/smj.4250120908

Harrell, F. E. (2001). *Regression modeling strategies: With applications to linear models, logistic regression, and survival analysis*. New York: Springer-Verlag.

Heide, J. B. (1994). Interorganizational governance in marketing channels. *Journal of Marketing*, *58*(1), 71–85. doi:10.2307/1252252

Heimer, C. A. (1985). Allocating information costs in a negotiated information order: Interorganizational constraints on decision making in Norwegian oil insurance. *Administrative Science Quarterly*, *30*(3), 395–417. doi:10.2307/2392671

Henderson, R., & Cockburn, I. (1994). Measuring competence? Exploring firm effects in pharmaceutical research. *Strategic Management Journal*, *15*(S1), 63–84. doi:10.1002/smj.4250150906

Hobbs, J. E. (1996). A transaction cost approach to supply chain management. *Supply Chain Management*, *1*(2), 15–27. doi:10.1108/13598549610155260

Hofman, D., & Aronow, S. (2012). *The Gartner supply chain top 25 for 2012*. Stamford, CT: Gartner, Inc.

Holland, J. H. (1995). *Hidden order: How adaptation builds complexity*. Reading, MA: Addison-Wesley Publishing Company.

Holland, J. H., & Miller, J. H. (1991). Artificial adaptive agents in economic theory. *The American Economic Review*, *81*(2), 365–370.

Holme, P., Kim, B. J., Yoon, C. N., & Han, S. K. (2002). Attack vulnerability of complex networks. *Physical Review E: Statistical, Nonlinear, and Soft Matter Physics*, *65*(5), 1–15. doi:10.1103/PhysRevE.65.056109

Hunt, S. D., & Morgan, R. M. (1995). The comparative advantage theory of competition. *Journal of Marketing*, *59*(2), 1–15. doi:10.2307/1252069

Huo, B. (2012). The impact of supply chain integration on company performance: An organizational capability perspective. *Supply Chain Management: An International Journal*, *17*(6), 596–610. doi:10.1108/13598541211269210

Ilachinski, A. (2000). Irreducible semi-autonomous adaptive combat (ISAAC), an artificial-life approach to land combat. *Military Operations Research*, *5*(3), 29–46. doi:10.5711/morj.5.3.29

Im, S., & Workman, J. P. (2004). Market orientation, creativity, and new product performance in high-technology firms. *Journal of Marketing*, *68*(2), 114–132. doi:10.1509/jmkg.68.2.114.27788

Inkpen, A. C., & Tsang, E. W. K. (2005). Social capital, networks, and knowledge transfer. *Journal of the Academy of Marketing Science*, *30*(1), 146–165.

Jayaram, J., & Pathak, S. (2013). A holistic view of knowledge integration in collaborative supply chains. *International Journal of Production Research*, *51*(7), 1958–1972. doi:10.1080/00207543.2012.700130

Jonsson, P., & Mattsson, S.-A. (2012). The value of sharing information in supply chains. *International Journal of Physical Distribution and Logistics Management*, *43*(4), 282–299.

Joshi, A. W., & Campbell, A. J. (2003). Effect of environmental dynamism on relational governance in manufacturer-supplier relationships: A contingency framework and an empirical test. *Journal of the Academy of Marketing Science*, *31*(2), 176–188. doi:10.1177/0092070302250901

Kahn, K. B. (1996). Interdepartmental integration: A definition with implications for product development performance. *Journal of Product Innovation Management*, *13*(2), 137–151. doi:10.1016/0737-6782(95)00110-7

Kim, S. W. (2003). *An investigation of information technology investments on buyer-supplier relationship and supply chain dynamics*. East Lansing, MI: Michigan State University.

Kirby, S. (2000). Syntax without natural selection: How compositionality emerges from vocabulary in a population of learners. In *The evolutionary emergence of language: Social function and the origins of linguistic form* (pp. 185–215). Academic Press. doi:10.1017/CBO9780511606441.019

Kogut, B. (2000). The network as knowledge: Generative rules and the emergence of structure. *Strategic Management Journal*, *21*(3), 405–425. doi:10.1002/(SICI)1097-0266(200003)21:3<405::AID-SMJ103>3.0.CO;2-5

Kogut, Z., & Zander, U. (1992). Knowledge of the firm, combinative capabilities, and the replication of technology. *Organization Science*, *3*(3), 383–397. doi:10.1287/orsc.3.3.383

Kohli, A. K., & Jaworski, B. J. (1993). Market orientation: The construct, research propositions, and managerial implications. *Journal of Marketing*, *54*(2), 1–18. doi:10.2307/1251866

Langlois, R. N. (2002). Modularity in technology and organization. *Journal of Economic Behavior & Organization*, *49*(1), 19–37. doi:10.1016/S0167-2681(02)00056-2

Law, A. M., & Kelton, W. D. (2000). *Simulation modeling and analysis* (3rd ed.). New York: McGraw-Hill Higher Education.

Lee, H. L., Padmanabhan, V., & Whang, S. (1997a). The bullwhip effect in supply chains. *Sloan Management Review*, *38*(3), 93–102.

Lee, H. L., Padmanabhan, V., & Whang, S. (1997b). Information distortion in a supply chain: The bullwhip effect. *Management Science*, *43*(4), 546–558. doi:10.1287/mnsc.43.4.546

Leiblein, M. J. (2003). The choice of organizational governance form and performance: Predictions from transaction cost, resource-based, and real options theories. *Journal of Management*, *29*(6), 937–961.

Limburg, K. E., O'Neill, R. V., Costanza, R., & Farber, S. (2002). Complex systems and valuation. *Ecological Economics*, *41*(3), 409–420. doi:10.1016/S0921-8009(02)00090-3

Lin, F.-R., Huang, S.-H., & Lin, S.-C. (2002). Effects of information sharing on supply chain performance in electronic commerce. *IEEE Transactions on Engineering Management*, *49*(3), 258–268. doi:10.1109/TEM.2002.803388

Liu, P., & Zhang, X. (2011). Research on the supplier selection of a supply chain based on entropy weight and improved ELECTRE-III method. *International Journal of Production Research*, *3*(1), 637–646. doi:10.1080/00207540903490171

Lummus, R. R., & Vokurka, R. J. (1999). Managing the demand chain through managing the information flow: Capturing 'moments of information'. *Production and Inventory Management Journal, 40*(1), 16–20.

Macher, J. T. (2006). Technological development and the boundaries of the firm: A knowledge-based examination in semiconductor manufacturing. *Management Science, 52*(6), 826–843. doi:10.1287/mnsc.1060.0511

Maisel, H., & Gnugnoli, G. (1972). *Simulation of discrete stochastic systems*. Chicago: Science Research Associates, Inc.

Markose, S. M. (2005). Computability and evolutionary complexity: Markets as complex adaptive systems (CAS). *The Economic Journal, 115*(504), F159. doi:10.1111/j.1468-0297.2005.01000.x

McElroy, M. W. (2000). Integrating complexity theory, knowledge management and organizational learning. *Journal of Knowledge Management, 4*(3), 195–203. doi:10.1108/13673270010377652

McEvily, S. K., & Chakravarthy, B. (2002). The persistence of knowledge-based advantage: An empirical test for product performance and technological knowledge. *Strategic Management Journal, 23*(4), 285–305. doi:10.1002/smj.223

McNamara, G., Vaaler, P. M., & Devers, C. (2003). Same as it ever was: The search for evidence of increasing hypercompetition. *Strategic Management Journal, 24*(3), 261–278. doi:10.1002/smj.295

Mena, C., Humphries, A., & Choi, T. Y. (2013). Toward a theory of multi-tier supply chain management. *Journal of Supply Chain Management, 49*(2), 58–77. doi:10.1111/jscm.12003

Moorman, C., & Miner, A. S. (1997). The impact of organizational memory on new product performance and creativity. *JMR, Journal of Marketing Research, 34*(1), 91–106. doi:10.2307/3152067

Morel, B., & Ramanujam, R. (1999). Through the looking glass of complexity: The dynamics of organizations as adaptive and evolving systems. *Organization Science, 10*(3), 278–293. doi:10.1287/orsc.10.3.278

Narasimhan, R., & Jayaram, J. (1998). Causal linkages in supply chain management: An exploratory study of North American manufacturing firms. *Decision Sciences, 29*(3), 579–605. doi:10.1111/j.1540-5915.1998.tb01355.x

Nelson, C. R. (1973). *Applied time series analysis for managerial forecasting*. San Francisco: Holden-Day, Inc.

Nickerson, J. A., & Zenger, T. R. (2004). A knowledge-based theory of the firm—The problem-solving perspective. *Organization Science, 15*(6), 617–632. doi:10.1287/orsc.1040.0093

Nielsen, B. B. (2005). The role of knowledge embeddedness in the creation of synergies in strategic alliances. *Journal of Business Research, 58*(9), 1194–1204. doi:10.1016/j.jbusres.2004.05.001

Ouchi, W. G. (1979). A conceptual framework for the design of organizational control mechanisms. *Management Science, 25*(9), 833–848. doi:10.1287/mnsc.25.9.833

Parunak, H. V. D. (1998). *The DASCh experience: How to model a supply chain*. Retrieved from http://www.erim.org/~vparunak/iccs98.pdf

Pathak, S. D. (2005). *An investigative framework for studying the growth and evolution of complex supply networks*. Nashville, TN: Vanderbilt University.

Pettit, T. J., Croxton, K. L., & Fiksel, J. (2012). Ensuring supply chain resilience: Development and implementation of an assessment tool. *Journal of Business Logistics, 34*(1), 46–76. doi:10.1111/jbl.12009

Pilbeam, C., Alvarez, G., & Wilson, H. (2012). The governance of supply networks: A systematic literature review. *Supply Chain Management: An International Journal*, *17*(4), 358–376. doi:10.1108/13598541211246512

Prahalad, C. K., & Hamel, G. (1990). The core competence and the corporation. *Harvard Business Review*, 71–90.

Prajogo, D., & Olhager, J. (2012). Supply chain integrationandperformance:The effects of long-term relationships, information technology and sharing, and logistics integration. *International Journal of Production Economics*, *135*, 514–522. doi:10.1016/j.ijpe.2011.09.001

Randall, W. S., Pohlen, T. L., & Hanna, J. B. (2010). Evolving a theory of performance-based logistics using insight from service-dominant logic. *Journal of Business Logistics*, *31*(2), 35–61. doi:10.1002/j.2158-1592.2010.tb00142.x

Rindfleisch, A., & Heide, J. B. (1997). Transaction cost analysis: past, present, and future applications. *Journal of Marketing*, *61*(4), 30–54. doi:10.2307/1252085

Rindfleisch, A., & Moorman, C. (2001). The acquisition and utilization of information in new product alliances: A strength-of-ties perspective. *Journal of Marketing*, *65*(2), 1–18. doi:10.1509/jmkg.65.2.1.18253

Roper, S., & Crone, M. (2003). Knowledge complementarity and coordination in the local supply chain: Some empirical evidence. *British Journal of Management*, *14*(4), 339–355. doi:10.1111/j.1467-8551.2003.00383.x

Roy, S., Sivakumar, K., & Wilkinson, I. F. (2004). Innovation generation in supply chain relationships: A conceptual model and research propositions. *Academy of Marketing Science Journal*, *32*(1), 61–79. doi:10.1177/0092070303255470

Schoenherr, T., Modib, S. B., Bentoncy, W. C., Carter, C. R., Choi, T. Y., & Larson, P. D. et al. (2011). Research opportunities in purchasing and supply management. *International Journal of Production Research*, *50*(6), 4556–4579.

Sethi, R., Smith, D. C., & Park, C. W. (2001). Cross-functional product development teams, creativity, and the innovativeness of new consumer products. *JMR, Journal of Marketing Research*, *38*(1), 73–85. doi:10.1509/jmkr.38.1.73.18833

Shannon, C. E. (1948). A mathematical theory of communication. *The Bell System Technical Journal*, *27*, 379–423, 623–656. doi:10.1002/j.1538-7305.1948.tb01338.x

Sivadas, E., & Dwyer, F. R. (2000). An examination of organizational factors influencing new product success in internal and alliance-based processes. *Journal of Marketing*, *64*(1), 31–49. doi:10.1509/jmkg.64.1.31.17985

Sorenson, O. (2003). Social networks and industrial geography. *Journal of Evolutionary Economics*, *13*(5), 513–527. doi:10.1007/s00191-003-0165-9

Stank, T. P., Keller, S. B., & Closs, D. J. (2001). Performance benefits of supply chain logistical integration. *Transportation Journal*, *41*(2-3), 32–46.

Stock, G. N., Greis, N. P., & Kasarda, J. D. (1999). Logistics, strategy and structure: A conceptual framework. *International Journal of Physical Distribution & Logistics Management*, *29*(4), 224–239. doi:10.1108/09600039910273948

Surana, A., Kumara, S., Greaves, M., & Raghavan, U. N. (2005). Supply-chain networks: A complex adaptive systems perspective. *International Journal of Production Research*, *43*(20), 4235–4265. doi:10.1080/00207540500142274

Tan, K. C., Kannan, V. R., & Handfield, R. B. (1998). Supply chain management: Supplier performance and firm performance. *International Journal of Purchasing and Materials Management, 34*(3), 2–9.

Van De Ven, A. H., & Poole, M. S. (1995). Explaining development and change in organizations. *Academy of Management Review, 20*(3), 510–540. doi:10.2307/258786

Varadarajan, P. R., & Cunningham, M. H. (1995). Strategic alliances: A synthesis of conceptual foundations. *Journal of the Academy of Marketing Science, 23*(4), 297–300. doi:10.1177/009207039502300408

Vargo, S. L., & Lusch, R. F. (2004a). Evolving to a new dominant logic for marketing. *Journal of Marketing, 68*(1), 1–17. doi:10.1509/jmkg.68.1.1.24036

Vargo, S. L., & Lusch, R. F. (2004b). The four service marketing myths: remnants of a goods-based, manufacturing model. *Journal of Service Research, 6*(4), 324–335. doi:10.1177/1094670503262946

Vargo, S. L., & Lusch, R. F. (2006). Service-dominant logic: reactions, reflections and refinements. *Marketing Theory, 6*(3), 281–288. doi:10.1177/1470593106066781

Vitasek, K. (2011). *The vested outsourcing manual: A guide for creating successful business and outsourcing agreements*. New York: Palgrave-MacMillan.

Wiggins, R. W., & Ruefli, T. W. (2005). Schumpeter's ghost: Is hypercompetition making the best of times shorter?. *Strategic Management Journal, 26*(S 887), 911.

Williamson, O. E. (1975). *Markets and hierarchies: Analysis and antitrust implications*. New York: The Free Press.

Williamson, O. E. (1986). *Economic organization*. New York: New York University Press.

Williamson, O. E. (1991). Comparative economic organization: The analysis of discrete structural alternatives. *Administrative Science Quarterly, 36*(2), 269–296. doi:10.2307/2393356

Williamson, O. E. (2008). Outsourcing: Transaction cost economics and supply chain management. *Journal of Supply Chain Management, 44*(2), 5–16. doi:10.1111/j.1745-493X.2008.00051.x

World Economic Forum. (2013). *Enabling trade: Valuing growth opportunities*. Geneva: World Economic Forum.

Zacharia, Z. G., Sanders, N. R., & Nix, N. W. (2011). The emerging role of the third-party logistics provider (3PL) as an orchestrator. *Journal of Business Logistics, 32*(1), 40–54. doi:10.1111/j.2158-1592.2011.01004.x

Zhou, K. Z., & Li, C. B. (2012). How knowledge affects radical innovation: Knowledge base, market knowledge acquisition, and internal knowledge sharing. *Strategic Management Journal, 33*, 1090–1102. doi:10.1002/smj.1959

Zhou, K. Z., & Wu, F. (2010). Technological capability, strategic flexibility, and product innovation. *Strategic Management Journal, 31*, 547–561.

ENDNOTES

1 Small companies were a special case where δ was based on the current capacity (C_t) and the size of the requested upgrade (C_U). In the case that $C_t<3$ and $C_U \leq 3C_t$, then δ=1; if $C_t<3$ and $C_U \leq 4C_t$, then δ equaled a constant.

APPENDIX A

Key:

EOS = Economies of Scale
EMhet = End Market (heterogeneous)
EMhom = End Market (homogeneous)
RTA = Rate of Technological Advance
see Tables A1, A2, A3, A4

Table A1. Manufacturer-to-assembler selection parameter estimates by strategy

		Low Price Strategy		*High Price Strategy*	
Parameter		*Estimate*	*Pr > ChiSq*	*Estimate*	*Pr > ChiSq*
TimeStep	1	4.96	<.0001	-0.35	0.4236
	2	-1.19	0.0015	-0.35	0.4236
	3	-2.49	<.0001	-0.35	0.4236
	4	-1.71	<.0001	0.16	0.7035
TimeStep*EOS	1	0.13	0.7312	0.22	0.5233
	2	0.39	0.2796	0.10	0.7691
	3	0.36	0.3588	0.10	0.7691
	4	-0.61	0.0658	-0.76	0.0087
TimeStep*EMhet	1	0.35	0.374	-0.24	0.5599
	2	-0.93	0.0135	-0.42	0.3242
	3	0.37	0.3566	-0.42	0.3242
	4	0.40	0.2137	0.30	0.3408
TimeStep*EMhom	1	.	.	-0.51	0.2272
	2	.	.	-0.69	0.1128
	3	.	.	-0.69	0.1128
	4	.	.	0.54	0.1439

Table A2. Assembler-to-manufacturer selection parameter estimates by strategy

		Low Price Strategy		*High Price Strategy*	
Parameter		*Estimate*	*Pr > ChiSq*	*Estimate*	*Pr > ChiSq*
TimeStep	1	1.73	<.0001	-1.13	0.0093
	2	-1.50	<.0001	-1.13	0.0093
	3	0.06	0.7513	-1.13	0.0093
	4	-0.40	0.0453	-0.24	0.4893
TimeStep*EMhet	1	2.03	<.0001	0.33	0.4136
	2	0.61	0.0032	-0.04	0.9283
	3	-1.29	<.0001	-0.04	0.9283
	4	-0.97	<.0001	-0.30	0.2947
TimeStep*EMhom	1	.	.	-0.09	0.8273
	2	.	.	-0.45	0.2896
	3	.	.	-0.45	0.2896
	4	.	.	-0.12	0.7428
TimeStep*RTA*EMhet	1	-0.82	0.0006	0.25	0.5312
	2	1.27	<.0001	-0.11	0.7897
	3	-0.19	0.3033	-0.11	0.7897
	4	-0.33	0.1123	0.17	0.6171

Table A3. Assembler-to-retailer selection parameter estimates by strategy

Parameter		Low Price Strategy Estimate	Low Price Strategy Pr > ChiSq	High Price Strategy Estimate	High Price Strategy Pr > ChiSq
TimeStep	1	-1.16	0.0073	-1.01	0.0195
	2	-1.16	0.0073	-1.01	0.0195
	3	-1.16	0.0073	-1.01	0.0195
	4	0.87	0.0035	-0.05	0.8868
TimeStep*EOS	1	0.33	0.436	0.27	0.4147
	2	0.33	0.436	0.33	0.3392
	3	0.33	0.436	0.20	0.564
	4	-0.95	0.0012	-0.78	0.0031
TimeStep*EMhom	1	.	.	-0.31	0.4371
	2	.	.	-0.61	0.1469
	3	.	.	-0.41	0.322
	4	.	.	0.46	0.1334

Table A4. Retailer-to-assembler selection parameter estimates by strategy

Parameter		Low Price Strategy Estimate	Low Price Strategy Pr > ChiSq	High Price Strategy Estimate	High Price Strategy Pr > ChiSq
TimeStep	1	1.73	<.0001	-1.13	0.0093
	2	-1.50	<.0001	-1.13	0.0093
	3	0.06	0.7513	-1.13	0.0093
	4	-0.40	0.0453	-0.24	0.4893
TimeStep*RTA	1	1.48	<.0001	0.04	0.9181
	2	-1.21	<.0001	0.04	0.9181
	3	-0.03	0.8663	0.04	0.9181
	4	0.03	0.8948	-0.12	0.7204
TimeStep*EMhet	1	2.03	<.0001	0.33	0.4136
	2	0.61	0.0032	-0.04	0.9283
	3	-1.29	<.0001	-0.04	0.9283
	4	-0.97	<.0001	-0.30	0.2947
TimeStep*EMhom	1	.	.	-0.09	0.8273
	2	.	.	-0.45	0.2896
	3	.	.	-0.45	0.2896
	4	.	.	-0.12	0.7428
TimeStep*RTA*EMhet	1	-0.82	0.0006	0.25	0.5312
	2	1.27	<.0001	-0.11	0.7897
	3	-0.19	0.3033	-0.11	0.7897
	4	-0.33	0.1123	0.17	0.6171

APPENDIX B: CREATE COMPUTATIONAL REPRESENTATION

In this section are included descriptions of variables and the equations utilized in the simulation.

Strategy

Each company used either a high quality or a low quality strategy. Companies following a high quality strategy did business exclusively with buyers and suppliers who also pursued a high quality strategy; companies following a low quality strategy similarly dealt exclusively with companies following a low quality strategy.

Life Points

Life points (λ) represented the health of the company. All companies started with the same number of life points. Companies gained or lost life points based on their capacity utilization (U). Companies making efficient use of existing capacity, as reflected by capacity utilization, were considered healthier and more resilient. *Capacity utilization* was calculated as a percent of a company's maximum capacity based on the company's sales, with the assumption that a company only produced to demand and sold all that it produced. If the capacity utilization fell below the survival threshold (θ), a life point was lost at a rate determined by a parameter called the *basis point division parameter* (β). The capacity utilization of 70% was in accordance with the recent U.S. Census Bureau plant capacity data (Bureau, 2005). Companies operating above this threshold gained a life point. Basis point division was arbitrarily set to 20; companies operating at capacity utilization between 50 and 70 lost one life point, companies operating at capacity utilization of 30-49 lose 2 life points, between 10-29 they lost 3 life points and at less than 10 they lost 4 life points. Thus a company with a long history of being successful that had fallen upon hard times would take much longer to die than a newer company or one with a history of being less successful.

Node Deaths

When a node died, all of the capacity it provided both upstream and downstream was removed from the system before calculation of node differentials (see the sub-section on *Capacity upgrades*), so any capacity lost due to node death will be supplemented by the system. The process for removing the capacity ensured that all requirements from downstream demand were balanced against capacity amongst the suppliers.

Demand Volume

Demand volume was determined by a function following a Gaussian distribution determined by the volume of peak demand (D_{max}) with a peak demand time step (τ) and a scale factor (s). Demand also used a random factor (p) that represented the percent variance from the value determined by the demand distribution function. This is shown below where r_t is a random number between 0 and 1 and p is the specified "random" percent of variance (and which is also shown in Figure 5).

$$D_t = D_{max}(e^{-((t-\tau)/s)^2})(1 + r_t p)$$

Price

Price started out the same for both high and low end markets (P_S). Demand was distinguished as being either high end demand (low price sensitivity) or low end demand (high price sensitivity) as determined by the established market price in each market. As demand increased, prices dropped in a linear fashion for both markets. It was implicitly assumed that price for both high and low end products would decline as the industry matured and price for low end product would always be less, or at most equal to, the price for high end product. As prices dropped, demand for low end product increased while demand for high end product decreased. The volume of high end demand was represented as:

$$D_H = D_t\left(\frac{P_L}{P_H}\right)$$

where P_L is the market price for the low end market and P_H is the market price in the high end market. Conversely, low end demand was represented as:

$$D_L = D_t\left(1 - \frac{P_L}{P_H}\right)$$

Economies of Scale

Production facilities of each company type (i.e., high end manufacturers, low end manufacturers, assemblers, and retailers) had an optimal economy of scale (κ), which increased as the industry matured, and changed linearly over time as determined by the optimum economy of scale size factor (ω). An optimum economy of scale size factor of one reflected a static growth of economies of scale, whereas ω=2 meant the economies of scale doubled over the course of the simulation.

Depending on the deviation of a plant's current capacity from the optimum economies of scale, the plant could experience economies or diseconomies of scale. The closer the current capacity was to optimal capacity, the more efficient the company. The effect of economies of scale primarily influenced the decision to increase capacity. Firms would uniformly decide to increase capacity until the optimal capacity was reached; the probability that a firm would increase capacity thereafter was reflected by a Gaussian distribution (see the section on *Capacity Upgrades*).

Technology and Capacity Upgrades

Companies faced a moving technological frontier. In order to keep up with the rate of technological change, companies would periodically upgrade their capacity. Competition amongst companies depended on having up-to-date production facilities—keeping up was assumed to be the cost of entry. The process for upgrading technology was incorporated with the process for capacity upgrades. As a result, early in the simulation companies tended to upgrade technology and capacity simultaneously, but as they reached their optimal economies of scale, technology upgrades would tend to occur without increases to capacity. The upgrade recovery time parameter (ζ) determined the pace at which plant capacity became obsolete and needed upgrading.

Upgrades also included increases to plant production capacity. The decision to upgrade depended on three factors: 1) recovery time since last upgrade, 2) relative size of upgrade, and 3) the economies of scale of the resulting upgrade. The decision to upgrade capacity occurred when the product of the three factors exceeded the company's upgrade decision threshold (υ). A brief description of each of the three factors follows:

1. **Recovery time since last upgrade (ρ):** This function was based upon the number of time steps since the last capacity upgrade. When the time since the last capacity upgrade reached the upgrade recovery time parameter ($t_U \geq \zeta$), then ρ=1; otherwise,

$$\rho = \frac{t_U}{\zeta}.$$

2. **Relative size of upgrade (δ):** Companies tend to avoid investing in trivial amounts of capacity upgrade; instead, they wait until it is worth their while. Companies also avoid making upgrades too rapidly lest they get ahead of the market. For example, an upgrade of only one unit was unlikely to occur, while an upgrade of 50% was likely to be much more useful. This function annotated current capacity with C_t, the size of a requested upgrade with C_U, and followed a Gaussian distribution designed so a requested capacity upgrade of 50% returned δ =1 (see equation below). Very small companies (with a capacity of 1 or 2) were handled differently with a "start up factor" allowing for slightly more drastic relative growth[1].

$$\delta = e^{-\left(\frac{C_t - 2C_U}{C_t + C_U}\right)^2}$$

3. **Economy of scale of the resulting upgrade (γ):** This function led the company to uniformly increase capacity until it reached the optimum economy of scale. The result of the function depended on the company capacity at the optimum economy of scale (κ) and the company's new capacity if the upgrade was implemented ($C_t + C_U$). The economy of scale of the resulting upgrade (γ) was determined by a Gaussian distribution based on κ:

$$\gamma = 1, C_t + C_U < \kappa$$

$$\gamma = e^{-\left(\frac{(C_t + C_U) - \kappa}{\kappa}\right)^2}, C_t + C_U < \kappa$$

A decision threshold (υ) determined when an individual company make an upgrade decision. The decision threshold was constant for the entire simulation system, with individual companies making their decision to upgrade when the product of the three factors was greater than the decision threshold (ρδγ> υ). A product function was used for calculating the decision threshold due to advancing economies of scale, which also affected the magnitude of the relative size of the upgrade; a product function kept all parameters to the same scale. The upgrade parameter (υ) was set based upon trial and error depending corresponding with the optimum economy of scale. Upgrading capacity went through the following recursion logic:

1. Beginning at the consumer end of the supply chain, the virtual transaction cost vector was created to map transaction costs with each active company to all other active companies of the same strategy in the upstream vector.
2. The virtual transaction cost vector was ordered from lowest to highest. In the case of declining demand, the vector was sorted highest to lowest.
3. The ordered transaction cost vector was searched by the downstream node for unused capacity which could be used if available. In the face of declining demand, relationships with the highest virtual transaction costs were eliminated first.
4. If no unused capacity was available, the downstream firm requested upgrades using the ordered virtual transaction cost vector to order additional capacity from a supplier.
5. If no supplier agreed to provide enough needed capacity, a new company was created for the unmet capacity. If the increase in demand was less than half the optimal economy of scale capacity for a retailer, then the demand was left unmet.

Virtual Transaction Costs

Virtual transaction costs were used to determine the relative cost associated with a transaction. It consisted of the sum of four transaction cost factors between buyer i and supplier j: total units, current units, economy of scale, and life of company. These virtual transaction "costs" were considered by each buyer as it selected a supplier with whom to do business. This reflected the movement of demand information up the supply chain. Buyers always selected suppliers in the order of lowest to highest virtual transaction cost in keeping with the desire to seek out the most efficient relationship available, with a bias toward suppliers that were familiar. Each virtual transaction cost factor is briefly described below:

1. **Total units cost (χ_{ij}):** Total units cost resulted from the total number of units ever exchanged between two companies (h_{ij}) such that

$$\chi_{ij} = \frac{100}{100 + h_{ij}}.$$

   This cost began at one and decreased over time as two companies continued to do business.
2. **Current units cost (ν_{ij}):** The units exchanged in the current cycle between two companies (n_{ij}) resulted in a cost that started at one and diminished as the volume of the current transaction increased:

$$\nu_{ij} = \frac{2}{2 + n_{ij}}$$

3. **Economy of scale cost (ι_j):** This virtual transaction cost depended on the buyer's capacity in relation to its optimum economy of scale. This cost ranged from 0 (optimum) to 0.1 (least optimum) based on current capacity (C_t) and the company capacity at the optimum economy of scale (κ):

$$\iota_j = \frac{1 - e^{-\left(\frac{C_t - \kappa}{\kappa}\right)^2}}{10}$$

4. **Life of company cost (φ_j):** This cost started at 0.5 and diminished as the company matured and was determined by how many time steps the buyer had been active (L):

$$\phi_j = \frac{5}{10 + L}$$

MODEL VERIFICATION AND VALIDATION

Verification relied on three techniques described by Law and Kelton (2000). The first technique entailed the writing and debugging of the program in modules or sub-programs. The process started with the main program consisting of the supply chain network interacting without transaction costs or capacity decisions. Then the sub-routine for transaction costs were implemented and debugged. Absolute transaction costs were originally envisioned; however, it quickly became apparent that the relative transaction costs between firms determined the outcomes of relationships. This agreed with Coase's vision of "an outside network of relative prices and costs" (1937, p. p. 389). A virtual transaction cost interaction vector was developed based on the exchange history (number of units exchanged), the current exchange (number of units), actual vs. optimal economies of scale, and the life of the company (number of cycles the company has been active). The virtual transaction cost resulted in the expected model behavior based upon TCA theory.

Capacity decision processes were implemented using a sub-program that incorporated two aspects of the capacity upgrade decision: 1) increasing the magnitude of capacity, and 2) upgrading product ensuing from the advancing technological frontier. The capacity decision process exhibited expected behavior based upon extant production literature with production capacity increasing and decreasing as the consequence of demand in the next level of the SCN (supply chain network) as well as following the increasing economies of scale and periodic upgrades to keep up with the technological frontier.

The second verification technique engaged more than one person using a "structured walk-through of the program" (Law & Kelton, 2000, p. p. 270). One author conducted a line-by-line walk-through with a mathematician who had no formal education or experience with TCA or market governance theory in order to assess two aspects of the computer program. First, the program's relationship validity was assessed to ensure the program accurately reflected the relationships between key variables as predicted by theory. Then the walk-through scrutinized the validity of the equations to model specific relationships between the variables.

The third verification technique checked robustness or sensitivity of the simulation to a variety of input parameters. The model was run under a variety of settings to authenticate these key simulation processes:

1. TCA processes,
2. Node death and birth processes,
3. Capacity decision processes,
4. Robustness checks for the simulation model as a whole over a range of parameter settings.

In all cases, the model behaved in accordance expectations. Random demand fluctuation led to the most unstable model response once random deviance from the demand curve reached 10%. At this point, the model exhibited excessively high node mortality throughout the simulation, in turn leading to underserved markets, low collaboration index scores, and generally unstable model behavior. For the rest of the parameter settings the model proved highly robust and certainly sufficient to model any realistic supply chain scenarios

Related References

To continue our tradition of advancing information science and technology research, we have compiled a list of recommended IGI Global readings. These references will provide additional information and guidance to further enrich your knowledge and assist you with your own research and future publications.

Abdullah, M., Ahmad, R., Peck, L. S., Kasirun, Z. M., & Alshammari, F. (2014). Benefits of CMM and CMMI-based software process improvement. In I. Management Association (Ed.), Software design and development: Concepts, methodologies, tools, and applications (pp. 1385-1400). Hershey, PA: Information Science Reference. doi: doi:10.4018/978-1-4666-4301-7.ch067

Abu-Shanab, E., & Ghaleb, O. (2012). Adoption of mobile commerce technology: An involvement of trust and risk concerns. *International Journal of Technology Diffusion*, *3*(2), 36–49. doi:10.4018/jtd.2012040104

Adapa, S. (2013). Electronic retailing of prepared food in Australia. In K. Tarnay, S. Imre, & L. Xu (Eds.), *Research and development in e-business through service-oriented solutions* (pp. 280–292). Hershey, PA: Business Science Reference.

Aklouf, Y., & Drias, H. (2011). An adaptive e-commerce architecture for enterprise information exchange. In I. Management Association (Ed.), Enterprise information systems: Concepts, methodologies, tools and applications (pp. 329-345). Hershey, PA: Business Science Reference. doi: doi:10.4018/978-1-61692-852-0.ch202

Al-Nawayseh, M. K., Alnabhan, M. M., Al-Debei, M. M., & Balachandran, W. (2013). An adaptive decision support system for last mile logistics in e-commerce: A study on online grocery shopping. *International Journal of Decision Support System Technology*, *5*(1), 40–65. doi:10.4018/jdsst.2013010103

Al-Somali, S. A., Clegg, B., & Gholami, R. (2013). An investigation into the adoption and implementation of electronic commerce in Saudi Arabian small and medium enterprises. In I. Association (Ed.), *Small and medium enterprises: Concepts, methodologies, tools, and applications* (pp. 816–839). Hershey, PA: Business Science Reference.

Al-Somali, S. A., Gholami, R., & Clegg, B. (2013). An investigation into the adoption of electronic commerce among Saudi Arabian SMEs. In M. Khosrow-Pour (Ed.), *E-commerce for organizational development and competitive advantage* (pp. 126–150). Hershey, PA: Business Science Reference. doi:10.4018/978-1-4666-3622-4.ch007

Alavi, S. (2013). Collaborative customer relationship management-co-creation and collaboration through online communities. *International Journal of Virtual Communities and Social Networking*, *5*(1), 1–18. doi:10.4018/jvcsn.2013010101

Alavi, S., & Ahuja, V. (2013). E-commerce in a web 2.0 world: Using online business communities to impact consumer price sensitivity. *International Journal of Online Marketing*, *3*(2), 38–55. doi:10.4018/ijom.2013040103

Alawneh, A., Al-Refai, H., & Batiha, K. (2011). E-business adoption by Jordanian banks: An exploratory study of the key factors and performance indicators. In A. Tatnall (Ed.), *Actor-network theory and technology innovation: Advancements and new concepts* (pp. 113–128). Hershey, PA: Information Science Reference. doi:10.4018/978-1-60960-587-2.ch412

Albuquerque, S. L., & Gondim, P. R. (2012). Applying continuous authentication to protect electronic transactions. In T. Chou (Ed.), *Information assurance and security technologies for risk assessment and threat management: Advances* (pp. 134–161). Hershey, PA: Information Science Reference.

Alfahl, H., Sanzogni, L., & Houghton, L. (2012). Mobile commerce adoption in organizations: A literature review and future research directions. *Journal of Electronic Commerce in Organizations, 10*(2), 61–78. doi:10.4018/jeco.2012040104

Almustafa, M. M., & Alkhaldi, D. (2014). Business intelligence and agile methodology for risk management in knowledge-based organizations. In I. Management Association (Ed.), Software design and development: concepts, methodologies, tools, and applications (pp. 1710-1735). Hershey, PA: Information Science Reference. doi: doi:10.4018/978-1-4666-4301-7.ch082

Aloini, D., Dulmin, R., & Mininno, V. (2013). E-procurement: What really matters in B2B e-reverse auctions. In P. Ordóñez de Pablos, J. Lovelle, J. Gayo, & R. Tennyson (Eds.), *E-procurement management for successful electronic government systems* (pp. 87–113). Hershey, PA: Information Science Reference.

Amer, M., & Gómez, J. M. (2010). Measuring B2C quality of electronic service: Towards a common consensus. In I. Lee (Ed.), *Encyclopedia of e-business development and management in the global economy* (pp. 135–143). Hershey, PA: Business Science Reference. doi:10.4018/978-1-61520-611-7.ch014

Amer, M., & Gómez, J. M. (2012). Measuring quality of electronic services: Moving from business-to-consumer into business-to-business marketplace. In E. Kajan, F. Dorloff, & I. Bedini (Eds.), *Handbook of research on e-business standards and protocols: Documents, data and advanced web technologies* (pp. 637–654). Hershey, PA: Business Science Reference. doi:10.4018/978-1-4666-0146-8.ch029

Andriole, S. J. (2010). Business technology strategy for an energy information company. *Journal of Information Technology Research, 3*(3), 19–42. doi:10.4018/jitr.2010070103

Andriole, S. J. (2010). Templates for the development of business technology strategies. *Journal of Information Technology Research, 3*(3), 1–10. doi:10.4018/jitr.2010070101

Antoniadis, D. N. (2013). Leadership style and the management of the effects of complexity. In S. Banerjee (Ed.), *Chaos and complexity theory for management: Nonlinear dynamics* (pp. 105–144). Hershey, PA: Business Science Reference.

Archer, N. (2010). Electronic marketplace support for B2B business transactions. In I. Management Association (Ed.), Electronic services: Concepts, methodologies, tools and applications (pp. 85-93). Hershey, PA: Information Science Reference. doi: doi:10.4018/978-1-61520-967-5.ch007

Archer, N. (2010). Management considerations for B2B online exchanges. In I. Management Association (Ed.), Business information systems: Concepts, methodologies, tools and applications (pp. 1740-1747). Hershey, PA: Business Science Reference. doi: doi:10.4018/978-1-61520-969-9.ch105

Arduini, D., Nascia, L., & Zanfei, A. (2012). Complementary approaches to the diffusion of innovation: Empirical evidence on e-services adoption in Italy. *International Journal of E-Services and Mobile Applications, 4*(3), 42–64. doi:10.4018/jesma.2012070103

Arh, T., Dimovski, V., & Blažic, B. J. (2011). ICT and web 2.0 technologies as a determinant of business performance. In M. Al-Mutairi, & L. Mohammed (Eds.), *Cases on ICT utilization, practice and solutions: Tools for managing day-to-day issues* (pp. 59–77). Hershey, PA: Information Science Reference. doi:10.4018/978-1-60960-783-8.ch312

Arikpo, I., Osofisan, A., & Eteng, I. E. (2012). Enhancing trust in e-commerce in developing IT environments: A feedback-based perspective. In A. Usoro, G. Majewski, P. Ifinedo, & I. Arikpo (Eds.), *Leveraging developing economies with the use of information technology: Trends and tools* (pp. 193–203). Hershey, PA: Information Science Reference. doi:10.4018/978-1-4666-1637-0.ch011

Aryanto, V. D., & Chrismastuti, A. A. (2013). Model for digital economy in Indonesia. In I. Oncioiu (Ed.), *Business innovation, development, and advancement in the digital economy* (pp. 60–77). Hershey, PA: Business Science Reference. doi:10.4018/978-1-4666-2934-9.ch005

Asim, M., & Petkovic, M. (2012). Fundamental building blocks for security interoperability in e-business. In E. Kajan, F. Dorloff, & I. Bedini (Eds.), *Handbook of research on e-business standards and protocols: documents, data and advanced web technologies* (pp. 269–292). Hershey, PA: Business Science Reference. doi:10.4018/978-1-4666-0146-8.ch013

Askool, S., Jacobs, A., & Nakata, K. (2013). A method of analysing the use of social networking sites in business. In I. Management Association (Ed.), IT policy and ethics: Concepts, methodologies, tools, and applications (pp. 794-813). Hershey, PA: Information Science Reference. doi: doi:10.4018/978-1-4666-2919-6.ch036

Association, I. R. (2013). *Enterprise resource planning: Concepts, methodologies, tools, and applications*. Hershey, PA: IGI Global. doi:10.4018/978-1-4666-4153-2

Azab, N., & Khalifa, N. (2013). Web 2.0 and opportunities for entrepreneurs: How Egyptian entrepreneurs perceive and exploit web 2.0 technologies. In N. Azab (Ed.), *Cases on web 2.0 in developing countries: Studies on implementation, application, and use* (pp. 1–32). Hershey, PA: Information Science Reference.

Azevedo, S. G., & Carvalho, H. (2012). RFID technology in the fashion supply chain: An exploratory analysis. In T. Choi (Ed.), *Fashion supply chain management: Industry and business analysis* (pp. 303–326). Hershey, PA: Information Science Reference.

Bagnato, A., Raiteri, F., Jung, C., & Elberzhager, F. (2014). Creating and applying security goal indicator trees in an industrial environment. In I. Management Association (Ed.), Software design and development: Concepts, methodologies, tools, and applications (pp. 999-1013). Hershey, PA: Information Science Reference. doi: doi:10.4018/978-1-4666-4301-7.ch048

Baporikar, N. (2013). ICT challenge for ebusiness in SMEs. *International Journal of Strategic Information Technology and Applications*, *4*(1), 15–26. doi:10.4018/jsita.2013010102

Barbin Laurindo, F. J., Monteiro de Carvalho, M., & Shimizu, T. (2010). Strategic alignment between business and information technology. In M. Hunter (Ed.), *Strategic information systems: Concepts, methodologies, tools, and applications* (pp. 20–28). Hershey, PA: Information Science Reference.

Barjis, J. (2012). Software engineering security based on business process modeling. In K. Khan (Ed.), *Security-aware systems applications and software development methods* (pp. 52–68). Hershey, PA: Information Science Reference. doi:10.4018/978-1-4666-1580-9.ch004

Barnes, D., & Hinton, M. (2011). The benefits of an e-business performance measurement system. In N. Shi, & G. Silvius (Eds.), *Enterprise IT governance, business value and performance measurement* (pp. 158–169). Hershey, PA: Information Science Reference.

Bask, A., Lipponen, M., & Tinnilä, M. (2012). E-commerce logistics: A literature research review and topics for future research. *International Journal of E-Services and Mobile Applications*, *4*(3), 1–22. doi:10.4018/jesma.2012070101

Bask, A., & Tinnilä, M. (2013). Impact of product characteristics on supply chains: An analytical literature review. *International Journal of Applied Logistics*, *4*(1), 35–59. doi:10.4018/jal.2013010103

Basu, S. (2012). Direct taxation and e-commerce: Possibility and desirability. In E. Druicã (Ed.), *Digital economy innovations and impacts on society* (pp. 26–48). Hershey, PA: Information Science Reference. doi:10.4018/978-1-4666-1556-4.ch003

Batista, Í. J., Varela, A. T., Andrade, E. P., Azevedo, J. V., Garcia, T. L., & Henrique da Silva, D. … Araújo, A. L. (2014). A mechatronic description of an autonomous underwater vehicle for dam inspection. In I. Management Association (Ed.), Robotics: Concepts, methodologies, tools, and applications (pp. 647-662). Hershey, PA: Information Science Reference. doi: doi:10.4018/978-1-4666-4607-0.ch033

Beckinsale, M. (2010). E-business among ethnic minority businesses: The case of ethnic entrepreneurs. In B. Thomas, & G. Simmons (Eds.), *E-commerce adoption and small business in the global marketplace: Tools for optimization* (pp. 187–207). Hershey, PA: Business Science Reference. doi:10.4018/978-1-60566-998-4.ch010

Beckinsale, M. (2011). eBusiness among ethnic minority businesses: Ethnic entrepreneurs' ICT adoption and readiness. In S. Sharma (Ed.), E-adoption and socio-economic impacts: Emerging infrastructural effects (pp. 168-189). Hershey, PA: Information Science Reference. doi: doi:10.4018/978-1-60960-597-1.ch009

Bedini, I., Gardarin, G., & Nguyen, B. (2011). Semantic technologies and e-business. In E. Kajan (Ed.), *Electronic business interoperability: Concepts, opportunities and challenges* (pp. 243–278). Hershey, PA: Business Science Reference. doi:10.4018/978-1-60960-485-1.ch011

Beedle, J., & Wang, S. (2013). Roles of a technology leader. In S. Wang, & T. Hartsell (Eds.), *Technology integration and foundations for effective leadership* (pp. 228–241). Hershey, PA: Information Science Reference.

Belhajjame, K., & Brambilla, M. (2013). Ontological description and similarity-based discovery of business process models. In J. Krogstie (Ed.), *Frameworks for developing efficient information systems: Models, theory, and practice* (pp. 30–50). Hershey, PA: Engineering Science Reference. doi:10.4018/978-1-4666-4161-7.ch002

Benou, P., & Bitos, V. (2010). Developing mobile commerce applications. In M. Khosrow-Pour (Ed.), *E-commerce trends for organizational advancement: New applications and methods* (pp. 1–15). Hershey, PA: Information Science Reference.

Berger, C. (2014). Cloud-based testing for context-aware cyber-physical systems. In I. Management Association (Ed.), Software design and development: Concepts, methodologies, tools, and applications (pp. 1851-1877). Hershey, PA: Information Science Reference. doi: doi:10.4018/978-1-4666-4301-7.ch088

Berisha-Namani, M. (2013). Information systems usage in business and management. In I. Oncioiu (Ed.), *Business innovation, development, and advancement in the digital economy* (pp. 48–59). Hershey, PA: Business Science Reference. doi:10.4018/978-1-4666-2934-9.ch004

Bermúdez, G. M., & Rojas, L. A. (2013). Model-driven engineering for electronic commerce. In V. Díaz, J. Lovelle, B. García-Bustelo, & O. Martínez (Eds.), *Progressions and innovations in model-driven software engineering* (pp. 196–208). Hershey, PA: Engineering Science Reference. doi:10.4018/978-1-4666-4217-1.ch007

Bernardi, S., & Merseguer, J. (2014). A meta-model-based approach to the definition of the analysis results of petri-net models. In M. Gribaudo, & M. Iacono (Eds.), *Theory and application of multi-formalism modeling* (pp. 104–116). Hershey, PA: Information Science Reference.

Bernardino, J. (2013). Open business intelligence for better decision-making. *International Journal of Information Communication Technologies and Human Development*, *5*(2), 20–36. doi:10.4018/jicthd.2013040102

Berzins, M. (2012). Scams and the Australian e-business environment. In K. Mohammed Rezaul (Ed.), *Strategic and pragmatic e-business: Implications for future business practices* (pp. 156–175). Hershey, PA: Business Science Reference. doi:10.4018/978-1-4666-1619-6.ch007

Bhattacharya, S. (2014). Model-based approaches for scanning keyboard design: Present state and future directions. In I. Association (Ed.), *Assistive technologies: Concepts, methodologies, tools, and applications* (pp. 1497–1515). Hershey, PA: Information Science Reference.

Binsaleh, M., & Hassan, S. (2011). Systems development methodology for mobile commerce applications. *International Journal of Mobile Computing and Multimedia Communications*, *3*(4), 36–52. doi:10.4018/jmcmc.2011100103

Binsaleh, M., & Hassan, S. (2013). Systems development methodology for mobile commerce applications. In I. Khalil, & E. Weippl (Eds.), *Contemporary challenges and solutions for mobile and multimedia technologies* (pp. 146–162). Hershey, PA: Information Science Reference.

Blake, R., Gordon, S., & Shankaranarayanan, G. (2013). The role of case-based research in information technology and systems. In P. Isaias, & M. Baptista Nunes (Eds.), *Information systems research and exploring social artifacts: approaches and methodologies* (pp. 200–220). Hershey, PA: Information Science Reference.

Boateng, R., Heeks, R., Molla, A., & Hinson, R. (2013). Advancing e-commerce beyond readiness in a developing country: Experiences of Ghanaian firms. In M. Khosrow-Pour (Ed.), *E-commerce for organizational development and competitive advantage* (pp. 1–17). Hershey, PA: Business Science Reference. doi:10.4018/978-1-4666-3622-4.ch001

Bonfatti, F., Monari, P. D., & Martinelli, L. (2011). Business document exchange between small companies. In E. Kajan (Ed.), *Electronic business interoperability: Concepts, opportunities and challenges* (pp. 482–510). Hershey, PA: Business Science Reference. doi:10.4018/978-1-60960-485-1.ch020

Boucadair, M., & Binet, D. (2014). Issues with current internet architecture. In M. Boucadair, & D. Binet (Eds.), *Solutions for sustaining scalability in internet growth* (pp. 1–16). Hershey, PA: Information Science Reference.

Bouras, A., Gouvas, P., & Mentzas, G. (2009). A semantic service-oriented architecture for business process fusion. In I. Lee (Ed.), *Electronic business: Concepts, methodologies, tools, and applications* (pp. 504–532). Hershey, PA: Information Science Reference.

Braun, P. (2011). Advancing women in the digital economy: eLearning opportunities for meta-competency skilling. In I. Management Association (Ed.), Global business: Concepts, methodologies, tools and applications (pp. 1978-1990). Hershey, PA: Business Science Reference. doi: doi:10.4018/978-1-60960-587-2.ch708

Brown, M., & Garson, G. (2013). Organization behavior and organization theory. In *Public information management and e-government: Policy and issues* (pp. 160–195). Hershey, PA: Information Science Reference. doi:10.4018/978-1-4666-3003-1.ch007

Brown, M., & Garson, G. (2013). The information technology business model. In *Public information management and e-government: Policy and issues* (pp. 76–98). Hershey, PA: Information Science Reference. doi:10.4018/978-1-4666-3003-1.ch004

Broy, M., Kirstan, S., Krcmar, H., & Schätz, B. (2014). What is the benefit of a model-based design of embedded software systems in the car industry? In I. Management Association (Ed.), Software design and development: Concepts, methodologies, tools, and applications (pp. 310-334). Hershey, PA: Information Science Reference. doi: doi:10.4018/978-1-4666-4301-7.ch017

Burete, R., Badica, A., Badica, C., & Moraru, F. (2011). Enhanced reputation model with forgiveness for e-business agents. *International Journal of Agent Technologies and Systems*, *3*(1), 11–26. doi:10.4018/jats.2011010102

Business Research and Case Center. (2011). *Cases on business and management in the MENA region: New trends and opportunities*. Hershey, PA: IGI Global.

Bwalya, K. J. (2011). E-commerce penetration in the SADC region: Consolidating and moving forward. In M. Cruz-Cunha, & J. Varajão (Eds.), *E-business managerial aspects, solutions and case studies* (pp. 235–253). Hershey, PA: Business Science Reference.

Cagliero, L. (2014). ConChi: Pattern change mining from mobile context-aware data. In K. Rızvanoğlu, & G. Çetin (Eds.), *Research and design innovations for mobile user experience* (pp. 263–285). Hershey, PA: Information Science Reference.

Charbaji, R., Rebeiz, K., & Sidani, Y. (2010). Antecedents and consequences of the risk taking behavior of mobile commerce adoption in Lebanon. In H. Rahman (Ed.), *Handbook of research on e-government readiness for information and service exchange: Utilizing progressive information communication technologies* (pp. 354–380). Hershey, PA: Information Science Reference.

Chaturvedi, N. (2013). Collaborative web for natural resources industries. In I. Association (Ed.), *Supply chain management: Concepts, methodologies, tools, and applications* (pp. 601–614). Hershey, PA: Business Science Reference.

Chelmis, C., Sorathia, V., & Prasanna, V. K. (2014). Enterprise knowledge preservation and management. In E. Nikoi, & K. Boateng (Eds.), *Collaborative communication processes and decision making in organizations* (pp. 25–45). Hershey, PA: Business Science Reference.

Chen, C., & Yang, S. C. (2008). E-commerce and mobile commerce application adoptions. In A. Becker (Ed.), *Electronic commerce: Concepts, methodologies, tools, and applications* (pp. 826–836). Hershey, PA: Information Science Reference.

Chen, Q., & Zhang, N. (2013). IT-supported business performance and e-commerce application in SMEs. *Journal of Electronic Commerce in Organizations, 11*(2), 41–52. doi:10.4018/jeco.2013040104

Chen, T. F. (2011). Emerging business models: Value drivers in e-business 2.0 and towards enterprise 2.0. In T. Chen (Ed.), *Implementing new business models in for-profit and non-profit organizations: Technologies and applications* (pp. 1–28). Hershey, PA: Information Science Reference. doi:10.4018/978-1-60960-587-2.ch801

Chen, T. F. (2011). The critical success factors and integrated model for implementing e-business in Taiwan's SMEs. In I. Management Association (Ed.), Global business: Concepts, methodologies, tools and applications (pp. 1109-1133). Hershey, PA: Business Science Reference. doi: doi:10.4018/978-1-60960-587-2.ch416

Chew, E., & Gottschalk, P. (2013). Critical success factors of IT strategy. In *Knowledge driven service innovation and management: IT strategies for business alignment and value creation* (pp. 185–220). Hershey, PA: Business Science Reference.

Chew, E., & Gottschalk, P. (2013). Strategic alignment and IT-enabled value creation. In *Knowledge driven service innovation and management: IT strategies for business alignment and value creation* (pp. 141–184). Hershey, PA: Business Science Reference.

Chew, E., & Gottschalk, P. (2013). Theories and models of service-oriented firms. In *Knowledge driven service innovation and management: IT strategies for business alignment and value creation* (pp. 1–34). Hershey, PA: Business Science Reference.

Chiang, L. (2010). Digital confidence in business: A perspective of information ethics. In M. Pankowska (Ed.), *Infonomics for distributed business and decision-making environments: Creating information system ecology* (pp. 288–300). Hershey, PA: Business Science Reference.

Chiprianov, V., Kermarrec, Y., & Rouvrais, S. (2014). Integrating DSLs into a software engineering process: Application to collaborative construction of telecom services. In I. Management Association (Ed.), Software design and development: Concepts, methodologies, tools, and applications (pp. 570-595). Hershey, PA: Information Science Reference. doi: doi:10.4018/978-1-4666-4301-7.ch028

Chollet, S., & Lalanda, P. (2014). A model-driven approach to service composition with security properties. In I. Management Association (Ed.), Software design and development: Concepts, methodologies, tools, and applications (pp. 2051-2071). Hershey, PA: Information Science Reference. doi: doi:10.4018/978-1-4666-4301-7.ch095

Chugh, R., & Gupta, P. (2011). A unified view of enablers, barriers, and readiness of small to medium enterprises for e-business adoption. In M. Cruz-Cunha, & J. Varajão (Eds.), *E-business issues, challenges and opportunities for SMEs: Driving competitiveness* (pp. 291–312). Hershey, PA: Business Science Reference. doi:10.4018/978-1-60960-587-2.ch719

Clear, F., Woods, A., & Dickson, K. (2013). SME adoption and use of ICT for networked trading purposes: The influence of sector, size and age of firm. In I. Association (Ed.), *Small and medium enterprises: Concepts, methodologies, tools, and applications* (pp. 774–791). Hershey, PA: Business Science Reference.

Connolly, R. (2013). eCommerce trust beliefs: Examining the role of national culture. In P. Isaias & M. Baptista Nunes (Eds.), Information systems research and exploring social artifacts: Approaches and methodologies (pp. 20-42). Hershey, PA: Information Science Reference. doi: doi:10.4018/978-1-4666-2491-7.ch002

Cormican, K. (2013). Collaborative networks: Challenges for SMEs. In I. Association (Ed.), *Small and medium enterprises: Concepts, methodologies, tools, and applications* (pp. 1638–1653). Hershey, PA: Business Science Reference.

Costa, L., Loughran, N., & Grønmo, R. (2014). Model-driven engineering, services and interactive real-time applications. In I. Management Association (Ed.), Software design and development: Concepts, methodologies, tools, and applications (pp. 178-202). Hershey, PA: Information Science Reference. doi: doi:10.4018/978-1-4666-4301-7.ch010

Costante, E., Petkovic, M., & den Hartog, J. (2013). Trust management and user's trust perception in e-business. In I. Management Association (Ed.), IT policy and ethics: Concepts, methodologies, tools, and applications (pp. 64-83). Hershey, PA: Information Science Reference. doi: doi:10.4018/978-1-4666-2919-6.ch004

Costin, Y. (2012). Adopting ICT in the mompreneurs business: A strategy for growth? In C. Romm Livermore (Ed.), *Gender and social computing: Interactions, differences and relationships* (pp. 17–34). Hershey, PA: Information Science Publishing.

Cox, S. (2013). E-business planning in morphing organizations: Maturity models of business transformation. In E. Li, S. Loh, C. Evans, & F. Lorenzi (Eds.), *Organizations and social networking: Utilizing social media to engage consumers* (pp. 286–312). Hershey, PA: Business Science Reference. doi:10.4018/978-1-4666-4026-9.ch015

Cruz-Cunha, M. M., Moreira, F., & Varajão, J. (2014). *Handbook of research on enterprise 2.0: Technological, social, and organizational dimensions*. Hershey, PA: IGI Global.

D'Aubeterre, F., Iyer, L. S., Ehrhardt, R., & Singh, R. (2011). Discovery process in a B2B emarketplace: A semantic matchmaking approach. In V. Sugumaran (Ed.), *Intelligent, adaptive and reasoning technologies: New developments and applications* (pp. 80–103). Hershey, PA: Information Science Reference. doi:10.4018/978-1-60960-595-7.ch005

Dabbagh, R. A. (2011). E-business: Concepts and context with illustrative examples of e-business and e-commerce in education. In A. Al Ajeeli, & Y. Al-Bastaki (Eds.), *Handbook of research on e-services in the public sector: E-government strategies and advancements* (pp. 450–462). Hershey, PA: Information Science Reference.

Dekker, S. W. (2013). Drifting into failure: Complexity theory and the management of risk. In S. Banerjee (Ed.), *Chaos and complexity theory for management: Nonlinear dynamics* (pp. 241–253). Hershey, PA: Business Science Reference.

Delgado, J. C. (2014). Objective-oriented modeling of enterprises under the service paradigm. In M. Cruz-Cunha, F. Moreira, & J. Varajão (Eds.), *Handbook of research on enterprise 2.0: Technological, social, and organizational dimensions* (pp. 572–598). Hershey, PA: Business Science Reference.

Demirkan, H., & Spohrer, J. C. (2012). Servitized enterprises for distributed collaborative commerce. In S. Galup (Ed.), *Technological applications and advancements in service science, management, and engineering* (pp. 70–83). Hershey, PA: Business Science Reference. doi:10.4018/978-1-4666-1583-0.ch005

Denno, P. (2013). Trade collaboration systems. In I. Association (Ed.), *Supply chain management: Concepts, methodologies, tools, and applications* (pp. 615–633). Hershey, PA: Business Science Reference.

Derghal, A., & Goléa, N. (2014). Multi-objective generation scheduling using genetic-based fuzzy mathematical programming technique. In P. Vasant (Ed.), *Handbook of research on novel soft computing intelligent algorithms: Theory and practical applications* (pp. 450–474). Hershey, PA: Information Science Reference.

Djoleto, W. (2011). E-business efficacious consequences the etiquettes and the business decision making. In O. Bak, & N. Stair (Eds.), *Impact of e-business technologies on public and private organizations: Industry comparisons and perspectives* (pp. 278–295). Hershey, PA: Business Science Reference. doi:10.4018/978-1-60960-501-8.ch017

Djoleto, W. (2013). Information technology and organisational leadership. In *Electronic commerce and organizational leadership: Perspectives and methodologies*. Hershey, PA: IGI Global.

Djoleto, W. (2013). Cloud computing and ecommerce or ebusiness: "The now it way" – An overview. In *Electronic commerce and organizational leadership: Perspectives and methodologies* (pp. 239–254). Hershey, PA: Business Science Reference. doi:10.4018/978-1-4666-2982-0.ch010

Djoleto, W. (2013). eCommerce and organisational leadership. In Electronic commerce and organizational leadership: Perspectives and methodologies (pp. 99-121). Hershey, PA: Business Science Reference. doi: doi:10.4018/978-1-4666-2982-0.ch005

Djoleto, W. (2013). eCommerce: An overview. In Electronic commerce and organizational leadership: Perspectives and methodologies (pp. 74-98). Hershey, PA: Business Science Reference. doi: doi:10.4018/978-1-4666-2982-0.ch004

Djoleto, W. (2013). Empirical analyses of ecommerce: The findings – A mixed methodology perspective. In *Electronic commerce and organizational leadership: Perspectives and methodologies* (pp. 150–189). Hershey, PA: Business Science Reference. doi:10.4018/978-1-4666-2982-0.ch007

Djoleto, W. (2013). Future endeavours. In *Electronic commerce and organizational leadership: Perspectives and methodologies* (pp. 269–280). Hershey, PA: Business Science Reference. doi:10.4018/978-1-4666-2982-0.ch012

Djoleto, W. (2013). Information technology: The journey. In *Electronic commerce and organizational leadership: Perspectives and methodologies* (pp. 32–54). Hershey, PA: Business Science Reference. doi:10.4018/978-1-4666-2982-0.ch002

Doolin, B., & Ali, E. I. (2012). Mobile technology adoption in the supply chain. In I. Management Association (Ed.), Wireless technologies: Concepts, methodologies, tools and applications (pp. 1553-1573). Hershey, PA: Information Science Reference. doi: doi:10.4018/978-1-61350-101-6.ch603

Dostál, P. (2014). The use of soft computing in management. In P. Vasant (Ed.), *Handbook of research on novel soft computing intelligent algorithms: Theory and practical applications* (pp. 294–326). Hershey, PA: Information Science Reference.

Duin, H., & Thoben, K. (2011). Enhancing the preparedness of SMEs for e-business opportunities by collaborative networks. In M. Cruz-Cunha, & J. Varajão (Eds.), *E-business issues, challenges and opportunities for SMEs: Driving competitiveness* (pp. 30–45). Hershey, PA: Business Science Reference. doi:10.4018/978-1-60960-587-2.ch202

Dulai, T., Jaskó, S., & Tarnay, K. (2013). IOTP and payments protocols. In K. Tarnay, S. Imre, & L. Xu (Eds.), *Research and development in e-business through service-oriented solutions* (pp. 20–56). Hershey, PA: Business Science Reference.

Dza, M., Fisher, R., & Gapp, R. (2013). Service-dominant logic and supply network management: An efficient business mix? In N. Ndubisi, & S. Nwankwo (Eds.), *Enterprise development in SMEs and entrepreneurial firms: Dynamic processes* (pp. 376–389). Hershey, PA: Business Science Reference.

Ehsani, E. (2011). Defining e-novation in action. In H. Pattinson, & D. Low (Eds.), *E-novation for competitive advantage in collaborative globalization: Technologies for emerging e-business strategies* (pp. 58–74). Hershey, PA: Business Science Reference. doi:10.4018/978-1-60566-394-4.ch005

Ekong, U. O., Ifinedo, P., Ayo, C. K., & Ifinedo, A. (2013). E-commerce adoption in Nigerian businesses: An analysis using the technology-organization-environmental framework. In I. Association (Ed.), *Small and medium enterprises: Concepts, methodologies, tools, and applications* (pp. 840–861). Hershey, PA: Business Science Reference.

Emens, S. (2010). The new paradigm of business on the internet and its ethical implications. In D. Palmer (Ed.), *Ethical issues in e-business: Models and frameworks* (pp. 15–27). Hershey, PA: Business Science Reference. doi:10.4018/978-1-61520-615-5.ch002

Eriksson, P., Henttonen, E., & Meriläinen, S. (2011). Managing client contacts of small KIBS companies: Turning technology into business. *International Journal of Innovation in the Digital Economy*, *2*(3), 1–10. doi:10.4018/jide.2011070101

Escofet, E., Rodríguez-Fórtiz, M. J., Garrido, J. L., & Chung, L. (2012). Strategic e-business/ IT alignment for SME competitiveness. In I. Management Association (Ed.), Computer engineering: Concepts, methodologies, tools and applications (pp. 1427-1445). Hershey, PA: Engineering Science Reference. doi: doi:10.4018/978-1-61350-456-7.ch604

Eze, U. C., & Poong, Y. S. (2013). Consumers' intention to use mobile commerce and the moderating roles of gender and income. In I. Lee (Ed.), *Strategy, adoption, and competitive advantage of mobile services in the global economy* (pp. 127–148). Hershey, PA: Information Science Reference.

Eze, U. C., & Poong, Y. S. (2013). The moderating roles of income and age in mobile commerce application. *Journal of Electronic Commerce in Organizations*, *11*(3), 46–67. doi:10.4018/jeco.2013070103

Fantechi, A. (2014). The role of formal methods in software development for railway applications. In I. Management Association (Ed.), Software design and development: Concepts, methodologies, tools, and applications (pp. 1103-1118). Hershey, PA: Information Science Reference. doi: doi:10.4018/978-1-4666-4301-7.ch054

Faulin, J., Lera-López, F., & Juan, A. A. (2013). Optimizing routes with safety and environmental criteria in transportation management in Spain: A case study. In J. Wang (Ed.), *Management innovations for intelligent supply chains* (pp. 144–165). Hershey, PA: Business Science Reference.

Fehér, P. (2012). Integrating and measuring business and technology services in the context of enterprise architecture. In V. Shankararaman, J. Zhao, & J. Lee (Eds.), *Business enterprise, process, and technology management: Models and applications* (pp. 148–163). Hershey, PA: Business Science Reference. doi:10.4018/978-1-4666-0249-6.ch008

Feja, S., Witt, S., & Speck, A. (2014). Tool based integration of requirements modeling and validation into business process modeling. In I. Management Association (Ed.), Software design and development: Concepts, methodologies, tools, and applications (pp. 285-309). Hershey, PA: Information Science Reference. doi: doi:10.4018/978-1-4666-4301-7.ch016

Fengel, J. (2012). Semantic alignment of e-business standards and legacy models. In E. Kajan, F. Dorloff, & I. Bedini (Eds.), *Handbook of research on e-business standards and protocols: Documents, data and advanced web technologies* (pp. 676–704). Hershey, PA: Business Science Reference. doi:10.4018/978-1-4666-0146-8.ch031

Ferreira, M. P. (2013). SMEs and e-business: Implementation, strategies and policy. In I. Association (Ed.), *Small and medium enterprises: Concepts, methodologies, tools, and applications* (pp. 97–117). Hershey, PA: Business Science Reference.

Fluvià, M., & Rigall-I-Torrent, R. (2013). Public sector transformation and the design of public policies for electronic commerce and the new economy: Tax and antitrust policies. In N. Pomazalová (Ed.), *Public sector transformation processes and internet public procurement: Decision support systems* (pp. 32-57). Hershey, PA: Engineering Science Reference. doi:10.4018/978-1-4666-2665-2.ch003

Franquesa, J., & Brandyberry, A. (2011). Organizational slack and information technology innovation adoption in SMEs. In I. Lee (Ed.), *E-business applications for product development and competitive growth: emerging technologies* (pp. 25–48). Hershey, PA: Business Science Reference.

Fries, T. P. (2014). Reengineering structured legacy system documentation to UML object-oriented artifacts. In I. Management Association (Ed.), Software design and development: Concepts, methodologies, tools, and applications (pp. 749-771). Hershey, PA: Information Science Reference. doi: doi:10.4018/978-1-4666-4301-7.ch036

Funabashi, M. (2014). Transdisciplinary science and technology and service systems. In M. Kosaka, & K. Shirahada (Eds.), *Progressive trends in knowledge and system-based science for service innovation* (pp. 101–127). Hershey, PA: Business Science Reference.

Galinski, C., & Beckmann, H. (2014). Concepts for enhancing content quality and eaccessibility: In general and in the field of eprocurement. In I. Association (Ed.), *Assistive technologies: Concepts, methodologies, tools, and applications* (pp. 180–197). Hershey, PA: Information Science Reference.

Gan, J., & Gutiérrez, J. A. (2011). Viable business models for m-commerce: The key components. In M. Cruz-Cunha, & F. Moreira (Eds.), *Handbook of research on mobility and computing: Evolving technologies and ubiquitous impacts* (pp. 837–852). Hershey, PA: Information Science Reference. doi:10.4018/978-1-60960-042-6.ch052

Garg, H., Rani, M., & Sharma, S. (2014). Predicting uncertain behavior and performance analysis of the pulping system in a paper industry using PSO and fuzzy methodology. In P. Vasant (Ed.), *Handbook of research on novel soft computing intelligent algorithms: Theory and practical applications* (pp. 414–449). Hershey, PA: Information Science Reference.

Garito, M. (2012). Mobile business and mobile TV: Available technologies, future opportunities and new marketing trends. In I. Management Association (Ed.), E-marketing: Concepts, methodologies, tools, and applications (pp. 1240-1251). Hershey, PA: Business Science Reference. doi: doi:10.4018/978-1-4666-1598-4.ch072

Génova, G., Llorens, J., & Morato, J. (2014). Software engineering research: The need to strengthen and broaden the classical scientific method. In I. Management Association (Ed.), Software design and development: Concepts, methodologies, tools, and applications (pp. 1639-1658). Hershey, PA: Information Science Reference. doi: doi:10.4018/978-1-4666-4301-7.ch079

Ghobakhloo, M., & Zulkifli, N. (2013). Adoption of mobile commerce: The impact of end user satisfaction on system acceptance. *International Journal of E-Services and Mobile Applications*, *5*(1), 26–50. doi:10.4018/jesma.2013010102

Gill, A. Q., & Bunker, D. (2014). SaaS requirements engineering for agile development. In I. Management Association (Ed.), Software design and development: Concepts, methodologies, tools, and applications (pp. 351-380). Hershey, PA: Information Science Reference. doi: doi:10.4018/978-1-4666-4301-7.ch019

Gimenez, J. (2014). Reflections of professional practice: Using electronic discourse analysis networks (EDANs) to examine embedded business emails. In H. Lim, & F. Sudweeks (Eds.), *Innovative methods and technologies for electronic discourse analysis* (pp. 327–345). Hershey, PA: Information Science Reference.

Gionis, G. A., Schroth, C., & Janner, T. (2011). Advancing interoperability for agile cross-organisational collaboration: A rule-based approach. In Y. Charalabidis (Ed.), *Interoperability in digital public services and administration: Bridging e-government and e-business* (pp. 238–253). Hershey, PA: Information Science Reference.

Gnoni, M. G., & Rollo, A. (2011). A content analysis for evaluating RFID applications in supply network management. In I. Mahdavi, S. Mohebbi, & N. Cho (Eds.), *Electronic supply network coordination in intelligent and dynamic environments: Modeling and implementation* (pp. 93–112). Hershey, PA: Business Science Reference.

Gonçalves, A., Serra, N., Serra, J., & Sousa, P. (2011). How to use information technology effectively to achieve business objectives. In M. Cruz-Cunha, & J. Varajao (Eds.), *Enterprise information systems design, implementation and management: Organizational applications* (pp. 21–37). Hershey, PA: Information Science Reference. doi:10.4018/978-1-61350-456-7.ch102

Gordini, N., & Veglio, V. (2014). Customer relationship management and data mining: A classification decision tree to predict customer purchasing behavior in global market. In P. Vasant (Ed.), *Handbook of research on novel soft computing intelligent algorithms: Theory and practical applications* (pp. 1–40). Hershey, PA: Information Science Reference.

Gottschalk, P. (2007). The CIO developing e-business. In P. Gottschalk (Ed.), *CIO and corporate strategic management: Changing role of CIO to CEO* (pp. 148–185). Hershey, PA: Idea Group Publishing.

Goutam, S. (2010). Analysis of speedy uptake of electronic and digital signatures in digital economy with special reference to India. In E. Adomi (Ed.), *Frameworks for ICT policy: Government, social and legal issues* (pp. 76–88). Hershey, PA: Information Science Reference. doi:10.4018/978-1-61692-012-8.ch005

Grieger, M., Hartmann, E., & Kotzab, H. (2011). E-markets as meta-enterprise information e systems. In I. Management Association (Ed.), Enterprise information systems: Concepts, methodologies, tools and applications (pp. 638-647). Hershey, PA: Business Science Reference. doi: doi:10.4018/978-1-61692-852-0.ch306

Grzonkowski, S., Ensor, B. D., & McDaniel, B. (2013). applied cryptography in Electronic Commerce. In I. Management Association (Ed.), IT policy and ethics: Concepts, methodologies, tools, and applications (pp. 368-388). Hershey, PA: Information Science Reference. doi: doi:10.4018/978-1-4666-2919-6.ch017

Guhathakurta, K., Banerjee, S., & Dan, P. K. (2013). Nonlinear dynamics of voltage fluctuation in power plants for strategic decisions. In S. Banerjee (Ed.), *Chaos and complexity theory for management: Nonlinear dynamics* (pp. 352–367). Hershey, PA: Business Science Reference.

Guhathakurta, K., Bhattacharya, S. N., Banerjee, S., & Bhattacharya, B. (2013). Nonlinear correlation of stock and commodity indices in emerging and developed market. In S. Banerjee (Ed.), *Chaos and complexity theory for management: Nonlinear dynamics* (pp. 63–87). Hershey, PA: Business Science Reference.

Ha, H. (2012). Online security and consumer protection in ecommerce an Australian case. In K. Mohammed Rezaul (Ed.), *Strategic and pragmatic e-business: Implications for future business practices* (pp. 217–243). Hershey, PA: Business Science Reference. doi:10.4018/978-1-4666-1619-6.ch010

Ha, H., Coghill, K., & Maharaj, E. A. (2012). Current measures to protect e-consumers' privacy in Australia. In I. Management Association (Ed.), Cyber crime: Concepts, methodologies, tools and applications (pp. 1728-1755). Hershey, PA: Information Science Reference. doi: doi:10.4018/978-1-61350-323-2.ch806

Halas, H., & Klobucar, T. (2011). Business models and organizational processes changes. In I. Management Association (Ed.), Global business: Concepts, methodologies, tools and applications (pp. 192-205). Hershey, PA: Business Science Reference. doi: doi:10.4018/978-1-60960-587-2.ch113

Hamid, S. S., Nasir, M. H., Sahibuddin, S., & Nor, M. K. (2014). Managing software projects with team software process (TSP). In I. Management Association (Ed.), Software design and development: Concepts, methodologies, tools, and applications (pp. 1550-1583). Hershey, PA: Information Science Reference. doi: doi:10.4018/978-1-4666-4301-7.ch076

Han, B. (2012). I play, I pay? An investigation of the user's willingness to pay on hedonic social network sites. *International Journal of Virtual Communities and Social Networking*, *4*(1), 19–31. doi:10.4018/jvcsn.2012010102

Harnesk, D. (2011). Convergence of information security in B2B networks. In E. Kajan (Ed.), *Electronic business interoperability: Concepts, opportunities and challenges* (pp. 571–595). Hershey, PA: Business Science Reference. doi:10.4018/978-1-60960-485-1.ch023

Harwood, T. (2012). Emergence of gamified commerce: Turning virtual to real. *Journal of Electronic Commerce in Organizations*, *10*(2), 16–39. doi:10.4018/jeco.2012040102

Haworth, J. L., Vallabhajosula, S., Tzetzis, G., & Stergiou, N. (2013). Optimal variability and complexity: A novel approach for management principles. In S. Banerjee (Ed.), *Chaos and complexity theory for management: Nonlinear dynamics* (pp. 328–351). Hershey, PA: Business Science Reference.

Hegazi, A. S., Ahmed, E. M., & Matouk, A. E. (2013). On dynamical behaviors and chaos control of the fractional-order financial system. In S. Banerjee (Ed.), *Chaos and complexity theory for management: Nonlinear dynamics* (pp. 34–49). Hershey, PA: Business Science Reference.

Heravi, B. R., & Lycett, M. (2012). Semantically enriched e-business standards development: The case of ebXML business process specification schema. In E. Kajan, F. Dorloff, & I. Bedini (Eds.), *Handbook of research on e-business standards and protocols: Documents, data and advanced web technologies* (pp. 655–675). Hershey, PA: Business Science Reference. doi:10.4018/978-1-4666-0146-8.ch030

Hill, D. S. (2012). An examination of standardized product identification and business benefit. In E. Kajan, F. Dorloff, & I. Bedini (Eds.), *Handbook of research on e-business standards and protocols: Documents, data and advanced web technologies* (pp. 387–411). Hershey, PA: Business Science Reference. doi:10.4018/978-1-4666-0146-8.ch018

Hohmann, S., & Zelewski, S. (2013). Effects of vendor-managed inventory on the bullwhip effect. In J. Wang (Ed.), *Management innovations for intelligent supply chains* (pp. 167–183). Hershey, PA: Business Science Reference.

Hoops, D. S. (2011). Legal issues in the virtual world and e-commerce. In B. Ciaramitaro (Ed.), *Virtual worlds and e-commerce: Technologies and applications for building customer relationships* (pp. 186–204). Hershey, PA: Business Science Reference.

Hu, W., Zuo, Y., Kaabouch, N., & Chen, L. (2010). A technological perspective of mobile and electronic commerce systems. In M. Khosrow-Pour (Ed.), *E-commerce trends for organizational advancement: New applications and methods* (pp. 16–35). Hershey, PA: Information Science Reference.

Hua, G. B. (2013). *Implementing IT business strategy in the construction industry*. Hershey, PA: IGI Global. doi:10.4018/978-1-4666-4185-3

Hua, S. C., Rajesh, M. J., & Theng, L. B. (2011). Determinants of e-commerce adoption among small and medium-sized enterprises in Malaysia. In S. Sharma (Ed.), *E-adoption and socio-economic impacts: Emerging infrastructural effects* (pp. 71–89). Hershey, PA: Information Science Reference. doi:10.4018/978-1-60960-597-1.ch005

Huang, J., & Dang, J. (2011). Context-sensitive ontology matching in electronic business. In E. Kajan (Ed.), *Electronic business interoperability: Concepts, opportunities and challenges* (pp. 279–301). Hershey, PA: Business Science Reference. doi:10.4018/978-1-60960-485-1.ch012

Hunaiti, Z., Tairo, D., Sedoyeka, E., & Elgazzar, S. (2010). Factors facing mobile commerce deployment in United Kingdom. In W. Hu, & Y. Zuo (Eds.), *Handheld computing for mobile commerce: Applications, concepts and technologies* (pp. 109–123). Hershey, PA: Information Science Reference. doi:10.4018/978-1-61520-761-9.ch007

Hung, W. J., Tsai, C., Hung, S., McQueen, R., & Jou, J. (2011). Evaluating web site support capabilities in sell-side B2B transaction processes: A longitudinal study of two industries in New Zealand and Taiwan. *Journal of Global Information Management*, *19*(1), 51–79. doi:10.4018/jgim.2011010103

Hunter, M. G. (2013). The duality of information technology roles: A case study. In C. Howard (Ed.), *Strategic adoption of technological innovations* (pp. 38–49). Hershey, PA: Information Science Reference.

Huq, N., Shah, S. M., & Mihailescu, D. (2012). Why select an open source ERP over proprietary ERP? A focus on SMEs and supplier's perspective. In R. Atem de Carvalho, & B. Johansson (Eds.), *Free and open source enterprise resource planning: Systems and strategies* (pp. 33–55). Hershey, PA: Business Science Reference.

Ichoua, S. (2014). Relief distribution networks: Design and operations. In I. Management Association (Ed.), Crisis management: Concepts, methodologies, tools and applications (pp. 360-375). Hershey, PA: Information Science Reference. doi: doi:10.4018/978-1-4666-4707-7.ch015

Ingvaldsen, J., & Gulla, J. (2010). Semantic business process mining of SAP transactions. In M. Wang, & Z. Sun (Eds.), *Handbook of research on complex dynamic process management: Techniques for adaptability in turbulent environments* (pp. 416–429). Hershey, PA: Business Science Reference. doi:10.4018/978-1-61692-852-0.ch320

Ingvaldsen, J., & Gulla, J. (2011). Semantic business process mining of SAP transactions. In I. Management Association (Ed.), Enterprise information systems: Concepts, methodologies, tools and applications (pp. 866-878). Hershey, PA: Business Science Reference. doi: doi:10.4018/978-1-61692-852-0.ch320

Ioannou, M. (2013). Customer relationship management (CRM): A one-size-fits-all philosophy? In H. Kaufmann, & M. Panni (Eds.), *Customer-centric marketing strategies: Tools for building organizational performance* (pp. 150–170). Hershey, PA: Business Science Reference.

Islam, M. S., & Scupola, A. (2013). E-service research trends in the domain of e-government: A contemporary study. In A. Scupola (Ed.), *Mobile opportunities and applications for e-service innovations* (pp. 152–169). Hershey, PA: Information Science Reference.

Jailani, N., Patel, A., Mukhtar, M., Abdullah, S., & Yahya, Y. (2010). Concept of an agent-based electronic marketplace. In I. Lee (Ed.), *Encyclopedia of e-business development and management in the global economy* (pp. 239–251). Hershey, PA: Business Science Reference. doi:10.4018/978-1-61520-611-7.ch024

Johns, R. (2011). Technology, trust and B2B relationships: A banking perspective. In O. Bak, & N. Stair (Eds.), *Impact of e-business technologies on public and private organizations: Industry comparisons and perspectives* (pp. 79–96). Hershey, PA: Business Science Reference. doi:10.4018/978-1-60960-501-8.ch005

Joshi, S. (2013). E-supply chain collaboration and integration: Implementation issues and challenges. In D. Graham, I. Manikas, & D. Folinas (Eds.), *E-logistics and e-supply chain management: Applications for evolving business* (pp. 9–26). Hershey, PA: Business Science Reference. doi:10.4018/978-1-4666-3914-0.ch002

Juárez, F. (2013). Chaos and complexity in financial statements. In S. Banerjee (Ed.), *Chaos and complexity theory for management: Nonlinear dynamics* (pp. 1–33). Hershey, PA: Business Science Reference.

Kamal, M., Qureshil, S., & Wolcott, P. (2013). Promoting competitive advantage in micro-enterprises through information technology interventions. In I. Association (Ed.), *Small and medium enterprises: Concepts, methodologies, tools, and applications* (pp. 581–606). Hershey, PA: Business Science Reference.

Kamel, S. (2012). Electronic commerce prospects in emerging economies: Lessons from Egypt. In I. Management Association (Ed.), Regional development: Concepts, methodologies, tools, and applications (pp. 1104-1115). Hershey, PA: Information Science Reference. doi: doi:10.4018/978-1-4666-0882-5.ch604

Kamoun, F., & Halaweh, M. (2012). User interface design and e-commerce security perception: An empirical study. *International Journal of E-Business Research, 8*(2), 15–32. doi:10.4018/jebr.2012040102

Kandjani, H., Bernus, P., & Wen, L. (2014). Enterprises as complex systems: Extended axiomatic design theory and its application in enterprise architecture practice. In P. Saha (Ed.), *A systemic perspective to managing complexity with enterprise architecture* (pp. 72–98). Hershey, PA: Business Science Reference.

Kapos, G., Dalakas, V., Nikolaidou, M., & Anagnostopoulos, D. (2014). An integrated framework to simulate SysML models using DEVS simulators. In P. Fonseca i Casas (Ed.), *Formal languages for computer simulation: Transdisciplinary models and applications* (pp. 305–332). Hershey, PA: Information Science Reference.

Karakaya, F. (2012). Business-to-consumers ecommerce: How companies use the internet in marketing products and services to consumers. In N. Delener (Ed.), *Service science research, strategy and innovation: Dynamic knowledge management methods* (pp. 227–244). Hershey, PA: Business Science. doi:10.4018/978-1-4666-0077-5.ch014

Karakaya, F. (2013). B2B ecommerce: Current practices. In I. Association (Ed.), *Supply chain management: Concepts, methodologies, tools, and applications* (pp. 497–510). Hershey, PA: Business Science Reference.

Karimov, F. P. (2013). Factors influencing e-commerce growth: A comparative study of central Asian transition economies. In S. Sharma (Ed.), *Adoption of virtual technologies for business, educational, and governmental advancements* (pp. 1 17). Hershey, PA: Information Science Reference.

Kart, F., Moser, L. E., & Melliar-Smith, P. M. (2010). An automated supply chain management system and its performance evaluation. *International Journal of Information Systems and Supply Chain Management, 3*(2), 84–107. doi:10.4018/jisscm.2010040105

Kawano, K. (2014). The design of service systems architecture for building smart public infrastructures. In M. Kosaka, & K. Shirahada (Eds.), *Progressive trends in knowledge and system-based science for service innovation* (pp. 220–235). Hershey, PA: Business Science Reference.

Kelarev, A. V., Brown, S., Watters, P., Wu, X., & Dazeley, R. (2011). Establishing reasoning communities of security experts for internet commerce security. In J. Yearwood, & A. Stranieri (Eds.), *Technologies for supporting reasoning communities and collaborative decision making: Cooperative approaches* (pp. 380–396). Hershey, PA: Information Science Reference.

Kerr, D., Gammack, J. G., & Boddington, R. (2011). Overview of digital business security issues. In D. Kerr, J. Gammack, & K. Bryant (Eds.), *Digital business security development: Management technologies* (pp. 1–36). Hershey, PA: Business Science Reference.

Kerr, D., Gammack, J. G., & Bryant, K. (2011). *Digital business security development: Management technologies*. Hershey, PA: IGI Global.

Kett, H. (2013). A business model approach for service engineering in the internet of services. In P. Ordóñez de Pablos, & R. Tennyson (Eds.), *Best practices and new perspectives in service science and management* (pp. 228–236). Hershey, PA: Business Science Reference. doi:10.4018/978-1-4666-3894-5.ch013

Khan, L., & Badar, R. (2014). Hybrid adaptive neurofuzzy bspline based SSSC damping control paradigm: Power system dynamic stability enhancement using online system identification. In P. Vasant (Ed.), *Handbook of research on novel soft computing intelligent algorithms: Theory and practical applications* (pp. 787–828). Hershey, PA: Information Science Reference.

Khurana, R., & Aggarwal, R. (2013). *Interdisciplinary perspectives on business convergence, computing, and legality*. Hershey, PA: IGI Global. doi:10.4018/978-1-4666-4209-6

Kim, G., & Suh, Y. (2012). Building semantic business process space for agile and efficient business processes management: Ontology-based approach. In V. Shankararaman, J. Zhao, & J. Lee (Eds.), *Business enterprise, process, and technology management: Models and applications* (pp. 51–73). Hershey, PA: Business Science Reference. doi:10.4018/978-1-4666-0249-6.ch004

King, K. P., & Foley, J. J. (2012). 21st century learning opportunities for SME success: Maximizing technology tools and lifelong learning for innovation and impact. In I. Management Association (Ed.), Human resources management: Concepts, methodologies, tools, and applications (pp. 731-752). Hershey, PA: Business Science Reference. doi: doi:10.4018/978-1-4666-1601-1.ch045

Kipp, A., & Schubert, L. (2011). E-business interoperability and collaboration. In E. Kajan (Ed.), *Electronic business interoperability: Concepts, opportunities and challenges* (pp. 153–184). Hershey, PA: Business Science Reference. doi:10.4018/978-1-60960-485-1.ch008

Klink, S., & Weiß, P. (2011). Social impact of collaborative services to maintain electronic business relationships. In I. Management Association (Ed.), Virtual communities: Concepts, methodologies, tools and applications (pp. 2011-2040). Hershey, PA: Information Science Reference. doi: doi:10.4018/978-1-60960-100-3.ch609

Köse, U., & Arslan, A. (2014). Chaotic systems and their recent implementations on improving intelligent systems. In P. Vasant (Ed.), *Handbook of research on novel soft computing intelligent algorithms: Theory and practical applications* (pp. 69–101). Hershey, PA: Information Science Reference.

Koumpis, A., & Protogeros, N. (2010). Doing business on the globalised networked economy: Technology and business challenges for accounting information systems. In M. Cruz-Cunha (Ed.), *Social, managerial, and organizational dimensions of enterprise information systems* (pp. 81–92). Hershey, PA: Business Science Reference. doi:10.4018/978-1-61692-852-0.ch702

Kritchanchai, D., Tan, A. W., & Hosie, P. (2010). An empirical investigation of third party logistics providers in Thailand: Barriers, motivation and usage of information technologies. *International Journal of Information Systems and Supply Chain Management*, *3*(2), 68–83. doi:10.4018/jisscm.2010040104

Kritchanchai, D., Tan, A. W., & Hosie, P. (2012). An empirical investigation of third party logistics providers in Thailand: Barriers, motivation and usage of information technologies. In J. Wang (Ed.), *Information technologies, methods, and techniques of supply chain management* (pp. 272–288). Hershey, PA: Business Science Reference. doi:10.4018/978-1-4666-0918-1.ch016

Kumar, M. (2011). Role of web interface in building trust in B2B e-exchanges. In S. Chhabra, & H. Rahman (Eds.), *Human development and global advancements through information communication technologies: New initiatives* (pp. 63–74). Hershey, PA: Information Science Reference. doi:10.4018/978-1-60960-497-4.ch005

Kumar, M., & Sareen, M. (2012). Trust theories and models of e-commerce. In *Trust and technology in B2B e-commerce: Practices and strategies for assurance* (pp. 58–77). Hershey, PA: Business Science Reference.

Kumar, M., Sareen, M., & Chhabra, S. (2013). Technology related trust issues in SME B2B e-commerce. In S. Chhabra (Ed.), *ICT influences on human development, interaction, and collaboration* (pp. 243–259). Hershey, PA: Information Science Reference.

Kung, M. T., & Zhang, Y. (2011). Creating competitive markets for small businesses with new media and e-business strategy. *International Journal of E-Business Research, 7*(4), 31–49. doi:10.4018/jebr.2011100103

Kuo, D., Wong, D., Gao, J., & Chang, L. (2013). A 2D barcode validation system for mobile commerce. In W. Hu, & S. Mousavinezhad (Eds.), *Mobile and handheld computing solutions for organizations and end-users* (pp. 1–19). Hershey, PA: Information Science Reference.

Kurosu, M. (2014). Concept of user experience and issues to be discussed. In M. Pańkowska (Ed.), *Frameworks of IT prosumption for business development* (pp. 31–47). Hershey, PA: Business Science Reference.

Kyobe, M. (2010). E-crime and non-compliance with government regulations on e-commerce: Barriers to e-commerce optimization in South African SMEs. In B. Thomas, & G. Simmons (Eds.), *E-commerce adoption and small business in the global marketplace: Tools for optimization* (pp. 47–66). Hershey, PA: Business Science Reference. doi:10.4018/978-1-60566-998-4.ch003

Lacuesta, R., Fernández-Sanz, L., & Romay, M. D. (2014). Requirements specification as basis for mobile software quality assurance. In I. Management Association (Ed.), Software design and development: Concepts, methodologies, tools, and applications (pp. 719-732). Hershey, PA: Information Science Reference. doi: doi:10.4018/978-1-4666-4301-7.ch034

Laporte, C., & Vargas, E. P. (2014). The development of international standards to facilitate process improvements for very small entities. In I. Management Association (Ed.), Software design and development: Concepts, methodologies, tools, and applications (pp. 1335-1361). Hershey, PA: Information Science Reference. doi: doi:10.4018/978-1-4666-4301-7.ch065

Lawrence, J. E. (2011). The growth of e-commerce in developing countries: An exploratory study of opportunities and challenges for SMEs. *International Journal of ICT Research and Development in Africa, 2*(1), 15–28. doi:10.4018/jictrda.2011010102

Lawrence, J. E. (2013). Barriers hindering ecommerce adoption: A case study of Kurdistan region of Iraq. In A. Zolait (Ed.), *Technology diffusion and adoption: Global complexity, global innovation* (pp. 152–165). Hershey, PA: Information Science Reference. doi:10.4018/978-1-4666-2791-8.ch010

Lee, I. (2012). B2B e-commerce, online auction, supply chain management, and e-collaboration. In *Electronic commerce management for business activities and global enterprises: Competitive advantages* (pp. 249–299). Hershey, PA: Business Science Reference. doi:10.4018/978-1-4666-1800-8.ch007

Lee, I. (2012). B2C online consumer behavior. In *Electronic commerce management for business activities and global enterprises: competitive advantages* (pp. 166–201). Hershey, PA: Business Science Reference. doi:10.4018/978-1-4666-1800-8.ch005

Lee, I. (2012). Introduction to e-commerce in the global economy. In *Electronic commerce management for business activities and global enterprises: Competitive advantages* (pp. 1–46). Hershey, PA: Business Science Reference. doi:10.4018/978-1-4666-1800-8.ch001

Lee, I. (2012). Mobile commerce. In *Electronic commerce management for business activities and global enterprises: Competitive advantages* (pp. 300–338). Hershey, PA: Business Science Reference. doi:10.4018/978-1-4666-1800-8.ch008

Lee, I. (2012). Online payment systems. In *Electronic commerce management for business activities and global enterprises: Competitive advantages* (pp. 340–365). Hershey, PA: Business Science Reference. doi:10.4018/978-1-4666-1800-8.ch009

Lee, J., Ma, S., Lee, S., Wu, C., & Lee, C. L. (2014). Towards a high-availability-driven service composition framework. In I. Management Association (Ed.), Software design and development: Concepts, methodologies, tools, and applications (pp. 1498-1520). Hershey, PA: Information Science Reference. doi: doi:10.4018/978-1-4666-4301-7.ch073

Leonard, L. N. (2010). C2C mobile commerce: Acceptance factors. In I. Lee (Ed.), *Encyclopedia of e-business development and management in the global economy* (pp. 759–767). Hershey, PA: Business Science Reference. doi:10.4018/978-1-61520-611-7.ch076

Lertpittayapoom, N., & Paul, S. (2010). The roles of online intermediaries in collective memory-supported electronic negotiation. In I. Management Association (Ed.), Electronic services: Concepts, methodologies, tools and applications (pp. 1831-1847). Hershey, PA: Information Science Reference. doi: doi:10.4018/978-1-61520-967-5.ch112

Levis, A. H., & Abu Jbara, A. (2014). Multi-modeling, meta-modeling, and workflow languages. In M. Gribaudo, & M. Iacono (Eds.), *Theory and application of multi-formalism modeling* (pp. 56–80). Hershey, PA: Information Science Reference.

Li, L., Liu, C., Zhao, X., & Wang, J. (2011). Transactional properties of complex web services. In H. Leung, D. Chiu, & P. Hung (Eds.), *Service intelligence and service science: Evolutionary technologies and challenges* (pp. 21–34). Hershey, PA: Information Science Reference.

Li, X., & Lin, J. (2011). Call u back: An agent-based infrastructure for mobile commerce. *International Journal of E-Entrepreneurship and Innovation*, *2*(2), 1–13. doi:10.4018/jeei.2011040101

Liao, Q., Luo, X., & Gurung, A. (2011). Trust restoration in electronic commerce. In S. Clarke, & A. Dwivedi (Eds.), *Organizational and end-user interactions: New explorations* (pp. 72–88). Hershey, PA: Information Science Reference. doi:10.4018/978-1-60960-577-3.ch003

Liberato, N. A., Varajão, J. E., Correia, E. S., & Bessa, M. E. (2011). Location based e-commerce system: An architecture. In M. Cruz-Cunha, & F. Moreira (Eds.), *Handbook of research on mobility and computing: Evolving technologies and ubiquitous impacts* (pp. 881–892). Hershey, PA: Information Science Reference. doi:10.4018/978-1-60960-042-6.ch055

Lim, S. Y., & Wong, S. F. (2012). Impact of applying aggregate query processing in mobile commerce. *International Journal of Business Data Communications and Networking*, *8*(2), 1–17. doi:10.4018/jbdcn.2012040101

Lin, C., & Jalleh, G. (2013). Key issues and challenges for managing and evaluating B2B e-commerce projects within the Australian pharmaceutical supply chain. In I. Association (Ed.), *Supply chain management: Concepts, methodologies, tools, and applications* (pp. 1083–1100). Hershey, PA: Business Science Reference.

Lin, C., Jalleh, G., & Huang, Y. (2013). E-business investment evaluation and outsourcing practices in Australian and Taiwanese hospitals: A comparative study. In K. Tarnay, S. Imre, & L. Xu (Eds.), *Research and development in e-business through service-oriented solutions* (pp. 244–266). Hershey, PA: Business Science Reference.

Lin, C., Lin, H. K., Jalleh, G., & Huang, Y. (2011). Key adoption challenges and issues of B2B e-commerce in the healthcare sector. In M. Cruz-Cunha, & F. Moreira (Eds.), *Handbook of research on mobility and computing: Evolving technologies and ubiquitous impacts* (pp. 175–187). Hershey, PA: Information Science Reference. doi:10.4018/978-1-60960-042-6.ch011

Liyanage, J. P. (2011). Copying with dynamic change: Collaborative business interfacing for SMEs under intergated eoperations. In M. Cruz-Cunha, & J. Varajão (Eds.), *E-business managerial aspects, solutions and case studies* (pp. 136–147). Hershey, PA: Business Science Reference.

Liyanage, J. P. (2012). Hybrid intelligence through business socialization and networking: Managing complexities in the digital era. In M. Cruz-Cunha, P. Gonçalves, N. Lopes, E. Miranda, & G. Putnik (Eds.), *Handbook of research on business social networking: organizational, managerial, and technological dimensions* (pp. 567–582). Hershey, PA: Business Science Reference.

Loeser, F., Erek, K., & Zarnekow, R. (2013). Green IT strategies: A conceptual framework for the alignment of information technology and corporate sustainability strategy. In P. Ordóñez de Pablos (Ed.), *Green technologies and business practices: An IT approach* (pp. 58–95). Hershey, PA: Information Science Reference.

Lukovic, I., Ivancevic, V., Celikovic, M., & Aleksic, S. (2014). DSLs in action with model based approaches to information system development. In I. Management Association (Ed.), Software design and development: Concepts, methodologies, tools, and applications (pp. 596-626). Hershey, PA: Information Science Reference. doi: doi:10.4018/978-1-4666-4301-7.ch029

Maamar, Z., Faci, N., Mostéfaoui, S. K., & Akhter, F. (2011). Towards a framework for weaving social networks into mobile commerce. *International Journal of Systems and Service-Oriented Engineering*, *2*(3), 32–46. doi:10.4018/jssoe.2011070103

Mahmood, M. A., Gemoets, L., Hall, L. L., & López, F. J. (2011). Building business value in e-commerce enabled organizations: An empirical study. In I. Management Association (Ed.), Global business: Concepts, methodologies, tools and applications (pp. 229-253). Hershey, PA: Business Science Reference. doi: doi:10.4018/978-1-60960-587-2.ch201

Mahran, A. F., & Enaba, H. M. (2013). Exploring determinants influencing the intention to use mobile payment service. In R. Eid (Ed.), *Managing customer trust, satisfaction, and loyalty through information communication technologies* (pp. 288–309). Hershey, PA: Business Science Reference.

Manalif, E., Capretz, L. F., & Ho, D. (2014). Fuzzy rules for risk assessment and contingency estimation within COCOMO software project planning model. In A. Masegosa, P. Villacorta, C. Cruz-Corona, M. García-Cascales, M. Lamata, & J. Verdegay (Eds.), *Exploring innovative and successful applications of soft computing* (pp. 88–111). Hershey, PA: Information Science Reference.

Marimuthu, M., Omar, A., Ramayah, T., & Mohamad, O. (2013). Readiness to adopt e-business among SMEs in Malaysia: Antecedents and consequence. In S. Sharma (Ed.), *Adoption of virtual technologies for business, educational, and governmental advancements* (pp. 18–36). Hershey, PA: Information Science Reference.

Mayes, P. (2014). Interactive advertising: Displays of identity and stance on YouTube. In H. Lim, & F. Sudweeks (Eds.), *Innovative methods and technologies for electronic discourse analysis* (pp. 260–284). Hershey, PA: Information Science Reference.

McGrath, T. (2012). The reality of using standards for electronic business document formats. In E. Kajan, F. Dorloff, & I. Bedini (Eds.), *Handbook of research on e-business standards and protocols: Documents, data and advanced web technologies* (pp. 21–32). Hershey, PA: Business Science Reference. doi:10.4018/978-1-4666-0146-8.ch002

McKee, J. (2013). Improving strategic plan documentation. In *Applying principles from IT architecture to strategic business planning* (pp. 60–85). Hershey, PA: Business Science Reference.

McKee, J. (2013). Strategic planning considerations. In *Applying principles from IT architecture to strategic business planning* (pp. 86–100). Hershey, PA: Business Science Reference.

McKee, J. (2013). Strategic planning, information systems alignment, and architecture. In *Applying principles from IT architecture to strategic business planning* (pp. 1–8). Hershey, PA: Business Science Reference.

Meredith, J., & Potter, J. (2014). Conversation analysis and electronic interactions: Methodological, analytic and technical considerations. In H. Lim, & F. Sudweeks (Eds.), *Innovative methods and technologies for electronic discourse analysis* (pp. 370–393). Hershey, PA: Information Science Reference.

Metzger, A., & Di Nitto, E. (2014). Addressing highly dynamic changes in service-oriented systems: Towards agile evolution and adaptation. In I. Management Association (Ed.), Software design and development: Concepts, methodologies, tools, and applications (pp. 164-177). Hershey, PA: Information Science Reference. doi: doi:10.4018/978-1-4666-4301-7.ch009

Millman, C., & El-Gohary, H. (2011). New digital media marketing and micro business: A UK perspective. *International Journal of Online Marketing*, *1*(1), 41–62. doi:10.4018/ijom.2011010104

Mishra, B., & Shukla, K. K. (2014). Data mining techniques for software quality prediction. In I. Management Association (Ed.), Software design and development: Concepts, methodologies, tools, and applications (pp. 401-428). Hershey, PA: Information Science Reference. doi: doi:10.4018/978-1-4666-4301-7.ch021

Misra, H., & Rahman, H. (2013). *Managing enterprise information technology acquisitions: Assessing organizational preparedness*. Hershey, PA: IGI Global. doi:10.4018/978-1-4666-4201-0

Mohammadi, S., Golara, S., & Mousavi, N. (2012). Selecting adequate security mechanisms in e-business processes using fuzzy TOPSIS. *International Journal of Fuzzy System Applications, 2*(1), 35–53. doi:10.4018/ijfsa.2012010103

Möhlenbruch, D., Dölling, S., & Ritschel, F. (2010). Interactive customer retention management for mobile commerce. In K. Pousttchi, & D. Wiedemann (Eds.), *Handbook of research on mobile marketing management* (pp. 437–456). Hershey, PA: Business Science Reference. doi:10.4018/978-1-61520-969-9.ch069

Molla, A., & Peszynski, K. (2013). E-business in agribusiness: Investigating the e-readiness of Australian horticulture firms. In S. Chhabra (Ed.), *ICT influences on human development, interaction, and collaboration* (pp. 78–96). Hershey, PA: Information Science Reference.

Monsanto, C., & Andriole, S. J. (2010). Business technology strategy for a major real estate and mortgage brokerage company. *Journal of Information Technology Research, 3*(3), 43–53. doi:10.4018/jitr.2010070104

Montes, J. A., Gutiérrez, A. C., Fernández, E. M., & Romeo, A. (2013). Reality mining, location based services, and e-business opportunities: The case of city analytics. In S. Nasir (Ed.), *Modern entrepreneurship and e-business innovations* (pp. 87–99). Hershey, PA: Business Science Reference.

Moqbel, A., Yani-De-Soriano, M., & Yousafzai, S. (2012). Mobile commerce use among UK mobile users: An experimental approach based on a proposed mobile network utilization framework. In A. Zolait (Ed.), *Knowledge and technology adoption, diffusion, and transfer: International perspectives* (pp. 78–111). Hershey, PA: Information Science Reference. doi:10.4018/978-1-4666-1752-0.ch007

Movahedi, B. M., Lavassani, K. M., & Kumar, V. (2012). E-marketplace emergence: Evolution, developments and classification. *Journal of Electronic Commerce in Organizations, 10*(1), 14–32. doi:10.4018/jeco.2012010102

Mugge, R., & Schoormans, J. P. (2010). Optimizing consumer responses to mass customization. In C. Mourlas, & P. Germanakos (Eds.), *Mass customization for personalized communication environments: Integrating human factors* (pp. 10–22). Hershey, PA: Information Science Reference.

Muller, G. (2014). High-level modeling to support software design choices. In I. Management Association (Ed.), Software design and development: Concepts, methodologies, tools, and applications (pp. 1440-1460). Hershey, PA: Information Science Reference. doi: doi:10.4018/978-1-4666-4301-7.ch070

Musso, F. (2012). Technology in marketing channels: Present and future drivers of innovation. *International Journal of Applied Behavioral Economics, 1*(2), 41–51. doi:10.4018/ijabe.2012040104

Mutula, S. M. (2010). Digital economy components. In S. Mutula (Ed.), *Digital economies: SMEs and e-readiness* (pp. 29–38). Hershey, PA: Business Science Reference.

Mutula, S. M. (2010). Trends and best practices in the digital economy. In S. Mutula (Ed.), *Digital economies: SMEs and e-readiness* (pp. 283–301). Hershey, PA: Business Science Reference.

Nabi, S. I., Al-Ghmlas, G. S., & Alghathbar, K. (2014). Enterprise information security policies, standards, and procedures: A survey of available standards and guidelines. In I. Management Association (Ed.), Crisis management: Concepts, methodologies, tools and applications (pp. 750-773). Hershey, PA: Information Science Reference. doi: doi:10.4018/978-1-4666-4707-7.ch036

Nachtigal, S. (2011). E-business: Definition and characteristics. In O. Bak, & N. Stair (Eds.), *Impact of e-business technologies on public and private organizations: Industry comparisons and perspectives* (pp. 233–248). Hershey, PA: Business Science Reference. doi:10.4018/978-1-60960-501-8.ch014

Nah, F. F., Hong, W., Chen, L., & Lee, H. (2012). Information search patterns in e-commerce product comparison services. In K. Siau (Ed.), *Cross-disciplinary models and applications of database management: Advancing approaches* (pp. 131–145). Hershey, PA: Information Science Reference.

Nair, P. R. (2010). Benefits of information technology implementations for supply chain management: An explorative study of progressive Indian companies. In S. Parthasarathy (Ed.), *Enterprise information systems and implementing IT infrastructures: Challenges and issues* (pp. 323–343). Hershey, PA: Business Science Reference. doi:10.4018/978-1-61520-625-4.ch021

Nasir, M. H., Alias, N. A., Fauzi, S. S., & Massatu, M. H. (2014). Implementation of the personal software process in academic settings and current support tools. In I. Management Association (Ed.), Software design and development: Concepts, methodologies, tools, and applications (pp. 807-838). Hershey, PA: Information Science Reference. doi: doi:10.4018/978-1-4666-4301-7.ch039

Ndou, V., Del Vecchio, P., Passiante, G., & Schina, L. (2013). Web-based services and future business models. In P. Papajorgji, A. Guimarães, & M. Guarracino (Eds.), *Enterprise business modeling, optimization techniques, and flexible information systems* (pp. 1–13). Hershey, PA: Business Science Reference.

Ndou, V., & Sadguy, N. (2013). Digital marketplaces as a viable model for SME networking. In I. Association (Ed.), *Supply chain management: Concepts, methodologies, tools, and applications* (pp. 275–288). Hershey, PA: Business Science Reference.

Newlands, D. J. (2012). Supply chain reengineering: A case study. In R. Valverde, & M. Rao Talla (Eds.), *Information systems reengineering for modern business systems: ERP, supply chain and e-commerce management solutions* (pp. 201–218). Hershey, PA: Business Science Reference. doi:10.4018/978-1-4666-0155-0.ch010

Ntoa, S., Margetis, G., Antona, M., & Stephanidis, C. (2014). Scanning-Based interaction techniques for motor impaired users. In G. Kouroupetroglou (Ed.), *Assistive technologies and computer access for motor disabilities* (pp. 57–89). Hershey, PA: Medical Information Science Reference.

Nunes de Castro, L., Xavier, R. S., Pasti, R., Maia, R. D., Szabo, A., & Ferrari, D. G. (2014). The grand challenges in natural computing research: The quest for a new science. In L. Nunes de Castro (Ed.), *Natural computing for simulation and knowledge discovery* (pp. 237–250). Hershey, PA: Medical Information Science Reference.

Ochara, N. M., & Krauss, K. (2012). Towards a collaborative e-business vision for Africa. In K. Mohammed Rezaul (Ed.), *Strategic and pragmatic e-business: Implications for future business practices* (pp. 396–414). Hershey, PA: Business Science Reference. doi:10.4018/978-1-4666-1619-6.ch018

Olanrewaju, O. S., & Kader, A. S. (2014). Applying the safety and environmental risk and reliability model (SERM) for Malaysian Langat River collision aversion. In O. Olanrewaju, A. Saharuddin, A. Ab Kader, & W. Wan Nik (Eds.), *Marine technology and sustainable development: Green innovations* (pp. 193–225). Hershey, PA: Information Science Reference.

Olanrewaju, O. S., Kader, A. S., & Saharuddin, A. H. (2014). Utilization of simulation for training enhancement. In O. Olanrewaju, A. Saharuddin, A. Ab Kader, & W. Wan Nik (Eds.), *Marine technology and sustainable development: Green innovations* (pp. 277–290). Hershey, PA: Information Science Reference.

Olszewska, M. P., & Waldén, M. (2014). Measuring the progress of a system development. In I. Management Association (Ed.), Software design and development: Concepts, methodologies, tools, and applications (pp. 1415-1439). Hershey, PA: Information Science Reference. doi: doi:10.4018/978-1-4666-4301-7.ch069

Oncioiu, I. (2013). *Business innovation, development, and advancement in the digital economy*. Hershey, PA: IGI Global. doi:10.4018/978-1-4666-2934-9

Ondimu, K. O., Muketha, G. M., & Ondago, C. O. (2013). E-business adoption framework in the hospitality industry: The case of Kenyan coast. In K. Tarnay, S. Imre, & L. Xu (Eds.), *Research and development in e-business through service-oriented solutions* (pp. 225–243). Hershey, PA: Business Science Reference.

Onwubiko, C. (2014). Modelling situation awareness information and system requirements for the mission using goal-oriented task analysis approach. In I. Management Association (Ed.), Software design and development: Concepts, methodologies, tools, and applications (pp. 460-478). Hershey, PA: Information Science Reference. doi: doi:10.4018/978-1-4666-4301-7.ch023

Ovaskainen, M., & Tinnilä, M. (2013). Megatrends in electronic business: An analysis of the impacts on SMEs. In S. Nasir (Ed.), *Modern entrepreneurship and e-business innovations* (pp. 12–27). Hershey, PA: Business Science Reference.

Özcan, O., & Reeves, K. A. (2011). The firm boundary decision for sustainability-focused companies. *International Journal of Applied Logistics*, *2*(2), 49–68. doi:10.4018/jal.2011040104

Öztayşi, B., & Kahraman, C. (2014). Quantification of corporate performance using fuzzy analytic network process: The case of e-commerce. In P. Vasant (Ed.), *Handbook of research on novel soft computing intelligent algorithms: Theory and practical applications* (pp. 385–413). Hershey, PA: Information Science Reference.

Palmer, D. E. (2010). The transformative nature of e-business: Business ethics and stakeholder relations on the internet. In D. Palmer (Ed.), *Ethical issues in e-business: Models and frameworks* (pp. 1–14). Hershey, PA: Business Science Reference. doi:10.4018/978-1-61520-615-5.ch001

Palomo-Duarte, M. (2014). Service composition verification and validation. In I. Management Association (Ed.), Software design and development: Concepts, methodologies, tools, and applications (pp. 2163-2182). Hershey, PA: Information Science Reference. doi: doi:10.4018/978-1-4666-4301-7.ch100

Pańkowska, M. (2014). Frameworks of IT prosumption for business development (pp. 1-347). doi:10.4018/978-1-4666-4313-0

Páscoa, C., & Tribolet, J. (2014). Change and information value in military organizations' transformation processes. In G. Jamil, A. Malheiro, & F. Ribeiro (Eds.), *Rethinking the conceptual base for new practical applications in information value and quality* (pp. 304–323). Hershey, PA: Information Science Reference.

Páscoa, C., & Tribolet, J. (2014). Maintaining organizational viability and performance: The organizational configuration map. In G. Jamil, A. Malheiro, & F. Ribeiro (Eds.), *Rethinking the conceptual base for new practical applications in information value and quality* (pp. 266–283). Hershey, PA: Information Science Reference.

Pelet, J. É., & Papadopoulou, P. (2013). The effect of e-commerce websites' colors on customer trust. In I. Lee (Ed.), *Mobile applications and knowledge advancements in e-business* (pp. 167–185). Hershey, PA: Business Science Reference.

Peña de Carrillo, C. I., Choquet, C., Després, C., Iksal, S., Jacoboni, P., Lekira, A., et al. (2014). Engineering and reengineering of technology enhanced learning scenarios using context awareness processes. In I. Management Association (Ed.), Software design and development: Concepts, methodologies, tools, and applications (pp. 1289-1313). Hershey, PA: Information Science Reference. doi: doi:10.4018/978-1-4666-4301-7.ch063

Pennington, R. (2012). Enhanced social presence through ebranding the consumer in virtual communities. In A. Kapoor, & C. Kulshrestha (Eds.), *Branding and sustainable competitive advantage: Building virtual presence* (pp. 189–206). Hershey, PA: Business Science Reference.

Peslak, A. R. (2012). Industry variables affecting ERP success and status. *International Journal of Enterprise Information Systems*, *8*(3), 15–33. doi:10.4018/jeis.2012070102

Peterson, D., & Howard, C. (2012). Electronic payment systems evaluation: A case study to examine system selection criteria and impacts. *International Journal of Strategic Information Technology and Applications*, *3*(1), 66–80. doi:10.4018/jsita.2012010105

Petkov, D., Alter, S., Wing, J., Singh, A., Petkova, O., Andrew, T., & Sewchurran, K. (2014). Project contexts and the possibilities for mixing software development and systems approaches. In I. Management Association (Ed.), Software design and development: Concepts, methodologies, tools, and applications (pp. 335-350). Hershey, PA: Information Science Reference. doi: doi:10.4018/978-1-4666-4301-7.ch018

Pflügler, C. (2012). Fostering networked business operations: A framework for B2B electronic intermediary development. *International Journal of Intelligent Information Technologies, 8*(2), 31–58. doi:10.4018/jiit.2012040103

Pillai, K., & Ozansoy, C. (2013). Web-based digital habitat ecosystems for sustainable built environments. In P. Ordóñez de Pablos (Ed.), *Green technologies and business practices: An IT approach* (pp. 185–199). Hershey, PA: Information Science Reference. doi:10.4018/978-1-4666-4852-4.ch031

Pinto, M., Rodrigues, A., Varajão, J., & Gonçalves, R. (2011). Model of funcionalities for the development of B2B e-commerce solutions. In M. Cruz-Cunha, & J. Varajão (Eds.), *Innovations in SMEs and conducting e-business: Technologies, trends and solutions* (pp. 35–60). Hershey, PA: Business Science Reference. doi:10.4018/978-1-60960-765-4.ch003

Pires, J. A., & Gonçalves, R. (2011). Constrains associated to e-business evolution. In M. Cruz-Cunha, & J. Varajão (Eds.), *E-business issues, challenges and opportunities for SMEs: Driving competitiveness* (pp. 335–349). Hershey, PA: Business Science Reference.

Polovina, S., & Andrews, S. (2011). A transaction-oriented architecture for structuring unstructured information in enterprise applications. In V. Sugumaran (Ed.), *Intelligent, adaptive and reasoning technologies: New developments and applications* (pp. 285–299). Hershey, PA: Information Science Reference. doi:10.4018/978-1-60960-595-7.ch016

Potocan, V., Nedelko, Z., & Mulej, M. (2011). What is new with organization of e-business: Organizational viewpoint of the relationships in e-business. In M. Cruz-Cunha, & J. Varajão (Eds.), *E-business issues, challenges and opportunities for SMEs: Driving competitiveness* (pp. 131–148). Hershey, PA: Business Science Reference. doi:10.4018/978-1-60960-587-2.ch804

Potocan, V., Nedelko, Z., & Mulej, M. (2011). What is new with organization of e-business: Organizational viewpoint of the relationships in e-business. In M. Cruz-Cunha, & J. Varajão (Eds.), *E-business issues, challenges and opportunities for SMEs: Driving competitiveness* (pp. 131–148). Hershey, PA: Business Science Reference. doi:10.4018/978-1-60960-587-2.ch804

Pucihar, A., & Lenart, G. (2011). eSME Slovenia: Initiative and action plan for the accelerated introduction of e-business in SMEs. In I. Management Association (Ed.), Global business: Concepts, methodologies, tools and applications (pp. 995-1022). Hershey, PA: Business Science Reference. doi: doi:10.4018/978-1-60960-587-2.ch409

Quan, J. (2011). E-business strategy and firm performance. In I. Management Association (Ed.), Global business: Concepts, methodologies, tools and applications (pp. 56-66). Hershey, PA: Business Science Reference. doi: doi:10.4018/978-1-60960-587-2.ch105

Quente, C. (2010). Brand driven mobile marketing: 5 theses for today and tomorrow. In K. Pousttchi, & D. Wiedemann (Eds.), *Handbook of research on mobile marketing management* (pp. 468–483). Hershey, PA: Business Science Reference.

Qureshil, S., Kamal, M., & Wolcott, P. (2011). Information technology interventions for growth and competitiveness in micro-enterprises. In M. Tavana (Ed.), *Managing adaptability, intervention, and people in enterprise information systems* (pp. 106–137). Hershey, PA: Information Science Reference. doi:10.4018/978-1-60960-529-2.ch006

Rabaey, M. (2014). Complex adaptive systems thinking approach to enterprise architecture. In P. Saha (Ed.), *A systemic perspective to managing complexity with enterprise architecture* (pp. 99–149). Hershey, PA: Business Science Reference.

Rahman, H., & Ramos, I. (2012). Trends of open innovation in developing nations: Contexts of SMEs. In H. Rahman, & I. Ramos (Eds.), *Cases on SMEs and open innovation: Applications and investigations* (pp. 65–80). Hershey, PA: Business Science Reference.

Rahman, H., & Ramos, I. (2013). Implementation of e-commerce at the grass roots: Issues of challenges in terms of human-computer interactions. *International Journal of Information Communication Technologies and Human Development, 5*(2), 1–19. doi:10.4018/jicthd.2013040101

Rajagopal, D. (2010). Customer value and new product retailing dynamics: An analytical construct for gaining competetive advantage. In I. Management Association (Ed.), Business information systems: Concepts, methodologies, tools and applications (pp. 1998-2014). Hershey, PA: Business Science Reference. doi: doi:10.4018/978-1-61520-969-9.ch121

Rajagopal, D. (2010). Internet, reengineering and technology applications in retailing. In I. Management Association (Ed.), Business information systems: Concepts, methodologies, tools and applications (pp. 1324-1342). Hershey, PA: Business Science Reference. doi: doi:10.4018/978-1-61520-969-9.ch082

Rajagopal, D. (2011). Marketing strategy, technology and modes of entry in global retailing. In I. Management Association (Ed.), Global business: Concepts, methodologies, tools and applications (pp. 1-27). Hershey, PA: Business Science Reference. doi: doi:10.4018/978-1-60960-587-2.ch101

Rajagopal, D. (2012). Convergence marketing. In *Systems thinking and process dynamics for marketing systems: Technologies and applications for decision management* (pp. 274–290). Hershey, PA: Business Science Reference. doi:10.4018/978-1-4666-0969-3.ch011

Rajagopal, D. (2012). Product development and market governance. In *Systems thinking and process dynamics for marketing systems: Technologies and applications for decision management* (pp. 88–117). Hershey, PA: Business Science Reference. doi:10.4018/978-1-4666-0969-3.ch004

Rajagopal, D. (2012). Systems thinking and cognitive process in marketing. In *Systems thinking and process dynamics for marketing systems: technologies and applications for decision management* (pp. 170–197). Hershey, PA: Business Science Reference. doi:10.4018/978-1-4666-0969-3.ch007

Rajagopal, D. (2013). Pricing for new products. In *Marketing decision making and the management of pricing: Successful business tools* (pp. 56–74). Hershey, PA: Business Science Reference. doi:10.4018/978-1-4666-4094-8.ch003

Rajesh, R., Pugazhendhi, S., & Ganesh, K. (2013). Genetic algorithm and particle swarm optimization for solving balanced allocation problem of third party logistics providers. In J. Wang (Ed.), *Management innovations for intelligent supply chains* (pp. 184–203). Hershey, PA: Business Science Reference.

Ramayah, T., Popa, S., & Suki, N. M. (2013). Key dimensions on B2C e-business: An empirical study in Malaysia. *International Journal of Human Capital and Information Technology Professionals*, *4*(2), 43–55. doi:10.4018/jhcitp.2013040104

Rambo, K., & Liu, K. (2011). Culture-sensitive virtual e-commerce design with reference to female consumers in Saudi Arabia. In B. Ciaramitaro (Ed.), *Virtual worlds and e-commerce: Technologies and applications for building customer relationships* (pp. 267–289). Hershey, PA: Business Science Reference.

Ratnasingam, P. (2010). The evolution of online relationships in business to consumer e-commerce. In M. Khosrow-Pour (Ed.), *E-commerce trends for organizational advancement: New applications and methods* (pp. 167–176). Hershey, PA: Information Science Reference.

Ratnasingam, P. (2010). The impact of e-commerce customer relationship management in business-to-consumer e-commerce. In M. Hunter (Ed.), *Strategic information systems: Concepts, methodologies, tools, and applications* (pp. 2099–2111). Hershey, PA: Information Science Reference. doi:10.4018/978-1-61520-969-9.ch095

Razavi, A. R., Krause, P., & Moschoyiannis, S. (2010). Digital ecosystems: Challenges and proposed solutions. In N. Antonopoulos, G. Exarchakos, M. Li, & A. Liotta (Eds.), Handbook of research on P2P and grid systems for service-oriented computing: Models, methodologies and applications (pp. 1003-1031). Hershey, PA: Information Science Reference. doi: doi:10.4018/978-1-61520-686-5.ch043

Regazzi, J. J. (2014). *Infonomics and the business of free: Modern value creation for information services*. Hershey, PA: IGI Global.

Regazzi, J. J. (2014). The beginnings of infonomics. In *Infonomics and the business of free: Modern value creation for information services* (pp. 1–10). Hershey, PA: Business Science Reference.

Regazzi, J. J. (2014). The shifting sands of the information industry. In *Infonomics and the business of free: Modern value creation for information services* (pp. 11–38). Hershey, PA: Business Science Reference.

Regazzi, J. J. (2014). Slippery slopes. In *Infonomics and the business of free: modern value creation for information services* (pp. 39–65). Hershey, PA: Business Science Reference.

Riaz, N., & Rehman, M. (2013). Negotiation by software agents in electronic business: An example of hybrid negotiation. In E. Li, S. Loh, C. Evans, & F. Lorenzi (Eds.), *Organizations and social networking: Utilizing social media to engage consumers* (pp. 327–349). Hershey, PA: Business Science Reference. doi:10.4018/978-1-4666-4026-9.ch017

Roberti, G., & Marinelli, A. (2012). Branding identity: Facebook, brands and self construction. In F. Comunello (Ed.), *Networked sociability and individualism: Technology for personal and professional relationships* (pp. 147–168). Hershey, PA: Information Science Reference.

Rodrigues, D. E. (2012). Cyberethics of business social networking. In M. Cruz-Cunha, P. Gonçalves, N. Lopes, E. Miranda, & G. Putnik (Eds.), *Handbook of research on business social networking: Organizational, managerial, and technological dimensions* (pp. 314–338). Hershey, PA: Business Science Reference.

Roos, G. (2013). The role of intellectual capital in business model innovation: An empirical study. In P. Ordóñez de Pablos, R. Tennyson, & J. Zhao (Eds.), *Intellectual capital strategy management for knowledge-based organizations* (pp. 76–121). Hershey, PA: Business Science Reference. doi:10.4018/978-1-4666-3655-2.ch006

Rowley, J., & Edmundson-Bird, D. (2013). Brand presence in digital space. *Journal of Electronic Commerce in Organizations*, *11*(1), 63–78. doi:10.4018/jeco.2013010104

Rusko, R. (2013). The redefined role of consumer as a prosumer: Value co-creation, coopetition, and crowdsourcing of information goods. In P. Renna (Ed.), *Production and manufacturing system management: Coordination approaches and multi-site planning* (pp. 162–174). Hershey, PA: Engineering Science Reference.

Sahi, G., & Madan, S. (2013). Developing a website usability framework for B2C e-commerce success. *International Journal of Information Communication Technologies and Human Development*, *5*(1), 1–19. doi:10.4018/jicthd.2013010101

Saidane, A., & Guelfi, N. (2014). Towards test-driven and architecture model-based security and resilience engineering. In I. Management Association (Ed.), Software design and development: Concepts, methodologies, tools, and applications (pp. 2072-2098). Hershey, PA: Information Science Reference. doi: doi:10.4018/978-1-4666-4301-7.ch096

Sainz de Abajo, B., de la Torre Díez, I., & López-Coronado, M. (2010). Analysis of benefits and risks of e-commerce: Practical study of Spanish SME. In I. Portela, & M. Cruz-Cunha (Eds.), *Information communication technology law, protection and access rights: Global approaches and issues* (pp. 214–239). Hershey, PA: Information Science Reference. doi:10.4018/978-1-61520-975-0.ch014

Samanta, I., & Kyriazopoulos, P. (2011). Can global environment influence B2B relationships? In P. Ordóñez de Pablos, M. Lytras, W. Karwowski, & R. Lee (Eds.), *Electronic globalized business and sustainable development through IT management: Strategies and perspectives* (pp. 54–69). Hershey, PA: Business Science Reference.

Sambhanthan, A., & Good, A. (2012). Implications for improving accessibility to e-commerce websites in developing countries: A study of hotel websites. *International Journal of Knowledge-Based Organizations*, *2*(2), 1–20. doi:10.4018/ijkbo.2012040101

Sampaio, L., & Figueiredo, J. (2011). E-sourcing electronic platforms in real business. In M. Cruz-Cunha, & J. Varajão (Eds.), *E-business managerial aspects, solutions and case studies* (pp. 185–205). Hershey, PA: Business Science Reference.

Savsar, M. (2013). Modeling of hybrid production systems with constant WIP and unreliable equipment. In J. Wang (Ed.), *Management innovations for intelligent supply chains* (pp. 235–256). Hershey, PA: Business Science Reference.

Seetharaman, A., & Raj, J. R. (2011). Evolution, development and growth of electronic money. In S. Sharma (Ed.), *E-adoption and socio-economic impacts: Emerging infrastructural effects* (pp. 249–268). Hershey, PA: Information Science Reference. doi:10.4018/978-1-60960-597-1.ch013

Segev, A. (2014). Multilingual crisis knowledge representation. In I. Management Association (Ed.), Crisis management: Concepts, methodologies, tools and applications (pp. 792-803). Hershey, PA: Information Science Reference. doi: doi:10.4018/978-1-4666-4707-7.ch038

Sengupta, A., & Glavin, S. E. (2013). Predicting volatile consumer markets using multi-agent methods: Theory and validation. In B. Alexandrova-Kabadjova, S. Martinez-Jaramillo, A. Garcia-Almanza, & E. Tsang (Eds.), *Simulation in computational finance and economics: Tools and emerging applications* (pp. 339–358). Hershey, PA: Business Science Reference.

Serpico, E., Aquilani, B., Ruggieri, A., & Silvestri, C. (2013). Customer centric marketing strategies: The importance and measurement of customer satisfaction – Offline vs. online. In H. Kaufmann, & M. Panni (Eds.), *Customer-centric marketing strategies: Tools for building organizational performance* (pp. 315–357). Hershey, PA: Business Science Reference.

Shareef, M. A., & Kumar, V. (2012). Prevent/control identity theft: Impact on trust and consumers' purchase intention in B2C EC. *Information Resources Management Journal, 25*(3), 30–60. doi:10.4018/irmj.2012070102

Sharma, N., Singh, K., & Goyal, D. (2014). Software engineering, process improvement, and experience management: Is the nexus productive? Clues from the Indian giants. In I. Management Association (Ed.), Software design and development: Concepts, methodologies, tools, and applications (pp. 1401-1414). Hershey, PA: Information Science Reference. doi: doi:10.4018/978-1-4666-4301-7.ch068

Sherringham, K., & Unhelkar, B. (2011). Business driven enterprise architecture and applications to support mobile business. In I. Management Association (Ed.), Enterprise information systems: Concepts, methodologies, tools and applications(pp. 805-816). Hershey, PA: Business Science Reference. doi: doi:10.4018/978-1-61692-852-0.ch316

Shin, N. (2011). Information technology and diversification: How their relationship affects firm performance. In N. Kock (Ed.), *E-collaboration technologies and organizational performance: Current and future trends* (pp. 65–79). Hershey, PA: Information Science Reference.

Sidnal, N., & Manvi, S. S. (2010). Service discovery techniques in mobile e-commerce. In I. Lee (Ed.), *Encyclopedia of e-business development and management in the global economy* (pp. 812–823). Hershey, PA: Business Science Reference. doi:10.4018/978-1-61520-611-7.ch081

Sidnal, N., & Manvi, S. S. (2013). English auction issues in mobile e-commerce. In K. Tarnay, S. Imre, & L. Xu (Eds.), *Research and development in e-business through service-oriented solutions* (pp. 208–223). Hershey, PA: Business Science Reference.

Simonette, M. J., & Spina, E. (2014). Enabling IT innovation through soft systems engineering. In M. Pańkowska (Ed.), *Frameworks of IT prosumption for business development* (pp. 64–72). Hershey, PA: Business Science Reference.

Singh, S. (2010). Usability techniques for interactive software and their application in e-commerce. In T. Spiliotopoulos, P. Papadopoulou, D. Martakos, & G. Kouroupetroglou (Eds.), *Integrating usability engineering for designing the web experience: Methodologies and principles* (pp. 81–102). Hershey, PA: Information Science Reference. doi:10.4018/978-1-60566-896-3.ch005

Söderström, E. (2010). Guidelines for managing B2B standards implementation. In E. Alkhalifa (Ed.), *E-strategies for resource management systems: Planning and implementation* (pp. 86–105). Hershey, PA: Business Science Reference. doi:10.4018/978-1-61692-016-6.ch005

Solemon, B., Sahibuddin, S., & Ghani, A. A. (2014). Requirements engineering process improvement and related models. In I. Management Association (Ed.), Software design and development: Concepts, methodologies, tools, and applications (pp. 203-218). Hershey, PA: Information Science Reference. doi: doi:10.4018/978-1-4666-4301-7.ch011

Sood, S. (2012). The death of social media in start-up companies and the rise of s-commerce: Convergence of e-commerce, complexity and social media. *Journal of Electronic Commerce in Organizations*, *10*(2), 1–15. doi:10.4018/jeco.2012040101

Soto-Acosta, P. (2010). E-business and the resource-based view: Towards a research agenda. In I. Lee (Ed.), *Encyclopedia of e-business development and management in the global economy* (pp. 336–346). Hershey, PA: Business Science Reference. doi:10.4018/978-1-61520-611-7.ch033

Sourouni, A. M., Mouzakitis, S., Kourlimpini, G., Askounis, D., & Psarras, J. (2010). Ontology-based registries: An e-business transactions' registry. In E. Alkhalifa (Ed.), *E-strategies for resource management systems: Planning and implementation* (pp. 106–117). Hershey, PA: Business Science Reference. doi:10.4018/978-1-61692-016-6.ch006

Srinivasan, S., & Barker, R. (2012). Global analysis of security and trust perceptions in web design for e-commerce. *International Journal of Information Security and Privacy*, *6*(1), 1–13. doi:10.4018/jisp.2012010101

Su, Q., & Adams, C. (2012). Consumers' attitudes toward mobile commerce: A model to capture the cultural and environment influences. In A. Scupola (Ed.), *Innovative mobile platform developments for electronic services design and delivery* (pp. 1–20). Hershey, PA: Business Science Reference. doi:10.4018/978-1-4666-1568-7.ch001

Swilley, E., Hofacker, C. F., & Lamont, B. T. (2012). The evolution from e-commerce to m-commerce: Pressures, firm capabilities and competitive advantage in strategic decision making. *International Journal of E-Business Research*, *8*(1), 1–16. doi:10.4018/jebr.2012010101

Swimm, N., & Andriole, S. J. (2010). Business technology strategy for an energy management company. *Journal of Information Technology Research*, *3*(3), 54–65. doi:10.4018/jitr.2010070105

Tadjouddine, E. M. (2011). E-commerce systems for software agents: Challenges and opportunities. In M. Cruz-Cunha, & J. Varajão (Eds.), *E-business issues, challenges and opportunities for SMEs: Driving competitiveness* (pp. 20–29). Hershey, PA: Business Science Reference.

Takaki, O., Seino, T., Izumi, N., & Hasida, K. (2014). User-centered business process modeling and pattern-based development for large systems. In I. Management Association (Ed.), Software design and development: Concepts, methodologies, tools, and applications (pp. 1014-1035). Hershey, PA: Information Science Reference. doi: doi:10.4018/978-1-4666-4301-7.ch049

Talla, M., & Valverde, R. (2014). Component-based modeling for information systems reengineering. In I. Management Association (Ed.), Software design and development: Concepts, methodologies, tools, and applications (pp. 220-234). Hershey, PA: Information Science Reference. doi: doi:10.4018/978-1-4666-4301-7.ch012

Taylor, P. R. (2014). Enterprise architecture's identity crisis: New approaches to complexity for a maturing discipline. In P. Saha (Ed.), *A systemic perspective to managing complexity with enterprise architecture* (pp. 433–453). Hershey, PA: Business Science Reference.

Tella, A. (2012). Determinants of e-payment systems success: A user's satisfaction perspective. *International Journal of E-Adoption*, *4*(3), 15–38. doi:10.4018/jea.2012070102

Terjesen, A. (2010). Anonymity and trust: The ethical challenges of e-business transactions. In D. Palmer (Ed.), *Ethical issues in e-business: Models and frameworks* (pp. 40–57). Hershey, PA: Business Science Reference. doi:10.4018/978-1-61520-615-5.ch004

Theron, P. (2014). ICT resilience as dynamic process and cumulative aptitude. In I. Management Association (Ed.), Crisis management: Concepts, methodologies, tools and applications (pp. 1384-1420). Hershey, PA: Information Science Reference. doi: doi:10.4018/978-1-4666-4707-7.ch070

Tijsen, R., Spruit, M., van de Ridder, M., & van Raaij, B. (2011). BI-FIT: Aligning business intelligence end-users, tasks and technologies. In M. Cruz-Cunha, & J. Varajao (Eds.), *Enterprise information systems design, implementation and management: Organizational applications* (pp. 162–177). Hershey, PA: Information Science Reference.

Toka, A., Aivazidou, E., Antoniou, A., & Arvanitopoulos-Darginis, K. (2013). Cloud computing in supply chain management: An overview. In D. Graham, I. Manikas, & D. Folinas (Eds.), *E-logistics and e-supply chain management: Applications for evolving business* (pp. 218–231). Hershey, PA: Business Science Reference. doi:10.4018/978-1-4666-3914-0.ch012

Tomaiuolo, M. (2014). Trust management and delegation for the administration of web services. In I. Portela, & F. Almeida (Eds.), *Organizational, legal, and technological dimensions of information system administration* (pp. 18–37). Hershey, PA: Information Science Reference.

Tran, Q., Huang, D., & Zhang, C. (2013). An assessment method of the integrated e-commerce readiness for construction organizations in developing countries. *International Journal of E-Adoption*, *5*(1), 37–51. doi:10.4018/jea.2013010103

Tsadimas, A., Nikolaidou, M., & Anagnostopoulos, D. (2014). Model-based system design using SysML: The role of the evaluation diagram. In P. Fonseca i Casas (Ed.), *Formal languages for computer simulation: Transdisciplinary models and applications* (pp. 236–266). Hershey, PA: Information Science Reference.

Tung, H., Kung, H., Lawless, D. S., Sofge, D. A., & Lawless, W. F. (2011). Conservation of information and e-business success and challenges: A case study. In M. Cruz-Cunha, & J. Varajão (Eds.), *E-business managerial aspects, solutions and case studies* (pp. 254–269). Hershey, PA: Business Science Reference. doi:10.4018/978-1-60960-587-2.ch106

Unhelkar, B. (2011). *Handbook of research on green ICT: Technology, business and social perspectives*. Hershey, PA: IGI Global.

Unhelkar, B., Ghanbary, A., & Younessi, H. (2010). Collaborative business process engineering (CBPE) model. In B. Unhelkar, A. Ghanbary, & H. Younessi (Eds.), *Collaborative business process engineering and global organizations: Frameworks for service integration* (pp. 98–120). Hershey, PA: Business Science Reference.

Unhelkar, B., Ghanbary, A., & Younessi, H. (2010). Emerging technologies for business collaboration. In B. Unhelkar, A. Ghanbary, & H. Younessi (Eds.), *Collaborative business process engineering and global organizations: Frameworks for service integration* (pp. 37–64). Hershey, PA: Business Science Reference.

Unhelkar, B., Ghanbary, A., & Younessi, H. (2010). Fundamentals of collaborative business. In B. Unhelkar, A. Ghanbary, & H. Younessi (Eds.), *Collaborative business process engineering and global organizations: Frameworks for service integration* (pp. 1–36). Hershey, PA: Business Science Reference.

Van Huy, L., Rowe, F., Truex, D., & Huynh, M. Q. (2012). An empirical study of determinants of e-commerce adoption in SMEs in Vietnam: An economy in transition. *Journal of Global Information Management*, *20*(3), 23–54. doi:10.4018/jgim.2012070102

Vannoy, S. A. (2011). A structured content analytic assessment of business services advertisements in the cloud-based web services marketplace. *International Journal of Dependable and Trustworthy Information Systems*, *2*(1), 18–49. doi:10.4018/jdtis.2011010102

Vasconcelos, V., & Campos, P. (2012). The role of social networks in distributed informal information systems for innovation. In J. Varajão, M. Cruz-Cunha, & A. Trigo (Eds.), *Organizational integration of enterprise systems and resources: Advancements and applications* (pp. 60–75). Hershey, PA: Business Science Reference. doi:10.4018/978-1-4666-1764-3.ch004

Venkatraman, R., Venkatraman, S., & Asaithambi, S. P. (2013). A practical cloud services implementation framework for e-businesses. In K. Tarnay, S. Imre, & L. Xu (Eds.), *Research and development in e-business through service-oriented solutions* (pp. 167–198). Hershey, PA: Business Science Reference.

Verkasalo, H. (2011). Analysis of the forces reshaping the mobile internet business. In M. Bartolacci, & S. Powell (Eds.), *Interdisciplinary and multidimensional perspectives in telecommunications and networking: Emerging findings* (pp. 19–45). Hershey, PA: Information Science Reference. doi:10.4018/978-1-60960-505-6.ch003

Verma, A. (2013). Effects of phishing on e-commerce with special reference to India. In R. Khurana, & R. Aggarwal (Eds.), *Interdisciplinary perspectives on business convergence, computing, and legality* (pp. 186–197). Hershey, PA: Business Science Reference. doi:10.4018/978-1-4666-4209-6.ch017

Walker, B., & Posey, E. (2013). Digital El Paso: A public-private business model for community wireless networks. In A. Abdelaal (Ed.), *Social and economic effects of community wireless networks and infrastructures* (pp. 94–111). Hershey, PA: Information Science Reference. doi:10.4018/978-1-4666-2997-4.ch006

Wan, Y., Clegg, B., & Dey, P. K. (2013). A framework for enabling dynamic e-business strategies via new enterprise paradigms and ERP solutions. In I. Association (Ed.), *Enterprise resource planning: Concepts, methodologies, tools, and applications* (pp. 1561–1595). Hershey, PA: Business Science Reference. doi:10.4018/978-1-4666-4153-2.ch083

Wang, F., Lupton, N., Rawlinson, D., & Zhang, X. (2012). EBDMSS: A web-based decision making support system for strategic e-business management. In P. Zaraté (Ed.), *Integrated and strategic advancements in decision making support systems* (pp. 265–284). Hershey, PA: Information Science Reference. doi:10.4018/978-1-4666-1746-9.ch019

Wenyin, L., Liu, A., Li, Q., & Huang, L. (2011). Business models for insurance of business web services. In H. Leung, D. Chiu, & P. Hung (Eds.), *Service intelligence and service science: Evolutionary technologies and challenges* (pp. 261–272). Hershey, PA: Information Science Reference. doi:10.4018/978-1-60960-587-2.ch215

Wiedmann, K., Reeh, M., & Schumacher, H. (2010). Employment and acceptance of near field communication in mobile marketing. In K. Pousttchi, & D. Wiedemann (Eds.), *Handbook of research on mobile marketing management* (pp. 190–212). Hershey, PA: Business Science Reference.

Williams, J. G., & Premchaiswadi, W. (2011). On-line credit card payment processing and fraud prevention for e-business. In I. Management Association (Ed.), Global business: Concepts, methodologies, tools and applications (pp. 699-717). Hershey, PA: Business Science Reference. doi: doi:10.4018/978-1-60960-587-2.ch312

Wilms, A., & Andriole, S. J. (2010). Business technology strategy for a specialty chemicals company. *Journal of Information Technology Research, 3*(3), 11–18. doi:10.4018/jitr.2010070102

Winkler, U., & Gilani, W. (2012). Business continuity management of business driven IT landscapes. In S. Reiff-Marganiec, & M. Tilly (Eds.), *Handbook of research on service-oriented systems and non-functional properties: Future directions* (pp. 381–399). Hershey, PA: Information Science Reference.

Wollenberg, A. (2013). Optimizing international joint venture (IJV) ownership structures: A technology and knowledge transfer-linked productivity growth perspective. In B. Christiansen, E. Turkina, & N. Williams (Eds.), *Cultural and technological influences on global business* (pp. 142–164). Hershey, PA: Business Science Reference. doi:10.4018/978-1-4666-3966-9.ch009

Wood, A. M., Moultrie, J., & Eckert, C. (2010). Product form evolution. In A. Silva, & R. Simoes (Eds.), *Handbook of research on trends in product design and development: Technological and organizational perspectives* (pp. 499–512). Hershey, PA: Business Science Reference. doi:10.4018/978-1-61520-617-9.ch027

Wresch, W., & Fraser, S. (2011). Persistent barriers to e-commerce in developing countries: A longitudinal study of efforts by Caribbean companies. *Journal of Global Information Management, 19*(3), 30–44. doi:10.4018/jgim.2011070102

Wresch, W., & Fraser, S. (2013). Persistent barriers to e-commerce in developing countries: A longitudinal study of efforts by Caribbean companies. In F. Tan (Ed.), *Global diffusion and adoption of technologies for knowledge and information sharing* (pp. 205–220). Hershey, PA: Information Science Reference.

Xiao, X., Liu, Y., & Zhang, Z. (2012). The analysis of the logistics mode decision to e-commerce. *Journal of Electronic Commerce in Organizations, 10*(4), 57–70. doi:10.4018/jeco.2012100105

Xing, B., & Gao, W. (2014). Used product delivery optimization using agent-based modelling and simulation. In *Computational intelligence in remanufacturing* (pp. 113–133). Hershey, PA: Information Science Reference.

Xing, B., & Gao, W. (2014). Used product remanufacturability evaluation using fuzzy logic. In *Computational intelligence in remanufacturing* (pp. 75–94). Hershey, PA: Information Science Reference.

Xu, L. (2010). Outsourcing and multi-party business collaborations modeling. In K. St.Amant (Ed.), *IT outsourcing: Concepts, methodologies, tools, and applications* (pp. 558–577). Hershey, PA: Business Science Reference.

Xu, M., Rohatgi, R., & Duan, Y. (2010). Engaging SMEs in e-business: Insights from an empirical study. In I. Management Association (Ed.), Business information systems: Concepts, methodologies, tools and applications (pp. 115-134). Hershey, PA: Business Science Reference. doi: doi:10.4018/978-1-61520-969-9.ch009

Yermish, I., Miori, V., Yi, J., Malhotra, R., & Klimberg, R. (2010). Business plus intelligence plus technology equals business intelligence. *International Journal of Business Intelligence Research, 1*(1), 48–63. doi:10.4018/jbir.2010071704

Yeung, W. L. (2013). Specifying business-level protocols for web services based collaborative processes. In A. Loo (Ed.), *Distributed computing innovations for business, engineering, and science* (pp. 137–154). Hershey, PA: Information Science Reference.

Zarour, M., Abran, A., & Desharnais, J. (2014). Software process improvement for small and very small enterprises. In I. Management Association (Ed.), Software design and development: Concepts, methodologies, tools, and applications (pp. 1363-1384). Hershey, PA: Information Science Reference. doi: doi:10.4018/978-1-4666-4301-7.ch066

Zavala, L., Mendoza, B., & Huhns, M. N. (2014). A test-driven approach to behavioral queries for service selection. In I. Management Association (Ed.), Software design and development: Concepts, methodologies, tools, and applications (pp. 381-400). Hershey, PA: Information Science Reference. doi: doi:10.4018/978-1-4666-4301-7.ch020

Zerenler, M., & Gözlü, S. (2012). Issues influencing electronic commerce activities of SMEs: A study of the Turkish automotive supplier industry. In I. Management Association (Ed.), Human resources management: Concepts, methodologies, tools, and applications (pp. 1035-1055). Hershey, PA: Business Science Reference. doi: doi:10.4018/978-1-4666-1601-1.ch064

Compilation of References

(1999). *Management Scientist*. New York, NY: South-Western Educational Publishing.

(2003). *QM Windows Manual*. New York, NY: Prentice Hall.

(2003). *WinQSB Manual*. New York, NY: Wiley.

(2013). *Excel: Basic Solve, Frontline Solvers*. New York, NY: Author.

(2013). *GAMS: A User's Guide*. Create Space Independent Publishing Platform.

4. TS Corporation Oy. (2013). *4TS dSeal device - Ensure the quality and integrity of your deliveries*. Retrieved April 5, 2013, from www.4ts.com

ABB Annual Report 2010. (2011). Retrieved October, 14, 2011 from http://www400.abbext.com/2010/ar/servicepages/downloads/files/download.php?file=annual_report_2010.pdf&cat=b

ABB. (2005, April 4). *ABB pioneers use of radio frequency identification technology in material logistics in Finland*. ABB.

Abel, T. (1998). Complex adaptive systems, evolutionism, and ecology within anthropology: Interdisciplinary research for understanding cultural and ecological dynamics. *Georgia Journal of Ecological Anthropology*, *2*, 6–29. doi:10.5038/2162-4593.2.1.1

About Mobile Aspects. (2010). *Mobile aspects*. Retrieved February 4, 2010, from www.mobileaspects.com

Accenture. (2012). *Achieving high performance in BPO*. Retrieved from http://www.accenture.com/us-en/Pages/insight-high-performance-bpo-eight-practices.aspx

Adshead, A., & Thomas, D. (2003, July 1). Web-based EDI forces suppliers to change tactics. *Computer Weekly*.

Aggarwal, S. P., & Jaggi, C. K. (1995). Ordering policies of deteriorating items under permissible delay in payments. *The Journal of the Operational Research Society*, *46*(5), 658–662.

Agresti, A. (1990). *Categorical data analysis*. New York: John Wiley and Sons.

Akkermans, H. (2001). *Emergent supply networks: System dynamics simulation of adaptive supply agents*. Paper presented at the 34th Hawaii International Conference on System Sciences. Hawaii, HI.

Akrich, M. (1992). The description of technical objects. In J. Law (Ed.), *Shaping technology/building society* (pp. 205–224). Cambridge, MA: MIT Press.

Alchian, A. A., & Demsetz, H. (1972). Production, information costs, and economic organization. *The American Economic Review*, *62*(5), 777–795.

Alcoba, J. (2012). The paradox of service industrialization and the creation of meaning. *International Journal of Service Science, Management, Engineering, and Technology*, *3*(2), 50–62. doi:10.4018/jssmet.2012040104

Al-Turki, Y. A. Y. (2000). Multi-product batch scheduling for lead-acid battery assembly line. In *Proceedings of the 11th Annual Production and Operations Management Society (POMS) Conference*. San Antonio, TX: POMS.

Alvarez-Gil, M., Berrone, P., Husillos, F. J., & Lado, N. (2007). Reverse logistics, stakeholders' influence, organizational slack, and managers' posture. *Journal of Business Research*, *60*(5), 463–473. doi:10.1016/j.jbusres.2006.12.004

Amaeshi, K., Osuji, O., & Nnodim, P. (2008). Corporate social responsibility in supply chains of global brands: A boundaryless responsibility? Clarifications, exceptions and implications. *Journal of Business Ethics*, *81*(1), 223–234. doi:10.1007/s10551-007-9490-5

An Internet of Objects-the Future of RFID Technology. (2009). Retrieved November 23, 2009, from http://www.cellular-news.com/story/37900.php

Anand, N., & Peterson, R. A. (2000). When market information constitutes fields: Sensemaking of markets in the commercial music industry. *Organization Science*, *11*(3), 270–284. doi:10.1287/orsc.11.3.270.12502

Anbumozhi, V., & Kanda, Y. (2005). Greening the production and supply chains in Asia: Is there a role for voluntarily initiatives? *IGES Kansai Research Center Discussion Paper*, KRC-2005 (6). Retrieved from www.iges.or.jp

Andersen, M., & Skjoett-Larsen, T. (2009). Corporate social responsibility in global supply chains. *Supply Chain Management*, *14*(2), 75–86. doi:10.1108/13598540910941948

Anderson, D., Sweeney, D., Williams, T., & Martin, R. (2007). *An introduction to management science: Quantitative approaches to decision making*. New York, NY: South-Western College Publisher.

Anderson, P. (1999). Complexity theory and organization science. *Organization Science*, *10*(3), 216–232. doi:10.1287/orsc.10.3.216

Angeles, R. (2000). Revisiting the role of internet-EDI in the current electronic commerce scene. *Logistics Information Management*, *13*(1), 45–57. doi:10.1108/09576050010306396

Ang, S., & Straub, D. (1998). Production and transaction economies and is outsourcing: A study of the US bank industry. *Management Information Systems Quarterly*, *22*(4), 535–352. doi:10.2307/249554

Appleyard, M. M. (1996). How does knowledge flow? Interfirm patterns in the semiconductor industry. *Strategic Management Journal*, *17*, 137–154.

Arbnor, I., & Björn, B. (1997). *Methodology for creating business knowledge* (2nd ed.). Thousand Oaks, CA: Sage Publications.

Ariely, D., & Berns, G. S. (2012). Neuromarketing: the hope and hype of neuroimaging in business. *Nature Reviews. Neuroscience*, *11*(4), 284–292. doi:10.1038/nrn2795 PMID:20197790

Ariely, D., Kamenica, E., & Prelec, D. (2008). Man's search for meaning: The case of Legos. *Journal of Economic Behavior & Organization*, *67*, 671–677. doi:10.1016/j.jebo.2008.01.004

Armstrong, C. E., & Shimizu, K. (2007). A review of approaches to empirical research on the resource-based view of the firm. *Journal of Management*, *33*(6), 959–986. doi:10.1177/0149206307307645

Arsham, H. (2013). An interior boundary pivotal solution algorithm for linear programs with the optimal solution-based sensitivity region. *International Journal of Mathematics in Operational Research.*

Arsham, H. (1990). Perturbation analysis of general LP models: A unified approach to senstivity parametric, tolerance, and more-for-less analysis. *Mathematical and Computer Modelling*, *13*, 79–102. doi:10.1016/0895-7177(90)90073-V

Arsham, H. (1996). On the more-for-less paradoxical situations in linear programs: A parametric optimization approach. *Journal of Information & Optimization Sciences*, *17*, 485–500. doi:10.1080/02522667.1996.10699297

Arsham, H. (1997a). An artificial-free simplex algorithm for general LP models. *Mathematical and Computer Modelling*, *25*, 107–123. doi:10.1016/S0895-7177(96)00188-4

Arsham, H. (1997b). Initialzation of the simplex algorithm: An artifical-free approach. *SIAM Review*, *39*, 736–744. doi:10.1137/S0036144596304722

Arsham, H. (2006). A big-m free soltuion algorithm for general linear programs. *Internation Journal of Pure and Applied Mathematics*, *32*, 549–564.

Arsham, H. (2007). A computationally stable solution algorithm for large-scalelinear programs. *Applied Mathematics and Computation*, *188*, 1549–1561. doi:10.1016/j.amc.2006.11.031

Arsham, H. (2008). A validation and verification tool for optimization solvers. *Journal of Information & Optimization Sciences*, *29*, 57–80. doi:10.1080/02522667.2008.10699791

Arsham, H. (2009). A simplified algebraic method for system of linear inequalities with LP applications. *Omega: International Journal of Management Science*, *37*, 876–882. doi:10.1016/j.omega.2008.06.001

Arsham, H. (2012). Foundation of linear programming: A managerial perspective from solving system of inequalities to software implementation. *International Journal of Strategic Decision Sciences*, *3*(3), 40–60. doi:10.4018/jsds.2012070104

Asif, Z., & Mandiviwalla, M. (2005). Integrating the supply chain with RFID: A technical and business analysis. *Communications of the Association for Information Systems*, (15): 393–427.

Attaran, M. (2007). RFID: An enabler of supply chain operations. *Supply Chain Management: An International Journal*, *12*(4), 249–257. doi:10.1108/13598540710759763

Aubert, B., Dussault, S., Patry, M., & Rivard, S. (1999). *Managing the risk of it sourcing*. Paper presented at the 32nd Hawaii International Conference of System Sciences (HICSS). Hawaii, HI.

Ausubel, D. P., Novak, J. D., & Hanesian, H. (2000). *Psicología educativa: Un punto de vista cognoscitivo*. Mexico: Trillas.

Avgerou, C. (1996). Transferability of information technology and organisational practices. In M. Odedra-Straub (Ed.), *Global information technology and socio-economic development* (pp. 106–115). Nashua, NH: Ivy League.

Avgerou, C. (2001). The significance of context in information systems and organisational change. *Information Systems Journal*, *11*(1), 43–63. doi:10.1046/j.1365-2575.2001.00095.x

Avgerou, C. (2008). Information systems in developing countries: A critical research review. *Journal of Information Technology*, *23*(3), 133–146. doi:10.1057/palgrave.jit.2000136

Bae, J., & Lawler, J. J. (2000). Organizational performance and HRM strategies in Korea: Impact on firm performance in an emerging economy. *Academy of Management Journal*, *43*(3), 502–517. doi:10.2307/1556407

Bahli, B., & Rivard, S. (2003). The information technology outsourcing risk: A transaction cost and agency theory-based perspective. *Journal of Information Technology*, *18*(3), 211–221. doi:10.1080/0268396032000130214

Baker, W. E., & Sinkula, J. M. (1999). The synergistic effect of market orientation and learning orientation on organizational performance. *Journal of the Academy of Marketing Science*, *27*(4), 411–427. doi:10.1177/0092070399274002

Baldwing, L. P., Irani, Z., & Love, P. (2001). Outsourcing information systems: Drawing lessons from a banking case study. *European Journal of Information Systems*, *10*(1), 15–24. doi:10.1057/palgrave.ejis.3000372

Ballou, R. H. (2004). *Business logistics/supply chain management: planning, organizing, and controlling the supply chain* (5th ed.). Upper Saddle River, NJ: Prentice Hall.

Bandyopadhyay, S., & Pathak, P. (2007). Knowledge sharing and cooperation in outsourcing projects - A game theoretic analysis. *Decision Support Systems*, *43*(2), 349–358. doi:10.1016/j.dss.2006.10.006

Banks, J. S. C. II, & Nelson, B. L. (1999). *Discrete-event system simulation* (2nd ed.). Upper Saddle River, NJ: Prentice Hall.

Barkley, B. T. (2007). *Project management in new product development*. New York, NY: McGraw-Hill.

Barney, J. (1991). Firm resources and sustained competitive advantage. *Journal of Management*, *17*, 99–120. doi:10.1177/014920639101700108

Bartel, A. P. (1994). Productivity gains from the implementation of employee training programs. *Industrial Relations*, *33*(4), 411–425. doi:10.1111/j.1468-232X.1994.tb00349.x

Barthelemy, J., & Adsit, D. (2003). The seven deadly sins of outsourcing. *The Academy of Management Executive*, *17*(2), 87–100. doi:10.5465/AME.2003.10025203

Bartlett, C. A., & Ghoshal, S. (1989). *Managing across borders: The transnational solution*. Cambridge, MA: Harvard Business School Press.

Beatriz, A., Jabbour, C., & Jabbour, J. (2009). Are supplier selection criteria going green? Case studies of companies in Brazil. *Industrial Management & Data Systems*, *109*, 477–495. doi:10.1108/02635570910948623

Bechini, A., Cimino, M., Marcelloni, F., & Tomasi, A. (2008). Patterns and technologies for enabling supply chain traceability through collaborative information e-business. *Information and Software Technology*, *50*(4), 342–359. doi:10.1016/j.infsof.2007.02.017

Behery, M. (2009). Person/organization job-fitting and affective commitment to the organization: Perspectives from the UAE. *Cross Cultural Management: An International Journal*, *6*(2), 179–196. doi:10.1108/13527600910953928

Belassi, W., & Tukel, O. I. (1996). A new framework for determining critical success/failure factors in projects. *International Journal of Project Management*, *14*(3), 141–151. doi:10.1016/0263-7863(95)00064-X

Bellifemine, F., Caire, G., & Greenwood, D. (2007). *Developing multi-agent systems with JADE*. Hoboken, NJ: Wiley. doi:10.1002/9780470058411

Bendavid, Y., & Cassivi, L. (2010). Bridging the gap between RFID/EPC concepts, technological requirements and supply chain e-business processes. *Journal of Theoretical and Applied Electronic Commerce Research*, *5*(3), 1–16. doi:10.4067/S0718-18762010000300002

Bendoly, E., Bharadwaj, J., & Bharadwaj, S. (2012). Complementary drivers of new product development performance: Cross-functional coordination, information system capability, and intelligence quality. *Production and Operations Management*, *21*(4), 653–667. doi:10.1111/j.1937-5956.2011.01299.x

Bettis, R. A., & Prahalad, C. K. (1995). The dominant logic: Retrospective extension. *Strategic Management Journal*, *16*(1), 5–14. doi:10.1002/smj.4250160104

Bharadwaj, A. S. (2000). A resource-based perspective on information technology capability and firm performance: An empirical investigation. *Management Information Systems Quarterly*, *24*(1), 169–196. doi:10.2307/3250983

Bhatnagar, R., Sohal, A. S., & Millen, R. (1999). Third-party logistics services: A Singapore perspective. *International Journal of Physical Distribution and Logistics Management*, *26*(9), 569–587.

Bititci, U., Mendibill, K., Martinez, V., & Albores, P. (2005). Measuring and managing performance in extended enterprises. *International Journal of Operations & Production Management*, *25*(4), 333–353. doi:10.1108/01443570510585534

Bottani, E., Montanari, R., & Volpi, A. (2010). The impact of RFID and EPC network on the bullwhip effect in the Italian FMCG supply chain. *International Journal of Production Economics*, *124*(2), 426–432. doi:10.1016/j.ijpe.2009.12.005

Boudourides, M. A. (2001, November 1-3). *Networks, fluids, chaos*. Paper presented at the International Conference Spacing and Timing, Rethinking Globalization & Standardization. Palermo, Italy.

Boukachour, J., Fredouet, C.-H., & Gningue, M. B. (2011). Building an expert-system for maritime container security risk management. *International Journal of Applied Logistics*, *2*(1), 35–56. doi:10.4018/jal.2011010103

Boulus, N., & Bjorn, P. (2010). A cross-case analysis of technology-in-use practices: ERP-adaptation in Canada and Norway. *International Journal of Medical Informatics*, *79*(6), 97–108. doi:10.1016/j.ijmedinf.2008.06.008 PMID:20053582

Bourlakis, M., Maglaris, G., & Fotopoulos, C. (2012). Creating a best value supply chain? Empirical evidence from the Greek food chain. *International Journal of Logistics Management*, *23*(3), 360–382. doi:10.1108/09574091211289228

Bowen, F. E. (2002). Does size matter? Organizational slack and visibility as alternative explanations for environmental responsiveness. *Business & Society*, *41*(1), 118–124. doi:10.1177/0007650302041001007

Bowen, F. E., Cousins, P. D., Lamming, R. C., & Faruk, A. C. (2001). The role of supply management capabilities in green supply. *Production and Operations Management*, *10*(2), 174–189. doi:10.1111/j.1937-5956.2001.tb00077.x

Bowersox, D. J., Closs, D. J., & Stank, T. P. (1999). 21st century logistics: Making supply chain integration a reality. *Supply Chain Management Review*, *3*(3), 17–28.

Bowersox, D. J., & Daugherty, P. J. (1995). Logistics paradigms: the impact of information technology. *Journal of Business Logistics*, *16*(1), 65–80.

Bowersox, D. J., Daugherty, P. J., Dröge, C. L., Rogers, D. S., & Wardlow, D. L. (1989). *Leading edge logistics: Competitive positioning for the 1990s*. Oak Brook, IL: Council of Logistics Management.

Bowersox, D. J., Daugherty, P. J., Droge, C. L., Rogers, D. S., & Wardlow, D. L. (1989). *Leading-edge logistics: Competitive positioning for the 90s. Oak Brook*. Council of Logistics Management.

Boynton, A. C., Zmud, R. W., & Jacobs, G. C. (1994). The influence of IT management practice on IT use in large organizations. *Management Information Systems Quarterly*, *18*(3), 299–318. doi:10.2307/249620

Boztepe, S. (2007). User value: Competing theories and models. *International Journal of Design*, *1*(2), 55–63.

Branzei, O., & Vertinsky, I. (2003). Eco-sustainability orientation in China and Japan: Differences between proactive and reactive firms. In *Research in corporate sustainability*. Northampton, MA: Edward Elgar Publishing.

Braubach, L., Pokahr, A., & Lamersdorf, W. (2005). Extending the capability concept for flexible BDI agent modularization. In *Proceedings of the 3rd International Workshop on Programming Multiagent Systems* (LNCS), (vol. 3862, pp. 139-155). Berlin: Springer.

Bresnahan, T., Brynjolfsson, E., & Hitt, L. (2002). Information technology workplace organization, and the demand for skilled labor, firm level evidence. *The Quarterly Journal of Economics*, *117*(1), 339–376. doi:10.1162/003355302753399526

Brooks, P., & Hestnes, B. (2010, March/April). User measures of quality of experience: Why being objective and quantitative is important. *IEEE Network*, 8–13. doi:10.1109/MNET.2010.5430138

Brown, T. J., & Dacin. (1997). The company and the product: Corporate associations and consumer product responses. *Journal of Marketing*, *61*(1), 68–84. doi:10.2307/1252190

Bruneteau, F., Bourhis, O., Cornish, R., & Hallauer, T. (2010). *European location study*. Ptolemus Consulting Group. Retrieved April 5, 2013, from www.ptolemus.com

Bucklin, L. P., & Sengupta, S. (1993). Organizing successful co-marketing alliances. *Journal of Marketing*, *57*(2), 32–46. doi:10.2307/1252025

Budhwar, P. S., Luthar, H. K., & Bhatnagar, J. (2006). The dynamics of HRM systems in Indian BPO firms. *Journal of Labor Research*, *27*(3), 339–360. doi:10.1007/s12122-006-1027-7

Building a Case for Active RFID ROI: Cost-Benefit Analysis. (2006). Retrieved November 18, 2012, from http://active-rfid.com/download/Building_a_Case_for_Active_RFID.pdf

Bureau, U. S. C. (2005). *Survey of plant capacity*. Author.

Burges, D. (2013). *Cargo theft, loss prevention, and supply chain security*. Waltham, MA: Butterworth-Heinemann.

Burton-Jones, A., & Gallivan, M. J. (2007). Toward a deeper understanding of system usage in organizations: A multilevel perspective. *Management Information Systems Quarterly*, *31*(4), 657–679.

Buysse, K., & Verbeke, A. (2003). Proactive environmental strategies: A stakeholder management perspective. *Strategic Management Journal*, *24*(5), 453–470. doi:10.1002/smj.299

Calderia, M., & Ward, J. (2003). Using resource-based theory to interpret the successful adoption and use of information systems and technology in manufacturing small and medium-sized enterprises. *European Journal of Information Systems*, *12*(2), 127–141. doi:10.1057/palgrave.ejis.3000454

Callon, M. (1986). Some elements of a sociology of translation: Domestication of the scallops and the fishermen of St Brieuc Bay. In J. Law (Ed.), *Power, action and belief: A new sociology of knowledge*. London, UK: Routledge.

Campbell, B., Kay, R., & Avison, D. (2005). Strategic alignment: A practitioner's perspective. *Journal of Enterprise Information Management*, *18*(6), 653–664. doi:10.1108/17410390510628364

Capaldo, A. (2007). Network structure and innovation: the leveraging of a dual network as a distinctive relational capability. *Strategic Management Journal*, *28*(6), 585. doi:10.1002/smj.621

Carlton, J. (2006, June 29). EU's environmental hurdles for electronics, rules to require mitigation of toxic materials that are common in many products. *Wall Street Journal*, p. B5.

Carr, N. (2003). IT doesn't matter. *Harvard Business Review*, *81*(5), 41–49. PMID:12747161

Carter, C. R., Ellram, & Ready, K. (1998). Environmental purchasing: Benchmarking our German counterparts. *Journal of Supply Chain Management*, *34*(4), 28–38.

Carter, C., Ellram, & Tate, W. (2007). The use of social network analysis in logistics research. *Journal of Business Logistics*, *28*(1), 137–170. doi:10.1002/j.2158-1592.2007.tb00235.x

Carter, C., & Dresner, M. (2001). Purchasing's role in environmental: Cross-functional development of grounded theory. *Journal of Supply Chain Management*, *37*, 12–27. doi:10.1111/j.1745-493X.2001.tb00102.x

Carter, C., & Jennings, M. (2004). The role of purchasing in corporate social responsibility: A structural equation analysis. *Journal of Business Logistics*, *25*(1), 145–186. doi:10.1002/j.2158-1592.2004.tb00173.x

Carter, C., Kale, R., & Grimm, C. (2000). Environmental purchasing and firm performance: An empirical investigation. *Transportation Research Part E, Logistics and Transportation Review*, *36*(3), 219–228. doi:10.1016/S1366-5545(99)00034-4

Carter, C., & Rogers, D. (2008). A framework of sustainable supply chain management: Moving toward new theory. *International Journal of Physical Distribution & Logistics Management*, *38*(5), 360–387. doi:10.1108/09600030810882816

Cavalieri, S., & Terzi, S. (2006). Proposal of a performance measurement system for the evaluation of scheduling solutions. *International Journal of Manufacturing Technology and Management*, *8*(1-3), 248–263. doi:10.1504/IJMTM.2006.008803

César, P., Rodrigues, C., & Oliveira, O. J. (2007). Intermediate stock management in the production process for automotive batteries: A case study. In *Proceedings of the 18th Annual POMS Conference*. Dallas, TX: POMS.

Chappell, D. (2011). *The benefits and risks of cloud platforms: A guide for business leaders*. David Chappell & Associates.

Chen, Z., & Dubinsky, A. J. (2003). A conceptual model of perceived customer value in e-commerce: A preliminary investigation. *Psychology and Marketing*, *20*(4), 323–347. doi:10.1002/mar.10076

Chesbrough, H. W., & Teece, D. J. (1996). When is virtual virtuous? Organizing for innovation. *Harvard Business Review*, *74*(1), 65–73.

Chiou, J. S. (2004). The antecedents of customer loyalty toward Internet service providers. *Information & Management*, *41*(6), 685–695. doi:10.1016/j.im.2003.08.006

Chiva-Gomez, R. (2003). The facilitating factors for organizational learning: Bringing ideas from complex adaptive systems. *Knowledge and Process Management*, *10*(2), 99–114. doi:10.1002/kpm.168

Choi, T. Y., Dooley, K. J., & Rungtusanatham, M. (2001). Supply networks and complex adaptive systems: Control versus emergence. *Journal of Operations Management*, *19*, 351–366. doi:10.1016/S0272-6963(00)00068-1

Chopra, S. & Meindl, P. (2001). Supply chain management: Strategy, planning and operation. *Prentice Hall,* USA.

Chopra, S., & Meindl, P. (2001). *Supply chain management: Strategy, planning, and operations*. Upper Saddle River, NJ: Prentice-Hall, Inc.

Chopra, S., & Meindl, P. (2012). *Supply chain management: Strategy, planning, & operations* (5th ed.). Upper Saddle River, NJ: Pearson.

Chow, W. S. (2004). An exploratory study of the success factors for extranet adoption in e-supply chain. *Journal of Global Information Management*, *12*(1), 60–67. doi:10.4018/jgim.2004010104

Chu, T. H., & Robey, D. (2008). Explaining changes in learning and work practice following the adoption of online learning: A human agency perspective. *European Journal of Information Systems*, *17*(1), 79–98. doi:10.1057/palgrave.ejis.3000731

Ciborra, C. U. (2004). Digital technologies and the duality of risk. London, UK: CARR, LSE.

CIPS. (2006). *Supply chain vulnerability*. Retrieved March 29, 2013 from www.cips.org

Clarck, T. D., Zmud, R. W., & McCray, G. E. (1995). The outsourcing of information services: Transforming the nature of business in the information industry. *Journal of Information Technology*, *10*(4), 221–237. doi:10.1057/jit.1995.26

Clarkson, M. (1995). A stakeholder framework for analyzing and evaluating corporate social performance. *Academy of Management Review*, *20*(1), 92–117.

Claver, E., Gonzales, R., Gasco, J., & Llopis, J. (2002). Information systems outsourcing: Reasons, reservations and success factors. *Logistics Information Management*, *15*(4), 294–308. doi:10.1108/09576050210436138

Cleland, D. I., Bidanda, B., & Chung, C. A. (1995). Human issues in technology integration – part 1. *Industrial Management (Des Plaines)*, *37*(Issue 4), 22.

Clemens, B., & Douglas, T. J. (2006). Does coercion drive firms to adopt 'voluntary' green initiatives? Relationships among coercion, superior firm resources, and voluntary green initiatives. *Journal of Business Research*, *59*, 483–491. doi:10.1016/j.jbusres.2005.09.016

Coase, R. H. (1937). The nature of the firm. *Economica*, *4*(16), 386–405. doi:10.1111/j.1468-0335.1937.tb00002.x

Code of Federal Regulations (CFR). (n.d.). *Title 14, Aeronautics and Space; Part 121; Operating Requirement: Domestic Flight and Supplemental Operations*. (This is a regulation that is periodically revised.)

Collier, J. E., & Bienstock, C. C. (2006). Measuring service quality in e-retailing. *Journal of Service Research*, *8*(3), 260–275. doi:10.1177/1094670505278867

Collins, J. (2005). School studies RFID's effect on Wal-Mart. *RFID Journal*. Retrieved November 22, 2009, from http://www.rfidjournal.com/article/articleview/1514/1/9/

Collins, J. (2006). RFID implementation is an art. *RFID Journal*. Retrieved November 18, 2009, from http://www.rfidjournal.com/article/articleview/2427/1/1/

Collins, C. M., Steg, & Koning. (2007). Customers' values, beliefs on sustainable corporate performance, and buying behavior. *Psychology and Marketing*, *24*(6), 555–577. doi:10.1002/mar.20173

Collis, D. (1994). Research note: How valuable are organizational capabilities. *Strategic Management Journal*, *15*(8), 143–152. doi:10.1002/smj.4250150910

Conner, K. R. (1991). A historical comparison of resource-based theory and five schools of thought within industrial organization economics: Do we have a new theory of the firm? *Journal of Management*, *17*(1), 121–154. doi:10.1177/014920639101700109

Contractor, F. J., & Mudambi, S. M. (2008). The influence of human capital investment on the exports of services and goods: An analysis of the top 25 services outsourcing countries. *Management International Review*, *48*(4), 433–445. doi:10.1007/s11575-008-0025-9

Cooke-Davies, T. J. (2001). *Towards improved project management practice*. (Unpublished PhD Thesis). Leeds Metropolitan University, Leeds, UK.

Cooke-Davies, T. (2002). The 'real' success factors on projects. *International Journal of Project Management*, *20*(3), 185–190. doi:10.1016/S0263-7863(01)00067-9

Covert, R. P., & Philip, G. C. (1973). An EOQ model for items with weibull distribution deterioration. *AIIE Transactions*, *5*, 323–336. doi:10.1080/05695557308974918

Cowles, D. L., Kiecker, P., & Little, M. W. (2002). Using key informant insights as a foundation for e-retailing theory development. *Journal of Business Research*, *55*(8), 629–636. doi:10.1016/S0148-2963(00)00203-4

Craddock, R. J., & Stansfield, E. V. (2005). Sensor fusion for smart containers. In *Proceedings of Signal Processing Solutions for Homeland Security*. IEE. doi:10.1049/ic:20050272

Crosby, P. (2012c). *Predictive indicators of success in science & engineering projects - Application to the SKA initiative* (Unpublished PhD thesis). Curtin University, Canberra, Australia.

Crosby, P. (2012a). Key success drivers: Meta-study findings applicable to large high-technology projects. *International Journal of Information Technology Project Management*, *3*(2), 1–20. doi:10.4018/jitpm.2012040101

Crosby, P. (2012b). Building resilience in large high-technology projects: Front end conditioning for success. *International Journal of Information Technology Project Management, 3*(4), 21–40. doi:10.4018/jitpm.2012100102

Crosby, P., & Boonstra, A.-J. (2011). *LOFAR project – Post project and lessons learned review (internal report).* ASTRON.

Csikszentmihalyi, M., & Rochberg-Halton, E. (1981). *The meaning of things: Domestic symbols and the self.* New York: Cambridge University Press. doi:10.1017/CBO9781139167611

Currall, S. C., & Inkpen, A. C. (2002). A multilevel approach to trust in joint ventures. *Journal of International Business Studies, 33*, 479–495. doi:10.1057/palgrave.jibs.8491027

Daily, B., & Bishop, J. (2003). TQM workforce factors and employee involvement: The pivotal role of teamwork. *Journal of Managerial Issues, 15*, 393–417.

Daily, B., & Huang, S. (2001). Achieving sustainability through attention to human resource factors in environmental management. *International Journal of Operations & Production Management, 21*(12), 1539–1552. doi:10.1108/01443570110410892

Damali, N., Jolley, G., & Handfield, R. (2008). Environmental management systems and green supply chain management: Complements for sustainability? *Business Strategy and the Environment, 17*(1), 30–45. doi:10.1002/bse.557

Daniel, S. E., Pappis, C. P., & Voutsinas, T. G. (2003). Applying life cycle inventory to reverse supply chains: A case study of lead recovery from batteries. *Resources, Conservation and Recycling, 37*(4), 251–281. doi:10.1016/S0921-3449(02)00070-8

Danneels, E. (2003). Tight-Loose coupling with customers: The enactment of customer orientation. *Strategic Management Journal, 24*(6), 559–576. doi:10.1002/smj.319

Darnall, N., & Edwards, D., Jr. (2004). *Predicting the cost of environmental system adoption.* Paper presented at Academy of Management National Meetings. New Orleans, LA.

Daugherty, P. J., Ellinger, A. E., & Gustin, C. M. (1996). Integrated logistics: Achieving logistics performance improvements. *Supply Chain Management, 1*(3), 25–33. doi:10.1108/13598549610155297

Davenport, T., & Brooks, J. (2004). Enterprise systems and the supply chain. *Journal of International Management, 17*(1), 8–19.

Davern, M. J., & Kauffman, R. J. (2000). Discovering potential and realizing value from information technology. *Journal of Management Information Systems, 16*(4), 121–143.

Dave, U., & Patel, L. K. (1981). Policy inventory model for deteriorating items with time proportional demand. *The Journal of the Operational Research Society, 32*(2), 137–142.

David, R. J., & Han, S.-K. (2004). A systematic assessment of the empirical support for transaction cost economics. *Strategic Management Journal, 25*, 39–58. doi:10.1002/smj.359

Davis, E. W., & Spekman, R. E. (2004). *The extended enterprise.* USA: Prentice Hall.

Davis, F. D. (1989). Perceived usefulness, perceived ease of use and user acceptance of information technology. *Management Information Systems Quarterly, 13*(3), 319–340. doi:10.2307/249008

Davis, J. P., Bingham, C. B., & Eisenhardt, K. M. (2007). Developing theory through simulation methods. *Academy of Management Review, 32*(2), 480–499. doi:10.5465/AMR.2007.24351453

Davydenko, I., Zhang, M., & Tavasszy, L. (2012). Assessment of SMART-CM platform benefits as the result of short transport time. In *Proceedings of e-Freight 2012e-Freight, International Conference on Paperless Freight Transport Logistics.* Delft, The Netherlands: e-Freight.

De Guinea, A. O., & Markus, M. L. (2009). Why break the habit of a lifetime? Rethinking the roles of intention, habit, and emotion in continuing information technology use. *Management Information Systems Quarterly, 33*(3), 433–444.

de Laet, M., & Mol, A. (2000). The Zimbabwe bush pump: Mechanics of a fluid technology. *Social Studies of Science, 30*(2), 225–263. doi:10.1177/030631200030002002

de Wit, A. (1988). Measurement of project success. *International Journal of Project Management, 6*(3), 164–170. doi:10.1016/0263-7863(88)90043-9

Dean, T., & Brown, R. (1995). Pollution regulation as a barrier to new firm entry: Initial evidence and implications for future research. *Academy of Management Journal, 38*, 288–303. doi:10.2307/256737

Debrito, M., Carbone, V., & Blanquart, C. (2008). Towards a sustainable fashion retail supply chain in Europe: Organisation and performance. *International Journal of Production Economics, 114*(2), 534–553. doi:10.1016/j.ijpe.2007.06.012

Deitrich, M. (2012). *Transaction cost economics and beyond: Toward a new economics of the firm*. London: Routledge.

Deloitte Consulting, L. L. P. (2007).Why Settle for Less? Deloitte Consulting 2008 Outsourcing Report http://www.deloitte.com/assets/Dcom-Estonia/Local%20Assets/Documents/ce_cons_outs_report_08(1).pdf

Deloitte Consulting, L. L. P. (2012). *2012* Global Outsourcing and Insourcing Survey Executive Summary. http://www.deloitte.com/assets/Dcom-UnitedStates/Local%20Assets/Documents/IMOs/Shared%20Services/us_sdt 2012GlobalOutsourcingandInsourcingSurveyExecutiveSummary_050112.pdf

Deloitte. (2003). *Chips with everything*. Retrieved March 2, 2010, from http://www.rfid4u.nl/downloads/deloitte-chipwitheverything.pdf

DeSanctis, G., & Poole, M. S. (1994). Capturing the complexity in advanced technology use: Adaptive structuration theory. *Organization Science, 5*(2), 121–147. doi:10.1287/orsc.5.2.121

Desarbo, W. S., Di Benedetto, C., Song, M., & Sinha, I. (2005). Revisiting the miles and snow strategic framework: Uncovering interrelationships between strategic types, capabilities, environmental uncertainty, and firm performance. *Strategic Management Journal, 26*(1), 47–74. doi:10.1002/smj.431

Deshpande, R., John, U. F., & Webster, F. E. (1993). Corporate culture, customer orientation, and innovativeness in Japanese firms: A quadratic analysis. *Journal of Marketing, 57*, 23–27. doi:10.2307/1252055

Devaraj, S., Fan, M., & Kohli, R. (2002). Antecedents of B2C channel satisfaction and preference: Validating e-commerce metrics. *Information Systems Research, 13*(3), 316–333. doi:10.1287/isre.13.3.316.77

Diller, S., Shedroff, N., & Rhea, D. (2008). *Making meaning: How successful businesses deliver meaningful customer experiences*. Berkeley, CA: New Riders.

DoE. (2011). Cost estimating guide. DOE G 413.3-21, Current Directives 400 series. Washington, DC: U.S. Department of Energy.

Dooley, K. J. (1997). A complex adaptive systems model of organization change. *Nonlinear Dynamics Psychology and Life Sciences, 1*(1), 69–97. doi:10.1023/A:1022375910940

Doonan, J., Lanoie, P., & Laplante, B. (2005). Analysis determinants of environmental performance in the Canadian pulp & paper industry: an assessment from inside the industry. *Ecological Economics, 55*, 73–84. doi:10.1016/j.ecolecon.2004.10.017

Downes, M., Thomas, A. S., & Singley, R. B. (2001). Predicting expatriate job satisfaction: The role of firm internationalization. *Career Development International, 7*(1), 24–36. doi:10.1108/13620430210414847

Droge, C., Vickery, S. K., & Jacobs, M. A. (2012). Does supply chain integration mediate the relationships between product/process strategy and service performance? An empirical study. *International Journal of Production Economics, 137*(2), 250–262. doi:10.1016/j.ijpe.2012.02.005

Drumwright, M. (1994). Socially responsible organizational buying: Environmental concern as a noneconomic buying criterion. *Journal of Marketing, 58*(3), 1–19. doi:10.2307/1252307

Duncan, R. B. (1972). Characteristics of organizational environments and perceived environmental uncertainty. *Administrative Science Quarterly, 17*(3), 313–327. doi:10.2307/2392145

Dunn, O. J. (1964). Multiple comparisons using rank sums. *Technometrics*, *6*, 241–252. doi:10.1080/00401706.1964.10490181

Dunstall, S., & Mills, G. (2002). Scheduling the charging of batteries. In *Proceedings of the Mathematics in Industry Study Group*. Canberra, Australia: University of South Australia.

Dunstall, S., & Mills, G. (2007). Robustness of cyclic schedules for the charging of batteries. *ANZIAM Journal*, *48*(4), 475–492. doi:10.1017/S1446181100003163

Dutta, A., Lee, H. L., & Whang, S. (2007). RFID and operations management: Technology, value and incentives. *Production and Operations Management*, *16*(5), 646–655. doi:10.1111/j.1937-5956.2007.tb00286.x

Dvir, D., Lipovetsky, S., Shemhar, A., & Tishler, A. (1998). In search of project classification: A non-universal approach to project success factors. *Research Policy*, *27*, 915–935. doi:10.1016/S0048-7333(98)00085-7

Dvir, D., Raz, T., & Shenhar, A. J. (2003). An empirical analysis of the relationship between project planning and project success. *International Journal of Project Management*, *21*(2), 89–95. doi:10.1016/S0263-7863(02)00012-1

Dyer, J. H. (2000). *Collaborative advantage*. Oxford University Press.

Dyer, J. H., & Nobeoka, K. (2002). Creating and managing a high-performance knowledge-sharing network: The Toyota case. *Strategic Management Journal*, *21*(3), 345–367. doi:10.1002/(SICI)1097-0266(200003)21:3<345::AID-SMJ96>3.0.CO;2-N

Dyer, J. H., & Singh, H. (1998). The relational view: Cooperative strategy and sources of interorganizational competitive advantage. *Academy of Management Review*, *23*(4), 660–679.

Economist (2011). *Slicing an apple*. Retrieved on June 24, 2013, from http://www.economist.com/blogs/dailychart/2011/08/apple-and-samsungs-symbiotic-relationship

Edwards, P., Peters, M., & Sharman, G. (2001). The effectiveness of information systems in supporting the extended supply chain. *Journal of Business Review*, *22*(1), 1–27.

Ehrgott, M., Reimann, F., Kaufmann, L., & Carter, C. (2011). Social sustainability in selecting emerging economy suppliers. *Journal of Business Ethics*, *98*, 99–119. doi:10.1007/s10551-010-0537-7

Eisenhardt, K., & Tabrizi, B. N. (1995). Accelerating adaptive processes: Product innovation in the global computer industry. *Administrative Science Quarterly*, *40*(1). doi:10.2307/2393701

Ekwall, D. (2009). The displacement effect in cargo theft. *International Journal of Physical Distribution & Logistics Management*, *39*(1), 47–62. doi:10.1108/09600030910929183

Elbanna, S. (2010). Strategic planning in United Arab Emirates. *International Journal of Commerce and Management*, *20*(1), 26–40. doi:10.1108/10569211011025934

Elimam, A. A., & Udayabhanu, V. (1999). Production planning for lead-acid batteries. In *Proceedings of IIE Industrial Engineering Research'99 Conference*. Phoenix, AZ: IIE.

Elimam, A. A., & Udayabhanu, V. (2001). Multi-level supply chain in lead-acid battery manufacturing. In *Proceedings of the First International Conference on Logistics and Supply Chain Management*. Coimbatore, India: IEEE.

Elimam, A. A., & Sartawi, M. M. (1986). Quality-control practices in lead-acid battery manufacturing to improve quality, cost, reliability. *IEEE Transactions on Reliability*, *35*(4), 369–374. doi:10.1109/TR.1986.4335471

Elimam, A. A., & Udayabhanu, V. (2011). Optimal and heuristic production planning in battery manufacturing. *California Journal of Operations Management*, *9*(1), 1–18.

Elmuti, D. (2003). The Perceived Impact of Outsourcing on Organizational Performance. *American Journal of Business*, *18*(Issue2), 33–42. doi:10.1108/19355181200300010

ElTayeb, T., Zailani, S., & Jayaraman, K. (2010). The examination on the drivers for green purchasing adoption among EMS 14001 certified companies in Malaysia. *Journal of Manufacturing Technology Management*, *21*(2), 206–225. doi:10.1108/17410381011014378

Enlow, S., & Ertel, D. (2006). Achieving outsourcing success: Effective relationship management. *Benefits Review*, *38*(3), 50. doi:10.1177/0886368706288216

EPC Information Services (EPCIS) Specification, EPC-Global Ratified Standard. (2007). Retrieved from http://www.gs1.org/gsmp/kc/epcglobal/epcis/epcis_1_0_1-standard-20070921.pdf

Epstein, M. (2008). *Making sustainability work: Best practices in managing and measuring social and environmental impacts*. Sheffield, UK: Greenleaf.

Erlich, P. R., & Erlich, A. H. (1991). *The population explosion*. New York: Touchstone.

Ernst & Whinney. (1987). *Corporate profitability & logistics*. USA: Council of Logistics Management.

Ernst, D. (2011, May 11). China's innovation policy is a wake-up call for America. *Asia Pacific Issues*, *100*, 2.

Espinosa, J., DeLone, W., & Lee, G. (2006). Global boundaries, task processes and IS project success: A field study. *Information Technology & People*, *19*(4), 345–370. doi:10.1108/09593840610718036

Estrin, S., Baghdasaryan, D., & Meyer, K. M. (2009). The impact of institutional and human resource distance on international entry strategies. *Journal of Management Studies*, *46*(7), 1171–1196. doi:10.1111/j.1467-6486.2009.00838.x

Ethiraj, S. K., & Levinthal, D. (2004). Modularlity and innovation in complex systems. *Management Science*, *50*(2), 159–173. doi:10.1287/mnsc.1030.0145

Europol. (2009). *Cargo theft report*. Europol. Retrieved April 5, 2013, from www.europol.europa.eu

Evans, P., & Wolf, B. (2005). Collaboration rules.[July –August]. *Harvard Business Review*, *83*(Issue 7/8), 96. PMID:16028820

Evaristo, R., Audy, J. L. N., Prikladnicki, R., & Avritchir, J. (2005). Wholly owned offshore subsidiaries for IT development: A program of research. In *Proceedings of the 38th Hawaii International Conference on System Sciences*. IEEE.

Ewell, C. D., & Chidester, T. (1996). American Airlines Converts CRM in Favor of Human Factors and Safety Training. *The Flightdeck, July/August*. Flight Department, American Airlines, DFW Airport.

Fabbe-Costes, N., & Jahre, M. (2008). Supply chain integration and performance: A review of the evidence. *International Journal of Logistics Management*, *19*(2), 130–154. doi:10.1108/09574090810895933

Faisal, M. (2010). Sustainable supply chains: A study of interaction among the enablers. *Business Process Management Journal*, *16*(3), 508–529. doi:10.1108/14637151011049476

Fawcett, S. E., & Magnan, G. M. (2005). Achieving world-class supply chain alignment: Benefits, barriers, and bridges. Center for Advanced Purchasing Studies, http://.www.capsresearch.org.

Federal Aviation Administration (1991). *AC 120-40B: Airplane Simulator Qualification* (7/29/91).

Federal Aviation Administration (2004). *AC 120-35C: Line Operational Simulations: Line Oriented Flight Training, Special Purpose Operational Training, Line Operational Evaluation* (09/27/04).

Federal Aviation Administration (2004). *AC 120-51E: Crew Resource Management* (01/22/04).

Federal Aviation Administration AC (1992). *120-45A: Airplane Flight Training Device Qualification* (2/5/92).

Feeny, D. F., & Willcocks, L. P. (1998). Core IS capabilities for exploiting information technology. *Sloan Management Review*, *39*(3), 9–21.

Feldman, M. S., & Orlikowski, W. J. (2011). Theorizing practice and practicing theory. *Organization Science*, *22*, 1240–1253. doi:10.1287/orsc.1100.0612

Fellows, J. D., & Alexander, J. K. (Eds.). (2010). *Decadal science strategy surveys: Report of a workshop*. Retrieved from http://www.nap.edu/catalog/11894.html

Fensel, D., & Bussler, C. (2002). The web service modelling framework. *Electronic Commerce Research and Applications*, *1*(2), 113–137. doi:10.1016/S1567-4223(02)00015-7

Fiksel, J. (2010). Evaluating supply chain sustainability. *Chemical Engineering Progress*, *106*(5), 28.

Fisher, J. R. (2010). *Large instrument development for radio astronomy: Astro2010 technology development white paper.* Observatory Technical Council, National Radio Astronomy Observatory.

Fisher, C. E., Chin, L., & Klitzman, R. (2010). Defining neuromarketing: Practices and professional challenges. *Harvard Review of Psychiatry, 18*(4), 230–237. doi:10.3109/10673229.2010.496623 PMID:20597593

Flauttau, J. (2011). Regaining control. *Aviation Week & Space Technology*, (38): 87–88.

Flin, R., Martin, L., et al. (2003, 2005). Development of the NOTECHS (non-technical skills) System for Assessing Pilots' CRM Skills. In Contemporary Issues in Human Factors and Aviation Safety (Harris, Don and Muir, Helen C., (Eds). (pp.133-154). Burlington, VT: Ashgate Publishing Co

Florida, R. (1996). Lean and green: The move to environmentally conscious manufacturing. *California Management Review, 39*, 80–105. doi:10.2307/41165877

Flyvbjerg, B. (2006). *Five misunderstandings about case-study research.* Sage Publications. Retrieved from http://qix.sagepub.com/content/12/2/219

Flyvbjerg, B. (2008). *What is a mega-project?* Retrieved from http://flyvbjerg.plan.aau.dk/whatisamegaproject.php

Fokianos, K., & Kedem, B. (2003). Regression theory for categorical time series. *Statistical Science, 18*(3), 357–376. doi:10.1214/ss/1076102425

Ford, R. M., & Coulston, C. (2008). *Design for electrical and computer engineers: Theory, concepts, and practice.* New York, NY: McGraw-Hill.

Forrester (2005). APAC study chain apps spending outlook, http://www.forrester.com.

Forrester, J. W. (1958). Industrial dynamics: A major breakthrough for decision makers. *Harvard Business Review*, 37–66.

Forrester, J. W. (1961). *Industrial dynamics.* Cambridge, MA: MIT Press.

Främling, K., Ala-Risku, T., Kärkkäinen, M., & Holmström, J. (2007). Design patterns for managing product life cycle information. *Communications of the ACM, 50*(6), 75–79. doi:10.1145/1247001.1247009

Främling, K., Parmar, S., Hinkka, V., Tätilä, J., & Rodgers, D. (2013). Assessment of EPCIS standard for interoperable tracking in the supply chain. In *Service orientation in holonic and multi agent manufacturing and robotics.* Berlin: Springer. doi:10.1007/978-3-642-35852-4_8

Freeman, E. (1984). *Strategic management: A stakeholder approach.* Boston: Pitman.

FreightWatch International. (2012). *FreightWatch international global threat assessment.* FreightWatch International. Retrieved April 5, 2013, from www.freightwatchintl.com

FreightWatch Security Net. (n.d.). *Geo F2 'pill bottle' tracker.* Retrieved April 5, 2013, from www.freightsecurity.net

Friedman, T. L. (2005). *The world is flat: A brief history of the twenty first century.* New York, NY: Farrar, Straus & Giroux.

Frohlich, M. T., & Westbrook, R. (2001). Arcs of integration: An international study of supply chain strategies. *Journal of Operations Management, 19*(2), 185–200. doi:10.1016/S0272-6963(00)00055-3

Frost & Sullivan. (2012). *Analysis of the real-time location systems (RTLS) market government and healthcare verticals drive growth.* NAE4-11.

Frost & Sullivan. (2012). *U.S. cloud user survey: Evolving usage of public, private, and hybrid clouds.* Author.

Frost, P. C. (1999). Developments in lead–acid batteries: A lead producer's perspective. *Journal of Power Sources, 78*(1–2), 256–266. doi:10.1016/S0378-7753(99)00042-7

Fukuyama, F. (1996). *Trust: The social virtues and the creation of prosperity.* New York Free Press.

Fung, V. K., Fung, W. K., & Wind, Y. (2007). *Competing in a flat world.* Wharton School Publishing.

Furneaux, B., & Wade, M. (2011). An exploration of organizational level information systems discontinuance intentions. *Management Information Systems Quarterly, 35*(3), 573–598.

Garicano, L., & Kaplan, S. N. (2001). The effects of business-to-business e-commerce on transaction costs. *The Journal of Industrial Economics, 49*(4), 463. doi:10.1111/1467-6451.00158

Garriga, E., & Mele. (2004). Corporate social responsibility theories: Mapping the territory. *Journal of Business Ethics, 53*(1/2), 51–71. doi:10.1023/B:BUSI.0000039399.90587.34

Gartner Webinars. (2012). *Integrated systems: High-tech Tuesday webinar: Cloud or fog: Clearing the air around cloud services.* Author.

Gartner. (2012a). *High-tech tuesday webinar: The AIM market: Beyond SOA and toward cloud.* Author.

Gartner. (2012b). *5 things private clouds are not!* Retrieved April 2, 2013, from http://www.businesscloud9.com/content/best-practice-5-things-private-clouds-are-not-according-gartner/11855

Gaski, R. M., & Etzel, M. J. (1986). The index of consumer sentiment toward marketing. *Journal of Marketing, 50*, 71–81. doi:10.2307/1251586

Gaukler, G. M., Özer, O., & Hausman, W. H. (2008). Order progress information: improved dynamic emergency ordering policies. *Production and Operations Management, 17*(6), 599–614. doi:10.3401/poms.1080.0066

Geffen, C., & Rothenberg, S. (2000). Suppliers and environmental innovation: The automotive paint process. *International Journal of Operations & Production Management, 20*, 166–186. doi:10.1108/01443570010304242

Gentner, D. (2003). Why we're so smart. In D. Gentner, & S. Goldin-Meadow (Eds.), *Language in mind: Advances in the study of language and thought.* Cambridge, MA: MIT Press.

Georgius, & Dibbern. (2005). Vendor strategies in the German market for information technology and business process outsourcing. In R. Hirschheim, A. Heinzl, & J. Dibbern (Eds.), *Information system outsourcing* (2nd Ed.), (pp. 381-386). New York: Springer-Verlag.

Gerbing, D. W., & Anderson, J. C. (1988). An updated paradigm for scale development incorporating unidimensionality and its assessment. *JMR, Journal of Marketing Research, 25*(2), 186–192. doi:10.2307/3172650

Ghare, P. M., & Schrader, G. F. (1963). A model for exponential decaying inventory. *Journal of Industrial Engineering, 14*(3), 238–243.

Ghijsen, P., Semeijn, J., & Wang, A. (2009). Modern 3PL services in China: The role of trust. *Journal of International Business and Economy, 10*(2), 103–135.

Ghosh, B. (2011). Cross-cultural knowledge management practices to support offshore outsourcing. In M. Al-Shammari (Ed.), *Knowledge management in emerging economies: Social, organizational and cultural implementation* (pp. 249–260). Denver, CO: Metropolitan State College of Denver. doi:10.4018/978-1-60960-783-8.ch623

Giddens, A. (1984). *The constitution of society: Outline of the theory of structure.* Berkeley, CA: University of California Press.

Gilmore, J. H., & Pine. (2007). *Authenticity: What consumers really want.* Cambridge, MA: Harvard Business School Press.

Gligor, D. M., & Holcomb, M. C. (2012). Understanding the role of logistics capabilities in achieving supply chain agility: A systematic literature review. *Supply Chain Management: An International Journal, 17*(4), 438–453. doi:10.1108/13598541211246594

Go'ncz, E., Skirke, U., Kleizen, H., & Barber, M. (2007). Increasing the rate of sustainable change: A call for a redefinition of the concept and the model for its implementation. *Journal of Cleaner Production, 15*(6), 525–537. doi:10.1016/j.jclepro.2006.05.018

Goel, A. (2010). The value of in-transit visibility for supply chains with multiple modes of transport. *International Journal of logistics: research and Applications, 13*(6), 475-492.

Goldman, S. L., Nagel, R. N., & Preiss, K. (1995). *Agile competitors and virtual organizations.* Van Nostrand Reinhold Publishing.

Goldstone, R. L., & Sakamoto, Y. (2003). The transfer of abstract principles governing complex adaptive systems. *Cognitive Psychology*, *46*(4), 414–466. doi:10.1016/S0010-0285(02)00519-4 PMID:12809681

Goles, T. (2001). *The impact of the client/vendor relationship on outsourcing success*. (Ph D dissertation). Department of Decision and Information Sciences, University of Houston, Houston, TX.

Gonçalves, A., & Figueiredo, J. (2009). Organizing competences: Actor-network theory in virtual settings. *International Journal of Networking and Virtual Organisations*, *6*(1), 23–35. doi:10.1504/IJNVO.2009.022481

Gonçalves, A., & Figueiredo, J. (2010a). How to recognize an immutable mobile when you find one: Translations on innovation and design. *International Journal of Actor-Network Theory and Technological Innovation*, *2*(2), 39–53. doi:10.4018/jantti.2010040103

Gonza'lez-Benito, J., & Gonza'lez-Benito. (2005). An analysis of the relationship between environmental motivations and ISO14001 certification. *British Journal of Management*, *16*(2), 133–148. doi:10.1111/j.1467-8551.2005.00436.x

Gonza'lez-Benito, J., & Gonza'lez-Benito. (2008). Operations management practices linked to the adoption of ISO 14001: An empirical analysis of Spanish manufacturers. *International Journal of Production Economics*, *113*(1), 60–73. doi:10.1016/j.ijpe.2007.02.051

Gonza'lez-Benito, J., & Gonza'lez-Benito. (2010). A study of determinant factors of stakeholder environmental pressure perceived by industrial companies. *Business Strategy and the Environment*, *19*(3), 164–181.

Gonzalez, R., Gasco, J., & Llopis, J. (2005). Information systems outsourcing: Reasons in the largest Spanish firms. *International Journal of Information Management*, *25*(2), 117–136. doi:10.1016/j.ijinfomgt.2004.10.002

Gopal, A., Sivaramakrishnan, K., Krishnan, M. S., & Mukhopadhyay, T. (2003). Contracts in offshore software development: An empirical analysis. *Information Technology & People*, *49*(12), 1671–1683.

Goranson, H. T. (1999). *The agile virtual enterprise*. Quorum Books.

Government Accountability Office (GAO). (1997). Human Factors: FAA Guidance and Oversight of Pilot Crew Resource Management training Can Be Improved. In GAO/RCED-98-7. GAO, Washington DC.

Govindu, R., & Chinnam, R. (2007). MASCF: A generic process-centered methodological framework for analysis and design of multi-agent supply chain systems. *Computers & Industrial Engineering*, *53*(4), 584–609. doi:10.1016/j.cie.2007.06.003

Gowen, C. R. III, & Tallon, W. J. (2003). Enhancing Supply Chain Practices through Human Resource Management. *Journal of Management Development*, *22*(Issue 1-2), 32. doi:10.1108/02621710310454842

Goyal, S. K. (1985). Economic order quantity under conditions of permissible delay in payments. *The Journal of the Operational Research Society*, *36*(3), 335–338.

Goyal, S. K., & Giri, B. C. (2001). Recent trends in modeling of deteriorating inventory. *European Journal of Operational Research*, *134*(1), 1–16. doi:10.1016/S0377-2217(00)00248-4

Granger, C. W. J., & Newbold, P. (1977). *Forecasting economic time series*. New York: Academic Press.

Grant, R. M. (1991). The resource-base theory of competitive advantage: Implications for strategy formulation. *California Management Review*, *33*(3), 114–135. doi:10.2307/41166664

Grant, R. M., & Baden-Fuller, C. (2004). A knowledge accessing theory of strategic alliances. *Journal of Management Studies*, *41*(1), 61–84. doi:10.1111/j.1467-6486.2004.00421.x

Gravier, M. J., & Farris, M. T. (2012). Integrative vs. disintegrative forces in the evolution of supply chain networks. *International Journal of Knowledge-Based Organizations*, *2*(3), 1–32. doi:10.4018/ijkbo.2012070101

Gravier, M. J., & Kelly, B. (2011). Measuring the cost of complexity in supply chains: Comparison of weighted entropy and the bullwhip effect index. In *Modelling value*. Berlin: Springer. doi:10.1007/978-3-7908-2747-7_13

Gravier, M. J., Randall, W. S., & Strutton, D. (2008). Investigating the role of knowledge in alliance performance. *Journal of Knowledge Management*, *12*(4), 117–130. doi:10.1108/13673270810884291

Gray, B. (1989). Negotiations: Arenas for reconstructing meaning. Working paper, *Pennsylvania State University, Center for Research in Conflict and Negotiation*, USA.

Gray, E. R., & Smeltzer. (1989). *Management: The competitive edge*. London: Collier MacMillan Publishers.

Green, H., & Khermouch, G. (2005). Bar codes better watch their backs. *BusinessWeek*. Retrieved December 8, 2012, from http://www.businessweek.com/magazine/content/03_28/b3841063.htm

Grenier, R., & Metes, G. (1995). *Going virtual*. USA: Prentice Hall.

Greve, H. R., Mitsuhashi, H., & Baum, J. A. C. (2012). Greener pastures: Outside options and strategic alliance withdrawal. *Organization Science*, *24*(1), 79–98. doi:10.1287/orsc.1110.0733

Grover, V., Cheon, M. J., & Teng, J. T. C. (1996). The effect of service quality and partnership on the outsourcing of information systems functions. *Journal of Management Information Systems*, *12*(4), 89–116.

Guide, P. M. B. O. K. (2008). A guide to the project management body of knowledge. The Project Management Institute, 4.

Guide, V., & VanWassenhove, L. N. (2001). Managing product returns for remanufacturing. *Production and Operations Management*, *10*(2), 142–155. doi:10.1111/j.1937-5956.2001.tb00075.x

Gunasekaran, A., & Ngai, E. W. T. (2004). Information systems in supply chain integration and management. *European Journal of Operational Research*, *159*(2), 269–295. doi:10.1016/j.ejor.2003.08.016

Gunasekaran, A., Patel, C., Ronald, E., & McGaughey, R. (2004). A framework for supply chain performance measurement. *International Journal of Production Economics*, *87*(3), 333–348. doi:10.1016/j.ijpe.2003.08.003

Gupta, S., & Palsule-Desai, O. (2011). Sustainable supply chain management: review and research opportunities. *IIMB Management Review*, *23*, 234–245. doi:10.1016/j.iimb.2011.09.002

Gustin, C. M., Daugherty, P. J., & Stank, T. P. (1995). The effects of information availability on logistics integration. *Journal of Business Logistics*, *16*(1), 1–21.

Haakansson, H., & Ford, D. (2002). How companies interact in business networks? *Journal of Business Research*, *55*(2), 133–139. doi:10.1016/S0148-2963(00)00148-X

Hackman, J. R. (1995). Teams, Leaders, Organizations. In Crew Resource Management; Weiner, E; Kanki, B; Helmreich, R (Eds) (p.52). Academic Press (now part of Elsevier), Maryland Heights, MO.

Hackman, M. (2004). *RFID could hack retail fraud*. Retrieved February 4, 2012, from http://www.eweek.com/article2/0,1759,1628696,00.asp

Hagel, J. (2002). Edging into web services. *The McKinsey Quarterly*, *4*, 29–38.

Hair, J. F. Jr, Black, W. C., Babin, B. J., Anderson, R. E., & Tatham, R. L. (2006). *Multivariate data analysis* (6th ed.). Upper Saddle River, NJ: Pearson Prentice Hall.

Halldórsson, A., Kotzab, H., Mikkola, J. H., & Skjøtt-Larsen, T. (2007). Complementary theories to supply chain management. *Supply Chain Management: An International Journal*, *12*(4), 284–296. doi:10.1108/13598540710759808

Hall, J. (2000). Environmental supply chain dynamics. *Journal of Cleaner Production*, *8*(6), 455–471. doi:10.1016/S0959-6526(00)00013-5

Hall, M., & Purchase, D. (2006). Building or bodging? Attitudes to sustainability in UK public sector housing construction development. *Sustainable Development*, *14*, 205–218. doi:10.1002/sd.265

Hamel, G. (1991). Competition for competence and interpartner learning within international strategic alliances. *Strategic Management Journal*, *12*(S1), 83–103. doi:10.1002/smj.4250120908

Handfield, R., Wallon, S., Seegers, L., & Melnyk, S. (1997). Green' Value Chain Practices in the Furniture Industry. *Journal of Operations Management*, *15*, 293–315. doi:10.1016/S0272-6963(97)00004-1

Hanfield, Robert B., & Nichols, Jr., Ernst L. (2002). Supply chain redesign, *Prentice Hall*, USA.

Han, S. H., Lee, J. N., & Seo, Y. W. (2008). Analyzing the impact of a firm's capability on outsourcing success: A process perspective. *Information & Management*, *45*, 31–42. doi:10.1016/j.im.2007.09.004

Hanseth, O., & Braa, K. (2000). Who's in control: Designers, managers, or technology? Infrastructure at north hydro. In C. U. Ciborra (Ed.), *From control to drift, the dynamics of corporate information infrastructures*. Oxford, UK: Oxford University Press.

Harrell, F. E. (2001). *Regression modeling strategies: With applications to linear models, logistic regression, and survival analysis*. New York: Springer-Verlag.

Harrington, H. James & Harrington, James S. (1995). Total quality management, McGraw-Hill, USA.

Harrington, H. James, Hefner, Mark B., & Cox, C. Keith (1995). Environmental change plans: Best practices for improvement planning and implementation. Harrington and Harrington's Total Improvement Management, McGraw-Hill, USA.

Harrington, S. J., & Guimaraes, T. (2005). Corporate culture, absorptive capacity and IT success. *Information and Organization*, *15*(1), 39–63. doi:10.1016/j.infoandorg.2004.10.002

Harris, E. (2009). *Strategic project risk appraisal and management*. London: Gower Publishing Limited.

Harrop, P. (2010). *HF RFID-the great leap forward*. Retrieved February 4, 2012, from http://www.rfidwizards.com.html

Harrop, P., Das, R., & Holland, G. (2009). RFID in healthcare: 2009-2019. *IDTechEx.com*. Retrieved February 4, 2010, from http://www.idtechex.com/research/reports/rfid_for_healthcare_and_pharmaceuticals_2009_2019_000146.asp

Hart, S. (1995). A natural-resource-based view of the firm. *Academy of Management Review*, *20*(4), 986–1014.

Hart, S., & Milstein, M. (2003). Creating sustainable value. *The Academy of Management Executive*, *17*(2), 56–69. doi:10.5465/AME.2003.10025194

Hatch, N. W., & Dyer, J. H. (2004). Human capital and learning as a source of sustainable competitive advantage. *Strategic Management Journal*, *25*(12), 1155–1178. doi:10.1002/smj.421

Hawkins, J., & Blakeslee, S. (2004). *On intelligence*. New York: Times Books. Henry Holt and Company.

Heath, J. (2006). Business ethics without stakeholders. *Business Ethics Quarterly*, *16*(3), 533–557. doi:10.5840/beq200616448

Heide, J. B. (1994). Interorganizational governance in marketing channels. *Journal of Marketing*, *58*(1), 71–85. doi:10.2307/1252252

Heimer, C. A. (1985). Allocating information costs in a negotiated information order: Interorganizational constraints on decision making in Norwegian oil insurance. *Administrative Science Quarterly*, *30*(3), 395–417. doi:10.2307/2392671

Heizer, J., & Render, B. (2008). *Principles of operations management* (7th ed.). Upper Saddle River, NJ: Prentice Hall.

He, J., & Wang, H. (2009). Innovative knowledge assetsand economic performance: The asymmetric roles of incentive and monitoring. *Academy of Management Journal*, *15*, 919–938. doi:10.5465/AMJ.2009.44633414

Helmreich, R. L., & Foushee, H. C. (1995). Why CRM? In Crew Resource Management; Weiner, E; Kanki, B; Helmreich, R (Eds) (p.38). Academic Press (now part of Elsevier)1995, Maryland Heights, MO.

Helmreich, R. L., Chidester, T. R., & Foushee, H. Clayton; Gregorich, Steve; Wilhelm, J.A.(1990). How Effective is Cockpit Resource Management Training? in Flight Safety Foundation's Flight Safety Digest, May 1990, 1-16, Alexandria, VA

Henderson, J. C., & Venkatraman, N. (1991). Understanding strategic alignment. *Business Quarterly*, *55*(3), 72–79.

Henderson, R., & Cockburn, I. (1994). Measuring competence? Exploring firm effects in pharmaceutical research. *Strategic Management Journal, 15*(S1), 63–84. doi:10.1002/smj.4250150906

Heng, K. J., Labban, J., & Linn, R. J. (1991). An order-level lot-size inventory model for deteriorating items with finite replenishment rate. *Computers & Industrial Engineering, 20*(2), 187–197. doi:10.1016/0360-8352(91)90024-Z

Henningsson, S., & Hedman, J. (2010). Industry-wide supply chain information integration: The lack of management and disjoint economic responsibilities. *International Journal of Information Systems and Supply Chain Management, 3*(1), 1–20.

Hinkka, V. (2012). Challenges for building RFID tracking systems across the whole supply chain. *International Journal for RF Technologies: Research and Applications, 3*(3), 201–218.

Hinkka, V., Holmström, J., & Främling, K. (2012). RFID tracking in the book supply chain: The transition from postponed to speculative tagging. *International Journal of Logistics: Research and Applications, 15*(3), 194–214. doi:10.1080/13675567.2012.709834

Hobbs, J. E. (1996). A transaction cost approach to supply chain management. *Supply Chain Management, 1*(2), 15–27. doi:10.1108/13598549610155260

Hoexter, M. (2006). *The concept of sustainability: Internal diversity and points of conflicts*. Retrieved from http://greenthoughts.us

Hofman, D., & Aronow, S. (2012). *The Gartner supply chain top 25 for 2012*. Stamford, CT: Gartner, Inc.

Holland, J. H. (1995). *Hidden order: How adaptation builds complexity*. Reading, MA: Addison-Wesley Publishing Company.

Holland, J. H., & Miller, J. H. (1991). Artificial adaptive agents in economic theory. *The American Economic Review, 81*(2), 365–370.

Hollier, R. H., & Mak, K. L. (1983). Inventory replenishment policies for deteriorating items in a declining market. *International Journal of Production Research, 21*(4), 813–826. doi:10.1080/00207548308942414

Holme, P., Kim, B. J., Yoon, C. N., & Han, S. K. (2002). Attack vulnerability of complex networks. *Physical Review E: Statistical, Nonlinear, and Soft Matter Physics, 65*(5), 1–15. doi:10.1103/PhysRevE.65.056109

Hoskisson, R. E., Eden, L., Lau, C. M., & Wright, M. (2000). Strategy in emerging economies. *Academy of Management Journal, 43*(3), 249–267. doi:10.2307/1556394

Hsiao, R. (2007). Misaligned market: The importance of industry context in technology-mediated exchanges. *Journal of Global Information Management, 15*(3), 69–87. doi:10.4018/jgim.2007070104

Hsu, C., & Hu, A. (2008). Green supply chain management in the electronic industry. *International Journal of Environmental Science and Technology, 5*(2), 205–216. doi:10.1007/BF03326014

Hsu, P., Kraemer, K. L., & Dunkle, D. (2006). Determinants of e-business use in U.S. firms. *International Journal of Electronic Commerce, 10*(4), 9–45. doi:10.2753/JEC1086-4415100401

Huan, S. H., Sheoran, S. K., & Wang, G. (2004). A review and analysis of supply chain operations reference (SCOR) model. *Supply Chain Management -. International Journal (Toronto, Ont.), 9*(1), 23–29.

Hu, G., Wang, L., Fetch, S., & Bidanda, B. (2008). A multi-objective model for project portfolio selection to implement lean and Six Sigma concepts. *International Journal of Production Research, 46*(23), 6611–6648. doi:10.1080/00207540802230363

Hunt, D., & At-Twaiijri, M. (1996). Values and the Saudi managers: An empirical investigation. *Journal of Management Development, 15*(5), 48–56. doi:10.1108/02621719610117259

Hunt, S. D., & Morgan, R. M. (1995). The comparative advantage theory of competition. *Journal of Marketing, 59*(2), 1–15. doi:10.2307/1252069

Hunt, S., & Davis, D. (2008). Grounding supply chain management in resource-advantage theory. *Journal of Supply Chain Management, 44*(1), 10–21. doi:10.1111/j.1745-493X.2008.00042.x

Huo, B. (2012). The impact of supply chain integration on company performance: An organizational capability perspective. *Supply Chain Management: An International Journal*, *17*(6), 596–610. doi:10.1108/13598541211269210

Hurst, I., & Hanessian, B. G. (1995). Navigating IT channels: Integrate or outsource? *The McKinsey Quarterly*, *3*, 103–110.

Hussain, Z., Wallace, J., & Cornelius, N. E. (2007). The use and impact of human resource information systems on human resource management professionals. *Information & Management*, *44*(1), 74–89. doi:10.1016/j.im.2006.10.006

Huston, T. W. (1996). *Production system design and cycle time reduction in a fuel cell manufacturing operation.* (Unpublished Master Thesis). MIT, Cambridge, MA.

ICCPM. (2011). *Complex project management: Global perspectives and the strategic agenda to 2025 (task force report).* International Centre for Complex Project Management.

Ichniowski, C., & Shaw, K. (1999). The effects of human resource management systems on economic performance: An international comparison of US, and Japanese plants. *Management Science*, *45*(5), 704–721. doi:10.1287/mnsc.45.5.704

IDC. (2012a). *Cloud transition series: Cloud definitions and market opportunities.* Author.

IDC. (2012b). *Converging the datacenter infrastructure: Why, how, so what?* Author.

Ika, L. A. (2009). Project success as a topic in project management journals. *Project Management Journal*, *40*(4), 6–19. doi:10.1002/pmj.20137

Ilachinski, A. (2000). Irreducible semi-autonomous adaptive combat (ISAAC), an artificial-life approach to land combat. *Military Operations Research*, *5*(3), 29–46. doi:10.5711/morj.5.3.29

Im, S., & Workman, J. P. (2004). Market orientation, creativity, and new product performance in high-technology firms. *Journal of Marketing*, *68*(2), 114–132. doi:10.1509/jmkg.68.2.114.27788

Inkpen, A. C., & Currall, S. C. (2004). The coevolution of trust, control, and learning in joint ventures. *Organization Science*, *15*(5), 586–599. doi:10.1287/orsc.1040.0079

Inkpen, A. C., & Tsang, E. W. K. (2005). Social capital, networks, and knowledge transfer. *Journal of the Academy of Marketing Science*, *30*(1), 146–165.

Intelliflex. (2013), *The true cause of $17 billion cold chain shrink.* Retrieved March 29, 2013, from www.intelliflex.com

Introna, L. D. (1991). The impact of information technology on logistics. *International Journal of Physical Distribution and Logistics Management*, *21*(5), 32–37. doi:10.1108/EUM0000000000387

Ives, B., Jarvenpaa, S. L., & Mason, R. O. (1993). Global business drivers: Aligning information technology to global business strategy. *IBM Systems Journal*, *32*(1), 143–161. doi:10.1147/sj.321.0143

Jabbour, A., Jabbour, C., & Govindan, K. et al. (2013). Factors affecting the adoption of green supply chain management practices in Brazil: Empirical evidence. *The International Journal of Environmental Studies.* PMID:24179250

Jackson, M. C. (2011). *Systems thinking: Creative holism for managers.* London: John Wiley & Sons.

Jaggi, C. K., Goel, S. K., & Mittal, M. (2011a). Economic order quantity model for deteriorating items with imperfect quality and permissible delay on payments. *International Journal of Industrial Engineering Computations*, *2*(2), 237–248. doi:10.5267/j.ijiec.2010.07.003

Jaggi, C. K., Goel, S. K., & Mittal, M. (2013). Credit financing in economic ordering policies for imperfect quality items with allowable shortages. *Applied Mathematics and Computation*, *219*(10), 5268–5282. doi:10.1016/j.amc.2012.11.027

Jaggi, C. K., Khanna, A., & Mittal, M. (2011b). Credit financing for deteriorating imperfect-quality items under inflationary conditions. *International Journal of Services Operations and Informatics*, *6*(4), 292–309. doi:10.1504/IJSOI.2011.045560

Jaggi, C. K., & Mittal, M. (2011). Economic order quantity model for deteriorating items with imperfect quality. *Revista Invetigacion Operacional*, *32*(2), 107–113.

Jaggi, C. K., Mittal, M., & Khanna, A. (2012). Effects of inspection on retailer's ordering policy for deteriorating items with time-dependent demand under inflationary conditions. *International Journal of Systems Science*, 1–9.

Jain, K., Manson, N., & Sankhe, S. (2005, February). The right passage to India. *The McKinsey Quarterly*. Retrieved March 2, 2006, from http://www.mckinseyquarterly.com/article_page.aspx?ar=1581&L2=16&L3=17

Jain, R. (2004). Quality of experience. *IEEE MultiMedia*, *11*(1), 95–96. doi:10.1109/MMUL.2004.1261114

Jain, V., Benyoucef, L., & Deshmukh, S. G. (2008). What's the buzz about moving from 'lean' to 'agile' integrated supply chains? A fuzzy intelligent agent-based approach. *International Journal of Production Research*, *46*(23), 6649–6678. doi:10.1080/00207540802230462

Jamal, A. M. M., Sarker, B. R., & Wang, S. (1997). An ordering policy for deteriorating items with allowable shortage and permissible delay in payments. *The Journal of the Operational Research Society*, *48*(8), 826–833.

Jani, A. (2010). Escalation of commitment in troubled IT projects: Influence of project risk factors and self-efficacy on the perception of risk and the commitment to a failing project. *International Journal of Project Management*, *29*(7), 934–945. doi:10.1016/j.ijproman.2010.08.004

Jarillo, P. (1988). On strategic networks. *Strategic Management Journal*, *9*(1), 31–41. doi:10.1002/smj.4250090104

Jarvenpaa, S. L., & Ives, B. (1991). Applications of global information technology: Key issues for management. *Management Information Systems Quarterly*, *15*(1), 33–50. doi:10.2307/249433

Jasperson, J. S., Carter, P. E., & Zmud, R. W. (2005). A comprehensive conceptualization of post-adoptive behaviors associated with information technology enabled work systems. *Management Information Systems Quarterly*, *29*(3), 525–558.

Jayaram, J., & Pathak, S. (2013). A holistic view of knowledge integration in collaborative supply chains. *International Journal of Production Research*, *51*(7), 1958–1972. doi:10.1080/00207543.2012.700130

Jayaratne, P. (2011). Sustainable supply and supply chain mapping - Sri Lankan tea supply chain. *HDR Student Conference, Paper 21*. University of Wollongong, NZ. Retrieved from http://ro.uow.edu.au/sbshdr/2011/papers/21

Jeyaraj, A., Rottman, J., & Lacity, M. (2006). A review of the predictors, linkages, and biases in IT innovation adoption research. *Journal of Information Technology*, *21*, 1–23. doi:10.1057/palgrave.jit.2000056

Jiang, B. (2009). The effects of interorganizational governance on supplier's compliance with SCC: An empirical examination of compliant and non-compliant suppliers. *Journal of Operations Management*, *27*(4), 267–280. doi:10.1016/j.jom.2008.09.005

Johanson, J., & Mattsson, L. G. (1987). Inter-organizational relations in industrial systems: A network approach compared with the transaction cost approach. *Inter-Organizational Studies of Management and Organization*, *17*(1), 34–48.

Johansson, O., & Pålsson, H. (2009). The impact of auto-ID on logistics performance: A benchmarking survey of Swedish manufacturing industries. *Benchmarking: An International Journal*, *16*(4), 504–522. doi:10.1108/14635770910972432

Johnson, P.R., Leenders, & Fearon. (2006). Supply's growing status and influence: A sixteen-year perspective. *The Journal of Supply Chain Management*, *13*, 33–43. doi:10.1111/j.1745-493X.2006.00010.x

Jones, P., Clarke-Hill, C., Comfort, D., Hillier, D., & Shears, P. (2005). Radio frequency identification and food retailing in the UK. *British Food Journal*, *107*(6), 350–360. doi:10.1108/00070700510602156

Jonsson, P., & Mattsson, S.-A. (2012). The value of sharing information in supply chains. *International Journal of Physical Distribution and Logistics Management*, *43*(4), 282–299.

Joshi, A. W., & Campbell, A. J. (2003). Effect of environmental dynamism on relational governance in manufacturer-supplier relationships: A contingency framework and an empirical test. *Journal of the Academy of Marketing Science, 31*(2), 176–188. doi:10.1177/0092070302250901

Jugdev, K., & Müller, R. A. (2005). Retrospective look at our evolving understanding of project success. *Project Management Journal, 36*(4), 19–31.

Kahn, K. B. (1996). Interdepartmental integration: A definition with implications for product development performance. *Journal of Product Innovation Management, 13*(2), 137–151. doi:10.1016/0737-6782(95)00110-7

Kalakota, R., & Whinston, A. B. (1996). *Frontiers of electronic commerce*. Reading, MA: Addison-Wesley.

Kamhawi, E. M. (2008). System characteristics, perceived benefits, individual differences and use intentions: A survey of decision support tools of ERP systems. *Information Resources Management Journal, 21*(4), 66–83. doi:10.4018/irmj.2008100104

Kannan, G., Sasikumar, P., & Devika, K. (2010). A genetic algorithm approach for solving a closed loop supply chain model: A case of battery recycling. *Applied Mathematical Modelling, 34*(3), 655–670. doi:10.1016/j.apm.2009.06.021

Kapoor, G., Zhou, W., & Piramuthu, S. (2009). Challenges associated with RFID tag implementations in supply chains. *European Journal of Information Systems, 18*(6), 526–533. doi:10.1057/ejis.2009.41

Karahanna, E., Straub, D. W., & Chervany, N. L. (1999). Information technology adoption across time: A cross-sectional comparison of pre-adoption and post-adoption beliefs. *Management Information Systems Quarterly, 23*(2), 183–204. doi:10.2307/249751

Kärkkäinen, M., Ala-Risku, T., & Främling, K. (2004). Efficient tracking for short-term multi-company networks. *International Journal of Physical Distribution & Logistics Management, 34*(7), 545–564. doi:10.1108/09600030410552249

Kathuria, R., Anandarajan, M., & Igbaria, M. (1999). Linking IT applications with manufacturing strategy: An intelligent decision support system approach. *Decision Sciences, 30*(4), 959–992. doi:10.1111/j.1540-5915.1999.tb00915.x

Katou, A., & Budhwar, P. (2006). The effect of human resource management systems on organizational performance: Test of a mediating model. *International Journal of Human Resource Management, 17*(7), 1223–1253. doi:10.1080/09585190600756525

Katz, J. P., & Townsend, J. B. (2000). The role of information technology in the fit between culture, business strategy and organizational structure of global firms. *Journal of Global Information Management, 8*(2), 24–35. doi:10.4018/jgim.2000040102

Kavcˇicˇ, K., & Tavcˇar, M. I. (2008). Planning successful partnership in the process of outsourcing. *Kybernetes, 37*(2), 241–249. doi:10.1108/03684920810851140

Kay, M. (2007). Closing the RFID knowledge gap in apparel and footwear. *RFID Journal*. Retrieved November 18, 2012, from http://www.rfidjournal.com/article/articleview/3267/1/82/

Kearney, A. T. (2005). *RFID will bring great benefits for retailers*. Retrieved November 24, 2009, from http://retailindustry.about.com/cs/it_rfid/a/bl_atk111003.htm

Kearns, G. S., & Lederer, A. L. (2000). The effect of strategic alignment on the use of IS-based resources for competitive advantage. *The Journal of Strategic Information Systems, 9*(4), 265–293. doi:10.1016/S0963-8687(00)00049-4

Kelly, S., Gibson, N., Holland, C. P., & Light, B. (1999). A business perspective of legacy information systems. *Communications of the Association for Information Systems, 2*(7), 1–27.

Kennedy, F. A., & Widener, S. K. (2008). A control framework: Insights from evidence on lean accounting. *Management Accounting Research, 19*(4), 301–319. doi:10.1016/j.mar.2008.01.001

Kern, T., & Blois, K. (2002). Norm development in outsourcing relationships. *Journal of Information Technology, 17*(1), 33–42. doi:10.1080/02683960210137174

Khadem, M., & Ali, A. (2008). Modeling and simulation for car battery manufacturing for cost effectiveness. In *Proceeding of the Intelligent Systems and Control Symposia: Computational Biology and Bioinform*. Orlando, FL: IEEE.

Khandwalla, P. (1977). *The design of organizations*. New York: Harcourt Brace Jovanovich.

Khan, S. U., Niazi, M., & Rashid, A. (2011a). Factors influencing clients in the selection of offshore software outsourcing vendors: An exploratory study using a systematic literature review. *Journal of Systems and Software*, *84*(4), 686–699. doi:10.1016/j.jss.2010.12.010

Kim, C., Yang, K. H., & Kim, J. (2008). A strategy for third-party logistics systems: A case analysis using the blue ocean strategy. *Omega*, *36*(4), 522–534. doi:10.1016/j.omega.2006.11.011

Kim, L., & Min, H. (2011). Measuring supply chain efficiency from a green perspective. *Management Research Review*, *34*(11), 1169–1189. doi:10.1108/01409171111178738

Kim, S. S., & Malhotra, N. K. (2005). A longitudinal model of continued IS use: An integrative view of four mechanisms underlying postadoption phenomena. *Management Science*, *51*(5), 741–755. doi:10.1287/mnsc.1040.0326

Kim, S. W. (2003). *An investigation of information technology investments on buyer-supplier relationship and supply chain dynamics*. East Lansing, MI: Michigan State University.

Kim, W. C., & Mauborgne, R. (2005). Blue ocean strategy: From theory to practice. *California Management Review*, *47*(3), 105–121. doi:10.2307/41166308

King, A., & Lenox, M. (2001). Lean and green? An empirical examination of the relationship between lean production and environmental performance. *Production and Operations Management*, *10*(3), 244–256. doi:10.1111/j.1937-5956.2001.tb00373.x

Kingshott, R. P. (2006). The impact of psychological contracts upon trust and commitment within supplier–buyer relationships: A social exchange view. *Industrial Marketing Management*, *35*(6), 724–739. doi:10.1016/j.indmarman.2005.06.006

King, W. R., & Sethi, V. (2001). Patterns in the organization of transnational information systems. *Information & Management*, *38*(4), 201–215. doi:10.1016/S0378-7206(00)00050-1

Kirby, S. (2000). Syntax without natural selection: How compositionality emerges from vocabulary in a population of learners. In *The evolutionary emergence of language: Social function and the origins of linguistic form* (pp. 185–215). Academic Press. doi:10.1017/CBO9780511606441.019

Kirkman, B. L., Rosen, B., Gibson, C. B., Tesluk, P. E., & McPherson, S. O. (2002). Five challenges to virtual team success: Lessons from Sabre, Inc. *The Academy of Management Executive*, *16*(Issue 3), 67. doi:10.5465/AME.2002.8540322

Klassen, R., & Vachon. (2003). Collaboration and evaluation in the supply chain: The impact on plant-level environmental investment. *Production and Operations Management*, *12*(3), 336–352. doi:10.1111/j.1937-5956.2003.tb00207.x

Kleindorfer, P. R., Singhal, K., & Van Wassenhove, L. N. (2005). Sustainable operations management. *Production and Operations Management*, *14*(4), 482–492. doi:10.1111/j.1937-5956.2005.tb00235.x

Klein, H. K., & Myers, M. D. (1999). A set of principles for conducting and evaluating interpretative field studies in information systems. *Management Information Systems Quarterly*, *23*(1), 67–94. doi:10.2307/249410

Klein, R., & Rai, A. (2009). Interfirm strategic information flows in logistics supply chain relationships. *Management Information Systems Quarterly*, *33*(4), 735–762.

Kogut, B. (2000). The network as knowledge: Generative rules and the emergence of structure. *Strategic Management Journal*, *21*(3), 405–425. doi:10.1002/(SICI)1097-0266(200003)21:3<405::AID-SMJ103>3.0.CO;2-5

Kogut, B., & Kulatilaka, N. (1994). Operating flexibility, global manufacturing, and the option value of a multinational network. *Management Science*, *40*(1), 123–139. doi:10.1287/mnsc.40.1.123

Kogut, Z., & Zander, U. (1992). Knowledge of the firm, combinative capabilities, and the replication of technology. *Organization Science*, *3*(3), 383–397. doi:10.1287/orsc.3.3.383

Kohli, A. K., & Jaworski, B. J. (1993). Market orientation: The construct, research propositions, and managerial implications. *Journal of Marketing*, *54*(2), 1–18. doi:10.2307/1251866

Koltai, T., & Terlaky, T. (2000). The difference between the managerial and mathematical interpretation of sensitivity analysis results in linear programming. *International Journal of Production Economics*, *65*, 257–274. doi:10.1016/S0925-5273(99)00036-5

Koong, L., & Lin, C. (2007). Evaluating the decision to adopt RFID systems using analytic hierarchy process. *Journal of American Academy of Business*, *11*(1), 72–77.

Kotzab, H., Seuring, S., Müller, M., & Reiner, G. (2005). *Research methodologies in supply chain management*. Heidelberg, Germany: Springer. doi:10.1007/3-7908-1636-1

KPMG. (2008). *Logistics in China: Property and infrastructure*. Retrieved March 2, 2010, from http://www.kpmg.com.hk/en/virtual_library/Property_Infrastructure/LogisticsChina.pdf

Kruskal, W. H., & Wallis, W. A. (1952). Use of ranks in one-criterion variance analysis. *Journal of the American Statistical Association*, *47*(260), 583–621. doi:10.1080/01621459.1952.10483441

Kuhn, T. S. (1962). *The Structure of Scientific Revolutions*. Chicago: University of Chicago Press.

Kuhn, T. S. (2000). *The Road Since Structure: Philosophical Essays, 1970-1993*. Chicago: University of Chicago Press.

Kumar, K., van Dissel, H. G., & Bielli, P. (1998). The merchant of Prato-revisited: Toward a third rationality of information systems. *Management Information Systems Quarterly*, *22*(2), 199–226. doi:10.2307/249395

Kumar, N. (2000). The power of trust in manufacturer-retailer relationships. Managing The Value Chain. *Harvard Business Review*.

Lacity, M., & Hirschheim, R. (1993). The information systems outsourcing bandwagon. *Sloan Management Review*, *35*(1), 73–86.

Lahiri, S., & Kedia, B. L. (2009). The effects of internal resources and partnership quality on firm performance: An examination of India BPO providers. *Journal of International Management*, *15*(20), 209–224. doi:10.1016/j.intman.2008.09.002

Lahiri, S., & Kedia, B. L., & Mukherjee. (2011). The impact of management capability on the resource–performance linkage: Examining Indian outsourcing providers. *Journal of World Business*, *47*(1), 145–155. doi:10.1016/j.jwb.2011.02.001

Lahtinen, T. (2010). *Supply chain integrity and efficiency – Creating a new service concept*. Lurea, Finland: Laurea University of Applied Sciences.

Lai, F., Hutchison, J., & Zhang, G. (2005). Radio frequency identification (RFID) in China: Opportunities and challenges. *International Journal of Retail and Distribution Management*, *33*(12), 905–916. doi:10.1108/09590550510634639

Lai, F., Zhao, X., & Wang, Q. (2006). The impact of information technology on the competitive advantage of logistics firms in China. *Industrial Management & Data Systems*, *106*(9), 1249–1271. doi:10.1108/02635570610712564

Lambert, D. M., & Cooper, M., & Pagh. (1998). Supply chain management: Implementation issues and research opportunities. *International Journal of Logistics Management*, *9*(2), 1–20. doi:10.1108/09574099810805807

Lambert, D., Stock, J., & Ellram, L. (1998). *Fundamentals of logistics management*. Boston, MA: Irwin/McGraw-Hill.

Lamming, R., & Hampson, J. (1996). The environment as a supply chain management issue. *British Journal of Management*, *7*, 45–62. doi:10.1111/j.1467-8551.1996.tb00147.x

Lancaster University. (2000). *Actor network resource: Thematic list*. Retrieved from http://www.lancs.ac.uk/fass/centres/css/ant/ant.htm

Langley, C. J., Jr. (2007). *2007 third-party logistics: results and findings of the 12th annual study*. Retrieved October 7, 2013, from http://www.3plstudy.com

Langley, C. J., Jr. (2008). *2008 third-party logistics: results and findings of the 13th annual study*. Retrieved October 7, 2013, from http://www.3plstudy.com

Langley, C. J., Jr. (2009). *2009 third-party logistics: results and findings of the 14th annual study*. Retrieved October 7, 2013, from http://www.3plstudy.com

Langley, C. J., Jr. (2010). *2010 third-party logistics: Results and findings of the 15th annual study*. Retrieved October 7, 2013, from http://www.3plstudy.com

Langley, C. J., Jr. (2012). *2012 third-party logistics: Results and findings of the 16th annual study*. Retrieved October 7, 2013, from http://www.3plstudy.com

Langley, C. J., Jr. (2013). *2013 third-party logistics: Results and findings of the 17th annual study*. Retrieved October 7, 2013, from http://www.3plstudy.com

Langlois, R. N. (2002). Modularity in technology and organization. *Journal of Economic Behavior & Organization, 49*(1), 19–37. doi:10.1016/S0167-2681(02)00056-2

Lao, Y, Hong, & Rao. (2010). Supply management, supply flexibility, and performance outcomes: An empirical investigation of manufacturing firms. *Journal of Supply Chain Management, 46*(3), 6–22. doi:10.1111/j.1745-493X.2010.03195.x

Latour, B. (1987). *Science in action: How to follow scientists and engineers through society*. Boston, MA: Harvard University Press.

Latour, B. (1990). Drawing things together. In M. Lynch, & S. Woolgar (Eds.), *Representation in scientific practice* (pp. 19–68). Cambridge, MA: MIT Press.

Latour, B. (1993). *We have never been modern*. Brighton, UK: Harvester Wheatsheaf.

Latour, B. (1994). On technical mediation. *Common Knowledge, 3*(2), 29–64.

Latour, B. (1997). On recalling ANT. In J. Law, & J. Hassard (Eds.), *Actor-network and after*. Oxford, UK: Blackwell.

Latour, B., Mauguin, P., & Teil, G. (1992). A note on socio-technical graphs. *Social Studies of Science, 22*(1), 33–57. doi:10.1177/0306312792022001002

Lau, K. H., & Ma, W. L. (2008). A supplementary framework for evaluation of integrated logistics provider. *International Journal of Information Systems and Supply Chain Management, 1*(3), 49–69. doi:10.4018/jisscm.2008070104

Lau, K. H., & Sirichoti, T. (2012). A study on RFID adoption in the grocery retailing industry of Thailand. *International Journal of Information Systems and Supply Chain Management, 5*(3), 58–77. doi:10.4018/jisscm.2012070104

Law, A. M., & Kelton, W. D. (2000). *Simulation modeling and analysis* (3rd ed.). New York: McGraw-Hill Higher Education.

Lawerence, J., & Pasternack, B. (2002). *Applied management science: Modeling, spreadsheet analysis, and communication for decision making*. New York, NY: Wiley.

Law, J. (1992). Notes on the theory of the actor-network: Ordering, strategy and heterogeneity. *Systems Practice, 5*(4), 379–393. doi:10.1007/BF01059830

Law, J., & Mol, A. (2000). *Situating technoscience: An inquiry into spatialities*. Lancaster, UK: Centre for Science Studies, Lancaster University.

Lee, H. L., Padmanabhan, V., & Whang, S. (1997a). The bullwhip effect in supply chains. *Sloan Management Review, 38*(3), 93–102.

Lee, H. L., Padmanabhan, V., & Whang, S. (1997b). Information distortion in a supply chain: The bullwhip effect. *Management Science, 43*(4), 546–558. doi:10.1287/mnsc.43.4.546

Lee, H., & Özer, Ö. (2007). Unlocking the value of RFID. *Production and Operations Management, 16*(1), 40–64. doi:10.1111/j.1937-5956.2007.tb00165.x

Lee, J. (2001). The impact of knowledge sharing, organizational capability and partnership quality on IS outsourcing success. *Information & Management, 38*(5), 323–335. doi:10.1016/S0378-7206(00)00074-4

Lee, S.-Y. (2008). Drivers for the participation of small and medium-sized suppliers in green supply chain initiatives. *Supply Chain Management: An International Journal*, *13*(3), 185–198. doi:10.1108/13598540810871235

Leiblein, M. J. (2003). The choice of organizational governance form and performance: Predictions from transaction cost, resource-based, and real options theories. *Journal of Management*, *29*(6), 937–961.

Lemley, D. (2009). The road to authentic brand is littered with design. In T. Lockwood (Ed.), *Design thinking: Integrating innovation, customer experience, and brand value*. New York: Allworth Press.

Leon, N., & Davies, A. (2008). Managed service paradox. *IBM Systems Journal*, *47*(1). doi:10.1147/sj.471.0153

Levina, N., & Ross, J. W. (2003). From the vendor's perspective: Exploring the value proposition in information technology outsourcing. *Management Information Systems Quarterly*, *27*(3), 331–364.

Levitt, T. (1976, September/October). The industrialization of service. *Harvard Business Review*, 63–74.

Lewin, A. Y., Massini, S., & Peeters, C. (2009). Why are companies offshoring innovation? The emerging global race for talent. *Journal of International Business Studies*, *40*(6), 901–925. doi:10.1057/jibs.2008.92

Li & Fung Research Centre. (2012). China's logistics industry update 2012. *China Distribution & Trading*, *96*, 1–46.

Li, D., Lai, F., & Wang, J. (2010). E-business assimilation in China's international trade firms: The technology-organization environment framework. *Journal of Global Information Management*, *18*(1), 39–65. doi:10.4018/jgim.2010091103

Lilienfeld, S. O., Lynn, S. J., Ruscio, J., & Beyerstein, B. L. (2010). *50 great myths of popular psychology: Shattering widespread misconceptions about human behavior*. Sussex, UK: Wiley-Blackwell.

Limburg, K. E., O'Neill, R. V., Costanza, R., & Farber, S. (2002). Complex systems and valuation. *Ecological Economics*, *41*(3), 409–420. doi:10.1016/S0921-8009(02)00090-3

Lim, S. H., & Koh, C. E. (2009). RFID implementation strategy: Perceived risks and organizational fits. *Industrial Management & Data Systems*, *109*(8), 1017–1036. doi:10.1108/02635570910991274

Lin, C. (2010). Computing shadow proces/costs of dengernerate LP problems with reduced simplex tables. *Expert Systems with Applications*, *37*, 5848–5855. doi:10.1016/j.eswa.2010.02.021

Lin, C. Y. (2007). Adoption of green supply chain practices in Taiwan's logistics industry. *Journal of International Management Studies*, *2*(2), 90–98.

Lin, C. Y., & Ho, Y. H. (2009). An empirical study on the adoption of RFID technology for logistics service providers in China. *International Business Research*, *2*(1), 23–36.

Lindgreen, A., Antioco, M., Harness, D., & Sloot, R. (2008). Purchasing and marketing of social and environmental sustainability for high-tech medical equipment. *Journal of Business Ethics*, *85*, 445–462. doi:10.1007/s10551-008-9740-1

LINDO. (2003). *LINDO users's manual*. New York, NY: Author.

Lindsay, J., Reade, W., & Roth, L. (2003). Retail RFID systems without smart shelves. *RFID Journal*. Retrieved November 18, 2009, from http://www.jefflindsay.com/rfid1.shtml

Lin, F.-R., Huang, S.-H., & Lin, S.-C. (2002). Effects of information sharing on supply chain performance in electronic commerce. *IEEE Transactions on Engineering Management*, *49*(3), 258–268. doi:10.1109/TEM.2002.803388

Linton, J. D., Klassen, & Jayaraman. (2007). Sustainable supply chains: An introduction. *Journal of Operations Management*, *25*(6), 1075–1082. doi:10.1016/j.jom.2007.01.012

Lipke, W. (2003). *Schedule is different*. The Measurable News.

Li, S., Godon, D., & Visich, J. K. (2010). An exploratory study of RFID implementation in the supply chain. *Management Research Review*, *33*(10), 1005–1015. doi:10.1108/01409171011084003

Li, S., Visich, J. K., Khumawala, B. M., & Zhang, C. (2006). Radio frequency identification technology: Applications, technical challenges and strategies. *Sensor Review*, *26*(3), 193–202. doi:10.1108/02602280610675474

Li, S., Wang, H., Hu, S. J., Lin, Y.-T., & Abell, J. A. (2011). Automatic generation of assembly system configuration with equipment selection for automotive battery manufacturing. *Journal of Manufacturing Systems*, *30*(4), 188–195. doi:10.1016/j.jmsy.2011.07.009

Liu, P., & Zhang, X. (2011). Research on the supplier selection of a supply chain based on entropy-weight and improved ELECTRE-III method. *International Journal of Production Research*, *3*(1), 637–646. doi:10.1080/00207540903490171

Locke, R., & Romis. (2007). Improving work conditions in a global supply chain. *MIT Sloan Management Review*, *48*(2), 54–62.

Locke, R. M., Fei, & Brause. (2007). Does monitoring improve labor standards? Lessons from NIKE. *Industrial & Labor Relations Review*, *61*(1), 3–31.

Lofaro, R. J. (1992). Workshop on Integrated Crew Resource Management (CRM). *In DOT/FAA/RD-95/5*. Springfield, VA: National Technical Information Service.

Lofaro, R. J., & Smith, K. M. (1993). The Role of LOFT in CRM Integration. In *Proceedings the Seventh International Symposium of Aviation Psychologists*. Columbus, OH: ISAP.

Lofaro, R. J., & Smith, K. M. (1998). Rising Risk? Rising Safety? The Millennium and Air Travel. *Transportation Law Journal*, *2*(25).

Lofaro, R. J., & Smith, K. M. (1999). Operational Decision-Making (ODM) and Risk Management (RM): Rising Risk, the Critical Mission Factors and Training. In *Proceedings of Tenth International Symposium on Aviation Psychology*. Columbus, OH: ISAP.

Lofaro, R. J., & Smith, K. M. (2011a) Similar Accidents: Lessons Learned and Countermeasures. In *Proceedings of the 16th International Symposium on Aviation Psychology*. Dayton, OH.

Lofaro, R.J., Adams, R.J., and C.N., (1992b). Workshop on Aeronautical Decision-Making (ADM) In *DOT/FAA/RD-92/14* (Vol. I, II). Springfield, VA: National Technical Information Service.

Lofaro, R. J., & Smith, K. M. (2011b). Collaborative decision-making between cockpit and dispatch for safety. *Technology and Innovation*, *3*(12), 225–231.

Lofaro, R. J., & Smith, K. M. (2011c). The Aviation Operational Environment: Integrating a Decision-Making Paradigm, Flight Simulator Training and an Automated Cockpit Display for Aviation Safety. In E. Abu-Taeih, A. El-Sheikk, & M. Jafari (Eds.), *Technology Engineering and Management in Aviation: Advancements and Discoveries* (pp. 241–282). Hershey, PA: IGI Global. doi:10.4018/978-1-60960-887-3.ch015

Lu, E. J. L., Tsai, R. H., & Chou, S. (2001). An empirical study of XML/EDI. *Journal of Systems and Software*, *58*(3), 271–279. doi:10.1016/S0164-1212(01)00044-9

Lummus, R. R., & Vokurka, R. J. (1999). Managing the demand chain through managing the information flow: Capturing 'moments of information'. *Production and Inventory Management Journal*, *40*(1), 16–20.

Macher, J. T. (2006). Technological development and the boundaries of the firm: A knowledge-based examination in semiconductor manufacturing. *Management Science*, *52*(6), 826–843. doi:10.1287/mnsc.1060.0511

Madu, C., Kuei, C., & Madu, I. (2002). A hierarchic metric approach for integration of green issues in manufacturing: A paper recycling application. *Journal of Environmental Management*, *64*, 261–272. doi:10.1006/jema.2001.0498 PMID:12040959

Mahmoud-Jouini, S. B. (2000). Coment passer d'une demarche réactive à une démarche proactive fondée sur des stratégies d'offres inovantes? In A. Bloch, & D. Manceau (Eds.), *Le cas des grandes entreprises générales de Bâtiment françaises, de l'idée au marché*. France: Vuibert.

Maisel, H., & Gnugnoli, G. (1972). *Simulation of discrete stochastic systems*. Chicago: Science Research Associates, Inc.

Majchrzak, A., Rice, R. E., Malhotra, A., King, N., & Ba, S. (2000). Technology adaptation: The case of a computer-supported inter-organizational virtual team. *Management Information Systems Quarterly*, *24*(4), 569–601. doi:10.2307/3250948

Maon, F., Lindgreen, & Swaen. (2009). Designing and implementing corporate social responsibility: An integrative framework grounded in theory and practice. *Journal of Business Ethics*, *87*, 71–89. doi:10.1007/s10551-008-9804-2

Marascuilo, L. A. (1966). Large-sample multiple comparisons. *Psychological Bulletin*, *65*(4), 280–290. doi:10.1037/h0023189 PMID:5325892

Mardi, A. G. (1992). Environmentally friendly development: Can the private sector succeed where others have failed? *The Columbia Journal of World Business*, *27*(3&4), 194–200.

Mark, K. L. (1982). A production lot-size inventory model for deteriorating items. *Computers & Industrial Engineering*, *6*(2), 309–317.

Markose, S. M. (2005). Computability and evolutionary complexity: Markets as complex adaptive systems (CAS). *The Economic Journal*, *115*(504), F159. doi:10.1111/j.1468-0297.2005.01000.x

Markus, M., Sia, S. K., & Soh, C. (2012). MNEs and information management: Structuring and governing IT resources in global enterprise. *Journal of Global Information Management*, *20*(1), 1–17. doi:10.4018/jgim.2012010101

Marshall, R. S., Cordano, M., & Silverman, M. (2005). Exploring individual and institutional drivers of proactive environmentalism in the US wine industry. *Business Strategy and the Environment*, *14*, 92–109. doi:10.1002/bse.433

Martin, S. F., Beimborn, D., Parikh, M. A., & Weitzel, T. (2008). Organizational Readiness for Business Process Outsourcing: A Model of Determinants and Impact on Outsourcing Success, *Proceedings of the 41st Hawaii International Conference on Systems Sciences*, IEEE, pp. 1-10.

Martín-de-Castro, G., Navas-López, J., López-Sáez, P., & Alama-Salazar, E. (2006). Organizational capital as competitive advantage of the firm. *Journal of Intellectual Capital*, *7*(3), 324–337. doi:10.1108/14691930610681438

Martin, K. (1999). *Large scale linear and integer optimization: A unified approach*. New York, NY: Springer. doi:10.1007/978-1-4615-4975-8

Mayasandra, R., Pan, S. L., & Leidner, D. (2011). Examining the strategic alignment and implementation success of a KMS: A subculture-based multi-level analysis. *Information Systems Research*, *22*(1), 39–59. doi:10.1287/isre.1080.0214

Maylor, H. (2006). Guest editorial - Special issue on rethinking project management. *International Journal of Project Management*, *24*, 635–637. doi:10.1016/j.ijproman.2006.09.013

McAdams, D. P. (2001). The psychology of life stories. *Review of General Psychology*, *5*(2), 100–122. doi:10.1037/1089-2680.5.2.100

McBurney, D., & White, T. (2009). *Research methods* (8th ed.). Belmont, CA: Academic Press.

McClellan, M. (2003). *Collaborative manufacturing*. St. Lucie Press.

McCormack, E., Jensen, M., & Hovde, A. (2010). Evaluating the use of electronic door seals (e-seals) on shipping containers. *International Journal of Applied Logistics*, *1*(4), 13–29. doi:10.4018/jal.2010100102

McCue, D. (2010). Sustainability and the triple bottom line. *World Trade*, *23*(10), 24.

McElroy, M. W. (2000). Integrating complexity theory, knowledge management and organizational learning. *Journal of Knowledge Management*, *4*(3), 195–203. doi:10.1108/13673270010377652

McEvily, S. K., & Chakravarthy, B. (2002). The persistence of knowledge-based advantage: An empirical test for product performance and technological knowledge. *Strategic Management Journal*, *23*(4), 285–305. doi:10.1002/smj.223

McFarlane, D., & Sheffi, Y. (2003). The impact of automatic identification on supply chain operations. *The International Journal of Logistics Management*, *14*(1), 1–17. doi:10.1108/09574090310806503

McIvor, R., Humphreys, P., & McCurry, L. (2003). Electronic commerce: Supporting collaboration in the supply chain. *Journal of Materials Processing Technology*, *139*(1-3), 147. doi:10.1016/S0924-0136(03)00196-1

McKenna, J. T. (1996). Carriers Hone CRM Programs. *Aviation Week & Space Technology*, (145): 146–147.

McKinsey, Q. (2001). Has outsourcing gone too far? https://www.mckinseyquarterly.com/Has_outsourcing_gone_too_far_1126.

McNamara, G., Vaaler, P. M., & Devers, C. (2003). Same as it ever was: The search for evidence of increasing hypercompetition. *Strategic Management Journal*, *24*(3), 261–278. doi:10.1002/smj.295

McWilliams, A., Siegel, & Wright. (2006). Corporate social responsibility, international perspectives. *The Journal of Business Strategy*, *23*(1), 1–12.

Melnyk, S. A., Sroufe, R. P., & Calantone, R. (2003). Assessing the impact of environmental management systems on corporate and environmental performance. *Journal of Operations Management*, *21*(2), 329–351. doi:10.1016/S0272-6963(02)00109-2

Mena, C., Humphries, A., & Choi, T. Y. (2013). Toward a theory of multi-tier supply chain management. *Journal of Supply Chain Management*, *49*(2), 58–77. doi:10.1111/jscm.12003

Menon, A., Sundar, G., Bharadwaj, P. T. A., & Stenen, W. E. (1999). Antecedent & consequences of marketing strategy making a model & a test. *Journal of Marketing*, *63*, 18–40. doi:10.2307/1251943

Mentzer, J., DeWitt, W., Keebler, J., Min, S., Nix, N., Smith, C., & Zacharia, Z. (2001). Defining supply chain management. *Journal of Business Logistics*, *22*(2), 1–25. doi:10.1002/j.2158-1592.2001.tb00001.x

Meyer-Larsen, N. (2009). Win-win solutions combining security and logistics. In *Proceedings of Second European Conference on ICT for Transport Logistics (ECITL)*. Venice, Italy: ECITL.

Middle East Battery Company. (2006). *MEBCO product standards manual*. Dammam, Saudi Arabia: MEBCO.

Miles, R. E., & Snow, C. C. (1992). Causes of Failure in Network Organizations. *California Management Review*, *34*, 53–72. doi:10.2307/41166703

Milne, M., Kearins, K., & Walton, S. (2006). Creating adventures in wonderland: The journey metaphor and environmental sustainability. *Organization*, *13*, 801–839. doi:10.1177/1350508406068506

Min, H., & Galle. (1997). Green purchasing strategies: Trends and implications. *International Journal of Purchasing and Materials Management*, *33*(3), 10–17. doi:10.1111/j.1745-493X.1997.tb00026.x

Min, H., & Galle, W. P. (2001). Green purchasing practices of US firms. *International Journal of Operations & Production Management*, *21*(9), 1222–1238. doi:10.1108/EUM0000000005923

Mitchell, R. K., Agle, B. R., & Wood, D. J. (1997). Toward a theory of stakeholder dentification and salience: Defining the principle of who and what really counts. *Academy of Management Review*, *22*(4), 853.

Mobile Aspects. (2007). *Is RFID ready for healthcare? Yes, if RFID is focused on addressing healthcare's needs* (White Paper). Retrieved November 24, 2009, from www.mobileaspects.com

Mobile Aspects. (2007). *Providing a more advanced automation strategy for medical device and supply utilization in preoperative care through RFID* (White Paper). Retrieved November 18, 2009, from www.mobileaspects.com

Mohapatra, S., & Srivastava, N. (2012). Sustaining competitive advantage in social entrepreneurship - A case study. *International Journal of Logistics Economics and Globalisation*, *4*, 197–220. doi:10.1504/IJLEG.2012.050207

Mol, A., & Law, J. (1994). Regions, networks and fluids: Anaemia and social topology. *Social Studies of Science*, *24*, 641–671. doi:10.1177/030631279402400402 PMID:11639423

Monteiro, E., & Hepsø, V. (2000). Infrastructure strategy formation: Seize the day at Statoil. In C. Ciborra (Ed.), *From control to drift: The dynamics of corporate information infrastructure*. Oxford, UK: Oxford University Press.

Monteiro, E., & Rolland, K. (2012). Trans-situated use of integrated information systems. *European Journal of Information Systems*, *21*(1), 1–13.

Moore, G. (2012). *Living on the fault line*. New York, NY: HarperCollins Publishers.

Moorman, C., & Miner, A. S. (1997). The impact of organizational memory on new product performance and creativity. *JMR, Journal of Marketing Research*, *34*(1), 91–106. doi:10.2307/3152067

Morel, B., & Ramanujam, R. (1999). Through the looking glass of complexity: The dynamics of organizations as adaptive and evolving systems. *Organization Science*, *10*(3), 278–293. doi:10.1287/orsc.10.3.278

Morris, M. H., & Joyce, M. L. (1988). How marketers evaluate price sensitivity. *Industrial Marketing Management*, *17*(2), 169–176. doi:10.1016/0019-8501(88)90019-3

Morris, P. W. G., Crawford, L., Hodgson, D., Shepherd, M. M., & Thomas, J. (2006). Exploring the role of formal bodies of knowledge in defining a profession – The case of project management. *International Journal of Project Management*, *24*, 710–721. doi:10.1016/j.ijproman.2006.09.012

Morris, P. W. G., & Hough, H. H. (1986). *Preconditions of success and failure in major projects*. Oxford, UK: The Oxford Centre for Management Studies.

Mouhalis, J. (2007). *The details behind a successful offshore BPO programme*. Oxford, UK: Zackin Publications.

Müller, R., & Turner, R. (2007). The influence of project managers on project success criteria and project success by type of project. *European Management Journal*, *25*(4), 289–309. doi:10.1016/j.emj.2007.06.003

Müller, R., & Turner, R. (2010). Leadership competency profiles of successful project managers. *International Journal of Project Management*, *28*, 437–448. doi:10.1016/j.ijproman.2009.09.003

Murphy, D. C., Baker, B. N., & Fisher, D. (1974). *Determinants of project success*. Boston: School of Management, Boston College.

Murphy, G. B., Trailer, J. W., & Hill, R. C. (1996). Measuring performance in entrepreneurship research. *Journal of Business Research*, *36*(1), 15–23. doi:10.1016/0148-2963(95)00159-X

Mustaffa, N., & Potter, A. (2009). Healthcare supply chain management in Malaysia: A case study. *Supply Chain Management: An International Journal*, *14*(3), 234–243. doi:10.1108/13598540910954575

Mwirigi, F. (2010). The challenge of building sustainable supply chain relationships among small firms in developing economies: The case of Kenya. *International Review of Business Research Papers*, *6*(4), 189–201.

Narasimhan, R., & Wook. (2001). Information system utilization strategy for supply chain integration. *Journal of Business Logistics*, *22*(2), 51–75. doi:10.1002/j.2158-1592.2001.tb00003.x

Narasimhan, R., & Jayaram, J. (1998). Causal linkages in supply chain management: An exploratory study of North American manufacturing firms. *Decision Sciences*, *29*(3), 579–605. doi:10.1111/j.1540-5915.1998.tb01355.x

NASA. (2011). ASK magazine: The NASA source for project management and engineering excellence. *NASA, 41*.

National Association of Software and Services Company. (2011). *Indian IT-BPO industry*. Retrieved December 12, 2001, from http://www.nasscom.org/indian-itbpo-industry

Naula, T., Ojala, L., Solakivi, T., Takalokastari, M., Rantanen, M., Kalske, M., ... Stenholm, P. (2006). Finland's state of logistics survey 2006. *Publications of the Ministry of Transport and Communications,* (35).

NCR Corporation (2003). *Fifty uses of RFID in retail*. Retrieved November 18, 2009, from http://www.usingrfid.com/features/read.asp?id=3

Nelson, C. R. (1973). *Applied time series analysis for managerial forecasting*. San Francisco: Holden-Day, Inc.

Neo, I. T. (2008). *Navigating economic uncertainty: Leveraging the services globalization advantage*. Retrieved November, 12, 2008, from http://www.neoadvisory.com/whitepapers.php#WPFeatured

Newland, G. (2005). *The potential impacts of RFID on the retail enterprise*. Retrieved November 18, 2012, from http://www.nrf-arts.org/presents/2004/retail/graham.pdf

Nickerson, J. A., & Zenger, T. R. (2004). A knowledge-based theory of the firm—The problem-solving perspective. *Organization Science*, *15*(6), 617–632. doi:10.1287/orsc.1040.0093

Niederman, F., Alhorr, H., Park, Y. H., & Tolmie, C. R. (2012). Global information management research: What have we learned in the past decade? *Journal of Global Information Management*, *20*(1), 18–56. doi:10.4018/jgim.2012010102

Nielsen, B. B. (2005). The role of knowledge embeddedness in the creation of synergies in strategic alliances. *Journal of Business Research*, *58*(9), 1194–1204. doi:10.1016/j.jbusres.2004.05.001

Nilsson, F., & Waidringer, J. (2010). Logistics management from a complexity perspective. *The ICFAI Journal of Operations Management*, *3*(2), 59–73.

Nonthaleerak, P., & Hendry, L. (2008). Exploring the six sigma phenomenon using multiple case study evidence. *International Journal of Operations & Production Management*, *28*(2), 279–303. doi:10.1108/01443570810856198

North, D. C. (1990). *Institutions, institutional change and economic performance*. Cambridge, MA: Cambridge University Press. doi:10.1017/CBO9780511808678

Novack, R. A., Langly, C. J. Jr, & Rinehart, L. M. (1995). *Creating logistics value: Themes for the future*. USA: Council on Logistics Management.

Nunnally, J. C. (1978). *Psychometric theory*. New York: McGraw-Hills Book Company, Inc.

Nurkka, P., Kujala, S., & Kemppainen, K. (2009). Capturing users' perceptions of valuable experience and meaning. *Journal of Engineering Design*, *20*(5), 449–465. doi:10.1080/09544820903158835

Nurmilaasko, J. M. (2008). Adoption of e-business functions and migration from EDI-based to XML-based e-business frameworks in supply chain integration. *International Journal of Production Economics*, *113*(2), 721–733. doi:10.1016/j.ijpe.2007.11.001

NZ Business Council for Sustainable Development Report. (2003). *Business guide to a sustainable supply chain*. Retrieved from http:// www.nzbcsd.org.nz/supplychain

O'Connor, M. C. (2004a). All CVS store to offer express pay. *RFID Journal*. Retrieved November 23, 2012, from http://www.rfidjournal.com/article/articleview/1297/1/11/

O'Connor, M. C. (2004b). Ending retail scams with RFID. *RFID Journal*. Retrieved November 23, 2012, from http://www.rfidjournal.com/article/articleview/1260/1/20/

OGC. (2002). *Contract management guidelines: principles for service contracts*. London: Office of Government Commerce, HMSO.

Oppenheim, J. Bonini, Bielak, Kehm, & Lacy. (2007). *Shaping the new rules of competition: UN global compact participant mirror*. McKinsey & Company. Retrieved from http://tinyurl.com/69shws

Orlady, H., & Foushee, H. C. (Eds.). (1986). Cockpit Resource Management Training. *NASA Conference Publication*, *2455*, ntrs.nasa.gov

Orlikowski, W. J. (1992). The duality of technology: Rethinking the concept of technology in organizations. *Organization Science*, *3*(3), 398–427. doi:10.1287/orsc.3.3.398

Orlikowski, W. J. (1996). Improvising organizational transformation over time: A situated change perspective. *Information Systems Research*, *7*(1), 63–92. doi:10.1287/isre.7.1.63

Orlikowski, W. J. (2000). Using technology and constituting structures: A practice lens for studying technology in organizations. *Organization Science*, *11*(4), 404–428. doi:10.1287/orsc.11.4.404.14600

Orlikowski, W. J., & Gash, D. C. (1994). Technological frames: Making sense of information technology in organizations. *ACM Transactions on Information Systems*, *12*(2), 174–207. doi:10.1145/196734.196745

Orlikowski, W. J., & Yates, J. (2002). It's about time: Temporal structuring in organizations. *Organization Science*, *13*(6), 684–701. doi:10.1287/orsc.13.6.684.501

Ouchi, W. G. (1979). A conceptual framework for the design of organizational control mechanisms. *Management Science*, *25*(9), 833–848. doi:10.1287/mnsc.25.9.833

Outsourcing Center. (2004). What Causes Outsourcing Failures? http://www.outsourcing-center.com/2004-08-what-causes-outsourcing-failures-article-37826.html.

Overview of RFID Solutions. (2010). Retrieved February 4, 2010, from http://www.hitachieu.com/mu/

Pache, G., & Spalanzani, A. (2007). *La gestion des chanes logistiques multi-acteurs: Perspectives stratgiques*. Presses Universitaires de Grenoble.

Pagell, M., Wu, & Wasserman. (2010). Thinking differently about purchasing portfolios: An assessment of sustainable sourcing. *Journal of Supply Chain Management*, *46*(1), 57–73. doi:10.1111/j.1745-493X.2009.03186.x

Papas, N. (2011). Combining EA techniques with bow-tie diagrams to enhance European port security. In *Proceedings of International Conference on Paperless Freight Transport Logistics*. Munich, Germany: IEEE.

Parasuraman, A., Zeithaml, V. A., & Berry, L. B. (1988). SERVQUAL: A multiple-item scale of measuring consumer perceptions of service quality. *Journal of Retailing*, *64*(1), 12–40.

Parkhe, A. (2007). International outsourcing of services: Introduction to the special issue. *Journal of International Management*, *13*(1), 3–6. doi:10.1016/j.intman.2006.12.004

Park, N., Park, M., & Peterson, C. (2010). When is the search for meaning related to life satisfaction? *Applied Psychology: Health and Well-Being*, *2*(1), 1–13.

Parunak, H. V. D. (1998). *The DASCh experience: How to model a supply chain*. Retrieved from http://www.erim.org/~vparunak/iccs98.pdf

Pathak, S. D. (2005). *An investigative framework for studying the growth and evolution of complex supply networks*. Nashville, TN: Vanderbilt University.

Paulraj, A. (2011). Understanding the relationships between internal resources and capabilities, sustainable supply management and organizational sustainability. *Journal of Supply Chain Management*, *47*(1), 19–37. doi:10.1111/j.1745-493X.2010.03212.x

Peattie, K. (2001). Towards sustainability: The third age of green marketing. *Marketing Review*, *2*(2), 129–146. doi:10.1362/1469347012569869

Peleg-Gillai, B., Bhat, G., & Sept, L. (2006). *Innovators in supply chain security: Better security drives business value*. Palo Alto, CA: Stanford University.

Peng, Y. S., & Lin, S. S. (2008). Local responsiveness pressure, subsidiary resources, green management adoption and subsidiary's performance: Evidence from Taiwanese manufactures. *Journal of Business Ethics*, *79*(1), 199–212. doi:10.1007/s10551-007-9382-8

Permala, A., Scholliers, J., & Granqvist, J. (2006). RFID-roadmap for business logistics. AINO Publications, (30).

Permala, A., Rantasila, K., & Pilli-Sihvola, E. (2012). RFID: From closed systems to improving visibility in the manufacturing supply chain. *International Journal of Applied Logistics*, *3*(2), 14–24. doi:10.4018/jal.2012040102

Peterson, C., Peterson, N., & Seligman, M. E. P. (2005). Orientations to hapiness and life satisfaction: The full life versus the empty life. *Journal of Happiness Studies*, *6*, 25–41. doi:10.1007/s10902-004-1278-z

Pettit, T. J., Croxton, K. L., & Fiksel, J. (2012). Ensuring supply chain resilience: Development and implementation of an assessment tool. *Journal of Business Logistics*, *34*(1), 46–76. doi:10.1111/jbl.12009

Phillips, N., & Hardy, C. (2002). *Discourse analysis: Investigating processes of social construction*. Thousand Oaks, CA: Sage Publications.

Pich, M. T., Loch, C. H., & De Meyer, A. (2002). On uncertainty, ambiguity, and complexity. *Project Management. Management Science*, *48*(8), 1008–1023. doi:10.1287/mnsc.48.8.1008.163

Piercy, N. F. (1981). British export market selection and pricing. *Industrial Marketing Management*, *10*, 287–297. doi:10.1016/0019-8501(81)90038-9

Pilbeam, C., Alvarez, G., & Wilson, H. (2012). The governance of supply networks: A systematic literature review. *Supply Chain Management: An International Journal*, *17*(4), 358–376. doi:10.1108/13598541211246512

Pilli-Sihvola, E., Rantasila, K., Permala, A., & Huhtaniemi, J. (2012). Solutions for monitoring multimodal supply chain. In *Proceedings of e-Freight 2012. e-Freight, International Conference on Paperless Freight Transport Logistics*. Delft, The Netherlands: e-Freight.

Pinto, J. K., & Slevin, D. P. (1989). Critical success factors in R&D projects. *Research Technology Management, 32*, 31.

Pison Teklogix Teams Up with Vocollect to Provide Voice Recognition to Work About Pro VIP Product Line. (2006). Retrieved February 5, 2010, from http://www.logisticsit.com/absolutenm/templates/articlevoice.aspx?articleid=2317&zoneid=39

PMBOK. (2008). *A guide to the project management body of knowledge* (4th ed.). Newtown Square, PA: Project Management Institute.

PMI. (2011). PM network magazine: Too big to handle - Megaprojects and meeting the triple constraints. *Project Management Institute, 25*(3), 22–23.

Podsakoff, P. M., MacKenzie, S. B., Lee, J., & Podsakoff, N. P. (2003). Common method biases in behavioral research: A critical review of the literature and recommended remedies. *The Journal of Applied Psychology, 88*(5), 879–903. doi:10.1037/0021-9010.88.5.879 PMID:14516251

Pokahr, A., Braubach, L., & Lamersdorf, W. (2005). Jadex: A BDI reasoning engine. In *Multi-agent programming: Languages, platforms and applications* (pp. 149–174). Berlin: Springer. doi:10.1007/0-387-26350-0_6

Poluha, R. G. (2007). *Application of the SCOR model in supply chain management*. Cambria Press.

Pons, D. (2008). Project management for new product development. *Project Management Journal, 39*(2), 82–97. doi:10.1002/pmj.20052

Pons, D. J., & Raine, J. (2005). Design mechanisms and constrains. *Research in Engineering Design, 16*, 73–85. doi:10.1007/s00163-005-0008-9

Porter, M. E. (1996, November-December). What is strategy? *Harvard Business Review*, 61–78. PMID:10158474

Posey, C., & Bari, A. (2009). Information sharing and supply chain performance: Understanding complexity, compatibility, and processing. *International Journal of Information Systems and Supply Chain Management, 2*(3), 67–76. doi:10.4018/jisscm.2009070105

Prahalad, C. K., & Hamel, G. (1990). The core competence and the corporation. *Harvard Business Review*, 71–90.

Prahalad, C. K., & Ramaswamy, V. (2004). Co-creation experiences: The next practice in value creation. *Journal of Interactive Marketing, 18*(3), 5–14. doi:10.1002/dir.20015

Prajogo, D., & Olhager, J. (2012). Supply chainintegratio nandperformance:The effects of long-term relationships, information technology and sharing, and logistics integration. *International Journal of Production Economics, 135*, 514–522. doi:10.1016/j.ijpe.2011.09.001

Premkumar, G. P. (2000). Inter-organizational systems and supply chain management: An information processing perspective. *Information Systems Management, 3*(17), 56–69.

Preston, D. S., & Karahanna, E. (2009). Antecedents of IS strategic alignment: A nomological network. *Information Systems Research, 20*(2), 159–179. doi:10.1287/isre.1070.0159

Preuss, L. (2009). Addressing sustainable development through public procurement. *Supply Chain Management: An International Journal, 14*(3), 213–223. doi:10.1108/13598540910954557

Project Management Institute. (2005). *Practice standard for earned value management* (2nd ed.). Newtown Square, PA: Author.

Provan, K. G., & Kenis, P. (2005). Modes of Network Governance: Structure, Management, and Effectiveness, *Public Management Research Association Conference*, September 29 – October 1, University of Southern California, Los Angeles, CA.

Pullman, M., Maloni, M., & Carter, C. (2009). Food for thought: Social versus environmental sustainability practices and performance outcomes. *J. Supply Chain Manag., 45*, 38–54. doi:10.1111/j.1745-493X.2009.03175.x

Pun, F., Chin, S., & Gill, R. (2001). Determinants of employee involvement practices in manufacturing enterprises. *Total Quality Management*, *12*(1), 95–109. doi:10.1080/09544120020010129

Raafat, F. (1991). Survey of literature on continuously deteriorating inventory model. *Journal of Operational Research*, *42*, 27–37.

Raafat, F., Wolfe, P. M., & Eldin, H. K. (1991). An inventory model for deteriorating items. *Computers & Industrial Engineering*, *20*(1), 89–94. doi:10.1016/0360-8352(91)90043-6

Rai, A., Maruping, L. M., & Venkatesh, V. (2009). Offshore information systems project success: The role of social embeddedness and cultural characteristics. *Management Information Systems Quarterly*, *33*(3), 617–641.

Raman, P., Wittmann, C. M., & Rauseo, N. A. (2006). The role of organizational capabilities in successful CRM implementation. *Journal of Personal Selling & Sales Management*, *26*(1), 39–54. doi:10.2753/PSS0885-3134260104

Ramanujan, V., Venkatraman, N., & Cammilus, J. C. (1986). Multi-objective assessment of effectiveness of strategic planning: A discriminant analysis approach. *Academy of Management Journal*, *29*(2), 347–372. doi:10.2307/256192

Ramkumar, N., Subramanian, P., Narendran, T. T., & Ganesh, K. (2011). Erratum to a genetic algorithm approach for solving a closed loop supply chain model: A case of battery recycling. *Applied Mathematical Modelling*, *35*(12), 5921–5932. doi:10.1016/j.apm.2011.05.026

Randall, W. S., Pohlen, T. L., & Hanna, J. B. (2010). Evolving a theory of performance-based logistics using insight from service-dominant logic. *Journal of Business Logistics*, *31*(2), 35–61. doi:10.1002/j.2158-1592.2010.tb00142.x

Ranji, U., Lundy, J., & Salganicoff, A. (2010). *US health-care costs: A background brief.* Retrieved February 4, 2010, from http://www.kaiseredu.org/topics_im.asp?imID=1&parentID=61&id=358

Rank, O. N., Robins, G. L., & Pattison, P. E. (2010). Structural logic of intraorganizational networks. *Organization Science*, *21*(3), 745–764. doi:10.1287/orsc.1090.0450

Rao, P. (2002). Greening the supply chain a new initiative in south East Asia. *International Journal of Operations & Production Management*, *22*, 632–655. doi:10.1108/01443570210427668

Rao, P. (2006). Greening of suppliers/in-bound logistics in the South East Asian context. In *Greening the supply chain*. London: Springer. doi:10.1007/1-84628-299-3_11

Rao, P., & Holt, D. (2005). Do green supply chains lead to competitiveness and economic performance? *International Journal of Operations & Production Management*, *25*(9), 898–916. doi:10.1108/01443570510613956

Ras, P., & Vermeulen, W. (2009). Sustainable production and the performance of South African entrepreneurs in a global supply chain: The case of South African table grape producers. *Sustainable Development*, *17*(5), 325–341. doi:10.1002/sd.427

Reed, R., & DeFillippi, R. J. (1990). Causal ambiguity, barriers to imitation and sustainable competitive advantage. *Academy of Management Review*, *15*(1), 88–102.

Rees, R. (2004). *ISO supply chain standards.* Retrieved from http://docbox.etsi.org//erm/open/rfidworkshop/rfid_20%20richard%20rees_bsi.pdf

Reichheld, F. F. (2003, December). The one number you need to grow. *Harvard Business Review*, 46–54. PMID:14712543

Remington, K., & Pollack, J. (2008). Are there special tools for complex projects? In *World congress project management to run*. IPMA.

Remington, K. (2011). *Leading complex projects*. London: Gower Publishing Ltd.

Rendleman. J. D., & Faulconer, J. W. (2011). Escaping the space acquisition death spiral, three part series. *High Frontier Journal*, *7*(4).

Research and Markets. (2009). *China logistics industry report, 2006-2010 brochure*. Retrieved March 2, 2010, from http://www.researchandmarkets.com/reports/357283/china_logistics_industry_report_2006_2010.pdf

Retzlaff, P. D., & Gibertini, M. (1987). Air Force Pilot Personality[Psychology Press/Taylor and Francis Group, Philadelphia, PA]. *Hard Data on the Right Stuff in Multivariate Behavioural Research*, *22*, 383–389. doi:10.1207/s15327906mbr2204_1

Reuter, C., Foerstl, Hartmann, & Blome. (2010). Sustainable global supplier management: The role of dynamic capabilities in achieving competitive advantage. *Journal of Supply Chain Management*, *26*(2), 45–63. doi:10.1111/j.1745-493X.2010.03189.x

Reynolds, M. A., & Yuthas. (2008). Moral discourse and corporate social responsibility reporting. *Journal of Business Ethics*, *78*, 47–64. doi:10.1007/s10551-006-9316-x

RFID and Voice Enhanced Productivity on Pick-to-Tote Operations. (2008). Retrieved November 23, 2012, from http://www.rfidproductnews.com/reader/83/articles/index.php

RFID in the Retail Industry. (2004). Retrieved November 18, 2012, from http://www.microsoft.com/industry/retail/RFIDoverview.mspx

RFID Industry Solutions. (2010). Retrieved November 18, 2012, from http://www.symbol.com/category.php?category=720

RFID Journal. (2003a). *RFID for your shopping cart*. Retrieved November 23, 2012, from http://www.rfidjournal.com/article/articleview/489/1/20/

RFID Journal. (2003b). *Wal-Mart CIO to speak on RFID*. Retrieved November 23, 2012, from http://www.rfidjournal.com/article/articleview/419/1/26/

RFID Journal. (2003c). *Wal-Mart details RFID requirement*. Retrieved November 23, 2012, from http://www.rfidjournal.com/article/articleview/642/1/1/

RFID Journal. (2005). *Wal-Mart tackles out-of-stocks*. Retrieved November 18, 2012, from http://www.rfidjournal.com/article/articleview/1551/4/160/.

RFID Journal. (2005a). *RFID in retail - The consumer in control*. Retrieved November 23, 2012, from http://www.semiconductors.philips.com/news/publications/content/file_1283.html

RFID Journal. (2005b). *Retailer tests RFID on garments*. Retrieved November 22, 2012, from http://www.rfidjournal.com/article/articleview/491/1/20/

RFID Journal. (2006). *Gen 2 EPC protocol approved as ISO 18000-6C*. Retrieved from http://www.rfidjournal.com/article/view/2481/1/1

RFID Journal. (2009). *The future of RFID*. Retrieved December 8, 2009, from http://www.rfidjournal.com/article/view/7268

RFID Journal. (2010a). *Frequently asked questions*. Retrieved February 4, 2010, from http://www.rfidjournal.com/faq#20

RFID Journal. (2010b). *Radioactive waste cleanup project becomes more efficient, greener*. Retrieved February 4, 2010, from http://www.rfidjournal.com/article/view/7435/3

RFID Lab Finland. (2009). *Case ABB*. Retrieved from http://www.rfidlab.fi/sites/rfidlab.fi/files/ABB_outbound_RFID_press_FI_090417.pdf

RFID Privacy and Security. (2010). Retrieved November 22, 2012, from http://www.rsa.com/rsalabs/node.asp?id=2115

Rimmington, M., Smith, J. C., & Hawkins, R. (2006). Corporate social responsibility and sustainable food procurement. *British Food Journal*, *108*(10/11), 824–837. doi:10.1108/00070700610702082

Rindfleisch, A., & Heide, J. B. (1997). Transaction cost analysis: past, present, and future applications. *Journal of Marketing*, *61*(4), 30–54. doi:10.2307/1252085

Rindfleisch, A., & Moorman, C. (2001). The acquisition and utilization of information in new product alliances: A strength-of-ties perspective. *Journal of Marketing*, *65*(2), 1–18. doi:10.1509/jmkg.65.2.1.18253

Ritter, L., Barrett, J. M., & Wilson, R. (2007). *Securing total transportation networks: A total security management approach*. New York, NY: McGraw-Hill.

Roberti, M. (2005a). Retailers say RFID will take time. *RFID Journal*. Retrieved November 22, 2009, from http://www.rfidjournal.com/article/articleview/1344

Roberti, M. (2005b). Wal-Mart begins RFID process changes. *RFID Journal*. Retrieved November 18, 2012, from http://www.rfidjournal.com/article/articleview/1385/1/20/

Roberti, M. (2005c). Wal-Mart to expand RFID tagging requirement. *RFID Journal*. Retrieved November 18, 2012, from http://rfidjournal.com/article/articleview/1930/1/9/

Roberts, P. W., & Dowling, G. R. (2002). Corporate reputation and sustained superior financial performance. *Strategic Management Journal*, *23*(12), 1077–1093. doi:10.1002/smj.274

Robinson, D. R., & Wilcox, S. (2008). The greening of the supply chain. *Logistics Management*, *47*(10), 67–72.

Rodrigues, A. (2009). *Methodology for implementing software as a service - The analysis of benefits management*. (Master's Thesis). Lisbon University Institute (ISCTE-IUL), Lisbon, Portugal.

Rodrigues, A. (2010). Effective measurement of time performance using earned value management – A proposed modified version for SPIU tested across various industries and project types. *PM World Today*, *12*(10).

Rodrigues, A., & O'Neill, H. (2010). *Framework based on benefits management and enterprise architecture*. Paper presented at CENTERIS 2010 – Conference on Enterprise Information Systems. Viana do Castelo, Portugal.

Rodrigues, A., & O'Neill, H. (2012). Framework based on benefits management and enterprise architecture. *Information Resources Management Journal*, *25*(2), 34–51. doi:10.4018/irmj.2012040103

Roh, J. J., Kunnathur, A., & Tarafdar, M. (2009). Classification of RFID adoption: An expected benefits approach. *Information & Management*, *46*(6), 357–363. doi:10.1016/j.im.2009.07.001

Roman, M. (2004, September/October). The industrialization of service: A personal journey. *Ivey Business Journal*, 1-7.

Roper, S., & Crone, M. (2003). Knowledge complementarity and coordination in the local supply chain: Some empirical evidence. *British Journal of Management*, *14*(4), 339–355. doi:10.1111/j.1467-8551.2003.00383.x

Rosemann, M., Recker, J., & Flender, C. (2008). Contextualisation of business processes. *International Journal of Business Process Integration and Management*, *3*(1), 47–60. doi:10.1504/IJBPIM.2008.019347

Ross, W. J., Weill, P., & Robertson, C. D. (2006). *Enterprise architecture as a strategy*. Cambridge, MA: Harvard Business School Press.

Rothenberg, S., Pil, F. K., & Maxwell, J. (2001). Lean, green, and the quest for superior performance. *Production and Operations Management*, *10*(3), 228–243. doi:10.1111/j.1937-5956.2001.tb00372.x

Roy, S., Sivakumar, K., & Wilkinson, I. F. (2004). Innovation generation in supply chain relationships: A conceptual model and research propositions. *Academy of Marketing Science Journal*, *32*(1), 61–79. doi:10.1177/0092070303255470

Rundh, B. (2008). Radio frequency identification: Invaluable technology or a new obstacle in the marketing process. *Marketing Intelligence & Planning*, *26*(1), 97–114. doi:10.1108/02634500810847174

Ryssel, R., Ritter, T., & Gemunden, H. G. (2004). The Impact of Information Technology Deployment on Trust, Commitment and Value Creation in Business Relationships. *Journal of Business and Industrial Marketing*, *19*(Issue 3), 197. doi:10.1108/08858620410531333

Saini, A., & Johnson, J. L. (2005). Organizational capabilities in e-commerce: An empirical investigation of e-brokerage service providers. *Journal of the Academy of Marketing Science*, *33*(3), 360–375. doi:10.1177/0092070305276150

Salmela, H., Toivonen, S., & Scholliers, J. (2010). Enhancing supply chain security with vulnerability management and new technology. *IET Intelligent Transport Systems*, *4*(4), 307–317. doi:10.1049/iet-its.2009.0124

Sambharya, R. B., Kumaraswamy, A., & Banerjee, S. (2005). Information technologies and the future of the multinational enterprise. *Journal of International Management*, *11*(2), 143–161. doi:10.1016/j.intman.2005.03.005

Samii, A. K. (2004). *Stratgie logistique, supply chain management: Fondements - méthodes - applications*. Dunod.

Sanders, E. B.-N., & Stappers, P. J. (2008). Co-creation and the new landscapes of design. *CoDesign*, *4*(1), 5–18. doi:10.1080/15710880701875068

Sarkis, J., & Tamarkin, M. (2005). Real options analysis for green trading: the case of greenhouse gases. *Eng. Econ.*, *50*, 273–294. doi:10.1080/00137910500227208

Sasikumar, P., & Haq, A. N. (2011). Integration of closed loop distribution supply chain network and 3PRLP selection for the case of battery recycling. *International Journal of Production Research*, *49*(11), 3363–3385. doi:10.1080/00207541003794876

Saunders, M., Lewis, P., & Thornhill, A. (2009). *Research methods for business students* (5th ed.). Essex, UK: Academic Press.

Savitz, A. W., & Weber, K. (2006). *The triple bottom line*. San Francisco: Jossey-Bass.

Scanlon, D. P., Swaminathan, S., Lee, W., & Chernew, M. (2008). Does competition improve health care quality? *Health Services Research*, *43*(6), 1931–1944. doi:10.1111/j.1475-6773.2008.00899.x PMID:18793214

Schacter, D. L., & Addis, R. L. (2007). The ghosts of past and future. *Nature*, *445*, 27. doi:10.1038/445027a PMID:17203045

Schacter, D. L., Addis, R. L., & Buckner, R. L. (2007). Remembering the past to imagine the future: The prospective brain. *Nature Reviews. Neuroscience*, *8*, 657–661. doi:10.1038/nrn2213 PMID:17700624

Schein, E. H. (1985). *Organizational culture and leadership*. Jossey-Baas Publishers.

Scherrer-Rathje, M., Boyle, T. A., & Deflorin, P. (2009). Lean, take two! Reflections from the second attempt at lean implementation. *Business Horizons*, *52*(1), 79–85. doi:10.1016/j.bushor.2008.08.004

Schlesinger, L. A., & Heskett, J. L. (1991). Breaking the cycle of failure in services. *Sloan Management Review*, *32*(3), 17–28.

Schmidt, R. C., Lyytinen, K., Keil, M., & Cule, P. (2001). Identifying software project risks: An international Delphi study. *Journal of Management Information Systems*, *17*(4), 5–36.

Schoenherr, T., Modib, S. B., Bentoncy, W. C., Carter, C. R., Choi, T. Y., & Larson, P. D. et al. (2011). Research opportunities in purchasing and supply management. *International Journal of Production Research*, *50*(6), 4556–4579.

Schultmann, F., Engels, B., & Rentz, O. (2003). Closed-loop supply chains for spent batteries. *Interfaces*, *33*(6), 57–71. doi:10.1287/inte.33.6.57.25183

Schultze, U., & Orlikowski, W. J. (2004). A practice perspective on technology-mediated network relations: The use of internet-based self-serve technologies. *Information Systems Research*, *15*(1), 87–106. doi:10.1287/isre.1030.0016

Schultz, M. (1994). *On studying organizational culture*. Walter de Gruyter Publishers.

Scott, W. R. (1995). *Institutions and organizations*. Thousand Oaks, CA: Sage.

Seidman, S. J., Brockman, R., & Lewis, B. M. (2010). In vitro tests reveal sample radio frequency identification readers inducing clinically significant electromagnetic interference to implantable pacemakers and implantable cardioverter-defibrillators. *Heart Rhythm: The Official Journal of the Heart Rhythm Society*, *7*(1), 99–107. doi:10.1016/j.hrthm.2009.09.071 PMID:20129290

Sen, F., & Shiel, M. (2006). From business process outsourcing (BPO) to knowledge process outsourcing (KPO), some issues. *Human Systems Management*, *25*, 145–155.

Sethi, R., Smith, D. C., & Park, C. W. (2001). Cross-functional product development teams, creativity, and the innovativeness of new consumer products. *JMR, Journal of Marketing Research*, *38*(1), 73–85. doi:10.1509/jmkr.38.1.73.18833

Setthasakko, W. (2009). Barriers to implementing corporate environmental responsibility in Thailand: A qualitative approach. *The International Journal of Organizational Analysis*, *17*(3), 169–183. doi:10.1108/19348830910974905

Seuring, S., & Müller, M. (2008). From a literature review to a conceptual framework for sustainable supply chain management. *Journal of Cleaner Production*, *16*(15), 1699–1710. doi:10.1016/j.jclepro.2008.04.020

Shah, N. H., & Shah, Y. K. (2000). Literature survey on inventory model for deteriorating items. *Economic Annals, 44*, 221–237.

Shah, Y. K. (1977). An order-level lot-size inventory model for deteriorating items. *AIIE Transactions, 9*(2), 108–112. doi:10.1080/05695557708975129

Shannon, C. E. (1948). A mathematical theory of communication. *The Bell System Technical Journal, 27*, 379–423, 623–656. doi:10.1002/j.1538-7305.1948.tb01338.x

Sharma, S., & Vredenburg. (1998). Proactive corporate environmental strategy and the development of competitively valuable organizational capabilities. *Strategic Management Journal, 19*(8), 729–753. doi:10.1002/(SICI)1097-0266(199808)19:8<729::AID-SMJ967>3.0.CO;2-4

Sharma, U. (2003). Implementing lean principles with the Six Sigma advantage: How a battery company realized significant improvement. *Journal of Organizational Excellence, 22*(3), 43–52. doi:10.1002/npr.10078

Sharpe, R. S., De Veaux, R. D., & Velleman, P. F. (2010). *Business statistics*. Boston: Addison Wesley.

Sheffi, Y. (2007). *The resilient enterprise: Overcoming vulnerability for competitive advantage*. The MIT Press.

Shenhar, A. J., & Wideman, R. M. (1996). *Improving PM: Linking success criteria to project type*. Project Management Institute, Symposium Creating Canadian Advantage through Project Management.

Shenhar, A. J., & Dvir, D. (2007). *Reinventing project management: The diamond approach to successful growth and innovation*. Boston: Harvard Business School Press.

Shenhar, A. J., Levy, O., & Maltz, A. C. (2001). Project success: A multi-dimensional strategic concept. *Long Range Planning, 34*, 699–725. doi:10.1016/S0024-6301(01)00097-8

Shoshanah, C., & Roussel, J. (2005). *Strategic supply chain management*. USA: McGraw-Hill.

Shrivastava, P. (1995). The role of corporations in achieving ecological sustainability. *Academy of Management Review, 20*(4), 936–960.

Sikdar, S. K. (2003). Sustainable development and sustainability metrics. *AIChE Journal. American Institute of Chemical Engineers, 49*(8), 1928–1932. doi:10.1002/aic.690490802

Sirico, L. (2010). *An RFID primer for manufacturers*. Retrieved February 4, 2010, from http://www.rfidwizards.com.html

Sivadas, E., & Dwyer, F. R. (2000). An examination of organizational factors influencing new product success in internal and alliance-based processes. *Journal of Marketing, 64*(1), 31–49. doi:10.1509/jmkg.64.1.31.17985

Smith, K. M., & Lofaro, R. J. (2001). A Paradigm for Developing Operational Decision-making (ODM). In *Proceedings of 2001 SAE World Aviation Congress (WAC) Conference*. Seattle, WA: SAE WAC

Smith, K. M., & Lofaro, R. J. (2003). The Finalized Paradigm for Operational Decision-Making (ODM) Paradigm: Components and Placement. In *Proceedings of the 12th International Symposium on Aviation Psychology*. Dayton, OH: ISAP.

Smith, A. D. (2005a). Reverse logistics and their affects on CRM and online behavior. *VINE: The Journal of Information and Knowledge Management, 35*(3), 166–181.

Smith, A. D. (2005b). Exploring the inherent benefits of RFID and automated self-serve checkouts in a B2C environment. *International Journal of Business Information Systems, 1*(1-2), 149–183. doi:10.1504/IJBIS.2005.007405

Smith, A. D. (2008). Evolution and acceptability of medical applications of RFID implants among early users of technology. *Health Marketing Quarterly, 24*(1-2), 121–155. doi:10.1080/07359680802125980 PMID:19042524

Smith, A. D., & Offodile, O. F. (2007). Exploring forecasting and project management characteristics of supply chain management. *International Journal of Logistics and Supply Management, 3*(2), 174–214.

Smith, A. D., & Offodile, O. F. (2008). Gauging web portals impact on supply chain management concepts. *International Journal of Logistics and Management, 4*(2), 184–206.

Smith, A. D., & Offodile, O. F. (2009). The perceived importance of major RFID-related technology initiatives among retail store managers. *International Journal of Services and Operations Management*, *5*(4), 520–547. doi:10.1504/IJSOM.2009.024583

Smith, G. E. (1995). Managerial pricing orientation: The process of making pricing decisions. *Pricing Strategy and Practice*, *3*(3), 28–39.

Soh, C., & Sia, S. (2004). An institutional perspective on sources of ERP package-organization misalignments. *The Journal of Strategic Information Systems*, *13*(4), 375–397. doi:10.1016/j.jsis.2004.11.001

Sohn, S. Y., Moon, T. H., & Seok, K. J. (2009). Optimal pricing for mobile manufacturers in competitive market using genetic algorithm. *Expert Systems with Applications*, *36*, 3448–3453. doi:10.1016/j.eswa.2008.02.051

Sonnemans, P. J. M., Geudens, W. H. J., & Brombacher, A. C. (2003). Organizing product releases in time-driven development processes: A probabilistic analysis. *IMA Journal of Management Mathematics*, *14*, 337–356. doi:10.1093/imaman/14.4.337

Soon, C.-B., & Gutiérrez, J. A. (2008). Effects of the RFID mandate on supply chain management. *Journal of Theoretical and Applied Electronic Commerce Research*, *3*(1), 81–91.

Sorenson, O. (2003). Social networks and industrial geography. *Journal of Evolutionary Economics*, *13*(5), 513–527. doi:10.1007/s00191-003-0165-9

Sroufe, R. (2003). Effects of environmental management systems on environmental management practices and operations. *Production and Operations Management*, *12*(3), 416–431. doi:10.1111/j.1937-5956.2003.tb00212.x

Stank, T. P., Keller, S. B., & Closs, D. J. (2001). Performance benefits of supply chain logistical integration. *Transportation Journal*, *41*(2-3), 32–46.

Starik, M., & Rands. (1995). Weaving an integrated web: Multilevel and multisystem perspectives of ecologically sustainable organizations. *Academy of Management Review*, *20*(4), 908–935.

Steger, M. F., & Kashdan, T. B. (2007). Search for meaning in life. *Encyclopedia of Social Psychology*. Retrieved 08-03, 2009, from http://sage-ereference.com/socialpsychology/Article_n469.html

Stern, N. (2007). *The Stern review*. Cambridge, MA: Cambridge University Press.

Stieb, J. (2009). Assessing Freeman's stakeholder theory. *Journal of Business Ethics*, *87*, 401–414. doi:10.1007/s10551-008-9928-4

Stock, G. N., Greis, N. P., & Kasarda, J. D. (1999). Logistics, strategy and structure: A conceptual framework. *International Journal of Physical Distribution & Logistics Management*, *29*(4), 224–239. doi:10.1108/09600039910273948

Strong, D., & Volkoff, O. (2010). Understanding organization-enterprise system fit: A path to theorizing the information technology artifact. *Management Information Systems Quarterly*, *34*(4), 731–756.

Sum, C. C., Teo, C. B., & Ng, K. K. (2001). Strategic logistics management in Singapore. *International Journal of Operations & Production Management*, *21*(9), 1239–1260. doi:10.1108/EUM0000000005926

Surana, A., Kumara, S., Greaves, M., & Raghavan, U. N. (2005). Supply-chain networks: A complex adaptive systems perspective. *International Journal of Production Research*, *43*(20), 4235–4265. doi:10.1080/00207540500142274

Swaminathan, S., Chernew, M., & Scanlon, D. P. (2008). Persistence of HMO performance measures. *Health Services Research*, *43*(6), 2033–2051. doi:10.1111/j.1475-6773.2008.00890.x PMID:18783460

Swinarski, M. E., Kishore, R., & Rao, H. R. (2006). *Impact of IT service provider process capabilities on service provider performance: An empirical study*. Paper presented at the 39th Hawaii International Conference on System Sciences (HICSS-39). Kauai, HI.

Tajima, M. (2007). Strategic value of RFID in supply chain management. *Journal of Purchasing and Supply Management*, *13*(4), 261–273. doi:10.1016/j.pursup.2007.11.001

Taleb, N. N. (2008). *El cisne negro: El impacto de lo altamente improbable*. Barcelona, Spain: Paidós.

Tan, J. J., & Litschert, R. J. (1994). Environment–strategy relationship and its performance implications: An empirical study of the Chinese electronics industry. *Strategic Management Journal*, *15*(1), 1–20. doi:10.1002/smj.4250150102

Tan, K. C., Kannan, V. R., & Handfield, R. B. (1998). Supply chain management: Supplier performance and firm performance. *International Journal of Purchasing and Materials Management*, *34*(3), 2–9.

Taylor, B. (2009). *Introduction to management science*. New York, NY: Prentice Hall.

Tencati, A., Russo, A., & Quaglia, V. (2010). Sustainability along the global supply chain: The case of Vietnam. *Social Responsibility Journal*, *6*(1), 91–107. doi:10.1108/17471111011024577

Teng, J. T., & Thompson, G. L. (1996). Optimal strategies for general price–quality decision models of new products with learning production costs. *European Journal of Operational Research*, *93*(3), 476–489. doi:10.1016/0377-2217(96)00028-8

Terzi, S., & Cavalieri, S. (2004). Simulation in the supply chain context: A survey. *Computers in Industry*, *53*(1), 3–16. doi:10.1016/S0166-3615(03)00104-0

Thomson, M., Davies, A., Jenkins, C. (2011). Three views of risk: Selecting and acquiring military equipment. *ASPI Special Report, 42*.

Tibben-Lembke, R. S. (2002). Life after death: reverse logistics and the product life cycle. *International Journal of Physical Distribution & Logistics Management*, *32*(3), 223–244. doi:10.1108/09600030210426548

To, P. L., Liao, C. C., & Liu, Y. C. (2004). The impact of the fit between global organizational configurations and information technology configurations on user information satisfaction. *Journal of International Management*, *11*(2), 109–138.

Topacan, U., Basoglu, A. N., & Daim, T. U. (2010). Exploring the adoption of technology driven services in the healthcare industry. *International Journal of Information Systems in the Service Sector*, *2*(1), 71–93. doi:10.4018/jisss.2010093005

Tsenga, F.-M., Chiub, Y.-J., & Chenc, J.-S. (2009). Measuring business performance in the high-tech manufacturing industry: A case study of Taiwan's large-sized TFT-LCD panel companies. *Omega*, *37*, 686–697. doi:10.1016/j.omega.2007.07.004

Tsoulfas, G. T., Pappis, C. P., & Minner, S. (2002). An environmental analysis of the reverse supply chain of SLI batteries. *Resources, Conservation and Recycling*, *36*(2), 135–154. doi:10.1016/S0921-3449(02)00016-2

Turner, J. R. (2004). Five conditions for project success. *International Journal of Project Management*, *22*, 349–350. doi:10.1016/j.ijproman.2004.02.001

Turnquist, M. A. (1991). University/industry cooperation on an integrated production planning software system. In *Proceedings of Technology Management: The New International Language*. Portland, OR: IEEE. doi:10.1109/PICMET.1991.183691

Tzafrir, S. S. (2005). The relationship between trust, HRM practices and firm performance. *International Journal of Human Resource Management*, *16*(9), 1600–1622. doi:10.1080/09585190500239135

U.S. Air Force: Air Force Instruction (AFI) (2005). 11-290: Cockpit/Crew Resource Management Training Program: Supplement 1 (03/22/2005).

U.S. Small Business Association (SBA). (2010). *Summary of size standards by industry*. Retrieved 7 October, 2013, from http://www.sba.gov/content/summary-size-standards-industry

Urciuoli, L. (2010). *Security in physical distribution networks- A survey study of Swedish transport operators*. Lund, Sweden: Lund University.

Vachon, S., & Klassen. (2006). Extending green practices across the supply chain: The impact of upstream and downstream integration. *International Journal of Operations & Production Management*, *26*(7), 795–821. doi:10.1108/01443570610672248

Vachon, S., & Mao, Z. (2008). Linking supply chain strength to sustainable development: A country-level analysis. *Journal of Cleaner Production*, *16*(15), 1552–1560. doi:10.1016/j.jclepro.2008.04.012

Vaidyanathan, G. (2005). A framework for evaluating third-party logistics. *Communications of the ACM, 48*(1), 89–94. doi:10.1145/1039539.1039544

Vaidyanathan, J. (2008). Business process outsourcing: An approach to gain access to world-class capabilities. *Business Process Management Journal, 14*(1), 23–38. doi:10.1108/14637150810849382

Valle, I. D. D., Castillo, M. A. S., & Rodriquez-Duarte, A. (2009). The effects of training on performance in service companies: A data panel study. *International Journal of Manpower, 30*(4), 393–407. doi:10.1108/01437720910973070

Van De Ven, A. H., & Poole, M. S. (1995). Explaining development and change in organizations. *Academy of Management Review, 20*(3), 510–540. doi:10.2307/258786

van den Engel, A. W., & Prummel, E. (2007). *Organised theft of commercial vehicles and their loads in the European Union*. Brussels, Belgium: European Parliament.

van Hoek, R. (1999). From reversed logistics to green supply chains. *Supply Chain Management: An International Journal, 4*(3), 129–135. doi:10.1108/13598549910279576

Varadarajan, P. R., & Cunningham, M. H. (1995). Strategic alliances: A synthesis of conceptual foundations. *Journal of the Academy of Marketing Science, 23*(4), 297–300. doi:10.1177/009207039502300408

Vargo, S. L., & Lusch, R. F. (2004a). Evolving to a new dominant logic for marketing. *Journal of Marketing, 68*(1), 1–17. doi:10.1509/jmkg.68.1.1.24036

Vargo, S. L., & Lusch, R. F. (2004b). The four service marketing myths: remnants of a goods-based, manufacturing model. *Journal of Service Research, 6*(4), 324–335. doi:10.1177/1094670503262946

Vargo, S. L., & Lusch, R. F. (2006). Service-dominant logic: reactions, reflections and refinements. *Marketing Theory, 6*(3), 281–288. doi:10.1177/1470593106066781

Vargo, S. L., Maglio, P., & Akaka, M. A. (2008). On value and value co-creation: A service systems and service logic perspective. *European Management Journal, 26*, 145–152. doi:10.1016/j.emj.2008.04.003

Venkatraman, N., & Ramanujam, V. (1986). Measurement of business performance in strategy research: A comparison of approaches. *Academy of Management Review, 11*(4), 801–814.

Vidgen, R., France, D., Powell, P., & Woerndl, M. (2004). Web service business transformation: Collaborative commerce opportunities in SMEs. *Journal of Enterprise Information Management, 17*(5), 372–381. doi:10.1108/17410390410561008

Vijayaraman, B. S., & Osyk, B. A. (2006). An empirical study of RFID implementation in the warehousing industry. *International Journal of Logistics Management, 17*(1), 6–20. doi:10.1108/09574090610663400

Visich, J. K., Li, S., Khumawala, B. M., & Reyes, P. M. (2009). Empirical evidence of RFID impacts on supply chain performance. *International Journal of Operations & Production Management, 29*(12), 1290–1315. doi:10.1108/01443570911006009

Vitasek, K. (2011). *The vested outsourcing manual: A guide for creating successful business and outsourcing agreements*. New York: Palgrave-MacMillan.

Vocollect Celebrates More Than Two Decades in Pittsburgh, Surpasses $100 Million in Revenue, Sees Future With New Markets. (2010). Retrieved February 4, 2010, from http://www.pghtech.org/news-and-publications/teq/article.aspx?Article=1855

Volkoff, O., Strong, D. M., & Elmes, M. B. (2007). Technological embeddedness and organizational change. *Organization Science, 18*(5), 832–848. doi:10.1287/orsc.1070.0288

Voss, M. D., Whipple, J. M., & Closs, D. J. (2009). The role of strategic security: Internal and external security measures with security performance implications. *Transportation Journal, 48*(2), 5–23.

Walker, H., & Brammer, S. (2009). Sustainable procurement in the United Kingdom public sector. *Supply Chain Management: An International Journal, 14*(2), 128–137. doi:10.1108/13598540910941993

Walker, H., Sisto, L. D., & McBain, D. (2008). Drivers and barriers to environmental supply chain management: Lessons from the public and private sectors. *Journal of Purchasing and Supply Management, 14*(1), 69–85. doi:10.1016/j.pursup.2008.01.007

Walsham, G. (1995). Interpretive cast studies in IS research: Nature and method. *European Journal of Information Systems, 4*(2), 74–81. doi:10.1057/ejis.1995.9

Walsham, G. (2006). Doing interpretive research. *European Journal of Information Systems, 15*(3), 320–330. doi:10.1057/palgrave.ejis.3000589

Walsh, J. P. (2005). Review essay: Taking stock of stakeholder management. *Academy of Management Review, 30*(2), 426–438. doi:10.5465/AMR.2005.16387898

Walton, S. V., Handfield, R. B., & Melnyk, S. A. (1998). The green supply chain: integrating suppliers into environmental management processes. *International Journal of Purchasing & Materials Management, 34*(2), 2–11. doi:10.1111/j.1745-493X.1998.tb00042.x

Wamba, S. F., & Chatfield, A. T. (2009). A contingency model for creating value from RFID supply chain network projects in logistics and manufacturing environments. *European Journal of Information Systems, 18*(6), 615–636. doi:10.1057/ejis.2009.44

Wang, Q., Lai, F., & Zhao, X. (2008). The impact of information technology on the financial performance of the third-party logistics firms in China. *Supply Chain Management: An International, 13*(2), 138–150. doi:10.1108/13598540810860976

Ward, J., & Daniel, E. (2007). *Benefits management delivering value from IS & IT investments*. London: Wiley.

Wautelet, Y. (2009). *Development of a multi-actor e-collaboration platform: business work unit 3, tasks 3.4.1 and 3.4.2*. Université catholique de Louvain.

Wautelet, Y., Achbany, Y., & Kolp, M. (2008). A service-oriented framework for mas modeling. In *Proceedings of the 10th International Conference on Enterprise Information Systems* (pp. 120-128). Barcelona, Spain: IEEE.

Wautelet, Y., Kolp, M., & Poelmans, S. (2013). Requirements-driven iterative project planning. In *Proceedings of Software and Data Technologies, 6th International Conference*. Berlin: Springer.

WCED (World Commission on Environment and Development). (1987). *Our common future*. Oxford, UK: Oxford University Press.

Wernerfelt, B. (1984). A resource-based view of the firm. *Strategic Management Journal, 5*(2), 171–180. doi:10.1002/smj.4250050207

Wessel, R. (2009, November 25). RFID drives up efficiencies at ABB. *RFID Journal.*

Whang, S. (2010). Timing of RFID adoption in a supply chain. *Management Science, 56*(2), 343–355. doi:10.1287/mnsc.1090.1121

What is RFID ? (2010). Retrieved February 4, 2010, from http://www.aimglobal.org/technologies/rfid/what_is_rfid.asp.html

White, A., Daniel, E. M., & Mohdzain, M. (2005). The role of emergent information technologies and systems in enabling supply chain agility. *International Journal of Information Management, 25*(5), 396–410. doi:10.1016/j.ijinfomgt.2005.06.009

Whiteman, G., & Cooper, W. H. (2000). Ecological embeddedness. *Academy of Management Journal, 43*(6), 1265–1282. doi:10.2307/1556349

Wiggins, R. W., & Ruefli, T. W. (2005). Schumpeter's ghost: Is hypercompetition making the best of times shorter?. *Strategic Management Journal, 26*(S 887), 911.

Wilcoxon, F. (1945). Individual comparisons by ranking methods. *Biometrics Bulletin, 1*(6), 80–83. doi:10.2307/3001968

Willcocks, L., & Lacity, M. (2006). *Global sourcing of business and IT services*. New York: Palgrave Macmillan. doi:10.1057/9780230288034

Willcocks, L., Lacity, M., & Fitzgerald, G. (1995). Information technology outsourcing in Europe and the USA: Assessment issues. *International Journal of Information Management, 15*(5), 333–351. doi:10.1016/0268-4012(95)00035-6

Williamson, O. (2008). Outsourcing: Transaction cost economics and supply chain management. *Journal of Supply Chain Management, 44*(1), 5–16. doi:10.1111/j.1745-493X.2008.00051.x

Williamson, O. E. (1975). *Markets and hierarchies: Analysis and antitrust implications*. New York: The Free Press.

Williamson, O. E. (1986). *Economic organization*. New York: New York University Press.

Williamson, O. E. (1991). Comparative economic organization: The analysis of discrete structural alternatives. *Administrative Science Quarterly, 36*(2), 269–296. doi:10.2307/2393356

Williams, T. (2004). Identifying the hard lessons from projects – Easily. *International Journal of Project Management, 22*, 273–279. doi:10.1016/j.ijproman.2003.11.001

Williams, T. (2008). How do organisations learn lessons from projects – And do they? *IEEE Transactions on Engineering Management, 55*(2). doi:10.1109/TEM.2007.912920

Williams, T., Samset, K., & Sunnevåg, K. J. (Eds.). (2009). *Making essential choices with scant information: Front end decision making in major projects*. London: Palgrave Macmillan. doi:10.1057/9780230236837

Winkler, J. K., Dibbern, J., & Heinzl, A. (2008). The impact of cultural differences in offshore outsourcing—Case study results from German–Indian application development projects. *Information Systems Frontiers, 10*(2), 243–258. doi:10.1007/s10796-008-9068-5

Winston, W. W. L., & Albright, S. (2011). *Practical management science*. New York, NY: Cengage Learning.

Winter, M., Smith, C., Morris, P., & Cicmil, S. (2006). Directions for future research in project management: The main findings of a UK government-funded research network. *International Journal of Project Management, 24*, 638–649. doi:10.1016/j.ijproman.2006.08.009

Wittmann, C.M., Hunt, & Arnett. (2009). Explaining alliance success: Competences, resources, relational factors, and resource-advantage theory. *Industrial Marketing Management, 38*, 743–756. doi:10.1016/j.indmarman.2008.02.007

Wolf, J. (2011). Sustainable supply chain management integration: A qualitative analysis of the German manufacturing industry. *Journal of Business Ethics, 102*(2), 221–235. doi:10.1007/s10551-011-0806-0

Wooldridge, M., Jennings, N., & Kinny, D. (2000). The gaia methodology for agent-oriented analysis and design. *Journal of Autonomous Agents and Multi-Agent Systems, 3*(3), 285–312. doi:10.1023/A:1010071910869

World Economic Forum. (2013). *Enabling trade: Valuing growth opportunities*. Geneva: World Economic Forum.

Wu, C. (2004). A readiness model for adopting web services. *Journal of Enterprise Information Management, 17*(5), 361–371. doi:10.1108/17410390410560991

Wu, F., Yeniyurt, S., Kim, D., & Cavusgil, T. (2006). The impact of information technology on supply chain capabilities and firm performance: A resource-based view. *Industrial Marketing Management, 35*(4), 1317–1323. doi:10.1016/j.indmarman.2005.05.003

Wyld, D. C. (2006). RFID 101: The next big thing for management. *Management Research News, 29*(4), 154–173. doi:10.1108/01409170610665022

Wynstra, J. Y. F. (2006). *Inkoop, leveranciers en innovatie: Van VOC tot space shuttle*. Erasmus Research Institute of Management. Retrieved from http://hdl.handle.net/1765/7439

Xua, D., Wijesooriyaa, C., Wanga, Y. G., & Beydounb, G. (2011). Outbound logistics exception monitoring: A multi-perspective ontologies' approach with intelligent agents. *Expert Systems with Applications, 38*(11), 13604–13611.

Yenradee, P. (1994). Application of optimized production technology in a capacity constrained flow shop: A case study in a battery factory. *Computers & Industrial Engineering, 27*(1-4), 217–220. doi:10.1016/0360-8352(94)90274-7

Yin, R. K. (2002). *Case study research: Design methods* (3rd ed.). Thousand Oaks, CA: Sage Publications.

Yin, R. K. (2009). *Case study research: Design and methods*. Thousand Oaks, CA: Sage Publications.

Yin, R. K. (2012). *Applications of case study research* (3rd ed.). Thousand Oaks, CA: Sage Publications.

Yiu, D., & Makino, S. (2002). The choice between joint venture and wholly owned subsidiary: An institutional perspective. *Organization Science, 13*(6), 667–683. doi:10.1287/orsc.13.6.667.494

Youndt, M. A., Snell, S. A., Dean, J. W., & Lepak, D. P. (1996). Human resource management, manufacturing strategy and firm performance. *Academy of Management Journal, 39*(4), 836–866. doi:10.2307/256714

Yu, E. (1995). *Modeling strategic relationships for process reengineering.* (Doctoral Dissertation). University of Toronto, Toronto, Canada.

Yu, E. (1997). Towards modeling and reasoning support for early-phase requirements engineering. In *Proceedings of the 3rd IEEE International Symposium on Requirements Engineering* (pp. 226-235). Annapolis, MD: IEEE.

Yu, E., Giorgini, P., Maiden, N., & Mylopoulos, J. (2011). *Social modeling for requirements engineering*. Cambridge, MA: MIT Press.

Yu, Z., Hong Yan, T. C., & Cheng, E. (2001). Benefits of information sharing with supply chain partnerships. *Industrial Management & Data Systems, 101*(3), 114–121. doi:10.1108/02635570110386625

Zacharia, Z. G., Sanders, N. R., & Nix, N. W. (2011). The emerging role of the third-party logistics provider (3PL) as an orchestrator. *Journal of Business Logistics, 32*(1), 40–54. doi:10.1111/j.2158-1592.2011.01004.x

Zailani, S., Jeyaraman, K., Vengadasan, G., & Premkumar, R. (2012). Sustainable supply chain management (SSCM) in Malaysia: A survey. *International Journal of Production Economics, 140*(1), 330–340. doi:10.1016/j.ijpe.2012.02.008

Zammuto, R. F., Griffith, T. L., Majchrzak, A., Dougherty, D. J., & Faraj, S. (2007). Information technology and the changing fabric of organization. *Organization Science, 18*(5), 749–762. doi:10.1287/orsc.1070.0307

Zhang, L., Hou, Z., & Qiu, X. (2012). *Third party logistics in Chinese automobile industry: A case study of a Chinese automotive manufacturer*. (Unpublished Thesis). Department of Industrial Development, IT and Land Management, Faculty of Engineering and Sustainable Development, Högskolan, I Gävle.

Zhou, K. Z., & Li, C. B. (2012). How knowledge affects radical innovation: Knowledge base, market knowledge acquisition, and internal knowledge sharing. *Strategic Management Journal, 33*, 1090–1102. doi:10.1002/smj.1959

Zhou, K. Z., & Wu, F. (2010). Technological capability, strategic flexibility, and product innovation. *Strategic Management Journal, 31*, 547–561.

Zhu, K., Dong, S., Xu, S. X., & Kraemer, K. L. (2006). Innovation diffusion in global contexts: Determinants of post-adoption digital transformation of European companies. *European Journal of Information Systems, 15*(6), 601–616. doi:10.1057/palgrave.ejis.3000650

Zhu, K., & Kraemer, K. L. (2002). E-commerce metrics for net-enhanced organizations: Assessing the value of e-commerce to firm performance in the manufacturing sector. *Information Systems Research, 13*(3), 275–295. doi:10.1287/isre.13.3.275.82

Zhu, K., & Kraemer, K. L. (2005). Post-adoption variations in usage and value of e-business by organizations: Cross-country evidence from the retail industry. *Information Systems Research, 16*(1), 61–84. doi:10.1287/isre.1050.0045

Zhu, Q. H., & Sarkis, J. (2004). Relationships between operational practices and performance among early adopters of green supply chain management practices in Chinese manufacturing enterprises. *Journal of Operations Management, 22*, 265–289. doi:10.1016/j.jom.2004.01.005

Zhu, Q., Sarkis, & Lai, K. (2008). Confirmation of a measurement model for green supply chain management practices implementation. *International Journal of Production Economics, 111*(2), 261–273. doi:10.1016/j.ijpe.2006.11.029

Zhu, Q., & Geng, Y. (2001). Integrating environmental issues into supplier selection and management: A study of large and medium-sized state-owned enterprises in China. *Greener Management International, 35*, 27–40.

Zhu, Q., Geng, Y., Fujita, T., & Hashimoto, S. (2010). Green supply chain management in leading manufacturers: Case studies in Japanese large companies. *Management Research Review, 33*(4), 380–392. doi:10.1108/01409171011030471

Zhu, Q., & Sarkis, J. (2006). An inter-sectoral comparison of green supply chain management in China: Drivers and practices. *Journal of Cleaner Production*, *14*, 472–486. doi:10.1016/j.jclepro.2005.01.003

Zhu, Q., Sarkis, J., & Lai, K. (2007). Initiatives and outcomes of green supply chain management implementation by Chinese manufacturers. *Journal of Environmental Management*, *25*(6), 179–189. doi:10.1016/j.jenvman.2006.09.003 PMID:17084502

Zhu, Q., Tian, Y., & Sarkis, J. (2012). Diffusion of selected green supply chain management practices: An assessment of Chinese enterprises. *Production Planning and Control*, *23*, 837–862. doi:10.1080/09537287.2011.642188

Zsidisin, G. A., & Siferd, S. P. (2001). Environmental purchasing: A framework for theory development. *European Journal of Purchasing and Supply Management*, *7*, 61–73. doi:10.1016/S0969-7012(00)00007-1

Zuo, K., Potangaroa, R., Wilkinson, S., & Rotimi, J. (2009). A project management prospective in achieving a sustainable supply chain for timber procurement in Banda Aceh, Indonesia. *International Journal of Managing Projects in Business*, *2*(3), 237–251. doi:10.1108/17538370910971045

Zuo, Y. (2010). Survivable RFID systems: Issues, challenges, and techniques. *IEEE Transactions on Systems, Man and Cybernetics. Part C, Applications and Reviews*, *40*(4), 406–418. doi:10.1109/TSMCC.2010.2043949

About the Contributors

John Wang is a professor in the Department of Information & Operations Management at Montclair State University, USA. Having received a scholarship award, he came to the USA and completed his PhD in Operations Research from Temple University. Due to his extraordinary contributions beyond a tenured full professor, Dr. Wang has been honored with a special range adjustment in 2006. He has published over 100 refereed papers and seven books. He has also developed several computer software programs based on his research findings. He is the Editor-in-Chief of International Journal of Applied Management Science, International Journal of Operations Research and Information Systems, and International Journal of Information Systems and Supply Chain Management. He is the Editor of Data Warehousing and Mining: Concepts, Methodologies, Tools, and Applications (six-volume) and the Editor of the Encyclopedia of Data Warehousing and Mining, 1st (two-volume) and 2nd (four-volume). His long-term research goal is on the synergy of operations research, data mining and cybernetics.

* * *

Hesham K. Alfares is Professor in the Systems Engineering Department of King Fahd University of Petroleum & Minerals (KFUPM). He has a BS in Electrical & Computer Engineering from the University of California, Santa Barbara, and an MS and a PhD in Industrial Engineering from the University of Pittsburgh and Arizona State University, respectively. Dr. Alfares research areas include employee scheduling, petrochemical industry optimization, maintenance modeling, production and inventory systems, and multi-criteria decision making. He has published more than 75 journal and conference papers in these areas.

Hossein Arsham is The Wright Distinguished Research Professor Emeritus of Decision Science at the Merrick School of Business, University of Baltimore. He is currently teaches graduate courses in Management Science at the Johns Hopkins University Carey Business School.

Maria Gisela Bardossy is an Assistant Professor at the Merrick School of Business, University of Baltimore, where she teaches business statistics and operations research. She earned her Ph.D. degree in Management Science from the Smith School of Business at the University of Maryland, College Park.

Phil Crosby works at CSIRO's Astronomy and Space Science Division, Australia, where he manages strategic business planning, industry engagement, and is a major projects specialist. Between 2009 and 2011 he was seconded to the SKA Program Development Office in the UK as Domain Specialist for the

SKA industry engagement strategy. Philip trained with ICL and BT, before operating his own medical and industrial electronics firm. Then followed 12 years with NATA in technical management standards; leading major field assessments including reviews of Antarctic science impacts, and ANSTO's operations. During 2005, Philip worked in Boeing Australia managing Industrial Participation. Apart from technical engineering qualifications, he holds a BA (Business Administration), is a graduate of the Australian Institute of Company Directors, and holds a PhD in mega-science project management. He is a member of the International Centre for Project Management, and an Associate Member of Engineers Australia.

M. Theodore (Ted) Farris II is a Professor in the Marketing and Logistics Department. He joined the University of North Texas faculty in 1997 to help grow the Professional Program in Logistics which is now ranked 22nd globally. He is presently a Charn Uswachoke International Scholar and was a Fulbright Scholar resident at the Fachhochschul Studiengang Produktions-und-Managementtechnik in Steyr, Austria. Dr. Farris received a B.S. in Business Administration and a Certificate in City and Regional Planning from Arizona State University, an M.B.A. in Materials Logistics Management from Michigan State University, an M.A.B.A. and a Ph.D. from The Ohio State University concentrating in Business Logistics and Management Information Systems. Prior to pursuing his Ph.D., Dr. Farris was employed by IBM Corporation and INTEL Corporation. Dr. Farris' research interests are Supply Chain Mapping, Disruptions in the Supply Chain, Cash-to-Cash, and Transportation regulation.

José Figueiredo, Professor at the Engineering and Management Department of IST – University of Lisbon, Researcher at CEG-IST, Engineering and Management Research Center, is an Electronics Engineer with an MBA in Information Management and a PhD in Industrial Engineering. Being always involved in university teaching he started two small companies in the information technologies sector. He currently teaches Project Management and communication skills. He published several papers in international journals, international edition book chapters, and conference proceedings.

Fernando Abreu Gonçalves, Professor at the Lisbon Politechnical School, Researcher at CEG-IST, Engineering and Management Research Center, PhD in Engineering and Management, MSC and MBA in Information Management, Engineering degree in Chemical Engineering, he has held different positions as Systems Engineer and worked at the Information Science and IO Department in different top Portuguese companies, and he is author of several papers in international journals, international edition book chapters, and conference proceedings.

Michael J. Gravier has been an Assistant Professor of Marketing at Bryant University since 2007. He holds a Ph.D. in Marketing and Logistics from the University of North Texas, an M.S. in logistics management from the Air Force Institute of Technology, and a B.A. with majors in Spanish and Anthropology from Washington University in St. Louis. Prior to his academic career he spent 12 years as an active duty logistics readiness officer in the United States Air Force and worked as a research assistant in the Washington University School of Medicine. Dr. Gravier holds professional certification status (C.T.L.) with the American Society of Transportation and Logistics. His research interests include supply chain network evolution, sustainable supply chains, risk mitigation, medical logistics, and transportation public policy. Dr. Gravier co-authored Rhode Island's Green Economy Road Map in 2010.

Daniel S. Hoops is an Associate Professor and Chair, Accounting and Taxation Department, Walsh College, Troy Michigan. Prof. Hoops received his Bachelor in Music (Trumpet Performance) from the University of Michigan, Juris Doctorate cum laude from the Detroit College of Law Michigan State University and Master of Law (Estate Planning) from the University of Miami.

Chandra K. Jaggi is Associate Professor in the Department of Operational Research, Faculty of Mathematical Sciences, University of Delhi, India. He earned his PhD, M.Phil. in Inventory Management and Masters Degree (Operational Research) from Department of Operational Research, University of Delhi. His teaching includes inventory modeling, and Financial Engineering. His research interest lies in modeling Inventory systems. He has more than thirty five Publications in reputed international as well as national journals viz. IJPE, JORS,EJOR,IJSS, AMO, IJCJPAS, TOP, OPSEARCH, JIOS etc.

Ronald John Lofaro, PhD is an aviation psychologist, currently an adjunct Associate Professor at Embry-Riddle Aernautical University/Worldwide. He spent 17 years with the FAA; his last FAA post (2002-2004; from which he retired from government service), was at Embry-Riddle Aeronautical University (ERAU), FL. There he was the FAA liaison, and visiting professor in Human Factors and Systems. Prior to that, he spent over 2 years as the sole FAA liaison to both the ten USAF Research Labs and the USAF Aeronautical Systems Center at WPAFB, OH. During that time, he also was a visiting professor at The Ohio State University and worked with NASA Glenn RC in Cleveland. Before Ron went to Ohio in late 1999, he spent 6 years at the FAA's Technical Center where he was a manager in the FAA's Airworthiness Assurance R&D Division. Prior to that, he spent almost 4 years at FAA HQ, where he was project manager on a variety of R&D efforts with NASA, USN, USAF and universities; these focused on human factors in flight crew training, performance and assessment as well as ATC selection. He came to the FAA from the Army Aviation Command, where his 5-plus year stint included R&D in helicopter pilot air-to-air combat training; pilot candidate selection and final placement in operational rotorcraft after initial training (this work resulted in an computer -based, operationally used, US Army Aviation, pilot selection/placement test: The Multitrack Algorithm); pilot performance assessment, and a year as a voting (doing on-site investigation as the human factors specialist) Board Member, of an Army Safety Center" go-team" for Class A aviation accident investigation. Ron has published over 80 articles and technical reports plus 4 book chapters; 10 articles were co-authored with Kevin Smith as well as two book chapters. Topics were: Aeronautical Decision-Making in the operational environment (ODM); Crew Resource Management (CRM), Line Oriented Flight Training (LOFT); use of ODM and Mission Performance Model in design and development of LOFTs; aviator selection and classification; aircrew/ pilot training and performance evaluation... aviation safety.

Isabelle Mirbel is associate professor at the University of Nice Sophia-Antipolis since 1998. Her research interests include information systems design, method engineering and requirement engineering. She has published papers at international level and she participated in the WIDE ESPRIT project (Workflow on Intelligent Distributed database Environment) and Chorochronos on the European projects: spatio-temporal databases (Training and Mobility of Researchers program). She has been part of the program committee in several international conferences, among which CAISE and ICEIS, and referee for several international journals. She is member of the IFIP 8.1 Working Group.

Mandeep Mittal is a research scholar in the department of Operational Research, Faculty of Mathematical Sciences, University of Delhi, India. He has completed his M.Sc.(Applied Mathematics)in 2000 from IIT Roorkee. At present he is pursuing his Ph.D. in operational research. His area of interest is "Inventory Control and Management." He has published one research paper in International Journal of Industrial Engineering Computations (IJIEC).

Antti Permala, born in 1952, obtained his MSc in 1976 and Licentiate of Technology in 1991 from Helsinki University of Technology (HUT). During 1977–1979 he worked as a researcher in Transport Planning for the City of Helsinki. In 1980–1984 he was in charge of administration and economics of the Road and Traffic Laboratory at VTT. Since 1985 he has worked in different research positions at VTT in freight transport and logistics. At the moment he is in the position of Chief Research Scientist in the Supply Chain Management research team at VTT. Main research focus during last years has been freight transport, logistics and telematics in domestic and European Union framework programme projects.

Eetu Pilli-Sihvola is a Research Scientist on Intelligent Transport Systems(ITS) at the Technical Research Centre of Finland (VTT). His areas of expertise include location-based services, mobile devices, ITS systems and architectures, the use of ICT in logistics, and impact assessment. The subject of his thesis was Road Traffic ITS as Critical Information Systems. Mr. Pilli-Sihvola has previously worked as a software designer at Foreca Ltd. testing tools for developing Web services. He has also worked as an ITS expert at the Finnish Road Administration specializing in transport-related information systems including road weather information systems as well as traffic control and management systems for tunnels, bridges and road sections.

Kenneth A. Saban is associate professor of marketing in the Palumbo/Donahue Schools of Business, Duquesne University. He possesses 36 years of manufacturing, consulting, and academic experience. Dr. Saban earned a BSBA from Youngstown State University, a MSJ from Northwestern University, and a PhD from the University of Pittsburgh. He has developed a particular expertise in the areas of business collaboration and technology management through a wide-range of research projects, grants and industry partnerships. He has published over 20 articles and given numerous academic and industry presentations. He is also working with the U.S. Air Force Research Labs to explore the impact of "human collaboration" on supply chain performance.

Johan Scholliers is working as Principal Scientist in the Vehicle ICT systems team at VTT Technical Research Centre of Finland. He received the MSc degree in electro-mechanical engineering in 1986 and PhD in 1992 from the Katholieke Universiteit Leuven in Belgium. His experience is in the use of monitoring and identification technologies for transport, logistics and traffic; and in the evaluation of ICT applications for logistics, mobility and vehicle safety applications. He has been involved in several national and international projects, and has been acting as coordinator of European research projects.

Karri Rantasila obtained his MSc (Econ.&Bus.Admin.) in 2010 from Turku School of Economics (TSE) with logistics as his major. In 2008-2010, he worked in several research projects at TSE, including EU funded C.A.S.H. project, and Finland State of Logistics 2009 and 2010 survey projects commissioned by the Ministry of Transport and Communications Finland. From June 2010 he has been employed at VTT as a research scientist on logistics, where he is involved in Several EUfunded (e.g.ENABLE, SKEMA, eFreight, SMART-CM), as well as domestic research and development projects in field of logistics and supply chain management. His research interests include transport systems, technologies enabling visibility in supply chains, and global logistics markets. The last one is closely related to the topic of his ongoing PhD dissertation work on measuring macro logistics costs at TSE.

D. K. Sharma is PhD in Project Management and works as a full time faculty member in the Merrick School of Business at University of Baltimore. He teaches Decision Science courses to undergraduate business students.

Alan D. Smith is presently University Professor of Operations Management in the Department of Management and Marketing at Robert Morris University, Pittsburgh, PA. Previously he was Chair of the Department of Quantitative and Natural Sciences and Coordinator of Engineering Programs at the same institution, as well as Associate Professor of Business Administration and Director of Coal Mining Administration at Eastern Kentucky University. He holds concurrent PhDs in Engineering Systems/ Education from The University of Akron and in Business Administration (OM and MIS) from Kent State University, as well as author of numerous articles and book chapters.

Sirra Toivonen MSc (Tech.) is a research scientist at Vulnerability Management Team at VTT technical Research Centre of Finland. She has worked at the Risk Management knowledge centre since 1995. MSc Toivonen has experience in radical innovation management, usabilityand safety development in both national research projects and confidential customer projects. Her main research focus areas in recent years have been supply chain security management and border security in national and in European research projects.

Yves Wautelet is an Assistant Professor of Information Systems at Katholieke Universiteit Leuven (Brussels campus), Belgium. He formerly has been an IT project manager and a postdoc fellow at the Université catholique de Louvain, Belgium. He completed a PhD thesis focusing on project and risk management issues in large enterprise software design. Dr. Wautelet also holds a Master of Management Sciences as well as a Master of Information Systems. His research interests include aspects of software engineering and enterprise information systems such as requirements engineering, agent-oriented programming and COTS-based development. He also focuses on the application of his research into industrial environments for the development of supply chain management and enterprise resource planning solutions.

Index

J

K

L

M

N

O

P

Q

R

S

T

U

V

W

www.ingramcontent.com/pod-product-compliance
Lightning Source LLC
LaVergne TN
LVHW060355300926
849162LV00006B/1270